A KEY TO VASCULAR PLANT SPECIES OF KERN COUNTY, CALIFORNIA

by

L. Maynard Moe

&

A FLORA OF KERN COUNTY, CALIFORNIA

by Ernest C. Twisselmann

Illustrated by

Eben and Gladys McMillan

AF693522

CALIFORNIA NATIVE PLANT SOCIETY

CALIFORNIA NATIVE PLANT SOCIETY

Kern County Chapter Officers, 1995

Scott Wilson, *President*
Debra Martin, *Vice-President*
Laura Stockton, *Secretary*
Lynne Hawk, *Treasurer*

CNPS Publications Committee, 1995

Phyllis Faber, *Chairman*
Wilma Follette
Joyce Hawley
Elly Bade
Carol Baird
Bob Berka
Steve Hartman
Harlan Kessel
Barbara Malloch Leitner
Carol LeNeve
Tom White

Copyright ©1995 California Native Plant Society
1722 J Street, Suite 17, Sacramento, CA 95814
(916) 447-2677

A Flora of Kern County, California by Ernest C. Twisselmann, reprinted with permission, was published originally by the University of San Francisco, San Francisco, California, 1967, as a reprint from *The Wasmann Journal of Biology*, Vol. 25, Nos. 1 and 2, 1967.

Publication of this book was made possible by financial support from the Kern County Chapter of CNPS and from the Lois B. and Rudolph E. Fink Publication Fund. Lois and Rudolph were native Californians with an abiding interest in California native plants, particularly in their native habitat. It is hoped that publication of this book will enable others to better experience that same special joy that added so much to the lives of Lois and Rudolph.

All Rights Reserved
First Printing
Printed in the United States of America
ISBN 0-943460-22-0

Cover: The Tehachapi Mountains in early spring. Photograph by L. Maynard Moe.

CONTENTS

KEY TO VASCULAR PLANT SPECIES OF KERN COUNTY, CALIFORNIA

INTRODUCTION ii

KEY . . . . 1 KEY

GLOSSARY . . . . 216 KEY

A FLORA OF KERN COUNTY, CALIFORNIA

NATURAL HISTORY 1

GEOGRAPHY 2

GEOLOGY AND SOILS 34

CLIMATE AND WEATHER 65

FLORISTICS 76

PLANT ASSOCIATIONS 87

NUMERICAL ANALYSIS OF THE FLORA 119

CENTERS OF DISTRIBUTION AND ORIGIN OF FLORA 125

RANGE LIMITS IN KERN COUNTY 126

BOTANICAL EXPLORATION 136

A NOTE ON THE ILLUSTRATONS 154

NOTES ON THE CATALOG OF PLANTS 154

ACKNOWLEDGMENTS 156

LITERATURE CITED 161

CATALOG OF PLANTS 165

INDEX 387

Douglas oak woodland near Caliente in early spring. Popcorn flower (*Plagiobothrys*) and fiddle-neck (*Amsinckia*) carpet the open hillside. Photograph by L. Maynard Moe.

A KEY TO VASCULAR PLANT SPECIES OF KERN COUNTY, CALIFORNIA

by

L. Maynard Moe

Key to Vascular Plant Species of Kern County, California

L. Maynard Moe
Bakersfield, California

Introduction

Ernest Twisselmann's *Flora of Kern County* (Twisselmann 1967) has been indispensable for field botanists for over twenty five years. It has very detailed descriptions of the geography, geology, soils, climate, floristics, plant associations, botanical explorations, and special plants of Kern County. However, this book has been generally of limited use to amateur botanists because it does not contain a key to assist in identification. Also, there have been new species added to the list for Kern County and there have been some nomenclatural changes since Twisselmann's *Flora* was published in 1967.

Extensive land development for housing, agriculture, land fills, mining, and recreation, has put increased pressures on the need for preserving what remains of our natural vegetation. Because State and Federal laws require that developers determine the potential impact their proposed plans would have on the natural resources, more and more botanists are in the field conducting plant surveys. In addition, membership in Kern chapters of The Native Plant Society, The Nature Conservancy, Sierra Club, and others is increasing; more and more people are interested in the local flora and need some assistance in identification of local plant species.

It is for these reasons that Twisselmann's *Flora* is being reissued and this key has been prepared. The key is based on my knowledge of the flora, experience with the keys in Munz's *A California Flora* (Munz and Keck 1968) and *A Flora of Southern California* (Munz 1974), Abrams' keys in *The Flora of the Pacific States* (Abrams and Ferris 1960), as well as over twenty years teaching California Flora courses. Most of the keys are modified from those of Munz.

The publication of *The Jepson Manual* (Hickman 1993) enabled me to include the most recent nomenclature. Where names of various taxa are different in *The Jepson Manual*, I have included names used by Twisselmann as synonyms in parentheses. The numbers in parentheses following each genus refer to page numbers in Twisselmann's *Flora*. Collector, numbers and locations are indicated for species not included in Twisselmann's *Flora*. Voucher specimens for most of these additions are deposited in the California Academy of Sciences (CAS), others are in the herbarium at California State University, Bakersfield unless otherwise indicated.

Since Twisselmann gave such detailed descriptions of locations, only those species not included in his *Flora* have location information in this key. Brief descriptions of families and genera are included in this key, but species descriptions are only indirectly obtained by piecing together choices made in the key. Information concerning the number of species in a genus is from Hickman (1993) where possible. Where not possible, the information was taken from Mabberley (1987). A complete list of Twisselmann's taxa indicating nommenclature changes and additions to the flora can be found in Moe (1994).

I have also tried to keep the technical terminology to a minimum but occasionally there are no simple alternatives for some terms. I have included a glossary of terms at the end of the keys. To keep the key small and easily used, I have used several abbreviations and symbols which are listed below.

I sincerely thank John Strother and Jim Shevock for reviewing an early version of this key and providing so much valuable assistance. Jim Shevock provided most of the additions to the list of species found in Kern County since publication of Twisselmann's *Flora*. Field work by many botanists continues to this date and range extensions for Kern County are being documented. Much work remains to be done, especially in the remote areas.

I also wish to thank the California Native Plant Society, especially the Kern County Chapter, whose efforts have made this publication possible.

Abbreviations and symbols

± : more or less
< : less than
> : more than
CA : California
ca. : about, nearly
caps. : capsule
cm. : centimeter
dm. : decimeter
fl. : flower
fr. : fruit
infl. :.inflorescence
invol. : involucre
lf. : leaf
lfts. : leaflets
lvs. : leaves
m. : meter
mm. : millimeter

References

Abrams, L. and R. S. Ferris. 1960. Illustrated Flora of the Pacific States. 4 vols. Stanford University Press, California.

Hickman, James C. 1993. The Jepson Manual of Higher Plants of California. University of California Press, Berkeley, California.

Mabberley, D. J. 1987. The Plant Book. Cambridge University Press, Cambridge.

Moe, L. M. 1994. Synonymized List of Vascular Plant Species in Kern County. *Crossosoma* 20 (1): 17-44.

Munz, P. A. 1974. A Flora of Southern California. University of California Press, Berkeley, California.

Munz, P. A. and D. D. Keck. 1968. A California Flora with Supplement. University of California Press, Berkeley, California.

Reveal, J. L. 1989. The Eriogonoid flora of California (Polygonaceae: Eriogonoideae). *Phytologia* 66(4): 295-414.

Twisselmann, E. C. 1967. A Flora of Kern County, California. *The Wassmann Journal of Biology*. 25 (1 and 2): 1-395.

Key to Artificial Groups of Families

1. Plants reproducing by spores borne in sporangia (ferns and fern allies)....... **Group A**
1. Plants reproducing by seeds.
  2. Seeds borne in cones; no true flowers; leaves needle or scale-like (gymnosperms). **Group B**
  2. Seeds in ripened pistils of true flowers; leaves various (angiosperms).
    3. Leaves mostly parallel-veined; flower parts mostly in 3's (monocots) .**Group C**
    3. Leaves mostly net-veined; flower parts mostly in 4's or 5's (dicots).
      4. Petals absent or not evident.
        5. Plants woody; well-developed trees or shrubs............................... **Group D**
        5. Plants herbaceous or woody at base only .... **Group E**
      4. Petals present, evident.
        6. Petals separate at base, not connate.
          7. Stamens numerous, more than fifteen... **Group F**
          7. Stamens few, not more than fifteen.. **Group G**
        6. Petals fused at base, connate.
          8. Ovary superior.. **Group H**
          8. Ovary inferior... **Group I**

GROUP A

(Ferns and fern allies; reproducing by spores)

1. Plants fern-like, terrestrial; one kind of spore in sporangia.
  2. Sporangia large, globular in stalked spike or panicle; fronds not coiled in development.. **Ophioglossaceae**
  2. Sporangia small, borne on lower surface of leaves; fronds coiled in development.
    3. Sporangia borne at or near margin.
      4. Blades to >100cm long, hairy underneath.......................... **Dennstaeditaceae**
      4. Blades <50cm long, glabrous... **Pteridaceae**
    3. Sporangia not marginal, borne in separate sori away from margin.
      5. Sporangia not in discrete sori, scattered along veins .... **Pteridaceae**
      5. Sporangia in discrete, round to elongate sori.
        6. Sori round to oval.
          7. Indusia absent, sori naked... **Polypodiaceae**
          7. Indusia present, though soon concealed by mature sporangia. **Dryopteridaceae**
        6. Sori oblong, linear to lunate, or horseshoe-shaped; fronds 6-20 dm. tall.
          8. Venation partly closed; sori 2-6 mm. long............................ **Blechnaceae**
          8. Venation entirely open; sori <1 mm. long.....................**Dryopteridaceae**
1. Plants not fern-like.
  9. Plants definitely aquatic or on mud.
    10. Small annual floating on water.
      11. Plants smooth, not moss-like; green (not a fern).......................... **Lemnaceae**

11. Plants wrinkeled, moss-like; often turning reddish........................ **Azollaceae**
10. Perennial rooted in mud.
12. Lvs. grass-like, stem subterranean corm.. **Isoetaceae**
12. Lvs. clover-like; leaflets 4; stem creeping rhizome rooting in mud. **Marsileaceae**
9. Plants terrestrial.
13. Stems upright, jointed, hollow; lvs. few, forming sheaths at nodes. **Equisetaceae**
13. Stems creeping, not jointed; lvs. numerous, overlapping......... **Selaginellaceae**

GROUP B
(Gymnosperms)

1. Internodes much longer than lvs.; lvs. brown scales.............................. **Ephedraceae**
1. Internodes shorter than lvs.; lvs. green.
2. Lvs. needle-like or linear.. **Pinaceae**
2. Lvs. scale-like; mostly decurrent and covering branches................. **Cupressaceae**

GROUP C
(Lvs. parallel veined; fl. part in 3's; monocots)

1. Plants tiny floating aquatics or stranded on mud, not differentiated into stems and lvs (see also Azollaceae) .... **Lemnaceae**
1. Plants larger, with stems and lvs.
2. Perianth well developed, at least inner segments petal-like in color and texture.
3. Carpels ± separate, 1-loculed, maturing into a whorl or bunch of achenes. **Alismataceae**
3. Carpels united into a 3-12-loculed ovary; fr. a capsule or berry.
4. Plants large "palm trees".. **Arecaceae**
4. Plants smaller; if tree-like then branched.
5. Ovary superior.
6. Floating aquatic plants with inflated petioles .... **Pontederiaceae**
6. Terrestrial plants .... **Liliaceae**
5. Ovary inferior.
7. Aquatic plants; submerged.. **Hydrocharitaceae**
7. Terrestrial plants.
8. Fls. regular; ovary 3-loculed; stem and lf.-bases flattened..... **Iridaceae**
8. Fls. irregular; ovary 1-loculed.. **Orchidaceae**
2. Perianth reduced to absent, not petal-like in color or texture.
9. Fls. enclosed by papery or husk-like bracts and mostly concealed by them; fls. in spikes, spikelets or heads.
10. Lf.-sheath split lengthwise on side opposite blade; lvs.usually 2-ranked; stems mostly hollow and rounded.. **Poaceae**
10. Lf.-sheath continuous around stem; lvs. mostly 3-ranked; stems often triangular... **Cyperaceae**
9. Fls. not enclosed by bracts.
11. Plants floating or submersed aquatics.

12. Fls. in spikes or heads.
  13. Fls. perfect, in peduncled axillary spikes................ **Potamogetonaceae**
  13. Fls. unisexual, in globose heads, the lower pistillate, the upper staminate. .... **Sparganiaceae**
12. Fls. axillary and mostly solitary.
  14. Lvs. alternate .... **Potamogetonaceae**
  14. Lvs. opposite.
    15. Lvs. entire .... **Zannichelliaceae**
    15. Lvs. with coarse teeth on margins .... **Hydrocharitaceae**
11. Plants of land or shallow water, mostly well emersed.
  16. Infl. a dense elongate spike.
    17. Plants small tufted herb to 0.4 m tall.............................. **Juncaginaceae**
    17. Plants tall, coarse, to 3 m tall.. **Typhaceae**
  16. Infl. of subglobose heads, racemes or open clusters.
    18. Fls. unisexual, the lower heads pistillate; perianth of papery scales. **Typhaceae**
    18. Fls. bisexual; perianth-segments not papery scales.
      19. Ovaries 6, separating at maturity................................ **Juncaginaceae**
      19. Ovary 1, forming a capsule .... **Juncaceae**

GROUP D

(Petals absent; woody trees or shrubs)

1. Lvs. simple.
  2. Plants parasitic on branches of trees and shrubs .... **Loranthaceae**
  2. Plants not parasitic.
    3. Fls. (at least staminate) in catkins.
      4. Lvs. opposite; plants dioecious.. **Garryaceae**
      4. Lvs. alternate.
        5. Staminate fls. with calyx.
          6. Fr. a nut or cone-like... **Betulaceae**
          6. Fr. an acorn with basal cup .... **Fagaceae**
        5. Calyx none; staminate fls. subtended by bracts.
          7. Fr. a capsule; seeds with tuft of hair .... **Salicaceae**
          7. Fr. 1-seeded; seeds without tuft of hair .... **Myricaceae**
    3. Fls. not in catkins.
      8. Plants normally trees with well-defined trunk.
        9. Lvs. entire, strongly aromatic when crushed.
          10. Ovary superior; stamens 9.. **Lauraceae**
          10. Ovary inferior; stamens numerous, colored white to red....... **Myrtaceae**
        9. Lvs. toothed or lobed, not aromatic.
          11. Lvs. 12-25 cm. broad, deeply 5-lobed; fls. in round heads. **Plantanaceae**
          11. Lvs. 2-5 cm. broad, ± serrate; fls. in small clusters or solitary. **Ulmaceae**
      8. Plants normally shrubs with several stems.
        12. Fls. yellowish, 1-6 cm. across; sepals petaloid...................... **Sterculiaceae**

12. Fls. not yellowish, smaller; sepals not petaloid.
13. Fr. an achene or utricle.
14. Style 1.
15. Style ± plumose in fr. .... **Rosaceae**
15. Style not plumose in fr. .... **Asteraceae**
14. Styles 2-3.
16. Fls. subtended by involucre .... **Polygonaceae**
16. Fls. lacking involucre of bracts .... **Chenopodiaceae**
13. Fr. not an achene or utricle.
17. Lvs. opposite.
18. Plants glabrous; fr. a blue-black drupe .... **Oleaceae**
18. Plants with silvery-scurfy branchlets; fr. red .... **Elaeagnaceae**
17. Lvs. alternate.
19. Lvs. palmate; fr. spiny .... **Euphorbiaceae**
19. Lvs. pinnate; fr. not spiny .... **Rhamnaceae**
1. Lvs. compound.
20. Woody vines; lvs. opposite; stamens numerous .... **Ranunculaceae**
20. Tree or shrub, not a vine.
21. Lvs. alternate; fr. a "nut" .... **Juglandaceae**
21. Lvs. opposite; fr. a samara.
22. Leaflets 3; samara pubescent .... **Aceraceae**
22. Leaflets 3-7; samara glabrous .... **Oleaceae**

GROUP E

(Petals absent; plants herbaceous)

1. Aquatic plants, submerged or on mud.
2. Lvs. dissected, whorled.
3. Ovary superior; lvs. 3-times forked .... **Ceratophyllaceae**
3. Ovary inferior; lvs. ± pinnatifid .... **Haloragaceae**
2. Lvs. entire, opposite, often crowded into terminal rosettes.
4. Ovary inferior; lvs. 15-25 mm. long. .... **Onagraceae**
4. Ovary superior; lvs. 3-8 mm. long .... **Callitrichaceae**
1. Land plants, sometimes in wet places.
5. Ovary superior, free from calyx.
6. Perianth absent; involucre often conspicuous.
7. Fls. in clusters often surrounded by invol. resembling a perianth; sap often milky .... **Euphorbiaceae**
7. Fls. in a spike subtended by conspicuous invol.; sap not milky... **Saururaceae**
6. Perianth present as a calyx (which is often petaloid).
8. Pistils >1, distinct; stamens >9 .... **Ranunculaceae**
8. Pistil 1; stamens various.
9. Fls. perigynous, floral tube present.
10. Stipules present; lvs. alternate .... **Rosaceae**
10. Stipules absent; lvs. opposite.
11. Stamens >10 .... **Aizoaceae**
11. Stamens 3-5 .... **Nyctaginaceae**

9. Fls. hypogynous; floral tube absent.

12. Style and stigma single.

13. Fls. in bractless racemes.. **Brassicaceae**

13. Fls. not in bractless racemes.

14. Calyx tubular, corolla-like ....**Nyctaginaceae**

14. Calyx not tubular.

15. Lvs. ovate, blades 10-20 mm. long.

16. Calyx mostly 4-parted; stamens 4; stinging hairs present, not stellate.. **Urticaceae**

16. Calyx 5-6-parted, stamens 6-7; stellate pubesence obvious. **Euphorbiaceae**

15. Lvs. linear, stiff, 3-6 mm. long..........................**Caryophyllaceae**

12. Styles and stigmas >1.

17. Lvs. deeply palmately 5-7-lobed or compound.

18. Tree... **Moraceae**

18. Herbs... **Cannabaceae**

17. Lvs. mostly entire or shallowly lobed.

19. Ovary 1-loculed.

20. Fr. a capsule, several-seeded....................................... **Aizoaceae**

20. Fr. an achene or utricle, 1-seeded.

21. Calyx mostly 6-parted (if 5-parted, stipular sheath present); stamens 3, 6 or 9.. **Polygonaceae**

21. Calyx mostly 5-parted; stipular sheath absent; stamens mostly 1, 4 or 5.

22. Bracts of infl. absent; stipules present, papery. **Caryophyllaceae**

22. Bracts of infl. herbaceous or papery; stipules absent.

23. Bracts of infl. herbaceous; plants mealy, scurfy or fleshy. **Chenopodiaceae**

23. Bracts of infl. papery; plants not scurfy or fleshy. **Amaranthaceae**

19. Ovary >1-loculed.

24. Lvs. alternate... **Euphorbiaceae**

24. Lvs. whorled or opposite.

25. Lvs. of each pair unequal; calyx sometimes colored.

26. Lvs. opposite; fr. circumscissile........................... **Aizoaceae**

26. Lvs. whorled; fr. splitting along length......... **Molluginaceae**

25. Lvs. of each pair equal; calyx not colored.....**Caryophyllaceae**

5. Ovary inferior.

27. Plants 1-2 m tall; dioecious... **Datiscaceae**

27. Plants <1 m tall

28. Stamens >10.. **Aizoaceae**

28. Stamens <10.

29. Lvs. alternate.

30. Lvs. entire; plants perennial... **Santalaceae**

30. Lvs. toothed to dissected; plants annual................................**Asteraceae**

29. Lvs. opposite or whorled.

31. Perianth 4-lobed; lvs. whorled .... **Rubiaceae**

31. Perianth 5-lobed or more; lvs. opposite.

32. Fls. regular, showy ....**Nyctaginaceae**

32. Fls. irregular, small; small annuals.............................. **Valerianaceae**

GROUP F

(Petals present, free; stamens numerous)

1. Ovary superior.

2. Plants woody shrubs or trees.

3. Plants without stipules ....**Papaveraceae**

3. Plants with evident stipules.

4. Shrub with recurved claw-like thorns; lvs. pinnate.......................... **Fabaceae**

4. Shrub not both thorny and with pinnate lvs.

5. Stamens united into a tube; lvs. often with stellate pubescence..**Malvaceae**

5. Stamens not united into a tube; stellate pubescence lacking......... **Rosaceae**

2. Plants herbaceous or woody at base only.

6. Sepals 2 or joined into a pointed cap.

7. Plant fleshy; sepals persistent... **Portulacaceae**

7. Plant not fleshy; sepals deciduous at anthesis............................**Papaveraceae**

6. Sepals >2.

8. Stamens united into a tube; lvs. rounded...**Malvaceae**

8. Stamens not united into a tube; leaves various.

9. Stamens perigynous; stipules well-developed............................... **Rosaceae**

9. Stamens hypogynous; stipules gland-like or absent.

10. Lvs. fascicled (clustered); fls. small with 2 petals................**Resedaceae**

10. Lvs. opposite or alternate; fls. larger.

11. Lvs. opposite; pistil 1 with 3 styles ....**Hypericaceae**

11. Lvs. mostly alternate; pistils >1................................. **Ranunculaceae**

1. Ovary inferior, at least partly so.

12. Ovary only partly inferior.

13. Fr. a fleshy pome... **Rosaceae**

13. Fr. a dry capsule... **Saxifragacaeae**

12. Ovary wholly inferior.

14. Petals numerous; plants succulent.

15. Plants not spiny; lvs. succulent...**Aizoaceae**

15. Plants spiny...**Cactaceae**

14. Petals few; plants not succulent.

16. Herbaceous; fr. a capsule ....**Loasaceae**

16. Tree or shrub; fr. fleshy.

17. Lvs. alternate; fr. a pome, <15 mm. long.................................... **Rosaceae**

17. Lvs. opposite; fr. a large leathery-skinned berry...................... **Punicaceae**

GROUP G

(Petals present, free; stamens few)

1. Pistils >1; nearly or quite separate.
  2. Plant fleshy; lvs. simple..**Crassulaceae**
  2. Plant not succulent; lvs. simple to compound.
    3. Stamens hypogynous; small herbs.. **Ranunculaceae**
    3. Stamens perigynous; hypanthium present.
      4. Hypanthium inconspicuous..**Crossosomataceae**
      4. Hypanthium well-developed.
        5. Stipules present; lvs. simple to compound, veins mostly pinnate.. **Rosaceae**
        5. Stipules absent; lvs. simple, veins mostly palmate................ **Saxifragaceae**
1. Pistil 1, of 1 or more ± united carpels.
  6. Tendrils present; lvs. palmate.
    7. Fr. a berry, borne in clusters..**Vitaceae**
    7. Fr. a pepo (melon or gourd), borne separately............................ **Cucurbitaceae**
  6. Tendrils absent or, if present, lvs. pinnate.
    8. Styles 2-5, separate to near base.
      9. Plant woody, well-developed shrub or tree.
        10. Lvs. scale-like, <4 mm long; fls. <4 mm wide.................... **Tamaricaceae**
        10. Lvs. well-developed.
          11. Ovary superior.
            12. Fr. not a samara; lvs. pinnate.......................................**Anacardiaceae**
            12. Fr. a samara; lvs. palmate..**Aceraceae**
          11. Ovary inferior.
            13. Fr. a berry; lvs. mostly palmate................................**Grossulariaceae**
            13. Fr. a pome; lvs. pinnate .... **Rosaceae**
      9. Plant herbaceous.
        14. Plant a submerged aquatic or on mud flats.
          15. Lvs. dissected..**Haloragaceae**
          15. Lvs. entire, opposite.. **Elatinaceae**
        14. Plant a normal terrestrial plant.
          16. Ovary superior.
            17. Lvs. compound, with 3 leaflets; stamens 10.................... **Oxalidaceae**
            17. Lvs. simple.
              18. Lvs. mostly alternate or basal.
                19. Sepals 2; plants ± fleshy... **Portulacaceae**
                19. Sepals 3-several; plants not fleshy.
                  20. Fls. irregular; stamens 3; petals 2..........................**Resedaceae**
                  20. Fls. regular, stamens >5; petals 5.
                    21. Ovary 2-5 celled...**Linaceae**
                    21. Ovary 1-celled..**Caryophyllaceae**
              18. Lvs. opposite.
                22. Stamens 4-7; plant of alkaline flats; corolla pink. **Frankeniaceae**
                22. Stamens 5 or 10; habitat and color various.......**Caryophyllaceae**
          16. Ovary inferior.
            23. Ovules several in each locule; lvs. simple.................... **Saxifragaceae**

23. Ovules solitary; fls. in umbels.
24. Fr. a berry; lvs. 3-15 dm. long .... **Araliaceae**
24. Fr. dry, splitting into 2 carpels; lf.-size various .... **Apiaceae**
8. Style 1, sometimes ± divided.
25. Ovary inferior.
26. Plants well-developed shrub or tree; fr. fleshy.
27. Lvs. opposite; stamens 4. .... **Cornaceae**
27. Lvs. alternate; stamens mostly 5 .... **Grossulariaceae**
26. Plants herbaceous; fr. dry.
28. Plant covered with short barbed hairs; petals 5....................... **Loasaceae**
28. Plant smooth; petals mostly 4... **Onagraceae**
25. Ovary superior.
29. Plant well-developed shrub or tree.
30. Fls. irregular.
31. Ovary 3-loculed; lvs. opposite............................... **Hippocastanaceae**
31. Ovary 1-loculed.
32. Fls. not papilionaceous; fr. a spiny 1-seeded pod..... **Krameriaceae**
32. Fls. papilionaceous; fr. a legume; lvs. mostly compound. **Fabaceae**
30. Fls. regular.
33. Lvs. compound.
34. Lvs. twice pinnate; fls. purpleish; tree .... **Meliaceae**
34. Lvs. once divided.
35. Fr. a samara; trees or shrubs.
36. Lvs. alternate; wing all around fr............................... **Rutaceae**
36. Lvs. opposite; wing of fr. terminal .... **Oleaceae**
35. Fr. not a samara.
37. Fr. fleshy; petals 3 plus 3; leaflets >3................ **Berberidaceae**
37. Fr. dry; petals in one series.
38. Ovary one-loculed, becoming a legume................. **Fabaceae**
38. Ovary >one-loculed.
39. Fr. <two times as long as broad; petals 5..**Zygophyllaceae**
39. Fr. >two times as long as broad; petals 4.
40. Fr. inflated.. **Capparaceae**
40. Fr. not inflated.. **Brassicaceae**
33. Lvs. simple.
41. Fr. a legume; fls. red-purple... **Fabaceae**
41. Fr. not a legume.
42. Lvs. palmate; trees... **Platanaceae**
42. Lvs. pinnate; shrubs or trees.
43. Pungent odor absent.
44. Stamens 4-5.. **Rhamnaceae**
44. Stamens 10.
45. Lvs. >2 cm. long; plant of boggy places .... **Ericaceae**
45. Lvs. <2 cm. long; plant of dry places................. **Rosaceae**
43. Lvs. with pungent odor... **Lauraceae**
29. Plant herbaceous.
46. Sepals 2 or 3.

47. Plants fleshy, succulent; sepals 2, persistent .... **Portulacaceae**
47. Plants not succulent.
48. Sepals deciduous, 2 or 3; stamens 6-12 .... **Papaveraceae**
48. Sepals not early deciduous; stamens 6; lvs. dissected.
49. Stamens in 2 sets of 3 each; petals 4, in 2 dissimilar pairs. **Papaveraceae**
49. Stamens in 1 set; petals 3, alike........................... **Limnanthaceae**
46. Sepals 4 or 5, sometimes more.
50. Fls. definitely irregular.
51. Fls. papilionaceous; fr. a legume .... **Fabaceae**
51. Fls. not papilionaceous; fr. a capsule .... **Violaceae**
50. Fls. mostly regular.
52. Lvs. compound.
53. Fr. 1-loculed, forming a legume; lvs. pinnate .... **Fabaceae**
53. Fr. 2-5-loculed; not a legume.
54. Leaflets 10-16; lvs. opposite............................ **Zygophyllaceae**
54. Leaflets 3-7; lvs. alternate.
55. Stamens 10; lvs. sour, not ill-smelling.............. **Oxalidaceae**
55. Stamens 2-6; lvs. acrid, ill-smelling................. **Capparaceae**
52. Lvs. simple, may be deeply lobed or divided.
56. Sepals and petals 4.
57. Lvs. opposite; fls. axillary....................................... **Lythraceae**
57. Lvs. alternate; fls. in terminal racemes................ **Brassicaceae**
56. Sepals and petals 5(6).
58. Ovary appearing inferior; fls. 4-6 merous.............. **Lythraceae**
58. Ovary appearing superior.
59. Sepals 4; filaments 6, 4 (rarely 2), distinct...... **Brassicaceae**
59. Sepals mostly 5, filaments 5, 8 or 10, often united at base.
60. Fr. a many-seeded capsule................................. **Ericaceae**
60. Fr. formed by separation of carpels into 1-seeded units.
61. Stipules scarious; carpels with long styles in mature fr. **Geraniaceae**
61. Stipules absent; carpels lacking long stout styles; lvs. opposite.. **Limnanthaceae**

GROUP H

(Petals connate; ovary superior)

1. Stamens >5.
2. Corolla not urn-shaped or tubular, petals united only at base.
3. Pistils 4-5; stamens 10; plants succulent.. **Crassulaceae**
3. Pistil 1; plants not or little succulent.
4. Corolla irregular .... **Papaveraceae**
4. Corolla regular.
5. Plant a shrub or tree; fls. <5 mm wide, in dense heads................. **Fabaceae**
5. Plant herbaceous; fls. larger.
6. Stamens >10; lvs. simple, palmate... **Malvaceae**

6. Stamens 10; lvs. 3-foliate .... **Oxalidaceae**
2. Corolla urn-shaped or tubular, petals markedly united.
7. Stamens inserted in corolla-throat .... **Lennoaceae**
7. Stamens nearly or quite free from corolla .... **Ericaceae**
1. Stamens not >5.
8. Plants parasitic, lacking chlorophyll.
9. Twining or trailing vines with small fls .... **Cuscutaceae**
9. Root parasites.
10. Corolla regular; stamens 5 .... **Lennoaceae**
10. Corolla irregular; stamens 4 .... **Orobanchaceae**
8. Plants with chlorophyll.
11. Corolla regular.
12. Pistils 2; plants with milky juice.
13. Styles united; stamens closely pressed around stigma .... **Apocynaceae**
13. Styles distinct; filaments fused and adnate to style column. **Asclepiadaceae**
12. Pistil 1; plants lacking milky juice.
14. Ovary 1-loculed, 1-ovuled; the seemingly colored corolla really a calyx inside a 1-flowered involucre .... **Nyctaginaceae**
14. Ovary and corolla not as above.
15. Stamens as many as the corolla-lobes and opposite them... **Primulaceae**
15. Stamens as many as or fewer than the corolla lobes and alternate with them.
16. Corolla small, dry-papery; lvs. usually basal .... **Plantaginaceae**
16. Corolla not dry-papery.
17. Ovary commonly 4-lobed, each lobe forming a nutlet when mature (though some may abort); infl. usually a scorpioid cyme. **Boraginaceae**
17. Ovary not 4-lobed; fr. various, not forming nutlets.
18. Plants shrubby; fls. 3-5 cm. long, white .... **Ericaceae**
18. Plants herbaceous or, if shrubby, fls. smaller.
19. Style 3-cleft; ovary 3-loculed; capsule 3-valved. **Polemoniaceae**
19. Style not 3-cleft; ovary 1-2-loculed.
20. Calyx 4-5-toothed or cleft; style 1, entire.
21. Ovary 1-loculed; stigmas 2; lvs. opposite ... **Gentianaceae**
21. Ovary 2-loculed; stigma 1; lvs. alternate .... **Solanaceae**
20. Calyx of 5 distinct sepals; styles 2 or 1, usually divided.
22. Stems and lvs. rigid; lvs. and bracts spine-tipped. **Polemoniaceae**
22. Stems and lvs. looser; lvs. not spine-tipped.
23. Plants twining or trailing; infl. not coiled. **Convolvulaceae**
23. Plants erect or diffuse; infl. cymose, often coiled. **Hydrophyllaceae**

11. Corolla ± irregular.
  24. Fr. of 2-4 nutlets; lvs. opposite.
    25. Ovary not lobed; style apical, entire .... **Verbenaceae**
    25. Ovary 4-lobed; style ± basal, cleft at apex .... **Lamiaceae**
  24. Fr. a capsule; lf. positions variable.
    26. Ovary 1-loculed.
      27. Plants aquatic; lvs. dissected; stamens 2..................... **Lentibulariaceae**
      27. Plants terrestrial; lvs. not dissected; stamens 4................. **Martyniaceae**
    26. Ovary >1-loculed.
      28. Stigma-lobes 3; lvs. spine-tipped; small annuals............ **Polemoniaceae**
      28. Stigma-lobes 2; lvs. not spine-tipped; annuals to perennials.
        29. Tree; seeds winged.. **Bignoniaceae**
        29. Shrub or herb; seeds not winged .... **Scrophulariaceae**

GROUP I

(Petals connate; ovary inferior)

1. Stamens >5.. **Ericaceae**
1. Stamens 5 or fewer.
  2. Stamens distinct.
    3. Lvs. alternate; stamens 5.
      4. Plants herbaceous.. **Campanulaceae**
      4. Plants woody shrubs .... **Grossulariaceae**
    3. Lvs. opposite or whorled.
      5. Stamens 1-3; fls. irregular; fr. 1-seeded.................................. **Valerianaceae**
      5. Stamens 4-5; fls. mostly regular; fr. 1 to many-seeded.
        6. Infl. many-flowered, cylindric, spiny..................................... **Dipsacaceae**
        6. Infl. few to many-flowered, open, not spiny.
          7. Lvs. opposite and with stipules, or whorled and without stipules. **Rubiaceae**
          7. Lvs. opposite or perfoliate, not whorled or with stipules. **Caprifoliaceae**
  2. Stamens united by anthers.
    8. Plants with tendrils; lvs. palmate.. **Cucurbitaceae**
    8. Plants without tendrils.
      9. Fls. in involucrate heads; stamens attached to corolla.................. **Asteraceae**
      9. Fls. not in involucrate heads; stamens free from corolla...... **Campanulaceae**

ACERACEAE

Acer (p. 280)

Trees with opposite simple or compound lvs.; fls. in drooping clusters, imperfect or perfect, with or without petals; pistils 1 with 2-lobed ovary and 2 styles; fr. 2-winged samara. About 120 species, 4 in CA.

1. Lvs. simple, palmately lobed; petals present.
   2. Shrub or small tree to 6 m tall; lvs. <6 cm broad...................... **A. glabrum** Pursh.
      3. Lvs. 3-5 cm broad; twigs reddish........................... var. **torreyi** (Greene) Smiley
      (Steep rocky canyon in Inspiration Point Botanical Area, Piute Mtns., Shevock 8705)
      3. Lvs. <3 cm broad; twigs whitish-gray............... var. **diffusum** (Greene) Smiley
      (Sierra crest south to Mt. Jenkins, Shevock 11215)
   2. Tree to 30 m tall; lvs. 10-25 cm broad..................................... **A. macrophyllum**
1. Lvs. pinnate with 3-5 lfts.; petals absent..................... **A. negundo** var. **californicum**

AIZOACEAE

Herbs or low shrubs, erect or prostrate, often fleshy; lvs. mostly opposite; fls. regular; sepals 4-8; petals 0 to many; stamens numerous; ovary superior or inferior; styles 3-many; fr. a circumscissile capsule.

1. Petals absent; calyx sometimes colored.
   2. Stipules present, papery; stamens 5-10... **Trianthema**
   2. Stipules absent; stamens >10; sepals rose-purple.................................. **Sesuvium**
1. Petals present... **Mesembryanthemum**

Mesembryanthemum

Low prostrate freely branching glandular succulent annuals with thickened lvs.; fls. solitary in axils; petals white, about 4 mm long. About 74 species mostly from Africa, 2 introduced into CA.

Lvs. linear, somewhat terete; plants covered with fine colorless vesicles.
M. nodiflorum L. (alkaline flats near Lokern,Yinger 158)
Lvs. spatulate, flat; plants covered with large vesicles...................... **M. crystalinum** L.
(alkaline flats near Buena Vista Lake, S. Cooley 234)

Sesuvium (p. 232)

Prostrate or decumbent glabrous perennial; lvs. opposite, papillate; fls. solitary, axillary; fl.-tube turbinate; sepals 5, purplish or rose-pink; petals 0; stamens 5-many; styles 3-5. About 5 species, one in CA. One species in Kern County: **S. verrucosum**

Trianthema (p. 232)

Prostrate diffusely-branched annuals; lvs. opposite, unequal; fls. solitary, axillary; sepals 5; petals 0; stamens 5-10, perigynous. About 20 species, one in CA. One species introduced in Kern County: **T. portulacastrum**

ALISMATACEAE

Mostly perennial aquatic or marsh plants with scapose stems and basal long-petioled lvs.; fls. in whorls, racemes or panicles; sepals 3, green; petals 3, white to pinkish; fr. a whorl or bunch of achenes.

1. Carpels in fr. spreading, long-beaked; petals toothed .... **Damasomium**
1. Carpels not spreading or long-beaked; petals entire.
    2. Upper fls. with stamens only; fruits winged .... **Sagittaria**
    2. Upper fls. perfect; fruits not winged.
        3. Fruits in a ring; stamens 6 .... **Alisma**
        3. Fruits in a dense head; stamens 6-30 .... **Echinodorus**

Alisma (p. 174)

Perennial herbs with erect or floating lvs.; fls. many in panicle of whorled branches; achenes 2-3-ribbed and curved on back. About 9 species, 3 in CA. One species in Kern County: **A. plantago-aquatica** L. (*A. triviale*)

Damasomium (p. 174)

Perennial scapose herb with erect or floating long-petioled lvs. with 3-5 veins; fls. white, in simple panicle; stamens 6, 2 opposite each sepal. About 5 species, one in CA. One species in Kern County: **D californicum**

Echinodorus (p. 175)

Annual to perennial herb with broadly ovate basal lvs. marked with little dots or lines; fls. white; fruiting heads subglobose, spinose by persistent firm beaks on achenes. About 50 species, one in CA. One species in Kern County: **E. berteroi**

Sagittaria (p. 175)

Perennial aquatic or marsh herbs with stolons or tubers and milky juice; outer lvs. long linear, inner lvs. petioled and shaped like arrowheads; fls. white, conspicuous; the pistillate below the staminate; carpels many, crowded, forming flat achenes. About 20 species, 5 in CA.

1. Sepals of pistillate fls. erect and growing with age.
        S. montevidensis Cham. & Schldl. ssp. **calycina** (Engelm.) Bogin(*S. calycina)*
1. Sepals of pistillate fls. reflexed or spreading.

2. Beak of achene on the side, well developed; fruiting heads 15-30 mm in diam. **S. latifolia**
2. Beak of achene terminal, minute; fruiting heads 8-15 mm in diam.... **S. longiloba**

AMARANTHACEAE

Annual to perennial subshrub with simple alt. or opp. lvs.; fls. minute, uni- or bisexual, arranged in clusters with 1-3 bracts per fl.; petals 0; fr. 1-seeded.

Stem-lvs. alt.; fl. unisexual; herbs...**Amaranthus**
Stem-lvs. opp.; fl. bisexual; subshrub..**Tidestromia**

Amaranthus (p. 230)

Usually coarse, erect to prostrate branched monoecious herbs; lvs. alt., entire, petioled; fls. 3-bracted, small, green to purplish, mostly staminate and pistillate intermixed. About 60 species, 15 in CA.

1. Sepals of pistillate fls. broadened upward; calyx ± urn-shaped................. **A. palmeri**
1. Sepals of pistillate fls. narrowed upward; calyx not urn-shaped.
   2. Infl. of terminal or axillary spikes; sepals 5.
      3. Sepals long-linear, at least one exceeding fr., all spreading at maturity. **A. retroflexus**
      3. Sepals lanceolate, not exceeding fr., erect .... **A. powellii**
   2. Infl. wholly of axillary compact clusters.
      4. Sepals 4-5; stems often purplish.................. **A. blitoides** Wats. (*A. graecizans*)
      4. Sepals 1-3.
         5. Sepals of pistillate fls. 2-3, only one well-developed; stems prostrate, much-branched, reddish. .... **A. californicus**
         5. Sepals of pistillate fls. 3, subequal; stems ascending, bushy-branched, pale. **A. albus**

Tidestromia

± mounded grayish-white subshrub with opposite (upper) entire, asymetric lvs.; hairs branched; fls. in clusters of 1-5 per lf. axil, subtended by invol. of 2-3 bract-like lvs. About 7 species, 2 in CA.One species in Kern County: **T. oblongifolia** (Wats.) Standley (Red Rock Canyon, Twisselmann T16,308)

ANACARDIACEAE

Trees or shrubs with acrid resinous sap; lvs. alt., compound; fls. small, in terminal or axillary clusters; fr. a dry berry-like drupe.

1. Stamens 5; lvs. 3-foliate; shrubs.
   2. Fls. whitish in loose axillary panicles; fr. glabrous, white........... **Toxicodendron**
   2. Fls. yellowish in sessile spikes; fr. hairy, red .... **Rhus**
1. Stamens 10; lvs. pinnately compound; tree ....**Schinus**

Rhus (p. 279)

Diffusely branched shrub, 8-14 dm tall; lvs. 3-foliate, pubescent. About 150 species, 3 in CA. One species in Kern County: **R. trilobata**

Schinus (p. 279)

Dioecious evergreen tree, 5-15 m tall; lvs. alt., pendent, odd-pinnate; 2-3 dm long; fr. reddish, 6-8 mm diam. About 25 species, one in CA. One species introduced into Kern County: **S. molle**

Toxicodendron (p. 279)

Erect shrub, 1-3 dm tall, deciduous; lvs. pinnately 3-foliate, bright green and shining above, aging red; fr. white, glabrous, 4-7 mm diam. About 15 species, one in CA. One species in Kern County: **T. diversilobum** (T. & G.) Greene (*Rhus diversiloba*)

APIACEAE
(Umbelliferae by some authors)

Aromatic herbs usually with hollow stems and alt. compound lvs. with basally dilated petioles; fls in simple or compound umbels or rarely a head; ovary inferior; styles 2 with swollen base (stylopodium); fr. dry, the 2 carpels separating at maturity.

1. Infl. a head, not an umbel.
    2. Fr. not winged; lvs., bracts and sepals spinose-tipped............................**Eryngium**
    2. Fr. winged; spinose tips absent .... **Cymopteris**
1. Infl. a distinct umbel, not a head.
    3. Ovaries and frs. bearing prickles, bristles or tubercles.
        4. Ovaries and frs. with spines, hooked bristles, or tubercles.
            5. Perennial; fls. perfect and staminate ....**Sanicula**
            5. Annual; fls. all perfect.
                6. Plants glabrous; lf.-divisions ± elongate, filiform; fr.-ribs papillose-roughened..**Apiastrum**
                6. Plants ± pubescent; lf.-divisions shorter; frs. with hooked bristles ... **Yabea**
        4. Ovaries and frs. with bristles which are not hooked, but may be barbed.
            7. Frs. several times longer than wide .... **Osmorhiza**
            7. Frs. not more than 2-times as long as wide; with some barbed bristles . **Daucus**
    3. Ovaries and frs. not bearing prickles, bristles or tubercles.
        8. Lvs. simple; umbels simple or infl. a spike.
            9. Ovaries and frs. glabrous; lvs. glabrous .... **Hydrocotyle**
            9. Ovaries and frs. with stellate hairs; lvs. stellate-pubescent............... **Bowlesia**
        8. Lvs. and umbels variously compound.
            10. Ribs of frs. not prominently winged; frs. rounded in cross- section or somewhat flattened laterally.
                11. Fls. yellow.

12. Involucel absent; lf.-divisions filiform; plants with anise odor. **Foeniculum**
12. Involucel present; lf.-divisions linear to ovate; plants lacking anise odor. .... **Tauschia**

11. Fls. not yellow.
13. Frs. elongate, several times longer than wide .... **Osmorhiza**
13. Frs. roundish to oblong, not more than 2-times as long as wide.
14. Plants with celery odor; annuals.......... **Apium**
15. Annual, 0.5 - 5 dm tall .... **Ciclospermum**
15. Perennial, 5 - 15 dm tall.......... **Apium**
14. Plants without celery odor; perennials.
16. Plant stemless, low; invol. absent.......... **Oreonana**
16. Plant caulescent with stems rather tall; invol. usually present.
17. Stems purple-dotted; lvs. decompound into small divisions. **Conium**
17. Stems not purple-dotted; lvs. pinnately or ternate-pinnately (in 3s) divided.
18. Lvs. all once-pinnate.......... **Berula**
18. Lvs. pinnately or ternate-pinnately divided.
19. Stylopodum prominent; fr.-ribs not corky; lf.-divisions few. **Perideridia**
19. Stylopodium none or low; fr.-ribs corky.
20. Styles 1/5 to 1/3 as long as fr.; fr. broadly ovoid to roundish.......... **Cicuta**
20. Styles about 1/2 as long as fr.; fr. sub-cylindric. **Oenanthe**

10. Ribs of frs. (all or some of them) winged; frs. ± dorsally compressed.
21. Lateral ribs winged, dorsal ribs filiform.
22. Stems present; lvs. pinnate into ovate divisions; fls. white; wet places. **Oxypolis**
22. Stems absent, lvs. pinnately or ternate-pinnately divided into linear divisions; fls. yellow or purple.......... **Lomatium**
21. Lateral, dorsal and intermediate ribs winged or prominent.
23. Stems present, leafy, 5-16 dm tall.
24. Umbellets not capitate .... **Angelica**
24. Umbellets capitate .... **Sphenosciadium**
23. Stems absent, to 4 dm tall .... **Cymopterus**

Angelica (p. 296)

Stout subglabrous perennial herb; lvs. pinnately compound in threes; sheaths often inflated and bladeless; infl. of loose compound umbels; fls. white to pinkish; fr. dorsally flattened, winged. About 50 species, 8 in CA.

Lf.-divisions linear, <1 cm broad, entire; frs. 10-15 mm long.......... **A. lineariloba**
Lf. divisions lance-shaped, 1.5-4 cm broad; frs. 3-5 mm long.
A. callii Mathias & Constance (Common along Shirley Creek, Shevock 10774)

Apiastrum (p. 296)

Smooth branching annual with finely dissected lvs.; corolla white; frs. ellipsoid heart-shaped with inconspicuous ribs. One species. One species introduced into Kern County: **A. angustifolium**

Apium (p. 296)

Glabrous perennial herb with pinnate lvs.; involucre and involucel absent; fls. whitish; frs. subglobose, glabrous. About 20 species, 2 introduced in CA. One species in Kern County: **A. graveolens**

Berula (p. 296)

Erect branched glabrous perennial; lvs. pinnately compound; involucre and involucels prominent; fls. white; frs. small, oval, glabrous. One species. One species introduced into Kern County: **B. erecta**

Bowlesia (p. 296)

Slender branching annual with stellate pubescence; lvs. opposite, simple; umbels axillary; frs. ovoid, stellate pubescent. About 16 species, one in CA. One species in Kern County: **B. incana**

Ciclospermum (p. 296)
(part of *Apium*)

Glabrous taprooted annual with lvs. pinnately dissected into linear segments; invols. absent; fr. 1.2-3 mm wide. About 3 species, one introduced in CA. One species in Kern County: **C. leptophyllum** (Pers.) Britton & Wilson (Lawn weed, CSUB, Moe 1143)

Cicuta (p. 296)

Glabrous branching stout perennial herb of wet places; lvs. 1-3 pinnate with serrate to incised lfts.; fls. greenish-white; frs. ovoid to subglobose, small. About 8 species, 2 in CA. One species in Kern County: **C. douglasii**

Conium (p. 297)

Tall biennial glabrous herb with spotted stems and pinnately more than once compound lvs.; invol. conspicuous; fls. white; frs. ovoid, laterally flattened, glabrous. About 6 species, one in CA. One species introduced into Kern County: **C. maculatum**

Cymopterus (p. 297)

Glabrous perennials from subterranean stems with tuft of lvs. and peduncles at surface of ground; lvs. 2-3 times pinnately divided; frs. oblong-ovoid, winged. About 50 species, 10 in CA.

Umbels congested and globose; fls. purple.
C. deserticola Brand. (Edwards AFB, CA City, Moe)
Umbels loose, rays evident; fls. greenish yellow .... **C. panamintensis**

Daucus (p. 297)

Pubescent annual or biennial; lvs. pinnately more than once compound; involucre and involucel present; fls. white; frs. oblong to ovoid, dorsally flattened. About 20 species, 2 in CA.

Bracts pinnately divided into short linear divisions; rays 4-40 mm long; central fl. of umbellet white.. **D. pusillus**
Bracts pinnately divided into elongate filiform divisions; rays 30-70 mm long; central fl. rose-purple.. **D. carota**

Eryngium (p. 297)

Glabrous perennial herb with pinnatifid spinulose lvs.; infl. capitate with spinose bracts; frs. ovoid, covered with white scales. About 200 species, 11 in CA. One species in Kern County: **E. spinosepalum**

Foeniculum (p. 298)

Erect perennial with strong anise odor; lvs. more than once compound into linear divisions; fls. yellow in large compound umbels; involucres absent; frs. oblong, glabrous. One species: **F. vulgare**

Hydrocotyle (p. 298)

Low perennials with creeping stems, rooting at nodes; lvs. long-petioled, not sheathing; fls. greenish-white, in axillary clusters. About 100 species; 5 in CA.

Lvs. peltate; ribs of fr. thick & corky ....**H. verticillata**
Lvs. round-reniform, not peltate; ribs of fr. filiform...................... **H. ranunculoides** L.
(shady slopes of Kern Canyon, Moe 1520)

Lomatium (p. 298)

Stemless or short-stemmed perennials with 1-several times compound lvs.; fls. white, yellow or purple in compound umbels; fr. strongly flattened dorsally. About 75 species, 35 in CA.

1. Peduncles conspicuously inflated and swollen at apex; fls. yellow........**L. nudicaule**
1. Peduncles not swollen at apex; fl.-color various.
  2. Lvs. 3 times or twice 3 times compound with large divisions; fls. yellow. **L. californicum**
  2. Lvs. more than once compound, dissected into many small divisions; fl.-color various.
    3. Ovaries and young fr. variously pubescent or roughened.
      4. Bractlets of involucel obovate; fls. yellow..... **L. utriculatum** (incl. *L. vaseyi*)
      4. Bractlets linear or reduced; fl.-color various.
        5. Bractlets with conspicuous papery margin, not tomentose or villous; fls. white. .... **L. nevadense**
        5. Bractlets not conspicuously papery-margined, ± villous or tomentose.
          6. Plants stemless, 1-3 dm tall; fls. purple............................. **L. mohavensis**
          6. Plants with leafy stem, 3-5 dm tall.
            7. Petals glabrous; frs. sparingly pubescent with long hairs; fls. white, yellow or purplish. .... **L. macrocarpum**
            7. Petals tomentose; frs. densely pubescent; fls. greenish-white to purplish. .... **L. dasycarpum**
              8. Pedicels longer than mature fr .... ssp. **dasycarpum**
              8. Pedicels shorter than mature fr. .... ssp. **tomentosum**
    3. Ovaries and fr. glabrous.
      9. Bractlets obovate to lanceolate, sometimes connate.
        10. Plants usually with several cauline lvs.; wings of frs. broader than body; fls. yellow. .... **L. utriculatum** (incl. *L.vaseyi*)
        10. Plants with 0-1 stem lf.; wings of frs. narrower than body; fl.-color various.
          11. Fls, yellow; leaflets linear .... **L. caruifolium**
          11. Fls. purple; leaflets ovate to elliptic, sharply dentate. **L. shevockii** Hartman & Const. (Owens Peak & Mt. Jenkins, eastern crest of Sierra)
      9. Bractlets filiform to lance-linear.
        12. Bractlets tomentose or villous; fls. white, yellow or purple. **L. macrocarpum**
        12. Bractlets mostly glabrous; fls. yellow to purplish. .... **L. dissectum**

Oenanthe (p. 299)

Succulent decumbent and ascending glabrous perennial of wet places; lvs. bipinnate with dentate to incised lfts.; fls. white; fr. oblong, purplish, glabrous. About 30 species, 2 in CA. One species introduced into Kern County: **O. sarmentosa**

Oreonana (p. 299)

Low tufted stemless perennial; lvs. pinnate to pinnate in threes; infl. globose, on elongated peduncle from tuft of lvs.; fls. white to purplish. Three species endemic to CA. One species in Kern County: **O. clementis**

Osmorhiza (p. 299)

Stout perennials from thick roots; lvs. pinnate in threes with ovate to lance-oblong lfts.; fls. yellowish; frs. linear oblong; crushed lvs. have odor of licorice. About 10 species, 5 in CA.

Involucels conspicuous, of several bractlets .... **O. brachypoda**
Involucels absent or rudimentary .... **O. chilensis**

Oxypolis (p. 299)

Slender erect glabrous perennial of wet places; lvs. pinnate with scalloped to serrate to incised lfts.; fls. white in loose compound umbels; frs. oval to oblong, glabrous. About 6 species, one in CA. One species in Kern County: **O. occidentalis**

Perideridia (p. 299)

Erect branching perennial herbs; lvs. compound in threes; infl. of loose compound umbels, invols. present; fls. white to pinkish; frs. laterally flattened. About 12 species, 11 in CA.

1. Basal lvs. 1-2 times compound, petioles and rachis not dilated, ultimate divisions not dimorphic.. **P. parishii**
1. Basal lvs. >2 times compound, petioles and rachis dilated, ultimate divisions dimorphic.
   2. Rays 10-20, 1-3 cm long; umbels ± compact .... **P. bolanderi**
   2. Rays 5-10, 3-8 cm long; umbels loose .... **P. pringlei**

Sanicula (p. 300)

Perennial, mostly stemless herbs with pinnately or palmately compound lvs.; sepals persistent, fls. yellowish to purple; frs. covered with tubercles or hooked hairs. About 40 species, 13 in CA.

1. Basal lvs. palmately divided; stem branched above the base; fls. yellow. **S. crassicaulis**
1. Basal lvs. pinnatifid or pinnately dissected.
   2. Lvs. with toothed margin on rachis; fls. yellow to purplish-red; fr. covered with prickles.. **S. bipinnatifida**
   2. Lvs. lacking margin on rachis; fls. yellow.
      3. Stems from tapering taproots; frs. prickly.

4. Lvs. 1-to 3-pinnate; staminate fls. few, inconspicuous................**S. bipinnata**
4..Lvs.pinnately compound in threes; staminate fls. several, conspicuous. **S.graveolens**
3. Lvs. from globose tuber; frs. lacking prickles .... **S. tuberosa**

Yabea (p. 296)
(*Caucalis*)

Hispid annual with pinnately dissected 2-3 times compound lvs.; corolla white; frs. oblong with hooked bristles. One species : **Y. microcarpa** (Hook. & Arn.) Koso-Polj.

APOCYNACEAE

Perennial herbs with milky juice and opposite entire lvs.; fls. regular, 5-merous; petals connate; stamens 5, alternate corolla-lobes; carpels 2, distinct or ± fused; fr. of 2 follicles.

Plant creeping or trailing; seeds without a tuft of hair.. **Vinca**
Plant not trailing; seeds with a tuft of hair ....**Apocynum**

Apocynum (p. 304)

Perennial herbs with small fls. <1 cm long; corolla cylindric or urn-shaped, whitish; stamens inserted at base of corolla; follicles 2; seeds comose. About 7 species, 2 in CA.

Corolla 4-6 mm long, twice the length of calyx; lvs. spreading. **A. androsaemifolium** L. (*A. medium* var. *floribundum*)
Corolla and calyx 2-3 mm long; lvs. ascending. **A. cannabinum** (incl. var. *glaberrimum*)

Vinca (p. 304)

Trailing perennial evergreen herb; fls. solitary in axil of every other lf., showy, 2-3 cm long, violet or blue; corolla with slender tube; follicles cylindric. About 6 species, one in CA. One species introduced into Kern County: **V. major**

ARECACEAE
(Palmae by some authors)

Washingtonia (p. 198)

Tall columnar unbranched tree clothed in dense thatch of dead drooping lvs. (often burned away); lvs. large, fan-shaped, in a tuft at top of tree. Two species, one in CA. One species in Kern County: **W. filifera**

ASCLEPIADACEAE

Asclepias (p. 304)

Perennial herbs with opposite or whorled lvs. and a milky juice; fls. perfect, regular, in axillary or terminal umbels; corolla 5-lobed, the lobes mostly reflexed and with a crown which is usually columnar and each of its 5-lobes hood-like, each with a horn-like projection, sometimes hornless; filaments connate into a tube attached to corolla tube, each anther winged and with a papery tip inflexed on summit of stylar disk; pollen grains united into sac-like pollinia; fr. a follicle; seeds with tuft of hair. About 100 species, 15 in CA.

Asclepias (p. 304)

1. Hoods hornless.
   2. Plants white-tomentose; corolla-lobes purplish, hoods dark maroon. **A. californica**
   2. Plants glabrous to ± puberulent, often tinged purple; corolla- and calyx-lobes red-purple; hoods purple. .... **A. cordifolia**
1. Hoods with horns on inner surface; corolla greenish-white, tinged purple.
   3. Lvs. narrow, <1 cm wide; horns slender, incurved .... **A. fascicularis**
   3. Lvs. broader, >1 cm wide; woolly.
      4. Lateral umbels sessile; horns included within hoods; corolla-lobes 6-7 mm long .... **A. vestita**
      4. Lateral umbels well peduncled; corolla-lobes 4-6 mm long.
         5. Lvs. whorled; horns mostly included within hoods .... **A. eriocarpa**
         5. Lvs. opposite; horns exserted .... **A. erosa**

ASTERACEAE

(Compositae by some authors)

Herbs or shrubs without stipules; fls. borne in a head on a common receptacle surrounded by an invol. (phyllaries); some fls. subtended by bracts or bristles, or not; corolla of fused petals, regular and tubular (disk) or irregular with strap-shaped limb (ray); calyx reduced to scales or bristles (pappus) surmounting inferior ovary; styles 2-lobed. A very large family subdivided into several tribes.

Key to Genera of ASTERACEAE

1. Plants with spiney lf.-margins and/or phyllary tips; florets all tubular with 5 long narrow lobes (thistles) .... **Group A**
1. Plants lacking spines on lf.-margins and/or phyllary tips.
   2. Florets all of ray type; plants with milky juice .... **Group B**
   2. Some or all florets of disk type; plants lacking milky juice.
      3. Pappus evident.
         4. Ray florets none or inconspicuous.
            5. Pappus of scales and/or awns .... **Group C**
            5. Pappus of capillary bristles .... **Group D**

4. Ray florets present.
   6. Pappus of firm awns or scales
      7. Rays yellow or orange .... **Group E**
      7. Rays white to reddish to purplish .... **Group F**
   6. Pappus of slender bristles .... **Group G**
3. Pappus none or reduced to a crown.
   8. Ray florets absent .... **Group H**
   8. Ray florets evident .... **Group I**

Group A

(spines present, florets tubular)

1. Achenes attached at their base; marginal corollas not enlarged.
   2. Lvs. white-mottled along veins; pappus not feathery .... **Silybum**
   2. Lvs. not white-mottled.
      3. Pappus of short scales in several series; annuals with yellow or orange fls. **Carthamus**
      3. Pappus feathery; fls. white, pink to purplish .... **Cirsium**
1. Achenes attached obliquely; marginal corollas ± enlarged.
   4. Lvs. lacking spines; heads not subtended by upper lvs .... **Centaurea**
   4. Lvs. prickly on margins; heads subtended by upper lvs .... **Cnicus**

Group B

(all florets ray-like, milky sap)

1. Pappus partly or completely of thin dry scales, or feathery, or with feathery awns.
   2. Pappus of thin dry scales, often awned above scale-like base.
      3. Coarse perennials, not scapose; corollas blue to whitish; pappus of short blunt scales .... **Cichorium**
      3. Scapose annuals (perennial in 1 species); corollas yellow to whitish; pappus-scales bristle-tipped.
         4. Heads erect in bud; pappus-scales 5, notched at tip .... **Uropappus**
         4. Heads nodding in bud; pappus-scales 5-10, notched or not.
            5. Perennial; fls. conspicuous; pappus-scales 5-10, not notched at tip. **Microseris**
            5. Annual; fls. inconspicuous, barely equaling phyllaries; pappus-scales 2-5.
               6. Heads strongly nodding in bud; pappus-scales broad, not notched at tip. **Microseris**
               6. Heads ± nodding in bud; pappus-scales narrowly lanceolate, notched at tip .... **Stebbinsoseris**
   2. Pappus conspicuously feathery.
      7. Achenes mostly with long slender beaks.
         8. Stems scapose; annuals .... **Hypochaeris**
         8. Stems leafy.
            9. Lvs. grass-like; fls. yellow; perennials .... **Tragopogon**
            9. Lvs. pinnately divided; fls. whitish; annuals .... **Rafinesquia**
      7. Achenes truncate.

10. Fls. pale yellow; phyllaries with reddish midrib; low scapose annuals. **Anisocoma**
10. Fls. rose to purplish; plants much-branched .... **Stephanomeria**
1. Pappus of capillary bristles, not feathery.
11. Achenes ± flattened.
12. Achenes beaked; lvs. spiny-toothed .... **Lactuca**
12. Achenes not beaked; lvs. spiny-serrate ....**Sonchus**
11. Achenes not flattened, either angled or round in cross section.
13. Fls. pink to rose; annuals .... **Prenanthella**
13. Fls. yellow to whitish.
14. Stems scapose.
15. Achenes truncate; pappus-bristles barbellate.............................**Microseris**
15. Achenes beaked; pappus-bristles not barbellate.
16. Achenes 10-15-ribbed, not spinulose above; pollen bright orange. **Agoseris**
16. Achenes 4-5-ribbed, spinulose above; pollen yellow........... **Taraxacum**
14. Stems leafy.
17. Achenes bumpy between the angles, beaked.
18. Infl. not glandular; achenes abruptly short-beaked....... **Glyptopleura**
18. Infl. with tack-shaped glands; achenes tapering to the beak. **Calycoseris**
17. Achenes with fine lines between the angles.
19. Pappus fragile (falls apart quickly).
20. Achenes beaked ....**Crepis**
20. Achenes truncate .... **Malacothrix**
19. Pappus persistent.
21. Achenes beaked; pappus white ....**Crepis**
21. Achenes truncate; pappus brownish .... **Hieracium**

Group C

(rays absent, pappus of scales or awns)

1. Receptacle with bracts or bristles at base of fls.
2. Pappus of plumose awns; plants nearly leafless shrubs...............................**Bebbia**
2. Pappus not of plumose awns; plants leafy.
3. Achenes with toothed wings; annuals with small nodding heads.............**Dicoria**
3. Achenes lacking wings.
4. Herbs; heads thistle-like but spineless .... **Centaurea**
4. Shrubs; heads not thistle-like.
5. Lvs. ovate to broadly lanceolate, ± 3-veined ....**Encelia**
5. Lvs. linear, 1-veined .... **Eastwoodia**
1. Receptacle lacking bracts at base of fls.
6. Pappus a mere crown which may be 2-lobed; phyllaries with thin dry margin. **Chamomilla**
6. Pappus of flat or awn-like scales; phyllaries not thin-margined.
7. Pappus of 20-30 silvery awns; plants shrubby .... **Acamptopappus**
7. Pappus with fewer awns; herbaceous or woody only at base.

8. Pappus-scales irregularly toothed or ciliate.
  9. Pappus-scales 11-17, deciduous in a ring .... **Orochaenactis**
  9. Pappus-scales fewer, deciduous separately; woolly annuals.
    10. Lvs. 3-lobed at tip; plants 3-8 cm high .... **Eriophyllum**
    10. Lvs. linear, entire to dentate; plants taller .... **Lembertia**
8. Pappus-scales mostly entire.
  11. Pappus of hyaline scales; annuals and perennials with dissected lvs.
    12. Fls. white to pinkish .... **Chaenactis**
    12. Fls. yellow .... **Eriophyllum**
  11. Pappus of 3-5 subulate awns; slender annuals with entire linear lvs. **Rigiopappus**

Group D

(rays absent, pappus of slender bristles)

1. Phyllaries completely hyaline or thin and dry; plants ± white-woolly.
  2. Plants taprooted; heads with outer pistillate and central perfect or staminate fls.
    3. Receptacle with bracts at base of outer fls.; small annual (to 30 cm tall)... **Filago**
    3. Receptacle lacking bracts; plants mostly larger .... **Gnaphalium**
  2. Plants fibrous-rooted, dioecious; low mat-like perennial .... **Antennaria**
1. Phyllaries herbaceous or only partly thin and dry or hyaline; plants mostly not white-woolly.
  4. Heads unisexual; plants mostly shrubby .... **Baccharis**
  4. Heads not strictly unisexual.
    5. Plants well-developed shrubs.
      6. Phyllaries 4-7, of equal length in single series .... **Tetradymia**
      6. Phyllaries more numerous.
        7. Phyllaries 9-18, in a single series; lvs. slender, resin-varnished. **Peucephyllum**
        7. Phyllaries in >one series; lvs. various.
          8. Lvs. scale-like, at least on flowering branches.
            9. Achenes silky; heads solitary .... **Aster**
            9. Achenes glabrous; heads clustered .... **Lepidospartum**
          8. Lvs. not scale-like.
            10. Phyllaries in ± distinct vertical ranks .... **Chrysothamnus**
            10. Phyllaries more irregularly arranged, not in vertical ranks.
              11. Fls. whitish to purplish, not yellow.
                12. Corolla purple; perennial to 12 dm tall, coarse .... **Pluchea**
                12. Corollas whitish to tinged red; much-branched shrubs. **Brickellia**
              11. Fls. yellow.
                13. Receptacle with bracts subtending fls.; branches slender and brittle .... **Bebbia**
                13. Receptacle naked; branches stouter.
                  14. Lvs. toothed.
                    15. Corolla abruptly expanded above tube .... **Isocoma**
                    15. Corolla gradually expanded above tube .... **Hazardia**

14. Lvs. entire.
16. Corolla abruptly expanded above tube...................... **Isocoma**
16. Corolla gradually expanded above tube.............. **Ericameria**
5. Plants herbaceous.
17. Annuals.
18. Plants white-woolly, aromatic; lvs. broad................................ **Psathyrotes**
18. Plants not white-woolly.
19. Phyllaries equal, in single series .... **Senecio**
19. Phyllaries unequal, in several series.
20. Outer corollas enlarged; plants often strong-smelling.......... **Lessingia**
20. Outer corollas mostly not enlarged.
21. Slender-stemmed annuals; heads solitary, fls. blue to pinkish white, thistle-like. .... **Centaurea**
21. Coarse annuals; heads several, with inconspicuous rays..... **Conyza**
17. Biennials to perennials.
22. Principal phyllaries ± equal, in single series.
23. Lvs. opposite .... **Arnica**
23. Lvs. alternate above.
24. Pappus-bristles plumose... **Raillardella**
24. Pappus of white soft bristles but not plumose......................... **Senecio**
22. Principal phyllaries ± equal in 2-several overlapping series.
25. Fls. red-purple to pinkish-white.. **Eupatorium**
25. Fls. yellow.
26. Phyllaries scarcely overlapping...**Erigeron**
26. Phyllaries plainly overlapping.
27. Phyllary-tips spreading at right angles..................**Machaeranthera**
27. Phyllary-tips not spreading.
28. Invol. bell-shaped or cylindric.
29 Lvs. narrowly oblanceolate; phyllaries in 5-7 series. **Heterotheca**
29. Lvs. broadly triangular; phyllaries in 2 series........... **Ageratina**
28. Invol. ovoid to ± cylindric, thistle-like.
30. Pappus-bristles not plumose; phyllary-tips fringed or spiny. **Centaurea**
30 Pappus-bristles plumose above; phyllary-tips not fringed or spiny. ....**Acroptilon**

Group E

(rays present, pappus of awns or scales, fls. yellow or orange)

1. Receptacles with bracts or bristles at base of fls.
2. Invol. of 1-several series, none of phyllaries enfolding ray-achenes.
3. Receptacles bristly ....**Gaillardia**
3. Receptacles with bracts.
4. Phyllaries in 2 distinct dissimilar series; achenes flattened perpendicular to radius of head.
5. Pappus of 2-6 firm, beaked awns...**Bidens**

        5. Pappus of 2 minute teeth.. **Coreopsis**
      4. Phyllaries in 2 or more similar series.
        6. Rays fertile, their achenes developing..**Wyethia**
        6. Rays neutral, their achenes not developing.
          7. Achenes ± thickened; pappus of 2 principal awns..................**Helianthus**
          7. Achenes quite flat.
            8. Low rounded shrub; awns 1-2, weak..**Encelia**
            8. Herbs; awns 2 with intermediate scales............................ **Helianthella**
  2. Invol. of 1 series; each phyllary partly or completely enclosing an achene.
    9. Ray-achenes completely enclosed by phyllary.
      10. Fls. red-brown; pappus silvery-papery...................................**Achyrachaena**
      10. Fls. yellow; pappus various.. **Layia**
    9. Ray-achenes half enclosed by phyllary...**Hemizonia**
1. Receptacle lacking bracts or bristles.
  11. Invols. of >3 series of graduated phyllaries.
    12. Herbs; heads large and gummy; lvs. 7-15 mm wide..........................**Grindelia**
    12. Shrubs; heads smaller; lvs. linear, <3 mm wide............................ **Gutierrezia**
  11. Invols. of 1-3 series of ± equal but not graduated phyllaries.
    13. Lvs. opposite; low annuals...**Lasthenia**
    13. Lvs. alternate; biennials to perennials.
      14. Short-lived perennials; lvs. petioled; pappus of 5-8 scales.............**Helenium**
      14. Robust perennials; lvs. sessile or clasping; pappus of 4 basally-united scales... **Hulsea**

Group F

(rays present, pappus of awns or scales, fls. white to purple)

1. Outer achenes ± enclosed by subtending phyllary.
  2. Disk-florets staminate (no achenes), glabrous; pappus none..................**Hemizonia**
  2. Disk-florets producing achenes, hairy; pappus feathery .... **Blepharizonia**
1. Outermost achenes not enclosed by phyllary.
  3. Perennials; phyllaries marked with oil glands; fls. purplish; plants with very strong odor..**Nicolletia**
  3. Annuals; phyllaries lacking obvious oil glands.
    4. Achenes compressed, 2-edged..**Perityle**
    4. Achenes ± thickened, 4-5-angled; very low annuals.
      5. Plants woolly; upper lvs. not subtending heads .... **Eriophyllum**
      5. Plants not woolly; upper lvs. subtending heads ....**Monoptilon**

Group G

(rays present, pappus of slender bristles)

1. Rays yellow
  2. Phyllaries many, usually ± overlapping (like shingles).
    3. Ray-achenes lacking pappus; tall leafy herbs................................. **Heterotheca**
    3. Ray-achenes with a pappus.
      4. Pappus in 1 series.

5. Plants shrubby.
6. Lvs. entire; phyllaries in 2-4 series..**Ericameria**
6. Lvs. toothed; phyllaries in 5-7 series..**Hazardia**
5. Plants herbaceous.
7. Lvs. toothed or lobed.
8. Heads <6 mm diam., often clustered on one side of infl.-branches. **Solidago**
8. Heads >6 mm diam., not in 1-sided clusters.
9. Basal rosette well-developed..**Pyrrocoma**
9. Basal rosette absent..**Hazardia**
7. Lvs. entire.
10. Heads 1-5 mm diam., in large panicles; achenes 1-2 mm long; non-alkaline habitats.
11. Lvs. <6mm wide, single-veined..**Euthamia**
11. Lvs. >6 mm wide, mostly 3-veined..**Solidago**
10. Heads 5-18 mm diam., in narrow clusters; achenes 2.5-6 mm long; alkaline habitats..**Pyrrocoma**
4. Pappus in 2 series, outer shorter than inner.
12. Lvs. linear, mostly basal..**Erigeron**
12. Lvs. wider, mostly cauline; rays yellow..**Heterotheca**
2. Phyllaries fewer, not imbricated, mostly in 1 series.
13. Phyllaries about 5, partly enclosing ray-achenes; pappus of 30-40 white bristles..**Syntrichopappus**
13. Phyllaries not at all enclosing ray-achenes.
14. Lvs. opposite..**Arnica**
14. Lvs. alternate or basal.
15. Receptacle flat; plants mostly perennial..**Senecio**
15. Receptacle strongly conic; delicate annual..**Crocidium**
1. Rays white to pink, blue or purple.
16. Pappus of ray-fls. much reduced, style-tips of disk-fls. ± hairy...........**Lessingia**
16. Pappus of both ray and disk-fls. well developed.
17. Heads nodding in bud; lvs. narrow; slender annuals.....................**Pentachaeta**
17. Heads erect in bud.
18. Rays filiform, scarcely exceeding disk-fls; annual weeds.................**Conyza**
18. Rays wider, much exceeding disk-fls.
19. Phyllaries subequal or ± overlapping, not leafy or with papery bases and green tips..**Erigeron**
19. Phyllaries subequal, outer ones leafy, or overlapping with papery bases and green tips.
20. Annual to perennial with taproot; lvs. toothed; phyllaries papery near base.
21. Heads <2 cm diam., clustered..**Machaeranthera**
21. Heads 3-6 cm diam., solitary..**Xylorhiza**
20. Perennials with fibrous roots; phyllaries herbaceous......................**Aster**

Group H

(rays absent, pappus none or low crown)

1. Heads unisexual, both sexes on same plant; invol. of pistillate heads winged to bur-like.
  2. Invol. of pistillate heads winged with broad papery scales; shrub with linear divided lvs. **Hymenoclea**
  2. Invol. of pistillate heads spiny, developing into a bur.
    3. Spines hooked; phyllaries of staminate heads distinct **Xanthium**
    3. Spines not hooked; phyllaries of staminate heads united **Ambrosia**
1. Heads not unisexual.
  4. Phyllaries in a single series.
    5. Lvs. opposite; low annuals.
      6. Plants green; fls. bright yellow-orange **Lasthenia**
      6. Plants white-woolly; fls. inconspicuous **Psilocarphus**
    5. Lvs. mostly alternate.
      7. Low white-woolly annuals.
        8. Bracts open, only subtending achenes **Hesperevax**
        8. Bracts sac-like, completely enclosing achene and falling away with it.
          9. Bracts throughout head, proper invol. absent.
            10. Bracts with hooked, spine-like tip **Ancistrocarphus**
            10. Bracts lacking hooked tips **Stylocline**
          9. Bracts lacking in center of head; invol. of about 5 phyllaries **Micropus**
      7. Larger annuals to perennials, not white-woolly.
        11. Annual herb; achenes flattened, winged **Dicoria**
        11. Perennial herb; achenes obovoid, not winged **Iva**
  4. Phyllaries in 2 or more overlapping series (like shingles).
    12. Heads in spikes, racemes or panicles; perennials with ± strong odor. **Artemisia**
    12. Heads solitary or in small clusters; annuals with mild or no obvious odor.
      13. Marginal achenes with little stem; receptacles mostly flat **Cotula**
      13. Marginal achenes sessile; receptacles conic.
        14. Heads conspicuous, terminal; corollas on all fls. **Chamomilla**
        14. Heads inconspicuous, bur-like in forks of stems; outer fls. apetalous. **Soliva**

Group I

(rays present, pappus none or low crown)

1. Receptacles with bracts or bristles subtending fls.
  2. Phyllaries in >1 series, overlapping like shingles.
    3. Lvs. opposite; heads small; ray corollas white **Eclipta**
    3. Lvs. alternate or basal.
      4. Ray corollas white; lvs. dissected.
        5. Heads large, few to solitary **Anthemis**
        5. Heads small, many, in flattish panicles **Achillea**
      4. Ray corollas yellow.

6. Invols. of 2 very unlike series, the lower spreading .... **Coreopsis**
6. Invols. of 2 or more similar series
7. Disk-achenes thickened, angled; stems scapose .... **Balsamorhiza**
7. Disk-achenes strongly flattened; stems ± shrubby......................... **Encelia**
2. Phyllaries in a single series.
8. Receptacle bristly; small desert annual.. **Eriophyllum**
8. Receptacle not bristly.
9. Ray-achenes enclosed by phyllaries.
10. Disk-fls. 6; ray-achenes laterally compressed.......................... **Lagophylla**
10. Disk-fls. 1 and ray-achenes laterally compressed, or more than 1 and ray-achenes not laterally compressed... **Madia**
9. Ray-achenes not completely enclosed by phyllaries.
11. Rays 3-toothed or -lobed, the lobes ± parallel; lvs. lacking tack-shaped glands.
12. Open pit glands on tips of upper lvs. and phyllaries............ **Holocarpha**
12. Open pit glands lacking..**Hemizonia**
11. Rays 3-parted into palmately spreading lobes; lvs. mostly with tack-shaped glands... **Calycadenia**
1. Receptacles naked, lacking bracts or bristles.
13. Lvs. opposite; low annuals.. **Lasthenia**
13. Lvs. alternate; annuals or perennials.
14. Rays with a tooth or lobe on inner side of throat opposite the ligule; annuals. **Monolopia**
14. Rays lacking a lobe on the inside; annuals or perennials.
15. Plants rather succulent; mostly glabrous except for tufts of wool in lf. axils and below heads. ....**Blennosperma**
15. Plants not succulent; hairs various.
16. Lvs. not white-woolly; annual herbs.................................**Dimorphotheca**
16. Lvs. white-woolly, at least when young; annuals or perennials.
17. Achenes ± flattened; corolla tube glabrous with a ring of hairs at junction with throat... **Pseudobahia**
17. Achenes angled; corolla-tube glandular or pubescent, lacking ring of hairs... **Eriophyllum**

Acamptopappus (p. 351)

Low much-branched desert shrub with white bark; lvs. linear to spatulate, 5-20 mm long, sessile; heads discoid, yellow, subglobose, 7-10 mm high; phyllaries broad, rounded, whitish with greenish tips; pappus of >30 silvery paleae and bristles.Two species, both in CA. One species in Kern County: **A. sphaerocephalus** var. **hirtellus**

Achillea (p. 373)

Perennial aromatic herbs with alternate pinnately dissected lvs.; heads several, small, in terminal corymbs; rays white. About 85 species. Recent treatments consider those of N. America to be in one highly polymorphic species. One species: **A. millefolium** L. (incl. *A. borealis, A. gigantea,* and *A. lanulosa*)

Achyrachaena (p. 365)

Annuals with lower lvs. opposite and clasping, upper lvs. alternate and linear; heads solitary, long-peduncled; invol. oblong, nearly equaling inconspicuous fls.; rays yellow turning red; pappus of papery scales. One species. One species in Kern County: **A. mollis**

Acroptilon (p. 378)
(part of *Centaurea*)

Erect, ± cobwebby perennial; lower lvs. 1-2 pinnately lobed, upper linear to toothed; heads in flet-topped clusters; rays absent; phyllaries in several series, entire, with widely papery tips; disk whitish to blue, pappus of slender bristles, barbed below, short-plumose above. One species: **A. repens** (L.) DC

Ageratina

Perennial with woody ± rhizomatose base; lvs. alt. with ± triangular, serrate blade; heads clustered; rays absent; disk whitish to blue; pappus of slender bristles. About 230 species, 4 in CA. One species in Kern County: **A. occidentalis** (Hook.) King & Robinson (On limestone - dolomites, Inspiration Point Botanical area, Piute Mtns. Shevock 10653)

Agoseris (p. 379)

Annual or perennial herbs with strong taproots and mostly basal rosette lvs.; infl. scapose, erect; fls. yellow to orange; achenes narrowed into long slender beak with pappus of capillary bristles. About 10 species, 7 in CA.

1. Annuals; invols. with purplish glandular hairs ....**A. heterophylla**
1. Perennials.
   2. Achene-beaks mostly not more than half as long as body.
      A. glauca (Pursh) Raf. var. **monticola** (greene) Jones (*A. glauca* var. *dasycephala)*
   2. Achene-beaks mostly 2-4 times times the length of body.
      3. Achenes abruptly beaked; ligules 6-14 mm long; pappus 14-18 mm long. **A. retrorsa**
      3. Achenes tapering gradually into beak; ligules 3-6 mm long; pappus 8-12 mm long. ....**A. grandiflora**

Ambrosia (p. 363)
(*Franseria*)

Annual or perennial herbs or small shrubs with mostly lobed or dissected lvs.; heads small, numerous, greenish and inconspicuous; staminate heads nodding in terminal spike-like racemes; pistillate heads in lf.-axils below the staminate heads, their invols.

armed with spines or prickles, becoming a bur. About 40 species, 12 in CA.

Low much branched shrub 2-6 dm high ....**A. dumosa** (Gray) Payne
Weedy annual..**A. acanthicarpa** Hook.

Ancistrocarphus (p. 362)
(part of *Stylocline*)

Low woolly annuals with alt. entire lvs.; heads in leafy-bracted groups; phyllary-like bracts enclosing pistillate fl. and falling with it; inner bracts with hooked tips. One species: **A. filagineus** Gray

Anisocoma (p. 380)

Low scapose annuals with basal rosettes of pinnately parted or toothed lvs. and several single-headed scapes; phyllaries with reddish dots, outer with purplish midrib and broad papery margins; fls. pale yellow; pappus white, feathery. One species. One species in Kern County: **A. acaulis**

Antennaria (p. 361)

Low mat-like silky-tomentose perennials with simple entire alternate and basal lvs.; dioecious; heads solitary on leafy stems 1-4 dm tall; phyllaries ± colorless to brownish; heads without rays. About 40 species, 14 in CA. One species in Kern County: **A. dimorpha**

Anthemis (p. 373)

Annual aromatic herbs with 2-3 times pinnately dissected lvs.; heads many; rays white, 5-11 mm long; receptacles convex to conic; pappus none. About 200 species, 3 introduced in CA. One species in Kern County: **A. cotula**

Arnica (p. 375)

Perennial herbs with simple mostly opposite lvs. on solitary upright stems; heads large with yellow rays to 18 mm long; receptacles flattish; pappus bristles barbellate. About 30 species, 17 in CA.

Stem-lvs. 5-10 pairs, 5-20 cm long (incl. petioles), toothed, the basal cordate.
A. chamissonis ssp. **foliosa**
Stem-lvs. 2-4 pairs, 2-9 cm long, ± entire, lanceolate......................**A. cordifolia** Hook.
(Tributary to Deep Creek near Tiger Flat, Greenhorn Mtns. Shevock 10527)

Artemisia (p. 373)

Herbs or shrubs with aromatic, entire to dissected alternate lvs.; heads small in paniculate clusters; rays lacking; phyllaries dry, overlapping, at least the inner with papery margins; pappus a short crown to none. About 300 species, 21 in CA.

1. Shrubs.
  2. Spiny shrubs intricately branched to 3 dm tall; lvs. crowded, 5-7-parted and divisions 3-lobed. .... **A. spinescens**
  2. Shrubs not spiny, taller.
    3. Lvs. mostly 3-lobed, those of fl.-shoots entire and linear; invol. 3-4 mm high; plants silvery hairy.......................... **A. tridentata** ssp. **tridentata**
    3. Lvs. once or twice palmately divided into linear segments; invol. 2-3 mm high; plants grayish.......................... **A. californica**
1. Herbaceous perennials.
  4. Lvs. mostly glabrous, linear, mostly entire, not very aromatic...... **A. dracunculus**
  4. Lvs. ± hairy and white-tomentose beneath, green above.
    5. Lvs. to 1 cm wide exclusive of lobes; plants rarely over 1 m tall; invol. tomentose.......................... **A. ludoviciana**
    5. Lvs. 1-5 cm wide exclusive of lobes; plants often to 1.5 m tall; invol. slightly hairy.......................... **A. douglasiana**

Aster (p. 351)

Late blooming rhizomatous herbaceous perennials or annuals with alternate entire to toothed lvs.; heads usually numerous, mostly raidate; ray fls. usually purplish blue or white; disk fls. yellow; pappus of numerous capillary bristles. About 250 species, 23 in CA.

1. Annuals; rays pinkish.
  2. Rays evident, longer than pappus; phyllaries with prominent papery margin.
    A. subulatus Mich. var. **ligulatus** Shinn. (*A. exilis*)
  2. Rays inconspicuous, same or shorter than pappus.......................... **A. frondosus**
1. Perennials.
  3. Rays absent.......................... **A. breweri** (Gray) Semple (*Chrysopsis breweri*)
  3. Rays present.
    4. Lvs. sessile, grass-like, glabrous; invols. glandular; rays <20, lavender to white.
      A. pauciflorus
    4. Lvs. often toothed, petioled; invols. not glandular ; rays 20-35.
      5. Hairs of stem in lines below lf.-bases; rays whitish to bluish.
        A. lanceolatus Willd. ssp. **hesperius** (Gray) Semple & Chmiel. (*A. hesperius*)
      5. Hairs of stem uniform.
        6. Invol. subequal, sometimes subtended by foliose bracts; rays pink to white.
          A. eatonii
        6. Invol. strongly graduated; foliose bracts absent; ; rays violet-purple to whitish.
          7. Bracts in infl. conspicuous.

8. Lvs. ± glabrous, hairs short or appressed............................**A. chilensis**
8. Lvs. with soft spreading hairs...................................... **A. bernardinus**
7. Bracts in infl. mostly lacking; rays violet .... **A. ascendens**

Baccharis (p. 352)

Dioecious shrubs or tall (1-3 m) perennial resinous herbs; lvs. alternate, entire to toothed; heads discoid, small, whitish or yellowish, in clusters; phyllaries overlapping (like shingles), whitish; pappus of copious capillary bristles. About 300 species, 9 in CA.

1. Stems herbaceous to base; lvs. 3-nerved; achenes pubescent..................**B. douglasii**
1. Stems woody; achenes glabrous.
2. Lvs. persistent, willow-like; heads clustered.
3. Lvs. linear to linear-lanceolate, entire to obscurely toothed, 5-15 cm long; achenes 5-nerved. ....**B. salicifolia** (Lopez & Pavon) Pers. (*B. glutinosa* and *B. viminea*)
3. Lvs. cuneate to oblong, toothed above middle, 3-veined; achenes 10-nerved. **B. emoryi**
2. Lvs. deciduous; plants broom-like with numerous erect grooved branches; heads scattered...**B. sergiloides**

Balsamorhiza (p. 364)

Perennial herb with leafless stems from thick caudex and mostly basal, deltateor sagittate entire to ± toothed lvs.; heads radiate, yellow, usually solitary; phyllaries herbaceous; achenes glabrous and without a pappus. About 12 species, 7 in CA.

Lvs. densely hairy, invols. woolly..**B. sagittata**
Lvs. sparsely hairy, green on both sides; invols. not woolly ....**B. deltoidea**

Bebbia

Rounded odorous desert shrub with green nearly leafless multibranched stems; heads solitary, discoid, about 4-8 mm high; corolla yellow; pappus of 15-20 ± plumose bristles. Two species, one in CA. One species in Kern County: **B. juncea** (Benth.) Greene var. **aspera** Greene (E-facing Sierra, Sand Canyon, M. A. Henry)

Bidens (p. 364)

Annual or perennial herbs with opposite simple to compound lvs.; heads mostly radiate, yellow to orange; invol. mostly in 2 series; pappus of 2-4 rigid barbed awns. About 230 species, 6 in CA.

Perennial; lvs. simple; heads radiate or discoid, pale yellow...........................**B. cernua**
Annual; lvs. compound; heads mostly discoid, orange .... **B. frondosa**

Blennosperma

Small annuals with alternate pinnately lobed lvs.; heads radiate, yellow; phyllaries in a single series, each with thin margin; achenes covered with minute bumps; pappus lacking. Three species, two in CA. One species in Kern County: **B. nanum** (Hook.) Blake var. **nanum** (Clay soils, Valley Grassland near Woody, Shevock 10159)

Blepharizonia (p. 365)

Stout fall-flowering xerophytic odorous annuals with white densely hirsute stems; lvs. mostly basal, linear, upper ones covered with yellow tack- shaped glands; rays white, veined with rose; pappus feathery. A single species: **B. plumosa** ssp. **viscida**

Brickellia (p. 351)

Mostly shrubby with simple mostly resinous-dotted lvs. with prominent veins; fls. white to pink-purple; heads discoid; phyllaries longitudinally nerved, in several ranks; pappus of several nearly smooth to subfeathery bristles. About 100 species, 15 in CA.

1. Heads 3-7-.fld; lvs. entire, 3-nerved.. **B. multiflora**
1. Heads 8-many-fld.
  2. Phyllaries with recurved or spreading tips; lvs. mostly sessile.
    3. Stems and lvs. white-hairy; phyllaries woolly...................................... **B. nevinii**
    3. Stems and lvs. greenish, glandular-villous .... **B. microphylla**
  2. Phyllaries erect; lvs. petioled.
    4. Heads small, 9-10 mm long; petioles 1-3 mm long...................... **B. desertorum**
    4. Heads larger, ± 14 mm long; petioles 5-20 mm long. .... **B. californica**

Calycadenia (p. 365)

Xerophytic annuals with linear grass-like lvs. which are basal and scattered in fascicles; tack-shaped glands at tips of upper lvs.; ray fls. few, with broad palmately 3-lobed white to yellow ligules; pappus of membranous scales. About 9 species, all in CA.

Bracts subtending heads flattened, nerved to tip, not concealing invol.; rays deep yellow to white, no red.................**C. multiglandulosa** DC (*C. hispida* forma *albiflora*)
Bracts subtending heads terete in upper third, concealing invol.; rays white fading rose.. **C. spicata**

Calycoseris (p. 380)

Annual, branched from base, glabrous below, with tack-shaped glands above; lvs. mostly basal, pinnately parted into linear segments; heads showy, yellow; achenes with short beak and pappus of white bristles. Two species, both in CA. One species in Kern County: **C. parryi**

Carthamus (p. 377)

Annuals with alternate spinose lvs.; heads terminal, with spreading leafy outer phyllaries and ± spiny inner ones; fls. red-orange; achenes white; pappus of numerous short scales. About 14 species, 4 introduced into CA. One species in Kern County: **C. tinctorius**

Centaurea (p. 378)

Annual herbs with entire to pinnately divided alternate lvs.; fls. tubular, the outer sometimes enlarged; invol. ovoid to ± cylindric, the phyllaries overlapping, sometimes spine-tipped; achenes smooth and shining. About 500 species, 12 introduced in CA.

1. Phyllaries fringed, not spiny at tip; fls. blue to white, outer often enlarged. **C. cyanus**
1. Phyllaries spiny at tip; fls. yellow.
   2. Phyllary-spines slender, purplish; plant branched above .... **C. melitensis**
   2. Phyllary-spines stout, yellowish; plant branched from base. .... **C. solstitialis**

Chaenactis (p. 369)

Annual to perennial herbs with mostly basal, some upper, pinnately dissected lvs.; heads large, terminal; fls. tubular, the outer often enlarged; pappus of 4-20 ± transparent scales. About 18 species, 13 in CA.

1. Perennials; pappus of 10 or more scales; fls. white to pinkish.
   2. Lvs. linear, with 10-15 pairs of short divisions .... **C. santolinoides**
   2. Lvs. broader, with few segments; alpine plants. **C. douglasii** (Hook.) Hook. & Arn. var. **douglasii** (*C. rubricaulis*)
1. Annuals
   3. Phyllaries narrow, linear, attenuate into reddish bristle-tip. **C. carphoclinia** var. **carphoclina** (incl. var. *attenuata*)
   3. Phyllaries acute or obtuse at apex, not bristle-tipped.
      4. Stamens included; corollas 10-14 mm long, pinkish .... **C. macrantha**
      4. Stamens exserted; corollas <10 mm long.
         5. Pappus-paleae 8 in 2 series; corollas white .... **C. xantiana**
         5. Pappus-paleae mostly 4 in 1 series.
            6. Corollas yellow .... **C. glabriuscula**
               7. Peduncles long, delicate; plants woolly .... var. **lanosa**
               7. Peduncles shorter, stouter.
                  8. Pappus-paleae 4; lvs. subentire to pinnately divided var. **glabriuscula** (inc. var. *curta*)
                  8. Pappus-paleae 4-5 with incomplete outer shorter whorl; some lvs. twice pinnately divided. .... var. **megacephala**
            6. Corollas white or tinged purplish.

9. Lvs. pinnate with few long linear divisions; plants mostly glabrous. **C. fremontii**
9. Lvs. pinnate or twice pinnate with blunt divisions; plants with tomentum. .... **C. stevioides** (incl. var. *brachypappa*)

Chamomilla (p. 375)
(*Matricaria*)

Annuals with alternate finely pinnately dissected lvs.; plants somewhat aromatic; heads many or solitary on ends of branches, dome-shaped; rays absent; fls. greenish; pappus reduced to a crown. About 40 species, 2 in CA.

Pappus-crown small, entire; heads mostly 6-9 mm high; plants much-branched; achenes deciduous............................ **C. suaveolens** (Pursh.) Rydb. (*M. matricarioides*)
Pappus with 2 small lobes; heads mostly >9 mm high, plants not much branched; achenes quite persistent on heads................................ **C. occidentalis** (E. Green) Rydb.

Chrysothamnus (p. 353)

Erect much-branched shrubs with alternate entire narrow lvs. (not in fascicles); heads numerous, mostly 5-flowered, yellow, in terminal clusters; rays lacking; phyllaries overlapping usually in 5 ranks; pappus of dense capillary bristles. About 16 species, 10 in CA.

1. Lvs. resinous-punctate, cylindrical.
 2. Phyllaries thin, lacking terminal gland.. **C. paniculatus**
 2. Phyllaries with conspicuous thick green terminal gland.................. **C. teretifolius**
1. Lvs. not resinous-punctate.
 3. Twigs not tomentose....................................... **C. viscidiflorus** (incl. ssp. *pumilus*)
 3. Twigs densely tomentose.
  4. Infl. racemose or spicate; phyllaries attenuate............ **C. parryi** ssp. **vulcanicus**
  4. Infl. cymose; phyllaries obtuse to ± attenuate .... **C. nauseosus**
   5. Phyllaries pubescent.
    6. Corolla-lobes lanceolate, >1 mm long. ssp. **albicaulis** (Nutt.) Hall & Clements (ssp. *occidentalis*)
    6. Corolla-lobes ovate, <1 mm long.........ssp. **hololeucus** (incl. ssp. *viscosus*)
   5. Phyllaries not hairy; lvs. not white-tomentose.
    7. Phyllaries with very slender recurved tip; corolla <7 mm long. ssp. **ceruminosus**
    7. Phyllaries without recurved mucronate tip.
     8. Invol. 6-9 mm high; phyllaries only slightly keeled.......... ssp. **consimilis**
     8. Invol. 9-10 mm high; phyllaries strongly keeled, in distinct vertical rows. ssp. **mohavensis**

Cichorium (p. 380)

Perennial herbs with mostly basal, toothed or divided lvs.; heads 1-3 in axils of reduced upper lvs., about 4 cm broad, blue to whitish; pappus-scales small. About 8 species, 2 introduced into CA. One species in Kern County: **C. intybus**

Cirsium (p. 378)

Annual to perennial herbs, spiny, with alternate pinnately divided lvs.; heads lacking rays; phyllaries numerous, overlapping, mostly spiny-tipped; corollas long-tubular; achenes glabrous; pappus of feathery bristles. About 200 species, 24 in CA. Many intermediates occur where 2 species overlap and hybridize.

1. Stems spinose-winged by leaf-bases which extend down the stem; lvs. coarsely hairy above; fls. purple. .... **C. vulgare**
1. Stems not spinose-winged.
  2. Heads mostly clustered, subtended by spiny leafy bracts.
    3. Plants 1-2 m tall; outer phyllaries pinnately spinose; corolla lavender-rose; heads lacking cobweb hairs. .... **C. crassicaule**
    3. Plants <50 cm tall; phyllaries not pinnately spinose; corolla whitish. **C. foliosum**
  2. Heads not much clustered, not subtended by leafy bracts.
    4. Phyllaries mostly glabrous, not with dense cobwebby hairs; stems 4-9 dm tall; corolla rose-purple. .... **C. andersonii**
    4. Phyllaries with dense cobwebby hairs, outer ones reflexed.
      5. Back of some phyllaries with a glandular spot or ridge.
        6. Plants from horizontal underground rootstock, perennials; corollas buff to off-white. .... **C. cymosum**
        6. Plants from taproot, winter annuals; corollas lavender to whitish. **C. occidentale** var. **californicum** (Gray) Kiel & Turner (*C. californicum*)
      5. Back of phyllaries without glandular spot or ridge.......... **C. occidentale**
        7. Heads mostly as broad as long; phyllaries mostly straight, covered with fine hair; corolla purplish red, rarely white.......... var. **occidentale**
        7. Heads mostly longer than broad; phyllaries mostly curved, cobweb-hairy; corolla crimson to rose, rarely white. var. **venustum** (Greene) Jeps. (*C. proteanum*)

Cnicus (p. 379)

Annual with wavy-dentate to pinnate lvs. with spiny lobes; heads large, terminal, solitary; fls. tubular, yellow; pappus of 2 series of awns, the outer yellow, the inner white. One species: **C. benedictus**

Conyza (p. 355)

Annual to perennial herbs with alternate entire to pinnately divided lvs.; heads small, numerous; rays absent or numerous and inconspicuous; phyllaries narrow, reflexed in age; pappus of few capillary bristles. About 50 species, 5 in CA.

1. Upper lvs. entire; ray-fls. numerous, small, white; plants to 2 m tall.
   2. Invol. mostly glabrous; phyllaries with translucent midvein. .... **C. canadensis**
   2. Invol. quite hairy; phyllaries lacking translucent midvein. .... **C. bonariensis**
1. Upper and lower lvs. coarsely dentate; ray-fls. lacking; plants less than 1 m tall. **C. coulteri**

Coreopsis (p. 364)

Annuals with opposite linear simple to compound mostly glabrous lvs.; heads showy, long-peduncled, yellow; invol. in 2 series, the outer ones spreading and herbaceous, the inner erect and membranous. About 115 species, 11 in CA.

1. Outer phyllaries linear. .... **C. bigelovii**
1. Outer phyllaries lanceolate to deltate.
   2. Disk-achenes ciliate, ray achenes glabrous; outer phyllaries deltate; pappus 2-awned. .... **C. calliopsidea**
   2. Disk- and ray-achenes not ciliate; outer phyllaries narrowly lanceolate; pappus reduced to a cup. .... **C. californica** var. **californica**

Cotula (p. 375)

Low diffuse rather strong-scented herbs with alternate clasping lobed to dissected lvs.; heads on long peduncles, yellow to greenish; rays absent; disk-fls. 4-toothed; pappus none. About 80 species, 2 introduced into CA.

Sparsely pubescent annual; lvs. finely pinnate; heads 2-5 mm broad. .... **C. australis**
Glabrous fleshy perennial; lvs. entire to divided; heads 8-10 mm broad. **C. coronopifolia**

Crepis (p. 380)

Perennial herbs with mostly basal deeply lobed lvs.; heads in terminal clusters; invol. cylindric, inner phyllaries in a single series with ± thickened midribs; outer phyllaries very small; fls. yellow, all ligulate; pappus of soft white bristles. About 200 species, 11 in CA.

Largest heads with 5-7 inner phyllaries. .... **C. acuminata**
Largest heads with 8-13 inner phyllaries. **C. occidentalis** (incl.vars. *occidentalis* & *pumila*)

Crocidium (p. 376)

Small slender annual with small entire to few-toothed mostly basal lvs.; heads solitary at ends of peduncle-like branches; tufts of woolly hair in lf.-axils; rays and disk yellow; pappus of several fragile white bristles. One species: **C. multicaule**

Dicoria (p. 375)

Diffusely branched annual with alternate ovate lvs.; heads small, numerous, nodding, arranged in spikes; corollas lacking in pistillate fls.; achene winged, lacking a pappus. About 4 species, one in CA. One species in Kern County: **D. canescens** (incl.ssp. *hispidula*)

Dimorphotheca (p. 375)

Loosely branched glandular pubescent annual with alternate dentate lvs.; heads large, solitary; rays orange-yellow. About 7 species, one introduced into CA. One species in Kern County: **C. sinuata**

Eastwoodia (p. 356)

Xerophytic white-stemmed shrub with linear alternate lvs.; heads yellow, solitary, subglobose; rays absent; invol. mostly 4-seriate; achenes silky; pappus of 5-6 linear segments. One species: **E. elegans**

Eclipta (p. 375)

Low annual with opposite lanceolate lvs.; heads inconspicuous with white rays; outer phyllaries somewhat larger than inner; achenes angled, hairy at tip; pappus obscure or none. About 4 species, one introduced in CA. One species in Kern County: **E. prostrata** (L.) L. (*E. alba*)

Encelia (p. 364)

Low branching shrubs with alternate lvs.; heads medium with yellow rays; invols. 2-3-seriate; achenes obovate, flat, hairy; pappus none or of 2 slender awns. About 13 species, 6 in CA.

Heads in clusters; lvs. white-hairy, clustered...... **E. farinosa** (Edwards AFB, Moe 782)
Heads solitary on pubescent peduncles; lvs. scattered, not white-hairy.
E. actoni Elmer (*E. virginensis* ssp. *actoni*)

Ericameria (p. 357)
(part of *Haplopappus*)

Shrubs with narrow to wedge-shaped entire lvs.; rays present or absent; phyllaries in 2-6 series, often resinous; disk corollas yellow; pappus of thin bristles. About 27 species, 18 in CA.

1. Lvs. oblong to obovate, generally >2.5 mm wide.
  2. Lvs. 2-25 mm long, oblanceolate to wedge-shaped.
    3. Rays 0-3, <5 mm long; lvs. wedge-shaped. **E. cuneata** (Gray) McClatchie var. **cuneata**
    3. Rays 4-6, 8-10 mm long; lvs. oblanceolate............ **E. gilmannii** (Blake) Nesom (Granitic rocky soils near summit of Owens Peak. Shevock 11811)
  2. Lvs. 20-70 mm long, slender....................................... **E. bloomeri** (Gray) Macbr.
1. Lvs. linear, cylindric, <2.5 mm wide.
  4. Invol. 8-14 mm wide .... **E. linearifolia** (DC) Urb. & Wussow
  4. Invol. 3-6 mm wide.
    5. Lvs. 30-60 mm long.. **E. arborescens** (Gray) Greene
    5. Lvs. <20 mm long.
      6. Invol. 4-5 mm long; rays 0-2...................................... **E. cooperi** (Gray) Hall
      6. Invol. 6-7 mm long; rays 1-6. **E. palmeri** (Gray) Hall var. **pachylepis** (Hall) Nesom

Erigeron (p. 356)

Biennial to perennial herbs with alternate to basal lvs.; heads few to many, many-fld.; phyllaries equal or somewhat overlapping; rays, when present, many, bluish to purple; disk-fls. yellow; pappus of capillary bristles. About 375 species, 51 in CA.

1. Rays absent.
  2. Pappus-bristles mostly <12, annual to biennial................................. **E. divergens**
  2. Pappus-bristles mostly 12-25; cushion-like perennial........... **E. compositus** Pursh (Lookout Point. Piute Mtns. Shevock 5575)
1. Rays present.
  3. Lvs. dissected into narrow lobes .... **E. compositus** Pursh (Lookout Point. Piute Mtns. Shevock 5575)
  3. Lvs. entire to toothed.
    4. Basal lvs. much different from stem lvs.; phyllaries ± equal.
      5. Stem lvs. many, sessile and ± clasping; rays pink to white.. **E. philadelphicus**
      5. Stem lvs. fewer, linear; rays bluish-lavender.
        6. Long-lived perennial; rays present, pale purple..................... **E. multiceps**
        6. Biennial; rays present or absent, light blue............................ **E. divergens**
    4. Basal lvs. not much different from stem lvs.; phyllaries much overlapping.
      7. Stems with minute glandular hairs.................................... **E. aequifolius** Hall (Granite rock outcrops, Owens Peak, Shevock)

7. Stems lacking glandular hairs.
  8. Stem-hairs dense, stiffly-spreading.
    E. brewerii Gray var. **covillei** (Greene) Nesom (*E. foliosus* var. *coveillei*)
  8. Stem-hairs sparse or lacking..**E. foliosus**
    9. Lvs. 2-4 cm long, glabrous to hairy, oriented variously.
      var. **foliosus** (incl. var. *stenophyllus*)
    9. Lvs. 4-6 cm long, hairy, ± oriented to one side..................var. **hartwegii**

Eriophyllum (p. 370)

Annual to perennial woolly herbs, some with ± woody base; lvs. entire, toothed or divided; heads solitary to clustered; rays yellow; invol. narrowly bell-shaped; phyllaries in one series, erect; pappus of few to several scales. About 14 highly variable species, all in CA.

1. Perennials, somewhat woody at base; lvs. divided.
  2. Heads 15-30 mm broad, long-peduncled...**E. lanatum**
    3. Tube of disk-corolla glandular or glandular-hairy........................var. **obovatum**
    3. Tube of disk-corolla glabrous...var. **hallii**
  2. Heads less than 15 mm broad, short-peduncled.
    E. confertiflorum var. **confertiflorum**
1. Small annuals to 20 cm high.
  4. Heads on small peduncles; rays present.
    5. Stems many from the base but little branched; pappus-scales opaque.
      E. wallacei
    5. Stems openly branched; pappus-scales translucent.......................**E. ambiguum**
      6. Invols. 4-5 mm high; phyllaries fused and ± thickened at base.
        var. **ambiguum**
      6. Invols. 5-7 mm high; phyllaries not thickened........................var. **paleaceum**
  4. Heads sessile in leafy-bracteate terminal clusters; rays absent.............**E. pringlei**

Euthamia (p. 360)
(part of *Solidago*)

Herbaceous perennial with ± whitish stems and alternate, sessile, linear-lanceolate, 3-5-veined, entire lvs.; heads in dense panicle-like ascending branches; rays yellow; invol. 3-5 mm high; pappus of long bristles. About 8 species, one in CA. One species in Kern County: **E. occidentalis** Nutt.

Filago (p. 361)

Low woolly annuals with entire alternate lvs.; heads small, clustered, lacking rays; outer pistillate fls. subtended by woolly bracts which ± enclose fl.; inner fls. lacking bracts and have pappus of capillary bristles. About 25 species, 5 in CA.

Upper lvs. much longer than heads, linear-pointed...**F. gallica**
Upper lvs. mostly as long as heads, wider.

Plants erect, 5-35 cm tall ....**F. californica**
Plants depressed, spreading, branched from base....................................**F. depressa**

Gaillardia (p. 371)

Leafy-stemmed annual with alternate entire to divided lvs.; heads solitary, large; rays 12-20 mm long, yellow with purple base or all purple; pappus of 5-10 awned scales. About 28 species, two introduced into CA. One species in Kern County: **G. aristata**

Glyptopleura (p. 380)

Depressed small winter annual with compact rosette of divided lvs. with whitish toothed margins; heads creamy to yellowish, large, on short peduncle; achenes about 4 mm long, abruptly beaked; pappus-bristles white. One species: **G. marginata** Eaton (*G. setulosa*)

Gnaphalium (p. 361)

Annual to perennial woolly herbs with alternate entire lvs.; heads clustered in spikes or panicles, rays absent; invol. ovoid to bell-shaped; phyllaries thin, dry, colored white, yellowish to tinted rose; outer fls. pistillate, inner ones perfect; pappus of capillary bristles. About 120 species, 11 in CA.

1. Lvs. (at least upper) green in age.. **G. californicum**
1. Lvs. gray, densely woolly.
  2. Perennial; basal lf.-tufts present at flowering.............................. **G. canescens** DC
    3. Heads 4-5 mm high; pistillate corollas 2-3 mm long.
      ssp. **thermale**(Nels.) Stebb. & Keil (*G. thermale*)
    3. Heads 5-6 mm high; pistillate corollas 3-4 mm long.
      4. Lvs. narrowly lanceolate, upper extending below point of attachment.
        ssp. **beneolens** (Davidson) Stebb. & Keil (*G. beneolens*)
      4. Lvs. oblong to spoon-shaped, not extending below point of attachment.
        ssp. **microcephalum** (Nutt.) Stebb. & Keil (*G. microcephalum*)
  2. Annual to biennial; basal lf.-tufts absent.
    5. Stems mostly <20 cm tall; phyllaries brown .... **G. palustre**
    5. Stems mostly >20 cm tall; phyllaries white to pale yellow.
      6. Pappus-bristles falling in clusters.. **G. luteo-album**
      6. Pappus-bristles falling singly ....**G. stramineum** Kunth (*G. chilense*)

Grindelia (p. 357)

Perennial herbs with alternate, punctate, mostly serrate ± clasping lvs.; heads medium to large, solitary at branch-tips; rays yellow; phyllaries thickish with appressed base and narrow spreading tips; pappus of 2-8 stiff awns. About 80 species, 6 in CA.

Outer phyllaries long-pointed, tightly coiled.
G. camporum Greene var. **bracteosum** (Howell) Lane (*G. camporum* var. *australis*)
Outer phyllaries reflexed but not tightly coiled **G. camporum** Greene var. **camporum** (incl. *G. procera* and *G. robusta*)

Gutierrezia (p. 357)

Small much-branched mostly glutinous shrubs with alternate entire slender lvs.; heads very small, numerous; rays yellow; phyllaries overlapping, tough, whitish; achenes ± hairy; pappus of short scales. About 25 species, 3 in CA.

1. Heads with only 2-3 fls.; invol. to 1.5 mm wide .... **G. microcephala**
1. Heads with more than 4 fls.; invol. more than 1.5 mm wide.
   2. Heads clustered at ends of branchlets; fls. 5-10 per head.................. **G. sarothrae**
   2. Heads mostly solitary at ends of branchlets; fls. mostly more than 10 per head. **C. californica** (DC) Torrey (*G. bracteata*)

Haplopappus (p. 357)

(see *Ericameria, Hazardia, Isocoma,* and *Pyrrocoma*)

Hazardia (p. 357)

(part of *Haplopappus*)

Perennial to shrubby with alt. toothed lvs.; rays present or absent; phyllaries in several overlapping series; pappus of several white to rusty bristles. About 13 species, 7 in CA.

1. Perennial or subshrub; rays present............**H. whitneyi** (Gray) Greene var. **whitneyi**
1. Shrub; rays absent.
   2. Fls. 4-8 per head; lvs. 15-25 mm long, 7-12 mm wide. **H. stenolepis** (Hall) Hoover
   2. Fls. 9-30 per head; lvs. 15-50 mm long, 10-20 mm wide. **H. squarrosa** (Hook. & Arn.) Greene var. **obtusa** (Greene) Jeps.

Helenium (p. 371)

Biennial or short-lived perennial herbs; lvs. extend below point of attachment; heads solitary on long peduncles; rays yellow to red-purple, reflexed; disk dome-shaped; invol. in 1-3 series, phyllaries soon reflexed; pappus of awn-tipped scales. About 35 species, 5 in CA.

Rays inconspicuous, 3-8 mm long; stems much branched above............ **H. puberulum**
Rays conspicuous, 8-22 mm long; stems few-branched above .... **H. bigelovii**

Helianthella (p. 364)

Perennial herb with many leafy stems from woody base; lvs. simple, entire, linear-lanceolate to 25 cm long, mostly basal; heads solitary on long naked peduncles; disk 1.5-2.2 cm wide; achenes compressed, no pappus. About 8 species, 2 in CA. One species in Kern County: **H. californica** var. **nevadensis**

Helianthus (p. 364)

Coarse annual to perennial herbs with simple lvs.; heads large, solitary; phyllaries ± herbaceous, overlapping; ray-fls. conspicuous, yellow; disk-fls. yellow to reddish; achenes glabrous, compressed front-to-back. About 65 species, 10 in CA.

Annual, 3-20 (or more) dm tall; disk-fls. reddish; disk 2-3.5 cm wide; lvs. 4-40 cm wide.......... **H. annuus**
Perennial to about 15 dm tall; disk-fls. yellow; disk 1-2.5 cm wide; lvs. 2-4 cm wide. **H. gracilentus**

Hemizonia (p. 365)

Annual or perennial glandular and aromatic fall-flowering xerophytic herbs; basal lvs. variously lobed; phyllaries half-enclosing ray-achenes; rays yellow, 2-3-lobed; disk-achenes with papery pappus, or none. About 25 species, 19 in CA.

1. Lvs. not spine-tipped.
   2. Ray-fls. 5; disk-fls. 6 .... **H. kelloggii**
   2. Ray- and disk-fls. more than 8.
      3. Pappus obvious; first lvs. lobed; rays pale yellow .... **H. pallida**
      3. Pappus reduced; first lvs. mostly entire; rays deep yellow.......... **H. arida**
1. Lvs. tipped with rigid spine.
   4. Pappus none; anthers yellow .... **H. pungens** ssp. **pungens**
   4. Pappus present; anthers black .... **H. fitchii**

Hesperevax (p. 361)
(*Evax*)

Low woolly diffusely branched annuals with entire alternate to ± whorled lvs.; heads axillary or crowded at tips of stems, consisting of perfect fls. with outer fls. pistillate, each subtended by a bract; rays absent; pappus none. About 3 species, 2 in CA.

Lf.-blades 7-15 mm wide; heads in terminal clusters........ **H. caulescens** (Benth.) Gray
Lf.-blades 3-5 mm wide; heads axillary; stems short. **H. acaulis** (Kellog) Greene var. **acaulis** (incl. var. *humilis*)

Heterotheca (p. 353, 359)
(incl. *Chrysopsis)*

Coarse erect annual or biennial herb with stout, unbranched stem below the terminal panicle of yellow fls.; infl. glandular and heavy-scented; lvs. densely hairy, the lower with a pair of stipule-like lobes; disk-achenes with double pappus of long inner and short outer bristles. About 30 species, 5 in CA.

1. Rays absent.
H. oregana (Nutt.) Shin. var. **scaberrima** (Gray) Semple (*Crysopsis o.* var. *s.*)
1. Rays present.
 2. Annual to biennial; ray achenes glabrous; ray pappus absent..........**H. grandiflora**

 2. Perennial; ray achenes hairy; ray pappus present.
H. sessiliflora (Nutt.) Shin. (*Chrysopsis villosa*)
  3. Upper lf.-margin flat to weakly wavy; hairs 2 mm long.
ssp. **echioides** (Benth.) Semple (South Fork Valley, Shevock)
  3. Upper lf.-margin distinctly wavy; hairs 1 mm long.
ssp. **fastigiata** (Greene) Semple

Hieracium (p. 380)

Perennial hairy herbs with mostly entire lvs.; heads in loose panicles; phyllaries linear to lanceolate, with a few small bract-like ones at the base; pappus-bristles dull white to brownish. About 250 species, 8 in CA.

Fls. white; stem and lvs. only slightly hairy ....**H. albiflorum**
Fls. yellow; stem and lvs. densely shaggy .... **H. horridum**

Holocarpha (p. 366)
(incl. in *Hemizonia*)

Glandular aromatic annuals; lvs. linear, the basal somewhat dentate, the upper with an apical gland; phyllaries half-enclosing ray-achenes and with terminal pit glands; fls. yellow, the rays 3-lobed; pappus none. About 4 species, all in CA.

Lvs. densely hairy above; infl. a raceme; phyllaries very glandular.
H. heermannii (Greene) Keck
Lvs. not densely hairy above; infl. a panicle; phyllaries only ± glandular.
H. obconica (Clausen & Keck) Keck

Hulsea

Densely glandular herbs with mostly basal, coarsely toothed lvs.; heads ± scapose, 15 mm wide; rays numerous, reddish-purple; achenes hairy with 1-3 pappus scales. About 8 species, 7 in CA. One species in Kern County: **H. heterochroma** Gray (Eastern crest of Owens Peak. Shevock 11200)

Hymenoclea (p. 363)

Low much-branched monoecious xerophytic shrubs; lvs. alternate, filiform, resinous, 2-5 cm long; heads small, numerous, the staminate above the pistillate; pistillate heads 1-fld., the phyllaries becoming papery horizontal wings, 3-8 mm wide. Two species, both in CA. One species in Kern County: **H. salsola** var. **salsola**

Hypochaeris (p. 380)

Annual with lvs. in a prominent rosette; heads on long naked stems 1-4 dm high; fls. yellow; achenes glabrous, upwardly scabrous; pappus of plumose bristles. About 50 species, 2 introduced into CA. One species introduced in Kern County: **H. glabra**

Isocoma (p. 357)
(part of *Haplopappus*)

Rounded to open shrub with spoon-shaped, entire to toothed, shiny lvs.; heads in clusters of 4-5; rays absent; disk corollas yellow; pappus of slender bristles. About 10 species, 3 in CA.

One species in Kern County: **I. acradenia** (Greene) Greene
Lvs. reduced upward, grading into bracts below heads. var. **bracteosa** (Greene) Nesom
Lvs. not reduced upward, uniform throughout.................................... var. **acradenia**

Iva (p. 363)

Herbaceous perennial spreading from slender rhizomes; lvs. oblanceolate, entire, thickish; heads solitary in axils, nodding; phyllaries united into a cup; fls. greenish white, inconspicuous. About 15 species, 3 in CA. One species in Kern County: **I. axillaris** Pursh. ssp. **robustior** (Hook.) Bassett (var. *pubescens*)

Lactuca (p. 381)

Leafy annual, prickly; lvs. spinose-denticulate, clasping the stems; heads in panicles, yellow often fading purplish; achenes with long beak and copious pappus of soft capillary bristles. About 100 species, 7 in CA. One species introduced into Kern County: **L. serriola** (incl. var. *integrata*)

Lagophylla (p. 367)

Late vernal to autumnal mesophytic or xerophytic annuals; basal lvs. linear-oblanceolate, lost before flowering; cauline lvs. with axillary fascicles; heads open in evening; rays yellow; phyllaries completely enfolding ray-achenes; pappus none. About 5 species, all in CA.

Rays 8-13 mm long, conspicuous, bright yellow .... **L. glandulosa**
Rays 3-6 mm long, inconspicuous, pale yellow ....**L. ramosissima** ssp. **ramosissima**

Lasthenia (p. 371)

Slender vernal annuals with opposite sessile entire to divided lvs.; heads small, terminal; invol. rounded at base; phyllaries distinct or united at base; rays yellow; pappus of small scales or none. About 17 species, 16 in CA.

1. Phyllaries distinct, not united at base.
  2. Receptacle pointed; phyllaries 3-6.
    3. Rays very short; disk-corollas most 4-lobed .... **L. microglossa**
    3. Rays >2 mm long; disk-corollas mostly 5-lobed .... **L. debilis**
  2. Receptacle conic to globose; phyllaries mostly >6.
    4. Lvs. pinnately lobed.
      5. Achenes >1.5 mm long.. **L. minor**
      5. Achenes <1.5 mm long...**L. fremontii**
    4. Lvs. all essentially entire.................................... **L. californica** (*L. chrysostoma)*
1. Phyllaries united below into a cup, the tips free.
    6. Achenes strongly flattened, with conspicuous marginal fringe of stiff hairs. **L. chrysantha**
    6. Achenes not much flattened, hairs not restricted to margins or glabrous.
      7. Achenes glabrous ....**L. glabrata** ssp. **coulteri**
      7. Achenes hairy .... **L. ferrisiae**

Layia (p. 367)

Vernal annuals; heads many-fld., on terminal peduncles; lower margins of phyllaries enfolding ray-achenes, the upper part flat; rays 3-lobed, white to yellow; ray-achenes compressed front to back; pappus of numerous bristles or scales. About 14 species, all in CA.

1. Pappus of paleae, not feathery.
  2. Rays white; anthers yellow; pappus white ....**L. leucopappa**
  2. Rays yellow with white tips; anthers black; pappus off-white, often with reddish base...**L. munzii**
1. Pappus of bristles or plumose feathery scales.
  3. Rays yellow with white tips; anthers black.. **L. platyglossa** (incl. ssp. *campestris*)
  3. Rays yellow or white; pappus feathery.
    4. Pappus of white deciduous bristles, feathery above the middle; stems hollow. **L. heterotricha**
    4. Pappus persistent; stems not much hollow.
      5. Basal lvs. dentate or lobed, cauline entire; pappus densely feathery; rays white... **L. glandulosa**
      5. Basal lvs. pinnately divided; pappus ± feathery; lvs. with strong odor. **L. pentachaeta**
        6. Rays yellow .... var. **pentachaeta**
        6. Rays white .... var. **albida**

Lembertia (p. 370)
(*Eatonella*)

Spring annual with tufts of woolly hairs and alternate lvs.; heads clustered, sessile or short-pedunculate; invol. with one row of 4-7 oval to obovate phyllaries; marginal fls. pistillate, usually with inconspicuous yellow rays; achenes with ciliate margins; pappus of thin scales. One species: **L. congdonii** (Gray) Greene

Lepidospartum (p. 376)

Rigid broom-like shrub 1-2 m tall with many slender straight stems and alternate entire ovate lvs. on young growth, small lf.-scales on older; heads numerous in small clusters in large panicles; rays absent; disk-fls. yellow; pappus of copious whitish capillary bristles. Three species, 2 in CA. One species in Kern County: **L. squamatum**

Lessingia (p. 359)

Summer to fall annuals to perennials, many with conspicuous tack-shaped or sessile glands;lvs. alternate; heads solitary or clustered; phyllaries overlapping in several series; true rays lacking but outer disk-corollas enlarged, 5-lobed; achenes hairy; pappus of numerous bristles. About 14 species, 13 in CA.

1. Rays present, violet to white.
   L. filaginifolia (Hook. & Arn.) Lane var. **filaginifolia** (*Corethrogyne filaginifolia*)
1. Rays absent.
   2. Stems very short, nearly absent, 1-2 cm tall; densely white-hairy; corollas rose to violet.. **L. nana**
   2. Stems taller.
      3. Corollas mostly yellow.
         4. Margins of stem-lvs. without glands; tips of inner phyllaries purple. **L. tenuis**
         4. Margins of stem-lvs. glandular; tips of inner phyllaries green.
            5. Phyllaries with glands on margins and back of tips; plants rigidly much-branched. ....**L. glandulifera** Gray var. **glandulifera** (*L. germanorum* vars. *tenuipes* and *vallicola*)
            5. Phyllaries with glands on margins only; plants mostly flexible. **L. lemmonii** Gray (vars. of *L. germanorum*)
               6. Phyllaries woolly....................................... var. **piersonii** (Howell) Ferris
               6. Phyllaries not woolly.
                  7. Stem-lvs. clasping base of invol.... var. **ramulosissima** (Nelson) Ferris
                  7. Stem-lvs. not clasping base of invol............................... var. **lemmonii**
      3. Corollas pink, lavender or whitish.
         8. Phyllaries woolly; stem lvs. with glandular dots; corollas lavender to purple. **L. leptoclada**
         8. Phyllaries not woolly; stem-lvs. with stipitate glands; corollas white with some lavender.. **L. nemaclada** (incl. var. *albiflora*)

Lygodesmia (p. 381)

(see *Prenanthella*)

Machaeranthera (p. 360)

Perennial herb or subshrub with alternate ± spiny-toothed lvs.; phyllaries in several series, herbaceous; rays shades of blue to white; pappus of numerous barbellate bristles. About 35 species, 8 in CA.

Rays present; herbaceous...**M. canescens** var. **canescens**
Rays none; subshrub ....**M. carnosa** (Gray) Nesom (*Aster intricatus*)

Madia (p. 368)

Annuals, usually glandular and heavy-scented, with linear to oblong lvs.; invol. ± angled due to shape of phyllaries which completely enclose laterally compressed ray-achenes; pappus none on ray-achenes. About 21 species, 19 in CA.

1. Disk-fls. more than 1, pubescent; phyllaries with empty flat tips.
  2. Ray-achenes prominently beaked; rays golden-yellow, remaining open in evening...**M. radiata**
  2. Ray-achenes not prominently beaked; rays closing in early evening.
    3. Disk-achenes sterile; rays conspicuous, bright yellow.......................**M. elegans**
      4. Anthers purple-black; rays yellow, 6-20 mm long.
        5. Stems simple to branched above, 2-8 dm high.
          6. Rays with maroon spot at base; flowering in late spring to summer. ssp. **elegans**
          6. Rays lacking maroon spot; flowering in spring. ....ssp. **vernalis**
        5. Stems stout, branching well above middle, to 25 dm tall; maroon spot usually present; flowering in late summer. ssp. **densifolia** (Greene) Keck (*M. densifolia*)
      4. Anthers yellow; rays 4-5 mm long, lacking maroon spot; flowering in summer...ssp. **wheeleri** (Gray) Keck (*M. wheeleri*)
    3. Disk-achenes fertile; rays inconspicuous, 3-8 mm long; anthers black. **M. gracilis**
1. Disk-fls. 1, glabrous; phyllary lacking empty flat tip; very tiny herbs.
  7. Heads on long thin divaricate peduncles; glands on phyllaries yellow... **M. exigua**
  7. Heads solitary or in small terminal clusters; glands on phyllaries black. **M. minima**

Malacothrix (p. 381)

Annual to perennial herbs with mostly basal toothed to pinnately divided lvs.; heads medium in size, yellow to white, often nodding in bud; phyllaries mostly in 1-2 series, with thin dry margins; rays truncate and 5-toothed; pappus of soft bristles. About 21 species, 14 in CA.

1. Annuals, 0.5-5 dm high.
  2. Phyllaries in 3-4 series, outer broadly ovate with conspicuous thin dry margins and dark central band; rays pale yellow .... **M. coulteri**
  2. Phyllaries in 1-2 series, outer shorter than inner, with narrow thin dry margins.
    3. Basal lvs. pinnatifid with long, linear, filiform lobes; rays 12-16 mm long, pale yellow.
      4. Stems simple, ending in solitary heads; invol. 12-15 mm high. **M. californica**
      4. Stems few-branched, each ending in a head; invol. 5-12 mm high. **M. glabrata**
    3. Basal lvs. with short, broad lobes; rays 2-10 mm long.
      5. Fls. white to pink.. **M. floccifera**
      5. Fls. yellow.
        6. Outer ligules exserted 1-4 mm.; corolla pale yellow............. **M. clevelandii**
        6. Outer ligules exserted 6-10 mm.; corolla lemmon yellow. **M. sonchoides** (Nutt.) T. & G. (Red Rock Canyon, Faull & Charlton POM)
1. Perennials, ± shrubby (not woody), 3-9 dm high; fls. white.....................**M. saxatilis**
  7. Lvs. fleshy, ± toothed; infl. condensed .... var. **saxatilis**
  7. Lvs. not fleshy, pinnatifid; infl. diffuse .... var. **tenuifolia** (Nutt.) Gray (*M. altissima*)

Micropus (p. 362)

Small woolly annuals with alternate entire lvs.; heads clustered; rays absent; outer fls. enclosed by in-folded, laterally compressed, woolly bracts; inner fls. small, staminate; pappus lacking. About 5 species, 2 in CA. One species in Kern County: **M californicus** var. **californicus**

Microseris (p. 382)

Annual to perennial herbs with mostly basal rosette of linear variously toothed lvs.; sap milky; heads on long peduncles, usually nodding in bud; invol. with lanceolate inner phyllaries and outer shorter ones; fls. yellow to white, often with reddish stripe; achene not beaked; pappus of scales which taper into bristle-like awns. About 15 species, 10 in CA.

1. Perennials; fls. conspicuous; pappus-parts 6-12 ..... **M. sylvatica** (incl. *M. montana*)
1. Annuals; fls. inconspicuous, barely equaling phyllaries; pappus-parts 2-5.
  2. Pappus-scales 5, straight and flat; awns hair-like................................. **M. elegans**
  2. Pappus-scales 2-5, curved; margins incurved; awns stouter.
    3. Pappus-scales smooth ....**M. campestris**
    3. Pappus scales hairy to scabrous ....**M. douglasii** ssp. **douglasii**

Monolopia (p. 372)

Woolly annuals with lvs. opposite below and alternate above; heads large, pedunculate, terminal; receptacle conic; phyllaries with black hairs at tips; rays bright yellow; pappus none. Four species, all in CA.

Rays 4-7 mm long, rounded apically ....**M. stricta**
Rays 10-20 mm long, truncate ....**M. lanceolata**

Monoptilon (p. 360)

Very small depressed desert annuals, branched from base; lvs. entire, hispid; heads daisy-like, subtended by upper lvs.; phyllaries mostly in one series, herbaceous; rays showy, white; pappus of short scales and a single bristle. Two species, both in CA. One species in Kern County: **M. bellidiforme**

Nicolletia (p. 372)

Perennial herb with deep taproot; lvs. alternate, glabrous, ± succulent, pinnately divided into bristle-tipped lobes, very strong-scented; heads ± sessile in upper axils; phyllaries acute with gland near tip; rays narrow, bright purple with pink stripe. Five species, one in CA. One species in Kern County: **N. occidentalis**

Orochaenactis (p. 372)

Low slender annual with linear entire alternate lvs.; heads in groups of 1-5; rays absent; phyllaries narrow, blunt, glandular-hairy; corollas greenish-yellow; achenes minutely glandular. One species: **O. thysanocarpha**

Pentachaeta (p. 353)
(*Chaetopappa)*

Diffusely branched glabrous annual with mostly basal lvs.; heads terminal on slender branches, 3-4-fld.; inner phyllaries lanceolate, the outer few minute; the inconspicuous fls. pink or rose; pappus of white capillary bristles. One species: **P. exigua** (Gray) Tomb

Peucephyllum

Rounded shrub to 3 m tall with alt., narrowly linear, thick, glabrous, resin-varnished lvs. 1-2 cm long; heads solitery, discoid; corollas pale yellow; achenes blackish; pappus of fine bristles. One species: **P. schottii** Gray (Red Rock Canyon. Faull & Charlton)

Perityle (p. 372)

Winter annuals with ascending brittle branches; lvs. broadly cordate to ovate, deeply toothed to palmately lobed; phyllaries in 2 series, fringed with hairs at tip; rays 10-13, white. About 75 species, 4 in CA. One species in Kern County: **P. emoryi**

Pluchea (p. 362)

Tall leafy herb or shrub with alternate lvs.; heads numerous, in terminal clusters; rays absent; corolla of marginal fls. 2-3-toothed, central fls. 5-toothed, purplish; pappus of capillary bristles. About 40 species, two in CA.

Annual to perennial herb, 3-12 dm tall .... **P. odorata** (L.) Cass (*P. purpurascens*)
Willow-like shrub, 1-4 m tall; lvs. silky .... **P. sericea**

Prenanthella (p. 381)
(*Lygodesmia*)

Diffusely branched glabrous annual with mostly basal lvs.; heads terminal on slender branches, 3-4-fld.; inner phyllaries lanceolate, the outer few minute; the inconspicuous fls. pink or rose; pappus of white capillary bristles. One species: **P. exigua** (Gray) Tomb.

Psathyrotes (p. 376)

Low compact scaly-tomentose aromatic herb with numerous spreading branches; lvs. alternate, broad, toothed; heads solitary in forks of branches; rays absent; fls. yellow, fading purple. Five species, two in CA. One species in Kern County: **P. annua**

Pseudobahia (p. 373)

Woolly spring annuals with alternate pinnately divided lvs.; heads solitary at ends of branches; phyllaries in one series, herbaceous; rays yellow; pappus mostly absent. Three species, all in CA.

Invol. 4-5 mm high; phyllaries united at base, with callous processes at sinuses. **P. heermannii**
Invol. 6-9 mm high; phyllaries free to base, lacking callous processes........ **P. peirsonii**

Psilocarphus (p. 362)

Small woolly annuals with opposite entire lvs.; heads small, round, subtended by pair of branches; true invol. lacking but heads subtended by small lvs.; outer pistillate fls.

loosely enclosed in saccate woolly bract with a small papery appendage below summit. Five species, 4 in CA.

1. Well developed receptacular bracts ca. 3 mm long. **P. brevissimus** var. **brevissimus**
1. Well developed receptacular bracts <3 mm long..**P. tenellus**
  2. Lvs. subtending heads oblong, 2-6 times as long as wide.................... var. **tenellus**
  2. Lvs. subtending heads broader, 1-3 times as long as wide. var. **globiferous** (DC) Morefield (var. *tenuisI*)

Pyrrocoma (p. 357)
(part of *Haplopappus*)

Herbaceous perennial from woody taproot; stems reddish-tinged; lvs. alt., simple, cauline clasping; heads in narrow clusters; rays 5-10 mm long, yellow; pappus of long bristles. About 10 species, 7 in CA. One species in Kern County: **P. racemosa** (Nutt.) T.&G. var. **paniculata** (Nutt.) Kart. & Gand.

Rafinesquia (p. 382)

Mostly ± glabrous branching annuals 2-15 dm tall with toothed or divided lvs.; lower lvs. petioled, upper reduced and auriculate-clasping; heads solitary at ends of branchlets; achenes beaked; pappus of plumose bristles. Two species.

Corollas white, exceeding invols. by about 5 mm; pappus brownish....... **R. californica**
Corollas white but veined rose-purple, exceeding invols. by 10-15 mm; pappus white. **R. neomexicana**

Raillardiopsis

Hairy, matted perennial herbs with simple opposite (below), alternate (above) lvs.; some heads scapose; pedicels 0-8 cm long; rays absent; corollas yellow; pappus-bristles about 8 mm long, plumose. Two species, both in CA. One species in Kern County: **R. muirii** (Gray) Rydb. (Granite rocky areas near summit of Owens Peak. Shevock)

Rigiopappus (p. 373)

Slender wiry annual with linear alternate entire lvs.; phyllaries in 2-3 series; rays yellow with some reddish, inconspicuous; pappus-scales often with purple spot at the ± united bases. One species. One species in Kern County: **R. leptocladus**

Senecio (p. 376)

Annual to perennial herbs or shrubs with alternate lvs.; rays usually present, yellow; phyllaries mostly equal, in 1 series with smaller bractlets at base; achenes beakless; pappus of white soft bristles. About 1500 species, 42 in CA.

1. Perennials.
   2. Lvs. divided into narrow segments, white-tomentose beneath, upper surface greenish; shrubby to ca. 1.5 m tall........................**S. flaccidus** Less. (*S. douglasii*)
      3. Shrubby to >1 m tall; invol. 8-10 mm high.
      var. **douglasii** (DC) Turner & Bark. (incl. var. *tularensis)*
      3. Less shrubby, 3-10 dm tall; invol. 7-8 mm high.
      var. **monoensis** (Greene) Turner & Bark.
   2. Lvs. not divided into narrow segments.
      4. Lvs. well-distributed along stems.
         5. Lvs. deeply lobed; phyllaries not black-tipped............................**S. clarkianus**
         5. Lvs. toothed; phyllaries with black tips ....**S. serra**
      4. Lvs. mostly basal, those on stem much reduced.
         6. Lvs. toothed, not lobed or divided; rays absent; woolly plants to 10 dm high. **S. astephanus**
         6. Lvs. lobed or divided, sometimes entire.
            7. Basal lvs. entire to lobed, not pinnately divided.
               8. Invol. 8-10 mm high; basal lvs. not cordate (heart-shaped). .**S. ionophyllus**
               8. Invol 5-8 mm high; basal lvs. subcordate to cordate. ...... **S. pseudaureus**
            7. Basal lvs. pinnately divided with large terminal lobe..................**S. breweri**
1. Annuals; phyllaries black-tipped.
   9. Rays conspicuous; lvs. narrow oblong, mostly toothed, most auriculate. **S. californicus**
   9. Rays none; lvs. coarsely divided, lower petioled and upper auriculate .. **S.vulgaris**

Silybum (p. 379)

Stout annual to biennial herb with large prickly lobed white-mottled lvs.; heads large, solitary, nodding, terminal; corollas. purple; phyllaries spiny-margined and -tipped; pappus of minutely barbed bristles. Two species, one in CA. One species introduced into Kern County: **S. marianum**

Solidago (p. 360)

Perennial herbs with leafy, usually simple stems with alternate entire to toothed lvs.; heads numerous, small, in racemose or panicled infl.; rays small, yellow; achenes short, pubescent; pappus of many white bristles. About 150 species, 10 in CA.

1. Well developed creeping rhizomes; lvs. not much reduced above; infl. open.
   2. Lvs. densely puberulent on both surfaces ....**S. californica**

2. Lvs. ± puberulent on nerves beneath, toothed .... **S. canadensis** ssp. **elongata**
1. Short woody rhizomes; lvs. reduced above, lower with clasping petioles; infl. dense; alkaline meadows.. **S. spectabilis**

Soliva

Small annuals with short stiff leafy branches; lvs. pinnately dissected; fls. greenish in small bur-like heads sessile in forks of stems; outer florets pistillate, apetalous; achenes flattened, winged or blunt-margined, pointed with hardened persistent style. Pappus none. About 5 species (S. America), one introduced in CA. One species introduced into Kern County: **S. sessilis** Lopez & Pavon (lawn weed, Hart Park, Cooley 615; CSUB, Moe 1257)

Sonchus (p. 382)

Leafy-stemmed, smooth coarse annuals with milky sap; lvs. spiny-denticulate, bases auriculate-clasping; phyllaries few, thin, with many shorter ones at base of invols.; fls. yellow; achenes flattened, thin-edged; pappus of soft white bristles. About 54 species, 4 introduced in CA.

Basal auricles of lvs. acute; achenes with fine lines .... **S. oleraceus**
Basal auricles of lvs. rounded; achenes 3-nerved .... **S. asper** ssp. **asper**

Stebbinsoceris (p. 382)
(part of *Microseris*)

Scapose annual with entire to pinnately-lobed lvs.; heads ± nodding in bud; invol. nearly concealing fls.; corollas yellow to whitish; pappus scales 5 with bristle arising from notched tip. Two species, both in CA. One species in Kern County: **S. heterocarpa** (Nutt.) Chambers

Stephanomeria (p. 383)

Annual or perennial mostly glabrous herbs with slender paniculately branched stems; lvs. alternate, linear to oblong, toothed to divided, uppermost much reduced; heads in panicles, rose to light lavender; invols. of several phyllaries; pappus of feathery bristles. About 24 species, 12 in CA.

1. Annuals.
   2. Pappus-bristles feathery throughout; plants mostly 5-30 dm high; branches ± leafless when flowering. .... **S. virgata** ssp. **virgata**
   2. Pappus-bristles naked at base, feathery above middle; plants <5 dm high; diffusely branched; lvs. clasping stem.................................. **S. exigua** ssp. **exigua**
1. Perennials.
   3. Phyllaries not much overlapping, small bractlets at base; lvs. mostly glabrous.
      4. Invols. 12-15 mm high, 10-14 fld... **S. parryi**

4. Invols. 8-10 mm high, 3-5-fld.; intricately branched bush. **S. pauciflora** var. **pauciflora**

3. Phyllaries overlapping; young lvs. woolly; plant erect, stout, 4-12 dm tall. **S. cichoriacea**

Stylocline (p. 362)

Low woolly annuals with alternate entire lvs.; heads terminal, subglobose, clustered; phyllaries completely enclosing marginal pistillate fls., ± boat-shaped and falling with achene; pappus none. About 6 species, all in CA.

Bracts of pistillate fls. with thin wing throughout their length........... **S. gnaphalioides**
Bracts of pistillate fls. with terminal thin wing-like appendage............ **S. micropoides**

Syntrichopappus (p. 373)

Low diffusely branched woolly annual with linear-spatulate lvs.; heads on short peduncles; invols. bell-shaped; phyllaries in 1 series; rays mostly 5, golden-yellow, 3-5 mm long; pappus of white barbellate bristles. Two species, both in CA. One species in Kern County: **S. fremontii**

Taraxacum (p. 383)

Perennial herbs with basal lvs. and solitary, showy yellow heads at ends of naked hollow scapes; lvs. pinnately divided to toothed; phyllaries in 2 series, outer ones short, reflexed, the inner erect, linear; achenes oblong, long-beaked; pappus of white bristles. About 50 species, 2in CA. One species introduced into Kern County: **T. officinale** (incl.*T. laevigatum*)

Tetradymia (p. 377)

Low to medium rigid shrubs with dense wool at least when young; lvs. alternate, entire, solitary or fascicled, the primary ones often modified into spines; heads 4-9-fld., yellow, usually in small racemes; invol. cylindric with 4-6 firm phyllaries; rays absent; pappus of soft capillary bristles. Ten species, 8 in CA.

1. Achenes with long ascending hairs which nearly conceal the pappus; primary lvs. stiff slender spines 2-3 cm long.
T. axillaris Nels.var. **longispina** (Jones) Strother (*T. spinosa* var. *longispina*)
1. Achenes glabrous or with much shorter hairs.
  2. Heads 4-fld., primary lvs. not changed into spines.
    3. Primary lvs. erect, slender, 5-10 mm long; secondary lvs. fascicled. **T. glabrata**
    3. Primary lvs. 20-30 mm long; secondary lvs. rarely fascicled.......... **T. canescens**
  2. Heads 5-fld.; primary lvs. modified into spines, 2-3 cm long, perpendicular to stems.. **T. stenolepis**

Tragopogon (p. 383)

Stout mostly glabrous perennial herb with entire elongate grass-like clasping lvs. and large solitary heads; invol. of mostly 8-9 lanceolate, tapering-pointed phyllaries about as long as chrome-yellow fls.; achenes long-beaked; pappus of feathery bristles. About 45 species, 3 introduced into CA. One species introduced into Kern County: **T. pratensis**

Uropappus (p. 382)
(part of *Microseris*)

Scapose annual with ± linear, entire to narrowly lobed, soft-hairy (esp. petiole) lvs.; heads erect; invol. 10-40 mm long, glabrous; corollas yellow; pappus of 5 deciduous scales with slender bristle from notched tip. One species: **U. lindleyi** (DC) Nutt.

Wyethia (p. 365)

Coarse perennial herbs with erect usually unbranched stems from a thick taproot; lvs. alternate, large, ovate to deltoid; heads solitary or few-clustered, yellow; phyllaries in 2-4 series; pappus reduced to a crown. About 14 species, 10 in CA.

Rays present; heads ± hidden by lvs., phyllaries in 2 series .... **W. ovata**
Rays absent (or 1-3 short ones); phyllaries in >2 series .... **W. invenusta**

Xanthium (p. 363)

Coarse weedy annuals with stout branching stems; lvs. alternate, lobed or toothed; staminate heads above pistillate in terminal and solitary clusters; invol. of pistillate heads enclosing 2 florets and becoming a 2-beaked bur with rigid hooked bristles. Two species, both in CA.

Lvs. lanceolate, with conspicuous 3-forked spines in axils .... **X. spinosum**
Lvs. broadly ovate or cordate, without spines .... **X. strumarium** (incl. var. *canadense*)

Xylorhiza (p. 360)
(part of *Machaeranthera*)

White-hairy subshrub with much-branched caudex; lvs. 3-10 cm long, spiny-dentate; heads solitary; peduncles 8-25 cm long; phyllaries in several series; rays light blue; pappus of many bristles. About 3 species, all in CA. One species in Kern County: **X. tortifolia** (T&G) Greene var. **tortifolia**

AZOLLACEAE
(Salviniaceae)

Azolla (p. 167)

Very small compact moss-like floating aquatic fern about 1-2.5 cm. long, reddish; lvs. ovate, 2-lobed, imbricate, the lower lobe submersed and forming sporocarp. About 6 species, one in CA. One species in Kern County: **A. filiculoides**

BERBERIDACEAE

Berberis (p. 239)

Evergreen shrub 3-18 dm high with pinnately compound lvs.; lfts. stiff-coriaceous, prominently veined with spinose teeth on margins; fls. yellow; fr. a blue-black berry. Over 600 species, 6 in CA. One species in Kern County: **B. aquifolium** Pursh. var. **dictyota** (Jeps.) Jeps. (*B. dictyota.*)

BETULACEAE

Alnus (p. 210)

Deciduous tree, 10-35 m tall with whitish bark and alt. simple serrate lvs. 5-11 cm long; fls. imperfect, apetalous, staminate in catkins, pistillate becoming small "cone". About 30 species, 4 in CA. One species in Kern County: **A. rhombifolia**

BIGNONIACEAE

Catalpa (p. 344)

Deciduous tree to 10 m tall with often whorled simple broadly cordate lvs.; panicles large; fls. about 5 cm broad, white with yellow lines and purplish-brown spots; pod 15-30 cm long, 1 cm or less wide. About 11 species, one in CA. One species introduced in Kern County: **C. bignonioides**

BLECHNACEAE

Woodwardia (p. 166)

Terrestrial ferns with large coarse pinnately compound fronds, pinnae deeply divided; fronds 1-2 m high, nearly erect; indusium nearly straight. About 15 species, one in CA. One species in Kern County: **W. fimbriata**

BORAGINACEAE

Annual or perennial usually rough-hairy herbs; lvs. simple, usually entire; fls. regular, mostly in coiled spikes or cymes; corolla tubular, sometimes crested in throat; stamens 5, alternate the corolla-lobes; ovary superior, 2-4-ovulate, becoming nutlets in fr.; style lobed or entire.

1. Style deeply cleft, each branch with capitate stigma .... **Tequilia**
1. Style entire, terminated by a single or obscurely lobed stigma.
   2. Style borne on summit of fr.; plants succulent................................. **Heliotropium**
   2. Style borne between the lobes of fr. (nutlets); plants not succulent.
      3. Nutlets widely spreading in fr., armed with barbed or hooked prickles.
         4. Nutlets flat, armed with hooked prickles on margins; fls. white; annuals. **Pectocarya**
         4. Nutlets subglobose, armed all over; fls. blue; perennial. .... **Cynoglossum**
      3. Nutlets erect, often unarmed.
         5. Fls. blue or pink; perennials.
            6. Nutlets smooth and shining..**Myosotis**
            6. Nutlets armed with hooked prickles .... **Hackelia**
         5. Fls. white, yellow or orange; plants mostly annuals.
            7. Fls. white.
               8. Nutlets attached laterally, with basal groove or slit.............. **Cryptantha**
               8. Nutlets with elevated keel and attachment scar. .... **Plagiobothrys**
            7. Fls. yellow or orange.
               9. Annuals .... **Amsinckia**
               9. Perennials... **Cryptantha**

Amsinckia (p. 321)

Pungent-bristly annuals; fls. in mostly naked coiled spikes; calyx 5-parted or with 1-several lobes ± united; corolla tubular to trumpet-shaped, yellow to orange; style filiform; stigma capitate, weakly 2-lobed; nutlets various. About 10 species, 9 in CA.

1. Nutlets smooth and shiny; lvs. coarsely pustulate.................................. **A. vernicosa**
   2. Corolla 8-12 mm long, 2-8 mm wide at top, yellow .... var. **vernicosa**
   2. Corolla 12-18 mm long, 8-14 mm wide at top, orange. var. **furcata** (Suksd.) Hoov. (*A. furcata*)
1. Nutlets roughened; lvs. not coarsely pustulate.
   3. Back of nutlets tessellate (like cobblestones); corolla 20-nerved (veins) below stamens, orange.
      4. Corolla 6-14 mm broad; calyx densely rusty-pubescent..............**A. douglasiana**
      4. Corolla 2-4 mm broad; calyx white-hirsute.............. **A. tessellata** var. **tessellata**
   3. Back of nutlets not tessellate; corolla 10-nerved below stamens.
      5. Corolla yellowish, 5-7 mm long, 2-3 mm broad, tube mostly included. **A. menziesii** var. **menziesii**
      5. Corolla orange, 8-18 mm long, tube exserted.

6. Corolla deep orange, 14-18 mm long.......................... **A. eastwoodae** Macbr. (*A. intermedia* var. *eastwoodae)*
6. Corolla light orange, 8-10 mm long. **A. menziesii** var. **intermedia** (Fisch. & Meyer) Ganders (*A. intermedia*)

Cryptantha (p. 322)

Small hispid annuals (one perennial) with slender stems; early lvs. opposite, others alternate; fls. in bractless coiled spikes; corolla small, white, the throat with intruded appendages; nutlets attached along a ventral slit in the pericarp to a columnar gynobase (prolonged receptacle). About 65 species, 52 in CA.

1. Perennial; corolla yellow....................................... **C. confertiflora** (Greene) Payson (dry rocky soils, Mt. Laura, Piute Mtns., Shevock 8071 - see Twisselmann p. 384)
1. Annuals; corollas white.
  2. Sepals falling in a ring when mature. **C. circumscissa** (incl. var. *hispida* and*C. similis*)
  2. Sepals not falling off, persistent.
    3. Receptacle subulate, protruding beyond nutlets; stems and root with purple dye. **C. micrantha** (incl. var. *lepida*)
    3. Receptacle shorter than nutlets; stems and roots with little or no purple dye.
      4. Nutlets roughened.
        5. Nutlet margins winged or knife-like.
          6. Nutlet margins decidedly winged; nutlets 4.
            7. Corolla conspicuous, 4-7 mm broad; nutlets alike............. **C. oxygona**
            7. Corolla inconspicuous, 1 mm broad; one nutlet unlike others. **C. pterocarya** (incl. vars. *cycloptera* and *purpusii*)
          6. Nutlet margins knife-like.
            8. Nutlets 1 or 2.. **C. utahensis**
            8. Nutlets 4... **C. angustifolia**
        5. Nutlet margins rounded or obtuse.
          9. Nutlets distinctly heteromorphous, one larger than the others. **C. dumetorum**
          9. Nutlets all alike.
            10. Nutlets lanceolate.
              11. Nutlets 1 or 2.............................. **C. decipiens** (incl. var. *corollata*)
              11. Nutlets 4 in normal fr.
                12. Corolla 2-8 mm broad; cismontane....................... **C. intermedia**
                12. Corolla 1-2 mm broad; deserts.
                  13. Calyx 6-12 mm long in fr.
                    14. Plants bristly, hairs spreading .... **C. barbigera**
                    14. Plants strigose, hairs appressed. **C. nevadensis** (incl. var. *rigida*)
                  13. Calyx 3-4 mm long in fr.................................. **C. angustifolia**
            10. Nutlets ovate to triangular-ovate.
              15. Nutlets with faint median ridge on back; corolla 1-6 mm broad. **C. muricata** (incl. vars. *denticulata* and *jonesii)*

15. Nutlets without median ridge; corolla <2 mm broad.
16. Plants closely strigose; spikes usually paired............ **C. simulans**
16. Plants spreading-hispid; spikes usually solitary........ **C. echinella**
4. Nutlets smooth and shiny.
17. Calyx with hooked or curved hairs.......... **C. flaccida** (incl. *C. sparsiflora*)
17. Calyx with straight hairs.
18. Nutlets with eccentric groove .... **C. affinis**
18. Nutlets with a centrally placed groove.
19. Corolla 3-7 mm broad.
20. Nutlets 2; cismontane.. **C. incana**
20. Nutlets 4; deserts... **C. mohavensis**
19. Corolla mostly ca. 1 mm broad.
21. Calyx 3-6 mm long in fr.; nutlets 4............................. **C. torreyana**
21. Calyx 1-3 mm long in fr.; nutlets 1-4.
22. Hairs of upper part of calyx-lobes retrorse............. **C. nemaclada**
22. Hairs of calyx-lobes spreading or ascending...... **C. microstachys**

Cynoglossum (p. 325)

Coarse tall perennial herb with spreading hairs; basal lvs. long-petioled, upper sessile; corolla blue, tinged pinkish, the tube closed by 5 scales stamens included; nutlets large, covered with short barbed prickles. About 80 species, 3 in CA. One species in Kern County: **C. occidentale**

Hackelia (p. 325)

Perennial herb with alternate narrow oblong lvs.; infl. ± compact of several few-fld. branches; pedicels recurved in fr.; fls. blue to pink with conspicuously crested throat; fr. armed with barbed prickles. About 40 species, 12 in CA. One species in Kern County: **H. velutina**

Heliotropium (p. 325)

Fleshy glaucous perennial herb from thick underground rootstocks; lvs. succulent; fls. in dense scorpioid cymes, white with yellow spots in throat, becoming purplish in age; fr. glabrous. About 250 species, 4 in CA. One species in Kern County: **H. curassavicum** (incl. var. *oculatum*)

Myosotis (p. 325)

Slender perennial herb; lvs. alternate, entire; corolla pale blue with yellow eye, salverform with prominent crests in throat; nutlets small, ovoid, smooth, shining and sharply margined. About 50 species, 6 in CA. One species in Kern County: **M. laxa**

Pectocarya (p. 325)

Low spreading annuals with slender stems and narrow lvs.; fls. in leafy-bracted racemes which constitute most of the plant; corolla white, small; nutlets linear to obovate, divaricate, usually margined, bearing hooked bristles or hairs. About 15 species, 8 in CA.

1. Nutlet margins entire; nutlet body obovate or rhomboidal.
  2. Nutlets equally divergent; calyx with hooked hairs .... **P. pusilla**
  2. Nutlets divergent in pairs; calyx with straight hairs.................................. **P. setosa**
1. Nutlet margins lacerate, bristly; nutlet body linear or oblong.
  3. Nutlet margins prickly most of their length.
    4 Nutlets straight or ± sickle-shaped.
      5. Stems prostrate; nutlet margin teeth distinct to base.
        P. linearis ssp. **ferocula** (Johnston) Thorne (var. *ferocula*)
      5. Stems more erect; nutlet margin teeth fused at base.
        P. platycarpa (Munz & Johnston) Munz & Johnston
        (Red Rock Canyon, Faull & Charlton POM)
    4. Nutlets conspicuously recurved .... **P. recurvata**
  3. Nutlet margins entire or undulate except for bristly tips.
    6. Nutlets all wing-margined... **P. penicillata**
    6. Nutlets dimorphic, with 2-winged and 2 mostly not winged....... **P. heterocarpa**

Plagiobothrys (p. 326)

Annual or perennial herbs with weak slender oppressed hairs; fls. in slender scorpioid racemes; corolla white with short tube; stamens included; nutlets mostly roughened, with elevated keel and attachment scar. About 65 species, 39 in CA.

1. Lvs. opposite, at least below (*Allocarya*).
  2. Scar large, 1/4-1/2 length of nutlet... **P. acanthocarpus**
  2. Scar small, <1/5 length of nutlet.
    3. Plants ± succulent; calyx-lobes strongly ribbed, 4-8 mm long..... **P. leptocladus**
    3. Plants not succulent; calyx not ribbed, 2-4 mm long.
      4. Scar basal, oblique to ventral keel... **P. bracteatus**
      4. Scar lateral, parallel with ventral keel... **P. hispidulus**
1. Lvs. alternate (*Plagiobothrys*).
  5. Scar elongate; nutlets tessellate (like cobblestones)................................. **P. jonesii**
  5. Scar ± round, at or below ventral keel; nutlets various.
    6. Calyx circumscissile (sepals falling along a line) in fr.; nutlets 1-2 (4).
      7. Infl. a long simple bracted raceme; corolla 2-3 mm broad. .... **P. arizonicus**
      7. Infl. forked, bracted only at base; corolla 3-9 mm broad.......... **P. nothofulvus**
    6. Calyx not circumscissile; nutlets mostly 4.
      8. Nutlets with annular scar, 2-3 mm long .... **P. infectivus**
      8. Nutlets with solid scar, less than 2 mm long.

9. Stems with purple dye; nutlets ovoid.
 10. Corolla 1-2 mm broad; calyx 2-3 mm long.............................. **P. torreyi**
 10. Corolla 2-4 mm broad; calyx 4-6 mm long......................... **P. canescens**
9. Purple dye absent; nutlets cruciform, constricted at base and apex.. **P. tenellus**

Tequilia (p. 322)
(*Coldenia*)

Low prostrate to matted canescent to hispid herbs; lvs. alternate, broad, with impressed veins; fls. clustered, usually at forks of stem; corolla short-tubed; style 2-cleft. About 27 species, 4 in CA.

Plant annual; corolla pink to white, 3-4 mm long................... **T. nuttallii** (Hook.) Rich.
Plant perennial; corolla blue or lavender, 4-6 mm long.............. **T. plicata** (Torr.) Rich.

BRASSICACEAE
(Cruciferae by many authors)

Herbs or shrubs with pungent juice and alternate lvs.; fls. regular, in terminal racemes; sepals and petals mostly 4, stamens 6 (rarely 2); ovary superior, with thin partition between two cells (valves); fr. 2-celled capsule, long and narrow (silique) or short (silicle), sometimes indehiscent.

1. Frs. indehiscent, 1-celled, 1-seeded.
 2. Silicles not winged, with hooked hairs.. **Athysanus**
 2. Silicles wing-margined, not with hooked hairs.............................. **Thysanocarpus**
1. Frs. dehiscent, usually more than 1-seeded.
 3. Frs. breaking cross-wise into seed bearing joints.
  4. Siliques 3-5 mm thick; pedicels 10-20 mm long; seeds in one row.... **Raphanus**
  4. Siliques >3 mm thick; pedicels 2-5 mm long; seeds in two rows...... **Chorispora**
 3. Frs. not breaking cross-wise, dehiscent by valves.
  5. Fr. a silique, several times longer than wide.
   6. Siliques with stalk between receptacle and fr..................................... **Stanleya**
   6. Siliques sessile.
    7. Siliques compressed perpendicular to suture lines on fr. ...**Tropidocarpum**
    7. Siliques not compressed perpendicular to narrow partition.
     8. Stigma lobes situated over valves (see young flower); anthers arrowhead-shaped at base.
      9. Calyx urn-shaped or flask-shaped.
       10. Siliques conspicuously beaked, pendent, strongly compressed. **Streptanthella**
       10. Siliques not conspicuously beaked, erect or deflexed.
        11. Siliques flattened; seeds flat, usually winged.......... **Streptanthus**
        11. Siliques cylindric or 4-sided; seeds wingless.
         12. Upper lvs. clasping; petals 5-30 mm long............. **Caulanthus**
         12. Upper lvs. not clasping; petals 3-9 mm long............. **Guillenia**

9. Calyx open, not urn-shaped.
   13. Siliques ± flattened.. **Thelypodium**
   13. Siliques ± 4-sided to cylindric.
      14. Fr. reflexed, not clustered; petals pinkish..................... **Guillenia**
      14. Fr. spreading, clustered; petals white..................... **Thelypodium**

8. Stigma-lobes situated over suture lines on fr.; anthers not arrowhead-shaped at base.
   15. Basal lvs. in definite rosettes.
      16. Siliques tetragonal-cylindric, 10-15 mm long.............. **Arabidopsis**
      16. Siliques flattened, mostly longer... **Arabis**
   15. Basal lvs. not in well-formed rosettes.
      17. Frs. with long indehiscent beak.
         18. Petals <10 mm long; seeds in 1 row per chamber.
            19. Fr.-valves with 1 midvein.. **Brassica**
            19. Fr.-valves with 3-7 veins.
               20. Fr. appressed to stem; beak-base swollen......... **Hirschfeldia**
               20. Fr. not appressed to stem; beak-base not swollen. **Thelypodium**
         18. Petals 15-20 mm long; seeds in 2 rows per chamber.......... **Eruca**
      17. Frs. dehiscent to tip.
         21. Pubescence simple or absent.
            22. Siliques strongly flattened.
               23. Plants with fleshy rootstocks; stems leafless in lower half. **Dentaria**
               23. Plants not as above.. **Arabis**
            22. Siliques cylindrical or 4-angled.
               24. Frs. ± 4-angled... **Barbarea**
               24. Frs. cylindric.
                  25. Frs. 3-10 cm long; dry places........................ **Sisymbrium**
                  25. Frs. 6-15 mm long; wet places. .... **Rorippa**
         21. Pubescence of forked or stellate hairs.
            26. Lvs. pinnatifid to pinnate....................................... **Descurainia**
            26. Lvs. entire or nearly so.
               27. Fls. yellow or orange.. **Erysimum**
               27. Fls. white to rose or purple.
                  28. Petals 3-15 mm long; stigmas not horned on back.. **Arabis**
                  28. Petals about 20 mm long; stigmas horned on back. **Matthiola**

5. Fr. a silicle, 1-2 times longer than wide.
   29. Fr. flattened.
      30. Fr. flattened parallel with partition.
         31. Stems scapose, 1-flowered.. **Idahoa**
         31. Stems not scapose, more than 1-flowered.
            32. Fr. tardily dehiscent; infl. with fls. along one side .... **Heterodraba**
            32. Fr. soon dehiscent; infl. not as above....................................... **Draba**
      30. Fr. flattened perpendicular to partition.

33. Frs. twin-like, the two locules rounded with a deep notch between them.
    34. Fr. 6-18 mm wide; pubescence stellate.............................. **Dithyrea**
    34. Fr. ca. 2 mm wide; pubescence not stellate.................... **Coronopus**
33. Fr. not twin-like.
    35. Fr. inverted-triangular; annuals.. **Capsella**
    35. Fr. orbicular to oval.
        36. Valves 1-seeded.. **Lepidium**
        36. Valves >1-seeded .... **Hutchinsia**

29. Fr. turgid or inflated, not flattened.
    37. Plants with leafy stems .... **Cardaria**
    37. Plants in ± scapose tufts, the lvs. mostly basal................................ **Draba**

Arabidopsis (p. 244)

Simple or branched annual with some forked hairs; basal lvs. in rosette, obscurely toothed; fls. small, white, in terminal racemes; siliques slender 10-15 mm long. About 12 species, one introduced in CA. One species in Kern County: **A. thaliana**

Arabis (p. 244)

Biennial or perennial herbs, glabrous or pubescent with simple forked or stellate hairs; lvs. mostly entire, cauline often auricled; infl. racemose; petals white to purple; siliques linear, straight or curved, flattened parallel to partition. About 120 species, 41 in CA.

1. Siliques 3-8 mm wide; seed-wing 1-3 mm wide.
    2. Lower stem-lvs. with winged petioles; petals scarcely longer than sepals. **A. repanda**
    2. All stem-lvs. sessile; petals exceeding sepals................................ **A. platysperma**
1. Siliques <3 mm wide; seed-wing <1 mm wide.
    3. Basal lvs. broadly oblanceolate.
        4. Siliques 4-10 cm long, 1 mm wide; seeds in 2 rows............................ **A. glabra**
        4. Siliques 3-5 cm long, 1.5-2 mm wide; seeds in one row... **A. davidsonii** Greene (Granite rock-faces, Spanish Needle & Owens Peak, Shevock 11212)
    3. Basal lvs. linear to linear-oblanceolate.
        5. Cauline lvs. arrowhead-shaped; seeds in one row.
            6. Siliques 6-12 cm long; basal lvs. 3-10 cm long ....**A. sparsiflora** var. **arcuata**
            6. Siliques 4-6 cm long; basal lvs. 2-3 cm long .... **A. inyoensis** Rollins (Summit of Owens Peak, Shevock 11188)
        5 Cauline lvs. not arrowhead-shaped ; seeds in 2 rows .... **A. pulchra**

Athysanus (p. 245)

Diffuse slender-stemmed annual, branched from near base, pubescent with forked hairs; lvs. basal, simple; fls. minute in one-sided racemes; pedicels recurved; silicles orbicular, indehiscent. One species: **A. pusillus**.

Barbarea (p. 245)

Glabrous biennial or perennial herbs with overwintering basal rosettes and stout angled stems; lvs. pinnatifid, the cauline with clasping bases; fls. racemose, yellow; siliques linear. About 20 species, 3 in CA. One species in Kern County: **B. orthoceras**

Brassica (p. 245)

Erect branched annual; basal lvs. pinnatifid to toothed; fls. racemose, showy, yellow; petals clawed; siliques elongate with stout indehiscent beak. About 35 species, ours introduced from Eurasia. One species in Kern County: **B. tournefortii** Gouan. (waste areas, Edwards AFB. Moe 1511)

Capsella (p. 245)

Low slender annual with forked pubescence; basal lvs. tufted; fls. racemose, small, white; silicle obcordate-triangular, flattened perpendicular to partition. About 4 species, one introduced in CA. One species in Kern County: **C. bursa-pastoris**

Cardamine (p. 246)
(*Dentaria*)

Glabrous perennial from thick rhizome; rhizomal lvs. 3-foliate with ovate lfts., cauline 2-5-foliate; infl. racemose; petals white, 9-14 mm long; siliques linear. About 170 species, 11 in CA. One species in Kern County: **C. californica** (T&G) Greene var. **californica**

Cardaria (p. 245)

Perennial rhizomatous pubescent herbs; lvs. dentate, upper clasping; fls. small, white; silicles ovoid, inflated, mostly indehiscent. Five species of Old World.

Silicles glabrous, broader than long.. **C. draba**
Silicles pubescent, longer than broad.. **C. pubescens**

Caulanthus (p. 245)

Annuals, glabrous or pubescent with simple hairs; cauline lvs. auriculate, basal not in conspicuous rosettes; fls. racemose, purple, yellow, or white; calyx flask-shaped; petals narrow; siliques slender, terete. About 18 species, 12 in CA..

1. Cauline lvs. sessile and auriculate at base.
  2. Stems conspicuously inflated .... **C. inflatus**
  2. Stems not conspicuously inflated.
    3. Stigma inconspicuously 2-lobed or entire.
      4. Fls. purplish; siliques erect or divaricate, 6-8 cm long. .... **C. amplexicaulis**
      4. Fls. yellow; siliques reflexed, 2-4 cm long .... **C. cooperi**
    3. Stigma distinctly 2-lobed.
      5. Siliques 2-4 cm long; filaments distinct .... **C. californicus**
      5. Siliques 5-13 cm long; at least two filaments united .... **C. coulteri**
        6. Siliques reflexed; pedicels 3-10 mm long .... var. **coulteri**
        6. Siliques erect; pedicels 10-20 mm long.
          var. **lemmonii** (Wats.) Munz (*C. lemmonii*)
1. Cauline lvs. sessile or petiolate, not auriculate at base .... **C. pilosus** Wats.
   (Near summit of Spanish Needle, eastern crest of Sierra. Shevock 11302)

Chorispora

Branched annual herb 1-3 dm tall; lower lvs. pinnately incised, middle and upper lanceolate, wavy-dentate; corollas purple, 10-12 mm long; siliques 3-4 cm long. About ten species, one introduced into CA. One species in Kern County: **C. tenella** (Pall.) DC (Occasional weed, Edison. Shevock 10169)

Coronopus

Strong-smelling prostrate annual, pubescent with simple hairs; lvs. pinnately divided; fls. minute, greenish white, in capitate clusters; silique with twin-like valves. About 10 species, 2 introduced in CA. One species in Kern County: **C. didymus** (L.) Sm. (Sidewalk weed in Southwest Bakersfield, Moe 731.)

Descurainia (p. 246)

Annual or biennial erect branched herbs; lvs. 1-3-pinnate, with fine divisions; pubescence stellate or forked; fls. small, racemose, yellow; siliques linear, curved. About 40 species, 7 in CA.

1. Upper lvs. 2-3-pinnate; siliques 10-30 mm long .... **D. sophia**
1. Upper lvs. pinnate with incised lfts., siliques <15 mm long.

2. Seeds in 1 row in each locule; style conspicuous; silique 9-15 mm long; stems with glandular hairs................**D. incana** (Fisch. & Meyer) Dorn (*D. richardsonii*)
2. Seeds in 2 rows in each locule; style obsolete; siliques club-shaped to oblong-elliptic. ....**D. pinnata**
   3. Racemes glandular-pubescent; petals 2-3 mm long...................... ssp. **menziesii**
   3. Racemes glabrous; petals ca. 2 mm long..ssp. **glabra**

Dithyrea (p. 247)

Stellate-pubescent annual with several decumbent stems; lvs. sinuate-dentate; fls. racemose, white; silicle strongly compressed perpendicular to partition, twin-like. Two species, both in CA. One species in Kern County: **D. californica**.

Draba (p. 247)

Low tufted annual with branched hairs; lvs.simple, entire to denticulate; fls. racemose, small white; silicle elliptic-oblanceolate. About 250 species, 22 in CA.

Lvs. all in basal rosette; pedicels 15-25 mm long; petals 2-lobed....................**D. verna**
Lv. also cauline; pedicels 2-5 mm long; petals not 2-lobed......... **D. cuneifolia** T. & G.
(On limestones, Erskine Creek Canyon. Shevock 10425)

Eruca (p. 247)

Rather succulent glabrous annual or biennial; lvs. pinnatifid to pinnately-lobed; fls. racemose; petals white, 15-18 mm long; siliques linear-oblong, long beaked. Three species, one introduced in CA. One species in Kern County: **E. vesicaria** (L.) Cav. ssp. **sativa** (Miller) Thell. (*E. sativa*)

Erysimum (p. 247)

Annual, biennial or perennial leafy-stemmed herbs with appressed 2-3-forked hairs; fls. yellow to orange, showy in terminal racemes; siliques linear, 4-sided or subterete. About 160 species, 9 in CA.

Basal lvs. lanceolate to oblanceolate; fr. mostly not narrowed between seeds; petals orange to yellow.......**E. capitatum** ssp. **capitatum** (*E. argillosum* and *E. moniliforme*)
Basal lvs. spoon-shaped; fr. narrowed between seeds; petals yellow.
E. capitatum (Dougl.) Greene ssp. **perenne** (Cov.) Price (*E. perenne*)

Guillenia (p. 251)
(part of *Thelypodium*)

Mostly glaucous annuals with basal rosette; upper lvs. reduced, subsessile; petals 2-9 mm long, white to pinkish; fr. cylindric, 1-7 cm long, ± straight. Three species, all in CA.

Fl. parts ± erect; sepals greenish; petals white to pale yellow.
G. lasiophylla (Hook. & Arn.) Greene
Fl. parts spreading; sepals pink to purplish; petals pink or whitish with pink veins.
G. lemmonii (Greene) Buck

Heterodraba (p. 248)

Diffuse slender-stemmed annual, branched from near base, pubescent with forked hairs; lvs. basal; fls. in lax racemes, minute, white; pedicels recurved; silicles round-oval, compressed, dehiscent. One species: **H. unilateralis**

Hirschfeldia (p. 245)
(part of *Brassica*)

Erect biennial to perennial herb, 2-6 dm tall; basal rosette with pinnately lobed lvs., cauline reduced, not clasping; petals yellow to white; fr. cylindric, valves 3-7-veined. Two species, one introduced in CA. One species in Kern County: **H. incana** (L.) Lagr.-Fossat (*B. geniculata*)

Hutchinsia (p. 248)

Low slender annual with forked pubescence; lvs. entire to pinnately lobed; fls. racemose, small, white; silicles compressed perpendicular to partition, elliptical. About 3 species, one in CA. One species in Kern County: **H. procumbens**

Idahoa (p. 248)

Low scapose glabrous annuals with basal lyrate lvs.; fls. solitary, small, white; silicle suborbicular, flattened parallel to broad partition. One species: **I. scapigera**

Lepidium (p. 248)

Annual to somewhat woody perennials, glabrous to hirsute with simple hairs; lvs. entire to bi-pinnate, sometimes clasping or surrounding stem; fls. minute, racemose; petals white, yellow or absent; stamens 2, 4 or 6; silicle ovate, obovate or elliptic, strongly compressed perpendicular to partition, usually notched or lobed at apex. About 175 species, 20 in CA.

1. Some lvs. surrounding stem or with clasping bases.

2. Fr. inflated, with conspicuously winged margin; cauline lvs. arrowhead-shaped at base. .... **L. campestre**
2. Fr. not inflated, not winged; bases of cauline lvs. surrounding stem. **L. perfoliatum**

1. Lvs. not surrounding stem or with clasping bases.
  3. Styles none or shorter than notch in fr.
    4. Fr. notched at apex but not or slightly winged.
      5. Pedicels terete or only slightly flattened.
        6. Sepals persistent; cauline lvs. pinnatifid to lobed........ **L. strictum**
        6. Sepals deciduous; cauline lvs. mostly entire.
          7. Petals wanting or shorter than sepals........ **L. densiflorum**
          7. Petals equalling or exceeding sepals........ **L. virginicum** var. **pubescens**
      5. Pedicels strongly flattened.
          8. Petals lacking; plant freely branched from base; stamens 2-4. **L. lasiocarpum**
          8. Petals evident; plant usually simple below; stamens 6 .... **L. nitidum**
            9. Fr.-tips ascending........ var. **oreganum**
            9. Fr.-tips erect.
              10. Stems densely pubescent; silicles minutely pubescent on margins. var. **howellii**
              10. Stems glabrous to hairy; fr. glabrous. var. **nitidum** (incl. var. *insigne*)
    4. Fr. plainly winged at apex with 2 prominent lobes........ **L. dictyotum**
      11. Silicles to 3 mm long, lobes <1 mm long, not divergent........ var. **dictyotum**
      11. Silicles mostly >4 mm long, lobes >1 mm long, divergent...... var. **acutidens**
  3. Styles well developed, exceeding notch in fr.
    12. Fls. white; plant perennial, shrubby........ **L. fremontii**
    12. Fls. yellow; plant annual........ **L. flavum**

Matthiola (p. 249)

Biennial or perennial herbs with a close stellate tomentum; lvs. oblong to linear, entire to dentate; fls. purple to white, racemose; petals ca. 2 cm long, clawed; siliques elongate, subterete, 8-30 cm long. About 50 species, one introduced in CA. One species in Kern County: **M. incana** (L.) R. Br. (*M. bicornis*)

Raphanus (p. 249)

Erect branched annual to biennial herbs; lvs. lyrate; fls. showy, purplish to white; petals long-clawed; siliques cylindric, 2-3-seeded. Three species, 2 introduced in CA. One species in Kern County: **R. sativus**

Rorippa (p. 249)

Glabrous herbs of wet places; lvs. pinnatifid; fls. in short racemes, yellow to white; siliques linear, cylindrical, mostly curved. About 75 species, 11 in CA.

1. Petals white; perennial with prostrate, freely rooting stems.
R. nasturtium-aquaticum (L.) Hayek (*Nasturtium officinale*)
1. Petals yellow; annual to ± perennial.
  2. Stems diffusely branched from base; pedicels 2-4 mm long........... **R. curvisiliqua**
  2. Stems erect, branched above; pedicels 3-8 mm long.
R. palustris (L.) Besser var. **occidentalis** (Wats.) Rollins (*R. islandica*

Sinapsis (p. 245)
(part of *Brassica*)

Erect leafy annual; basal lvs. irregularly lobed, not in basal rosette; infl. terminal raceme; petals yellow, ± 10 mm long; fr. cylindric, beaked. six species, two introduced in CA. One species in Kern County: **S. arvensis** L. (*B. kaber*)

Sisymbrium (p. 250)

Erect branched annual or perennial herbs; lvs. dentate to pinnatifid; pubescence of simple hairs; fls. yellow or white, small, racemose; siliques cylindric to 10 cm long. About 90 species, 6 introduced in CA.

1. Fruiting pedicels about as thick as silique; petals 6-9 mm long.
  2. Upper lvs. pinnatifid; siliques ca. 1 mm wide.................................. **S. altissimum**
  2. Upper lvs. entire to hastate; siliques ca. 2 mm wide............................**S. orientale**
1. Fruiting pedicels more slender than siliques; petals 3-4 mm long.................... **S. irio**

Stanleya (p. 250)

Suffrutescent glaucous perennial; lower lvs. pinnatifid; infl. becoming 2-6 dm long in fr., many-fld.; petals yellow, 12-16 mm long; siliques 3-8 cm long, long-stipitate (a 1-2 cm long stalk between receptacle and fr.). About 6 species, 3 in CA. One species in Kern County: **S. pinnata** var. **pinnata**

Streptanthella (p. 250)

Glabrous annuals with slender branched stems; lvs. entire to pinnatifid; fls. small, racemose; sepals with lateral pair basally saccate; petals narrow, yellowish; siliques pendent. One species: **S. longirostris** (incl. var. *derelicta*)

Streptanthus (p. 250)

Annual to short-lived perennial, glabrous; lvs. entire to lyrate-pinnatifid, the upper auriculate-clasping; calyx flask-shaped; petals narrow; siliques narrow, sessile. About 40 species, 24 in CA.

Stem lvs. purple; infl. with some leafy bracts.................................**S. farnsworthianus**
Stem lvs. green; infl. without leafy bracts.............................**S. cordatus** var. **piutensis**

Thelypodium (p. 251)

Annual to biennial herbs, glabrous or with simple hairs; infl. racemose; petals linear to oblanceolate, clawed, white to lilac; siliques stipitate to subsessile. About 20 species, 8 in CA. One species in Kern County: **T. integrifolium**

Thysanocarpus (p. 251)

Erect slender simple or branched annuals; lvs. subentire to pinnatifid; fls. minute, white in slender racemes; silicle indehiscent, compressed, round, winged. About 5 species, 4 in CA.

Stem lvs. auriculate, basal lvs. hirsute..**T. curvipes**
Stem lvs. not auriculate, basal lvs. glabrous ....**T. laciniatus**

Tropidocarpum (p. 252)

Slender branched diffusely spreading annual with simple and branched hairs; lvs. pinnatifid; fls. racemose, yellow; silique elongate, flattened perpendicular to narrow partition. Two species, both in CA. One species in Kern County: **T. gracile**

CACTACEAE

Perennial succulent plants with columnar globose terete or flattened stems; lvs. mostly absent; branches, spines and fls. on areoles; fls. solitary, sessile; perianth with or without a tube and consists of numerous outer segments (sepals) intergrading with inner segments (petals); stamens many; ovary inferior; fr. a berry.

1. Stems jointed, flat or cylindrical; areoles with numerous minute barbed bristles. **Opuntia**
1. Stems not jointed; areoles without minute barbed bristles
   2. Fr. and stem-tips densely woolly; petals yellow ....**Echinocactus**
   2. Fr. and stems not woolly; petals magenta ....**Sclerocactus**

Echinocactus (p. 288)

Stems cylindric, oblong or globose, not jointed, with spiniferous ribs; fls. short-tubular, campanulate; fr. spineless, woolly. About 6 species, 1 in CA. One species in Kern County: **E. polycephalus**

Opuntia (p. 289)

Somewhat woody plants with short-jointed stems; joints cylindrical or flattened, often tubercled but not ribbed; lvs. small, fleshy, subulate, soon shed; areoles with many short, easily detached bristles and usually longer stouter spines; fr. fleshy or dry. About 200 species, 22 in CA.

1. Joints terete, cylindrical; plants mostly >3 dm tall; fls. yellow-green.
   2. Spines solitary; branches slender, 6-10 mm thick .... **O. ramosissima** Engelm. (Creosote Bush Association, Edwards AFB. Moe 836)
   2. Spines more than 1; branches stouter.. **O. echinocarpa**
1. Joints flattened, pad-like; plants <3 dm high; fls. orchid to rose.............. **O. basilaris**
   3. Areoles mostly lacking spines .... var. **basilaris**
   3. Areoles mostly bearing spines 7-30 mm long. var. **treleasei** (Coulter) Toum. (*O. treleasei*)

Sclerocactus (p. 288)
(part of *Echinocactus*)

Stem cylindric to ovoid, mostly single, ribbed; central spines 9-11, most hooked at tip; fl. magenta; fr. scaley. About 19 species, 2 in CA. One species in Kern County: **S. polyancistrus** (Eng. & Bigel.) Britt. & Rose

CALLITRICHACEAE

Callitriche (p. 279)

Aquatic herb with slender tufted stems; lvs. opposite, to about 8 mm long; fls. small, axillary; perianth none. About 40 species, 6 in CA.

Fruits pedicelled; lvs. spatulate .... **C. marginata**
Fruits sessile; submerged lvs. linear, terminal lvs. obovate in a rosette. **C. heterophylla** var. **bolanderi**

CAMPANULACEAE
(incl. Lobeliaceae)

Herbaceous annuals often with milky sap; lvs. simple, alternate; fls. 5-merous, regular or irregular; ovary inferior, fr. a capsule (includes Lobeliaceae of Twisselmann).

1. Corolla regular; anthers distinct.

2. Lvs. broad to roundish; caps. laterally dehiscent .... **Heterocodon**
2. Lvs. narrow, oblong; caps. terminally dehiscent .... **Githopsis**

1. Corolla irregular.
3. Anthers distinct; fls. minute; inconspicuous .... **Nemacladus**
3. Anthers united into a tube; fls. showy.
4. Fls. sessile in lf.-axils, the ovary linear and resembling a pedicel ..... **Downingia**
4. Fls. pedicelled; ovary much shorter .... **Porterella**

Downingia (p. 349)

Low glabrous soft-stemmed annuals with sessile lvs.; fls. blue, bilabiate, on long slender inferior twisted ovaries which "invert" fls.; anthers fused, the 2 smaller ones with terminal tuft of bristles. About 13 species, all in CA.

1. Anther-tube 2.5-3.5 mm long, prominently exserted; lower corolla-lip with 3 basal purple spots. .... **D. pulchella**
1. Anther-tube 1.5-2.5 mm long, not much exserted.
2. Lower corolla-lip without any purple spots .... **D. cuspidata**
2. Lower corolla-lip with 2 or 3 basal purple spots .... **D. bella**

Githopsis (p. 349)

Small annual with angled lower stem; lvs. small, withering early; fls. in upper axils and terminal, corolla bright blue, sepals linear-subulate; fr. a capsule. About 4 species, all in CA.

Corolla funnel-shaped, >3 mm long .... **G. specularioides**
Corolla cylindric, <3 mm long.. **G. tenella** Morin
(East slopes of Greenhorn Mtns. Hardham 3673)

Heterocodon (p. 349)

Delicate annual with small roundish dentate sessile lvs.; fls. solitary, axillary, the upper ones blue, lower ones set fruit without blooming. One species: **H. rariflorum**

Nemacladus (p. 350)

Small annuals with capillary diffusely branched stems; basal lvs. in rosette, upper lvs. reduced to bracts; fls. minute on capillary pedicels; stamens fused by filaments with anthers stellately spreading; filaments between the 3 ovary glands with appendages. About 13 species, all in CA.

1. Corolla-lobes united at base, tube as long as lobes.
2. Plants 0.5-1 cm tall; bract longer than pedicel; ovary glands absent. **N. twisselmannii**
2. Plants 3-12 cm tall; pedicels longer than bracts; ovary glands absent.

3. Basal lvs. roundish, mostly entire; pedicels mostly very thin, forming a double curve........ **N. sigmoideus**
3. Basal lvs. narrower, dentate to finely divided, pedicels various.
   4. Racemes with fls. on both sides; pedicels finely capillary, forming a double curve........ **N. gracilis**
   4. Racemes with fls. along one side; pedicels ± stouter, somewhat sigmoid (in age)........ **N. secundiflorus**

1. Corolla-lobes scarcely united at base.
   5. Upper corolla-lobes ciliate on margins; stamens exserted.
      6. Corolla yellow with purple markings........ **N. rubescens**
      6. Corolla white with lobes ± purplish at tips........ **N. glanduliferus**
   5. Upper corolla-lobes not ciliate; stamens not much exserted.
      7. Base of fr. rounded; corolla pale lilac with red-violet blotches or whitish with pink tinge........ **N. interior**
      7. Base of fr. more pointed; corolla white........ **N. capillaris**

Porterella (p. 351)

Erect glabrous simple to few-branched annual; fls. blue with yellow or whitish spot, bilabiate, on slender pedicels 5-15 mm long. One species: **P. carnosula**

CANNABACEAE
(part of Moraceae)

Cannabis (p. 210)

Stout erect annual; lvs. with 5-11 coarsely toothed lfts.; fls. greenish, axillary, staminates panicled, pistillates spicate; fr. an achene. One species, introduced in Kern County: **C. sativa**

CAPPARACEAE
(Capparidaceae)

Ill-smelling herb or shrub with alternate palmately-compound lvs.; fls. perfect, racemose; sepals and petals 4; stamens 6; fr. a stipitate capsule.

1. Shrubs with conspicuously inflated capsule; petals >1 cm long........ **Isomeris**
1. Herbs with caps. not much inflated; petals mostly <1 cm long.
   2. Fr. twin-like, 2-celled, 2-seeded........ **Wislizenia**
   2. Fr. not twin-line, 1-celled, several-seeded........ **Cleomella**

Cleomella (p. 252)

Erect or diffuse annuals with petioled trifoliate lvs.; fls. yellow, small, mostly racemose; caps. broader than long. About 10 species, 4 in CA.

Stipe much longer than caps.; pedicels 6-12 mm long; petals 5-6 mm long. **C. obtusifolia**
Stipe shorter than caps.; pedicels 10-25 mm long; petals 1-2 mm long..... **C. parviflora**

Isomeris (p. 252)

Ill-scented widely-branched shrubs with trifoliate lvs.; fls. yellow, racemose; petals 10-16 mm long; caps. large, inflated, pendulous on long stipe. One species: **I. arborea** (incl. vars. *angustata* and *globosa*)

Wislizenia (p. 252)

Erect branched ill-scented annual with trifoliate lvs.; fls. yellow, racemose; caps. 2-seeded, twin-like, stipitate. One species: **W. refracta** Engelman ssp. **californica** (Greene) Kell.

CAPRIFOLIACEAE

Shrubs, trees or vines with opposite lvs.; fls. regular or irregular, mostly 5-merous; ovary inferior; fr. fleshy.

1. Corolla rotate, regular, white; shrub with pinnately compound lvs............ **Sambucus**
1. Corolla tubular to funnelform, irregular.
   2. Berry white, 1-2-seeded; corolla subregular; erect shrub............. **Symphoricarpos**
   2. Berry yellowish or red, few seeded; corolla irregular; twining shrub....... **Lonicera**

Lonicera (p. 347)

Twining shrubs with simple entire lvs., uppermost pair often connate-perfoliate; corolla tubular, 10-14 mm long, irregular, ± gibbous at base; fr. a few-seeded berry. About 200 species, 10 in CA.

Upper pair of lvs. only slightly connate; corolla yellowish, glandular-pubescent; berry yellowish or red.......... **L. subspicata** Hook. & Arn. var. **denudata** Reh. (*L. johnstonii*)
Upper pair of lvs. connate; corolla yellowish, glabrous; berry red............**L. interrupta**

Sambucus (p. 348)

Large shrubs with opposite odd-primate lvs. and serrate lfts.; fls. small, whitish, in compound cymes; corolla rotate; fr. a berry-like drupe. About 20 species, 3 in CA. One species in Kern County: **S. mexicana** (incl. *S. caerulea*)

Symphoricarpos (p. 348)

Shrubs with simple opposite deciduous lvs.; fls. white to rose-pink; corolla campanulate; fr. a white berry. About 10 species, 4 in CA.

1. Corolla short-bell-shaped, the lobes at least as long as tube, hairy inside.
   2. Plant erect; corolla 5-6 mm long.
   S. albus (L.) Blake var. **laevigatus** (Fern.) Blake (*S. rivularis)*
   2. Plant trailing, spreading; corolla 3-5 mm long .... **S. mollis** Nutt. (*S. acutus*)
1. Corolla elongate-bell-shaped, 6-7 mm long, the lobes shorter than the tube, glabrous inside...................**S. rotundifolius** Gray var. **parishii** (Rydb.) Dempster (*S. parishii*)

CARYOPHYLLACEAE
(incl. Illecebraceae)

Annual or perennial herbs with mostly opposite, entire, simple lvs. which are often connected at base by transverse line; nodes swollen; fls. regular, 4-5-merous, some apetalous; styles 2-5; fr. a capsule.

1. Plants apetalous; fls. small, greenish.
   2. Fr. a 1-seeded utricle; fls. in axillary clusters (Illecebraceae in Twisselmann). **Herniaria**
   2. Fr. a several-seeded caps; fls. terminal or axillary.
      3. Style 1; plant glandular-pubescent.. **Loeflingia**
      3. Styles as many as sepals.
         4. Annual; styles alternate sepals .... **Sagina**
         4. Perennial; styles opposite sepals.. **Stellaria**
1. Plants with petals; fls. larger, white.
   5. Sepals united; petals clawed and borne on carpophore (stipe of ovary)......... **Silene**
   5. Sepals distinct; petals without claws; ovary sessile.
      6. Stipules absent.
         7. Caps. ovoid or ellipsoid.
         7. Caps. cylindrical, often bent at summit.. **Cerastium**
            8. Styles usually 3, opposite sepals.
               9. Petals deeply notched or bifid; valves of caps. 2-parted.
                  10. Petal-lobes <1/3 petal length; plants glandular-hairy. **Pseudostellaria**
                  10. Petal-lobes >1/2 petal length; plants not glandular .... **Stellaria**
               9. Petals mostly entire; valves of caps. mostly entire.
                  11. Capsule-valves 3; annuals .... **Minuartia**
                  11. Capsule-valves 6; perennials .... **Arenaria**
            8. Styles 4-5, alternate sepals.. **Sagina**
      6. Stipules present, scarious.
         12. Lf. oblanceolate, ± petioled; plants glabrous .... **Polycarpon**
         12. Lf. linear to awl-shaped; plants glandular hairy.
            13. Styles 3.. **Spergularia**
            13. Style 1, 3-cleft; petals minute... **Loeflingia**

Arenaria (p. 233)

Low branched tufted or matted annual to perennial herbs; lvs. sessile, without stipules; often subulate; fls. small, white; 5-merous; styles usually 3; caps. globose to oblong, valves 6. About 150 species, 8 in CA. One species in Kern County: **A. macradenia**

Key to varieties

1. Sepals >5 mm long; stems woody at base.
   2. Cauline lvs. ascending, ca. 1 mm broad. var. **macradenia** (incl. var. *parishiorum*)
   2. Cauline lvs. strongly arcuate, 1-2 mm broad.................................... var. **arcuifolia**
1. Sepals 4-5 mm long; stems scarcely woody at base .... var. **ferrisiae**

Cerastium (p. 234)

Pubescent erect sticky annual; fls. in terminal dichotomous cyme; petals 2-cleft, white; caps. slender, slightly curved. About 60 species, 5 in CA. One species introduced in Kern County: **C. glomeratum**

Herniaria (p. 234)

Much branched mat-forming prostrate annual, leafy throughout; fls. in nearly all axils; petals absent; sepals with short stiff hairs; fr. a 1-seeded utricle. About 20 species, one introduced in CA. One species in Kern County: **H. hirsuta** L. ssp. **cinerea** (DC) Cout. (*H. cinerea*)

Loeflingia (p. 234)

Low spreading rigid glandular-pubescent annuals; lvs. small, subulate; stipules setaceous; fls. small, sessile, axillary; petals none or 3-5, small, white; stigmas 3. About 7 species, 1 in CA. One species in Kern County: **L. squarrosa**

Key to varieties:

Sepals 3-6 mm long, unequal, tip strongly recurved..... var. **squarrosa** (incl. *L. pusilla*)
Sepals 2-3 mm long, ± equal, tip ± straight. var. **artemisiarum** (Barn. & Twiss.) Dorn (SE Kern County, Twisselmann)

Minuartia (p. 233)
(part of *Arenaria*)

Simple or branched annuals with slender awl-shaped lvs.; fls. small, white; capsule with 3 teeth. About 120 species, 11 in CA.

1. Petals shorter than sepals... **M. pusilla** (Wats.) Mattf.
1. Petals equal to or longer than sepals.
   2. Lvs. ± lanceolate, 2-5 mm long.................................**M. californica** (Gray) Mattf.
   2. Lvs. filiform, 5-30 mm long... **M. douglasii** (T&G) Mattf.

Polycarpon

Matted or tufted annual with opposite to ± whorled lvs.with scarious stipules; lf.-blades obovate, 4-12 mm long; fr. a capsule. About 16 species, 2 in CA. One species introduced into Kern County: **P. tetraphyllum** (L.) L. (Weed in Bakersfield landscaping flowerbed. L. Cooley 158)

Pseudostellaria (p. 236)
(part of *Stellaria*)

Branched glandular-hairy perennial with lanceolate lvs. 1.5-10 cm long; petals white, shallowly lobed; fr. a globose caps. About 16 species, one in CA. One species in Kern County: **P. jamesiana** (Torr.) Weber & Hartman

Sagina (p. 234)

Low tufted or matted annual or perennial herbs; lvs. filiform or subulate, papery-connate at base; fls. small, whitish, terminal; petals 4 or 5 or 0, entire; styles as many as sepals, alternate them; caps. many-seeded. About 25 species, 5 in CA.

1. Plants annual, without sterile basal rosettes; stems filiform.
   2. Upper lvs. minutely ciliate near base; petals none .... **S. apetala**
   2. Upper lvs. glabrous; petals 5.
      S. decumbens (Elliott) T.&G. ssp. **occidentalis** (Watson) Crow (Red Rock Canyon, Faull & Charlton POM)
1. Plants perennial, with sterile basal rosettes; stems ± fleshy.................. **S. saginoides**

Silene (p. 234)

Annual or perennial herbs, usually sticky; stipules absent; fls. cymose; calyx cylindrical to ovoid, 5-toothed; petals clawed, usually with scale-like appendage at base of blade which is usually cleft or toothed; styles 3; ovary stipitate. About 500 species, 29 in CA.

1. Plants annual, weedy.
   2. Plants subglabrous; upper internodes with sticky bands .... **S. antirrhina**
   2. Plants densely pubescent .... **S. gallica**
1. Plants perennial.
   3. Corolla bright red.. **S. californica**
   3. Corolla white to pink.
      4. Fls. nodding at anthesis; stamens long-exserted.............................. **S. lemmonii**
      4. Fls. mostly erect.
         5. Basal lvs. <2 cm long and 1-4 mm wide; plants tufted, not >2 dm high; calyx 13-16 mm long.............................. **S. bernardina** Wats. (incl. *S. montana*)
         5. Basal lvs. mostly >3 cm long and >4 mm wide; plants mostly >3 dm high; calyx 10-12 mm long.. **S. verecunda** ssp **platyota**

Spergularia (p. 235)

Low annual to perennial herbs, usually of saline areas; lvs. linear, fleshy, ± terete; fls. whitish to pink in terminal cymes; petals entire; styles 3. About 40 species, 9 in CA.

1. Lvs. densely clumped; stipules lance-acuminate; plants mostly perennial.
  2. Sepals 5-10 mm long; styles >1 mm long .... **S. macrotheca** var. **leucantha**
  2. Sepals 3-5 mm long; styles <1 mm long .... **S. rubra**
1. Lvs. not densely clumped; stipules deltoid; plants annual.
  3. Seeds shiny black.. **S. atrosperma**
  3. Seeds brown.
    4. Stamens 6-10.. **S. bocconii**
    4. Stamens 2-5... **S. marina**

Stellaria (p. 236)

Low diffuse annuals or perennial herbs; fls. white, terminal; sepals 4-5; petals 0 or 4-5, deeply 2-cleft; styles 3; caps. 1-celled. About 120 species, 12 in CA.

1. Plants annual.
  2. Internodes with longitudinal line of hair... **S. media**
  2. Internodes glabrous... **S. nitens**
1. Plants perennial.
  3. Petals equal to or longer than sepals................................ **S. longipes** var. **longipes**
  3. Petals shorter than sepals.
    4. Lf.-margins serrulate-ciliate; lvs. lanceolate.
      S. borealis Bigel. ssp. **sitchana** (Steudel) Piper (*S. sitchana* var. *bongardiana*)
    4. Lf.-margins smooth, glabrous, crisped; lvs. ovate to oblong................. **S. crispa**

CERATOPHYLLACEAE

Ceratophyllum (p. 236)

Submerged aquatic with branching stem and whorled lvs. dissected into filiform divisions; monoecious with staminate and pistillate fls. on separate nodes, sessile, without perianth (but with calyx-like invol.); fr. an achene. About 3 species, 1 in CA. One species in Kern County: **C. demersum**

CHENOPODIACEAE

Herbs or shrubs, often succulent or scurfy, frequently of saline or subsaline places; lvs. simple, mostly alternate, sometimes reduced to scales; fls. perfect or imperfect, small, greenish, usually in small cymose glomerules; calyx persistent, mostly enclosing fr., sometimes reduced in pistillate fls.; petals 0; styles 2-3; fr. one-seeded, dry.

1. Sepals strongly overlapping; papery; lvs. opposite, united at base............ **Nitrophila**
1. Sepals not or only slightly overlapping, fleshy when young; lvs. mostly alternate.

2. Lvs. foliaceous, flattened (± fleshy in *Grayia*).
  3. Fls. mostly perfect, all with calyx, not enclosed in a pair of bracts.
    4. Sepals 3-5; stamens 4-5.
      5. Glabrous biennial herb; lvs. large, >1 dm long .... **Beta**
      5. Mealy, glandular or pilose herbs; lvs. smaller.
        6. Sepals with hooked spines; plants pilose .... **Bassia**
        6. Sepals without hooked spines; plants mealy or glandular.. **Chenopodium**
    4. Sepal 1; stamen 1 .... **Monolepis**
  3. Fls. imperfect, the pistillate enclosed in 2 bracts that grow with fr.
    7. Fr. not hairy; lf.-margins not rolled backward.
      8. Bracts compressed, margins never wholly united; lvs. ± mealy...... **Atriplex**
      8. Bracts obcompressed, wholly united, dorsally winged in fr.; lvs. glabrate **Grayia**
    7. Fr. conspicuously hairy; lvs. linear, margins rolled backward. **Krascheninnikovia**
2. Lvs. fleshy and sublinear, or scaly and spiny, not much flattened.
  9. Lvs. scale-like; stems and branches fleshy.
    10. Branches opposite .... **Salicornia**
    10. Branches alternate .... **Allenrolfea**
  9. Lvs. not scale-like, but fleshy and sublinear, or spiny-tipped.
    11. Fls. imperfect, staminate in spikes, pistillate solitary; large shrub. **Sarcobatus**
    11. Fls. perfect or perfect and pistillate.
      12. Lvs. tipped with spine or bristle; fruiting sepals winged on back... **Salsola**
      12. Lvs. not tipped with spine or bristle.
        13. Fruiting calyx transversely winged .... **Kochia**
        13. Fruiting calyx not transversely winged .... **Suaeda**

Allenrolfea (p. 224)

Much-branched glabrous erect succulent shrub with alternate green branches; lvs. scale-like; fls. sessile, in axils of fleshy peltate bracts and forming cylindrical spikes; fr. ovoid. One species: **A. occidentalis**

Atriplex (p. 225)

Herbs or shrubs, usually grayish or whitish, scurfy; lvs. flattened; plants monoecious or dioecious, the fls. small, green, in axillary clusters or in panicled spikes; staminate fls. with 3-5-parted calyx, bractless; pistillate fls. a naked pistil enclosed between a pair of foliaceous appressed bracts which enlarge in fr. About 250 species, 38 in CA.

1. Plants herbaceous.
  2. Lvs. green or greenish on both surfaces.
    3. Fruiting bracts united basally; lvs. hastate (arrowhead-shaped).
      4. Bracts hastate to rounded at base, tips close together. **A. triangularis** Willd. (*A. hastata*)

4. Bracts mostly with rounded ear-like lobes near base, tips far apart. **A. phyllostegia**
3. Fruiting bracts united to above the middle; lvs. not hastate. .... **A. serenana**
2. Lvs. gray or whitish with a fine scurf.
5. Bracts thickened, fleshy, becoming red .... **A. semibaccata**
5. Bracts not thickened or fleshy.
6. Bracts broadest below middle.
7. Lvs. coarsely toothed; bracts hard, often warty............................... **A. rosea**
7. Lvs. entire; bracts not becoming hard.
8. Stems simple or with few branches.
9. Lvs. cordate at base (heart-shaped).
10. Lvs. 5-15 mm long; fruiting bracts 4-5 mm long.........**A. cordulata**
10. Lvs. 2-7 mm long; fruiting bracts 2-3.5 mm long.......... **A. vallicola**
9. Lvs. rounded at base; bracts 2.5-3.5 mm long..................**A. tularensis**
8. Stems intricately branched throughout ....**A. parishii** (not in Twisselmann but likely to be in western Mojave)
6. Fruiting bracts broadest at or above middle.
11. Bracts becoming very hard... **A. rosea**
11. Bracts not becoming hard.
12. Lvs. narrowly ovate to elliptic, narrowed at base, 3-10 mm wide; bracts mostly 3-4 mm long.. **A. coronata**
12. Lvs. cordate- to lance-ovate, broad and often subhastate, mostly 10-40 mm wide; bracts 4-8 mm long... **A. argentea**
1. Plants shrubby, definitely woody.
13. Fruiting bracts with conspicuous extra lateral wings which are often coarsely dentate. ....**A. canescens** ssp. **canescens**
13. Fruiting bracts without extra lateral wings.
14. Plants not spiny.
15. Lf.-margins entire, flat or crisped but not toothed.
16. Lf.-blades oblong, <5 mm wide.. **A. polycarpa**
16. Lf.-blade ovate to deltate, >10 mm wide. **A. lentiformis** (Tor.) Wats.ssp. **lentiformis** (*A. breweri*)
15. Lf.-margins dentate...**A. hymenelytra**
14. Plants spiny (with sharp-pointed twigs).
17. Bracts 2-4 mm long.
18. Bracts suborbicular; lvs. truncate or triangular at base, blades 15-40 mm long...**A. lentiformis**
19. Twigs sharply angled.................. ssp. **torreyi** (Wats.) Hall & Clements (Edwards AFB, Moe 1547)
19. Twigs terete, not sharply angled .... ssp. **lentiformis**
18. Bracts with broad summit; lvs. cordate at base, blades 5-15 mm long. **A. parryi** Wats. (Edwards AFB, Moe 1582)
17. Bracts 6-15 mm long.
20. Body of bract small, not constricted beneath the free terminal wings; lvs. entire.. **A. confertifolia**
20. Body of bract large, thick, constricted to a neck beneath the free terminal wings; lvs. sometimes ± arrowhead-shaped at base................. **A. spinifera**

Bassia (p. 227)

Grayish densely pilose prostrate annual with alternate entire lvs.; fls. minute, sessile, axillary, perfect and pistillate; calyx-lobes ovate, villous, each with a stout hooked spine. About 5 species, 1 in CA. One species introduced in Kern County: **B. hyssopifolia**

Beta (p. 227)

Glabrous biennial herb with large fleshy root ("beet"); basal lvs. large, long-petioled, reddish; calyx 5-parted with ± succulent base containing the ovary. About 5 species, 1 introduced in CA. One in Kern County: **B. vulgaris**

Chenopodium (p. 227)

Annual or perennial mostly mealy (farinose) herbs; lvs. alternate, entire to lobed; fls. perfect, without bracts; calyx 5-parted, persistent and enveloping fr.; stamens 5; styles mostly 3. About 150 species, 27 in CA.

1. Plants ± glandular-pubescent, not mealy.
  2. Fls. in glomerules, these in short spikes .... **C. ambrosioides**
  2. Fls. solitary in small cymes, these in elongated panicles .... **C. botrys**
1. Plants mostly mealy, not glandular-pubescent.
  3. Seeds horizontal in fr.
    4. Lvs. shining on upper surface, ± rhombic; plants branched from base. **C. murale**
    4. Lvs. dull on upper surface; plants variously branched.
      5. Pericarp attached to seed; at least lower lvs. conspicuously sinuate-denate.
        6. Pericarp and seeds essentially smooth.
          7. Seeds mostly 1.2-1.5 mm wide; sepals largely covering fr. .... **C. album**
          7. Seeds 0.9-1.2 mm wide; sepals exposing mature fr. **C. strictum** var. **glaucophyllum**
        6. Pericarp and seeds reticulate .... **C. berlandieri**
      5. Pericarp not attached to seed, or if attached, the lower lvs. linear or entire or 3-lobed.
        8. Lf.-blades narrow to linear, mostly 1-3-veined.
          9. Plants low, diffusely branched from base, 1-3 dm high; lvs. 8-20 mm long. .... **C. desiccatum**
          9. Plants erect, 3-12 dm high, branched from base, lvs. 15-35 mm long. **C. incognitum**
        8. Lf.-blades lance-ovate to ovate, pinnately veined.
          10. Plants ill-scented; calyx-lobes rounded on back .... **C. vulvaria**
          10. Plants not ill-scented; calyx-lobes keeled on back .... **C. fremontii**
  3. Seeds vertical; lvs. sinuate-dentate.
    11. Plants annual, glabrous, reddish .... **C. rubrum**
    11. Plants perennial, mealy on upper parts .... **C. californicum**

Grayia (p. 229)

Stiff much branched scurfy-pubescent spinose shrub with alternate entire sessile lvs.; dioecious; staminate fls. small, bractless; pistillate fls. with 2 orbicular flattened bracts which are dorsally winged in fr. and turn reddish. One species: **G. spinosa**

Kochia (p. 229)

Low densely grayish pubescent perennial herb with spreading flattish lvs.; calyx 5-lobed, incurved and winged in fr. About 20 species, 3 in CA. One species in Kern County: **K. californica**

Krascheninnikovia (p. 228)
(*Eurotia*)

Low stellate-tomentose shrub with alternate or clumped entire lvs.; staminate fls. bractless in dense axillary spikes, calyx 4-parted; pistillate fls. with a pair of bracts which form a silky-hairy sac in fr. About 8 species, 1 in CA. One species in Kern County: **K. lanata** (Pursh) Meeuse & Smit

Monolepis (p. 229)

Low branched succulent mealy annual; lvs. triangular-lanceolate, hastate; fl.-clusters dense, reddish; calyx of 1 persistent sepal; stamen 1 or 0; styles 2. About 3 species, all in CA. One species in Kern County: **M. nuttalliana**

Nitrophila (p. 229)

Low glabrous perennial herbs with opposite linear fleshy sessile lvs.; fls. axillary; calyx papery-textured, 5-parted; stamens 5; styles 2. About 8 species, 2 in CA. One species in Kern County: **N. occidentalis**

Salicornia (p. 229)

Annual to somewhat shrubby herbs with succulent leafless jointed stems, opposite branches and short internodes; lvs. opposite, reduced to scales; fls. in cylindric fleshy spikes with sunken fls.; calyx fleshy; stamens 1 or 2. About 13 species, 5 in CA.

Perennial from creeping rootstocks .... **S. subterminalis**
Annual .... **S. europaea** (*S. depressa*)

Salsola (p. 229)

Densely and intricately branched annual forming a round bushy clump; lower lvs. fleshy, terete, linear; bracts ovate, prickly-pointed; calyx broadly winged in fr., often reddish. About 100 species, 4 in CA.

Sepal wings 0.5-2.5 mm long in fr.; lvs. ± green..... **S. tragus** L.(*S. kali* var. *tenuifolia*)
Sepal wings 2.5-4.5 mm long in fr.; lvs. ± yellow green..................... **S. paulsenii** Litv.
(Red Rock Canyon, Twisselmann)

Sarcobatus

Much-branched spinescent shrubs; lvs. linear, fleshy, sessile; monoecious or dioecious; staminate fls. in terminal condensed spikes; stamens 2-3, covered by a peltate scale; pistillate fls. sessile, 1 or 2 in lf.-axile. One species: **S. vermiculatus** (Hook.) Torr. (Edwards AFB, Moe 1632)

Suaeda (p. 230)

Somewhat shrubby perennials with alternate ± fleshy lvs.; fls. small, clustered in lf.-axils in elongate infls.; calyx 5-lobed, globose. About 115 species, 5 in CA. One species in Kern County: **S. moquinii** (Torr.) Greene (incl. *S. torreyana* and *S. fruiticosa*)

CONVOLVULACEAE

Annual or perennial mostly trailing or twining herbs with alternate lvs.; fls. regular, 4-5-merous, axillary; petals connate, bell- to urn-shaped; ovary superior; fr. a 2-celled capsule.

1. Erect herbs; fls. <1 cm long; styles 2.. **Cressa**
1. Trailing herbs; fls. >1 cm long; styles 1.
   2. Stigmas capitate.. **Ipomoea**
   2. Stigmas filiform to oblong-cylindric.
      3. Stigmas filiform, the stigmatic area and the style ± continuous; ovary with a complete septum.. **Convolvulus**
      3. Stigmas oblong, ± cylindrical, the stigmatic area and the style distinct; ovary with incomplete septum... **Calystegia**

Calystegia (p. 305)
(part of *Convolvulus*)

Twining or trailing perennial herbs; lvs. alternate, often heart-shaped or arrowhead-shaped; peduncles axillary, usually with a pair of bracts just below calyx; fls. showy, white to rose; corolla 2-4 cm long, funnelform to bell-shaped, pleated; ovary 2-celled; stigmas 2, oblong-cylindric with blunt apices, the stigmatic area and style distinct;

caps. 1-loculed with incomplete septum. About 150 species, 12 in CA.

1. Bractlets >4 mm wide, attached just below calyx and ± concealing it. **C. malacophylla** ssp. **pedicellata**
1. Bractlets <4 mm wide, attached >1 mm below calyx.
   2. Plants glabrous.......... **C. longipes**
   2. Plants pubescent (esp. peduncle & lf.-sinuses) ... **C. occidentalis** (Gray) Brummitt
      3. Bractlets entire, not lf.-shaped.......... ssp. **occidentalis** (*C. tomentellus*)
      3. Bractlets toothed or lobed, lf.-shaped. ssp. **fulcrata** (Gray) Brummitt (*C. fulcrata*)

Convolvulus (p. 305)

Similar to *Calystegia* but with stigmas linear with ± acute tips, the stigmatic area and style ± continuous; ovary 2-celled with complete septum. About 250 species, 3 in CA.

Plants annual; corolla ca. 6 mm long, deeply cleft.......... **C. simulans**
Plants perennial; corolla 2-6 cm long, not cleft.......... **C. arvensis**

Cressa (p. 306)

Low much-branched nontwining tufted woolly-villous perennial herbs; lvs. alternate, sessile, entire; fls. solitary in upper axils, 5-merous; corolla funnelform, white; stamens 5, exserted; fr. a capsule. About 5 species, 1 in CA. One species in Kern County: **C. truxillensis** (incl. var. *vallicola*)

Ipomoea (p. 306)

Twining or trailing herbs similar to *Calystegia* but with capitate stigmas and corolla purple, blue, pink or white. About 500 species, 5 in CA.

Sepals acute, ca. 12 mm long; lvs. mostly entire.......... **I. purpurea**
Sepals long-accuminate, 20-25 mm long; lvs. 3-lobed.......**I. nil** (L.) Rot (*I. hederacea*)

CORNACEAE

Cornus (p. 300)

Trees or shrubs with opposite entire lvs.; (lf.-veins remain attached when blade is gently torn);.fls. 4-merous; ovary inferior, fr. a drupe. About 50 species, 5 in CA.

1. Fls. in bractless cyme, appearing with or after lvs.
   2. Lvs. with 3-4 veins on each side of midrib; lf.-blades 3-5 cm long; drupes white to bluish.......... **C. glabrata**

2. Lvs. with 4-7 veins; lf.-blades 5-10 cm long; drupes white.
C. sericea L. ssp. **sericea** (*C. stolonifera*)

1. Fls. in a head subtended by petaloid white bracts at least 5 cm long; drupes red.
C. nuttallii

CRASSULACEAE

Succulent mostly glabrous herbs with entire lvs.; petals 4-5, distinct or ± united at base; stamens as many or twice as many as petals; carpels as many as sepals, usually distinct, becoming follicles.

1. Annuals; stamens 3-5.
   2. Fls. 1-several, axillary; petals 1-2 mm long .... **Crassula**
   2. Fls. in terminal cymes; petals 2-4 mm long .... **Parvisedum**
1. Perennials; stamens twice the number of petals.
   3. Flowering stems lateral, from axils of basal rosette lvs .... **Dudleya**
   3. Flowering stems terminal .... **Sedum**

Crassula (p. 253)
(*Tillaea*)

Very small branched glabrous annuals with opposite minute lvs.; fls. axillary, mostly 4-merous. About 300 species, 4 in CA.

Fls. clustered; carpels 1-2-seeded; dry places.
C. connata (Lopez & Pavon) Berger (*T. erecta*)

Fls. solitary; carpels several seeded; mud flats .... **C. aquatica**

Dudleya (p. 253)

Perennial herbs with fleshy ovate to linear lvs. in basal rosettes; flowering stems axillary; fls. in terminal paniculate clusters; corolla cylindric, erect; stamens 10; carpels 5. About 45 species, 25 in CA.

1. Petals yellow to ± tinged red; flowering stems 1-2 dm tall; cauline lvs. rounded at base.
   2. Lf. oblanceolate to spoon-shaped, to 5 cm wide .... **D. cymosa**
   2. Lf. lanceolate, to 1.5 cm wide .... **C. calcicola** Bartel & Shevock (occasional to common, Kernville-Lake Isabella area eastward to Sierra crest, Shevock)
1. Petals orange or pale green with red tinge; flowering stems 2-6 dm tall; cauline lvs. cordate-sagittate at base .... **D. lanceolata**

Parvisedum (p. 253)

Very small slender-stemmed diffusely branched annual with small fleshy lvs.; fls. in cymes; petals bright yellow with reddish median line. About 4 species, all in CA. One species in Kern County: **P. congdonii**

Sedum (p. 253)

Perennial herbs with prominent basal rosettes of spatulate lvs. ;fls. in terminal cyme; petals 4-5, yellow (rarely orange or white), lanceolate. About 350 species, 15 in CA. One species in Kern County: **S. spathulifolium**

CROSSOSOMATACEAE

Glossopetalon

Erect shrub <2 m tall with thorny branches and small, entire, deciduous, alternate lvs.; stipules bristle-like; hypanthium short; petals 5, white; stamens 6-10; fr. 1-2 follicles, 3-5 mm long. About 5 species, 2 in CA. One species in Kern County: **G. spinescens** Gray (Limestone-marble outcrops, Laura Peak, Erskine Creek & Long Canyon of Piute Mtns. Purpus 5520 (UC, 1897); Shevock 8047, 8070, 8250, 8272)

CUCURBITACEAE

Annual or perennial herbs with trailing stems; lvs. alternate, broad, palmately veined or lobed; tendrils opposite the lvs.; monoecious; fls. regular, 5-merous; ovary inferior; fr. a pepo (melon or gourd).

1. Fr. fleshy with hard, firm rind, indehiscent; fls. yellow.
  2. Corolla bell-shaped, petals markedly connate, 5-lobed, 5-12 cm long; perennial. **Cucurbita**
  2. Corolla ± wheel-shaped, petals barely connate, <2 cm long, annual.
    3. Anther-connective with terminal appendage; seeds not margined. .... **Cucumis**
    3. Anther-connective without terminal appendage; seeds with obtuse margin. **Citrullus**
1. Fr. not fleshy, spinose, dehiscent; fls. white or cream .... **Marah**

Citrullus (p. 348)

Hairy annual with long trailing stems with branched tendrils; lvs. ovate, 3-8 cm long, pinnately divided into 3-4 pairs of lobes; fr. a watermelon. About 4 species, one introduced in CA. One species in Kern County: **C. colocynthis** (L.) Schrader var. **lanatus** (Thun.) Mats. & Nakai (*C. lanatus*)

Cucumis (p. 348)

Herbaceous scabrous annual with simple tendrils; lvs. rounded-obtuse, 3-7-lobed; pepo globose, medium orange, smooth, 5-6 cm in diam., fragrant. About 40 species, 2 in CA. One species introduced into Kern County: **C. melo** L. var. **dudaim** (L.) Naudin

Cucurbita (p. 348)

Perennial trailing herb with well developed tendrils; fls. large, yellow, bell-shaped; pepo with 1-5 cavities. About 30 species, 3 in CA.

Lf.-blades triangular-ovate, 1-2.5 dm long; fr. dull green with cream-white stripes and mottling on slender pedicel without thickened ridges .... **C. foetidissima**
Lf.-blades palmately 5-lobed; less than 1 dm long; fr. dull light green with broad, ill-defined bands and splashes of greenish-white on pedicel with thickened ridges. **C. palmata**

Marah (p. 349)

Climbing or trailing herbs with annual stems from large underground tubers; lvs. orbicular, ± heart-shaped, 5-7-lobed; tendrils branched; pistillate fls. solitary, staminate fls. in raceme or panicle in same axil as pistillate; fr. dehiscent with large, irregular spines. About 7 species, 5 in CA.

Mature corolla wheel-shaped; fr. globose, 4-5 cm long. **M. fabaceus** (incl. var. *agrestis*)
Mature corolla bell-shaped, white; fr. cylindrical, 9-15 cm long................ **M. horridus**

CUPRESSACEAE

Evergreen trees with opposite or whorled scale-like lvs. thickly covering branchlets; staminate cones small; pistillate cones woody or fleshy, small.

1. Cone woody and dry at maturity; lvs. in opposite pairs.
   2. Cones oblong, 2-2.5 cm long; branchlets flattened .... **Calocedrus**
   2. Cones globose, 2-3 cm long and wide; branchlets 4-sided .... **Cupressus**
1. Cone berry-like .... **Juniperus**

Calocedrus (p. 171)

Tree 25-40 m tall with thick cinnamon-brown fibrous bark and flattened branchlets; cones with 3 pairs of woody scales abruptly pointed at tips. Three species, one in CA. One species in Kern County: **C. decurrens**

Cupressus (p. 171)

Erect tree to about 10 m tall with broad crown and gray-brown fibrous bark; cones with 6-8 peltate scales. About 30 species, 10 in CA. One species in Kern County: **C. arizonica** Greene ssp. **nevadensis** (Abrams) Murray (*C. nevadensis*)

Juniperus (p. 172)

Trees or large shrubs with thin shredding bark; cones ovoid to subglobose, berry-like. About 60 species, 4 in CA.

Berries blue or blue-black, 6-8 mm long; tree to 20 m tall...................... **J. occidentalis**
Berries reddish-brown, 12-18 mm long; shrub to 6 m tall......................... **J. californica**

CUSCUTACEAE
(part of Convolvulaceae)

Cuscuta (p. 306)

Slender-stemmed twining, yellow to orange parasitic herb fastened to hosts by haustoria; lvs. reduced to small scales; fls. small, white, in lateral clusters, 4-5-merous; corolla campanulate to urn-shaped; ovary 2-celled, styles 2. About 150 species, 16 in CA.

1. Caps. globose, ± depressed.
   2. Corolla with scale-like appendages attached to tube below stamens; stems yellow. **C. Pentagona** Engelm. (*C. campestris*)
   2. Corolla lacking scale-like appendages; stems bright orange-yellow. **C. californica** var. **californica**
1. Caps. ovoid, conic or beaked, longer than wide.
   3. Calyx-lobes round, obtuse, denticulate; stems pale yellow.............. **C. denticulata**
   3. Calyx-lobes acute to acuminate, entire.
      4. Calyx shorter than corolla-tube; fls. ca. 5 mm long. **C. subinclusa** Durand (*C. ceanothi*)
      4. Calyx as long as or longer than corolla-tube; fls. smaller.
         5. Corolla-lobes long-attenuate; stems pale; on plants of meadows. **C. californica** var. **breviflora** Engelm. (*C. suksdorfii* var. *subpedicellata*)
         5. Corolla-lobes not markedly long-attenuate; stems orange; on plants of saline places.. **C. salina**

CYPERACEAE

Grass-like or rush-like mostly perennial herbs with mostly solid stems; lvs. mostly basal, commonly 3-ranked, with closed sheaths; infl. various, of spikelets composed of scales or bracts subtending individual fls.; perianth represented by several bristles or a papery bract or absent; fr. an achene.

1. Fls. all unisexual; achene surrounded by a sac-like bract .... **Carex**
1. Fls. perfect or perfect and pistillate; achenes lacking a sac-like bract.
  2. Scales of spikelets spirally overlapping.
    3. Bristles subtending ovary present.
      4. Style deciduous from tip of achene; invol. lvs. present .... **Scirpus**
      4. Style persistent as tubercle at tip of achene; invol. lvs.absent......... **Eleocharis**
    3. Bristles absent.
      5. Thin perianth segment present between fl. and rachis of spikelet. **Lipocarpha**
      5. Thin perianth segment absent .... **Fimbristylis**
  2. Scales of spikelets 2-ranked .... **Cyperus**

Carex (p. 193)

Perennial mostly tufted herbs with mostly triangular solid stems; lvs. 3-ranked with closed sheaths; uppermost lvs. scale-like or leaf-like; fls. unisexual, borne in axils of scale-like bracts and congested into spikelets which may be solitary but mostly arranged into various infl.-types; each staminate fl. consists of 3 stamens; each pistillate fl. consists of a 2- or 3-parted style on a pistil enclosed in an urn-shaped bract--the stigma protrudes through an opening at the top of the bract. About 1000 species, 134 in CA.

1. Spikelet one on each stem; stems ± rounded; lowest pistillate scales 1-4 cm long. **C. multicaulis**
1. Spikelets several on each stem.
  2. Styles 2; achenes lens-shaped.
    3. Spikelets sessile, short, with relatively few pistillate fls.
      4. Staminate spikelets above pistillate, or unisexual and plants ± dioecious.
        5. Infl. simple, the spikelets in heads.
          6. Plants open or loosely tufted, with creeping rootstocks; spikelets in elongate heads or spikes.
            7. Beak of pistillate bract >half as long as body.
              8. Stems smooth, <2 dm long; lvs. folded .... **C. douglasii**
              8. Stems rough, >2 dm long; lvs. flat or grooved........... **C. praegracilis**
            7. Beak <half as long as body; lvs. flat.................................... **C. simulata**
          6. Plants densely tufted on short rootstocks .... **C. jonesii**
        5. Infl. ± compound, some of the spikes on branches.
          9. Infl. to 20 cm long, the branches apparent .... **C. alma**
          9. Infl. to 5 cm long, more compact, the branches not very apparent. **C. vicaria** (incl. *C. densa*)
      4. Staminate spikelet below pistillate.
        10. Margin of pistillate bract winged, bract not spongy at base.
          11. Infl. not conspicuously leafy-bracteate.
            12. Lf.-sheaths white-translucent on inside.
              13. Lf.-sheaths not conspicuously prolonged.

14. Pistillate bract small, 2-3.5 mm long. **C. subfusca** (incl. *C. teneraeformis*)
14. Pistillate bract larger, 3.5-8 mm long.
15. Pistillate bracts strongly nerved on inner surface; lvs. 2-3 mm wide.. **C. mariposana**
15. Pistillate bracts smooth on inner surface; lvs. 1-2 mm wide. **C. gracilior**
13. Lf.-sheaths conspicuously prolonged into a fragile translucent appendage.. **C. fracta**
12. Lf.-sheaths green-striped to top on inside................................ **C. feta**
11. Infl. densely capitate, conspicuously leafy-bracteate, the bracts exceeding infl.. **C. athrostachya**
10. Margin of pistillate bract rounded or sharp-edged but not winged; bract ± spongy thickened at base.
16. Plants densely tufted, 1-6 dm tall, ± as long as lvs.; lvs. 1-2 mm wide; pistillate bracts spreading.................. **C. echinata** Murray (*C. angustior*)
16. Plants loosely tufted, 2-9 dm tall, longer than lvs.; lvs. 2-5 mm wide; pistillate bracts appressed. ....**C. bolanderi**
3. Spikelets stalked, the staminate 1 or 2 broadly linear, the pistillate 2-5; the lowest bract leaf-like.
17. Lf.-sheaths not becoming filamentous when breaking.
18. Beak of pistillate bract >0.5 mm long, entire; lvs. 1.5-4 mm wide; plants to 6 dm tall...**C. nebrascensis**
18. Beak of pistillate bract <0.5 mm long, entire; lvs. 1.5-4 mm wide; plants to 6 dm tall...**C. lenticularis** Mich.
19. Pistillate bracts roundish, short-stalked. var. **impressa** (Bailey) Stand. (*C. paucicostata*)
19. Pistillate bracts ovate, long-stalked. var. **lipocarpa** (Holm.) Stand. (*C. kelloggii*)
17. Lf.-sheaths breaking and becoming filamentous.
20. Beak of pistillate bract 0.5 mm long, toothed, minutely bristly. **C. barbarae**
20. Beak of pistillate bract <0.5 mm long, entire, not bristly............. **C. senta**
2. Styles 3; achenes triangular.
21. Pistillate bract pubescent.
22. Plants to 3 dm tall; staminate spikelet terminal, 5-15 mm long; lowest bract not leaf-like.
23. Flowering stems shorter than basal lvs.; plants not "tuft-like". **C. brainerdii**
23. Flowering stems longer than basal lvs.; plants "tuft-like"... **C. rossii** Boott (common grass-like mats, Piute Mtns., Owens Peak, Kern Plateau. Shevock 8714)
22. Plants 3-10 dm tall; staminate spikelets 1-3 cm long; lowest bracts leaf-like.
24. Lvs. longer than flowering stem; staminate spikelet 2-6 cm long. **C. lanuginosa**
24. Lvs. shorter than flowering stem; staminate spikelets 1-3 cm long. **C. sartwelliana**

21. Pistillate bracts glabrous.
  25. Plants stoloniferous, rank; flowering stems 5-10 dm tall; not exceeding lvs.; lf.-blades 8-18 mm wide; staminate spikelets <1 cm long; pistillate spikelets 3-14 cm long; lowest bracts leaf-like .... **C. amplifolia**
  25. Plants clumped or tufted on short rootstocks; flowering stems exceeding lvs.; lf.-blades 1.4-5 mm wide; staminate spikelets >1 cm long.
    26. Lower bracts leaf-like, with short sheaths.......................... **C. serratodens**
    26. Lower bracts not leaf-like, with long sheaths......................... **C. lemmonii**

Cyperus (p. 195)

Perennial or annual herbs with unbranched triangular upright stems; lvs. mostly at base; invol. lvs. much exceeding infl.; infl. umbellate or capitate; spikelets several-fld., scales 2-ranked. About 600 species, 16 in CA.

1. Styles 2; achenes lens-shaped; infl. capitate.
  2. Achenes laterally flattened; lf.-sheaths reddish; invol. lvs. 3..................... **C. niger**
  2. Achenes dorsally flattened; invol. lvs. 1-2 .... **C. laevigatus**
1. Styles 3; achenes triangular.
  3. Spikelets breaking above the basal pair of scales into 1-seeded joints; edges of rachis nearly concealing fls.. **C. odoratus** L. (*C. ferax*)
  3. Spikelets persistent on the spike.
    4. Rachis not winged.
      5. Perennial with short rhizomes; invol. lvs. 5-8, to 50 cm long; spikelets 10-20 mm long. .... **C. eragrostis**
      5. Annual with fibrous roots.
        6. Low tufted plants, 1-20 cm tall; lvs. celery-scented. **C. squarrosus** L. (*C. aristatus*)
        6. Taller, 15-50 cm tall; lvs. not celery-scented; weedy................ **C. difformis**
    4. Rachis winged with translucent appendages at each node.
      7. Perennial, stoloniferous with tubers; invol. lvs. 2-6; infl. umbellate, rays 0-12 cm long.. **C. esculentus**
      7. Annual; invol. lvs. 4-10; infl. compound umbel with rays 1-30 cm long. **C. erythrorhizos**

Eleocharis (p. 196)

Perennial, rarely annual, herbs with upright rounded stems; lvs. reduced to basal sheaths; spikelets solitary, terminal; scales spirally overlapping; bristles subtending ovary 1-8 (0), downwardly barbed; style-base persistent forming a tubercle on tip of achene. About 250 species, 15 in CA.

1. Achenes lens-shaped; styles 2; plants 5-10 dm tall; spikelets 5-25 mm long. **E. macrostachya**
1. Achenes plump or triangular; styles 3.
  2. Stems slender, to 40 cm tall.
    3. Spikelets 6-12 mm long, 10-20-flowered.

4. Spikelets ovoid to oblong, obtuse; bristles 2-6. .... **E. montevidensis**
4. Spikelets linear, acute; bristles 6-7 .... **E. parishii**
3. Spikelets 2-7 mm long, 2-10-flowered.
5. Scales broad with green midrib and translucent margins.
6. Perennial with stolons or rhizomes; plants 2-20 cm tall
7. Stems thin; stamens 3 .... **E. acicularis** var. **acicularis**
7. Stems spongy; stamens 2 ....**E. radicans**
6. Annual with fibrous roots; plants 2-6 cm tall.
E. acicularis var. **bella** (*E. bella*)
5. Scales lanceolate, purplish-brown ....**E. pauciflora**
2. Stems wiry, coarse, 25-150 cm tall; scales brownish with translucent margins.
E. rostellata

Fimbristylis (p. 197)

Annual or perennial sedge-like herbs; invol. lvs. few; infl. a loose umbel or capitate cluster of spikelets; scales spirally overlapping; fls. perfect; bristles none; stamens 1-3; achene lens-shaped. About 200 species, 3 in CA.

Dwarf annual; infl. a simple capitate cluster .... **F. vahlii**
Taller (20-70 cm tall) perennial; infl. a loose umbel ....**F. thermalis**

Lipocarpha (p. 197)
(*Hemicarpha*)

Dwarf tufted annual with basal lvs.; edges rolled inward; invol. lvs. 1-3, one much exceeding infl.; infl. of 1-3 spikelets; scales with brown sides and green midrib extending into a curved awn. About 5 species, 3 in CA. One species in Kern County: **L. occidentalis** (Gray) Tucker

Scirpus (p. 197)

Perennial (one annual) herbs of meadows, marshes and other wet habitats; stems triangular to round in cross section; invol. of several blade-bearing lvs. or reduced to few; infl. an open umbel or cluster of spikelets, rarely solitary; fls. perfect; bristles 1-6, barbed or ciliate. About 200 species, 23 in CA.

1. Invol. lvs. 2-5, exceeding infl.
2. Spikelets small, 3-6 mm long; stamens 2 .... **S. microcarpus**
2. Spikelets larger, 10-25 mm long; stamens 3 .... **S. maritimus**
(incl. *S. paludosus* and *S. tuberosus*)
1. Invol. lf. solitary, often appearing as a continuation of stem.
3. Stems leafy, triangular; spikelets few, 1-12.
4. Perennials with rhizomes; stems not thread-like.
5. Lf.-blades longitudinally rolled, to 18 cm long; involucre-lf.. 3-10 cm long.
S. pungens Vahl. (*S. americanus*)

5. Lf.-blades flat, to 13 cm long; invol.-lf. 1-3 cm long. **S. americanus** Pers. (*S. olneyi*)
4. Annual; stems slender, rounded .... **S. saximontanus**
3. Stem-lvs. reduced to basal sheaths with short blades to 8 cm long; stems rounded; spikelets numerous in umbels.
7. Bristles slender, barbed; plants to 5 m tall. **S. acutus** var. **occidentalis** (Wats.) Beetle
7. Bristles broad, plumose with dense soft hairs; plants to 4 m tall.. **S. californicus**

DATISCACEAE

Datisca (p. 288)

Stout glabrous perennial herb with alt. (above) pinnately dissected lvs. 1-2 dm long; fls. dioecious, clustered in axils of a leafy raceme; petals absent; staminate calyx 4-9 lobed; pistillate calyx 3-toothed; fruit a capsule. Two species, one in CA. One species in Kern County: **D. glomerata**

DENNSTAEDTACEAE
(part of Polypodiaceae)

Pteridium (p. 166)

Coarse fern with erect 3-15 dm tall fronds; petioles straw-colored; fronds 3 times pinnate; indusium double. About 8 species, one in CA. One species in Kern County: **P. aquilinum** var. **pubescens**

DIPSACACEAE

Dipsacus

Tall stout herbs with opposite lvs., the stem-leaves united at base; fls. pinkish in dense terminal oblong heads; bracts subtending the fls. rigid and spine-like; calyx and corolla 4-lobed; stamens 4. About 15 species, 2 introduced in CA. One speciesin Kern County: **D. fullonum** L. (Pettit Meadow, Greenhorn Mtns. Possibly introduced during seeding of Red Mountain Burn. Shevock 8628)

DRYOPTERIDACEAE
(part of Polypodiaceae)

Terrestrial ferns with pinnately compound fronds with sori away from margins; indusium present, mostly round to kidney-shaped.

1. Petioles slender, <1.5 mm diameter; blade of frond 1-1.5 dm long; indusium hood-like.. **Cystopteris**

1. Petioles coarser, 2-4 mm. diam.; blade of frond more than 2 dm. long.
   2. Indusium kidney-shaped and circular in outline .... **Dryopteris**
   2. Indusium curved, elongate.. **Athyrium**

Athyrium (p. 165)

Large stout fern with erect arching fronds 6-15 dm. long and 1-2 dm. broad; petioles straw-colored; fronds 2-3 times pinnate, lfts. scalloped-dentate; sori curved, elongate. About 100 species, 2 in CA. One species in Kern County: **A. felix-femina** var. **cyclosporum** Rupr. (var. *californicum*)

Cystopteris (p. 165)

Delicate fern with few to several fronds 5-20 cm. long; petioles slender, brittle, greenish; fronds nearly twice-pinnate, thin; lfts. decurrent on margined rachis. About 10 species, one in CA. One species in Kern County: **C. fragilis**

Dryopteris (p. 166)

Woodland fern with several erect fronds 3-8 dm. long; petioles stout, shorter than blades, scaly; fronds 2 times pinnate with serrate lfts.; sori in 2 rows. About 100 species, 3 in CA. One species in Kern County: **D. arguta**

ELAEAGNACEAE

Shepherdia (p. 290)

Erect spiny dioecious shrub, 2-6 m tall; lvs. opposite, deciduous, 2-6 cm long, silver-scurfy on both sides; fls. clumped at nodes. Three species, one in CA. One species in Kern County. **S. argentea**

ELATINACEAE

Low herbs with opposite or whorled simple lvs.; fls. small, regular, axillary, perfect; sepals and petals 3-5; stamens hypogynous; fr. a capsule.

Plants glabrous, aquatic or semiaquatic.. **Elatine**
Plants glandular-pubescent, terrestrial.. **Bergia**

Bergia (p. 285)

Diffuse or ascending glandular-pubescent annual; lvs. ovate, 1-2 cm long; fls. 5-merous, white, solitary. About 25 species, 1 in CA. One species in Kern County: **B. texana**

Elatine (p. 285)

Dwarf glabrous annual in water or on mud flats, often rooting at nodes; fls. 2-4-merous, small. About 25 species, 6 in CA. One species in Kern County: **E. brachysperma**

EPHEDRACEAE

Ephedra (p. 172)

Much-branched shrubs with jointed, grooved greenish stems and scale-like lvs.; dioecious with small axillary cones; staminate cone with 2-8 stamens; pistillate cone to about 1 cm long with small lower bracts. About 40 species, 7 in CA.

1. Lvs. and bracts in 3's; branchlets to 4 mm thick; shrub to 1 m tall; cones sessile. **E. californica**
1. Lvs. and bracts in 2's; branchlets to 2 mm thick; shrubs to 1.5 m tall; cones stalked.
    2. Lf.-bases brown; branchlets yellow-green, ± parallel and broom-like.... **E. viridis**
    2. Lf.-bases gray; branchlets gray-green, ± divergent .... **E. nevadensis**

EQUISETACEAE

Equisetum (p. 167)

Rush-like to branching plants with creeping rhizomes; upright stems cylindrical, simple or with whorled branches, segmented; spores in terminal cones. About 15 species, 5 in CA.

1. Upright stems all green.
    2. Lf.-sheaths longer than broad, with narrow black band at tip, green. **E. laevigatum**
    2. Lf.-sheaths as broad as long, with 2 black bands, gray. **E. hyestaminate** ssp. **affine** (Engelm.) Calder & Taylor
1. Upright stems of 2 kinds, fertile ones yellowish, sterile ones green with whorls of thin branches.. **E. arvense**

ERICACEAE
(incl. Pyrolaceae)

Perennial herbs or shrubs with simple lvs. without stipules; chlorophyll present or absent; fls. regular, in axillary or terminal racemes; petals free or connate , 4-5; stamens hypogynous; anthers 2-celled, opening by slits or terminal pores; ovary 1-5-celled; fr. a capsule or berry-like.

1. Plants not green; lvs. scale-like.
    2. Fr. a berry; plant pinkish, 2-10 cm tall.. **Hemitomes**
    2. Fr. a capsule; plant pink to red, to 17 dm tall.

3. Petals fused; plants not striped.
  4. Plants pinkish, slender, to 17 dm tall .... **Pterospora**
  4. Plants red, stout, <5 dm tall.. **Sarcodes**
3. Petals free; plants with maroon stripes.. **Allotropa**

1. Plants green; lvs. expanded.
  5. Plants mostly herbaceous; lvs. basal or on short stems; petals free.
    6. Fls. 1-few on umbel-like raceme .... **Chimophylla**
    6. Fls. few-many in elongate receme .... **Pyrola**
  5. Plants woody; petals connate.
    7. Fr. a dry capsule; corolla large, funnel-shaped; stamens 5 .... **Rhododendron**
    7. Fr. fleshy; corolla small, urn-shaped; stamens 10...................... **Arctostaphylos**

Allotropa

Nongreen glabrous perennial with scale-like lvs.; Infl. white with reddish stripes; corolla cup-shaped, white; fruit a capsule. One species: **A. virgata** Gray (Near Tiger Flat, Greenhorn Mtns. Shevock 10521 , Hammerly 553)

Arctostaphylos (p. 301)

Woody evergreen shrubby plants with crooked branches and smooth thin reddish exfoliating bark; lvs. simple, alternate, tough; calyx 4-5-lobed; corolla small, urn-shaped, white to rose-pink; stamens 10 (8); fr. berry-like. About 60 species, 57 in CA.

1. Branches prostrate or procumbent, usually <6 dm high.
  2. Hairs of branchlets and infl. not glandular.
    3. Lvs. mostly broadest above the middle; fr. 5-6 mm in diameter...**A. nevadensis**
    3. Lvs. mostly broadest at or below the middle; fr. 8-12 mm in diam. **A. parryana**
  2. Hairs of branchlets and infl. glandular.. **A. patula**
1. Branches erect or ascending, taller.
  4. Bracts of infl. as long as or exceeding pedicels, foliaceous..................... **A. glauca**
  4. Bracts of infl. shorter than pedicels, deltoid to nearly subulate, not foliaceous.
    5. Pedicels glabrous.
      6. Branchlets and rachises of infl. not glandular pubescent.
        7. Fr. globose or depressed-globose, nutlets ± separable .... **A. pungens**
        7. Fr. somewhat ovoid, nutlets united into solid stone................. **A. parryana**
      6. Branchlets and rachises glandular pubescent .... **A. patula**
    5. Pedicels ± glandular pubescent.
      8. Fr. pulpy, 6-10 mm in diam., depressed globose ....**A. viscida**
        9. Lvs. glabrous, smooth; branchlets mostly glabrous; calyx-lobes ciliate. ssp. **viscida**
        9. Lvs. glabrous or glandular, scabrous; branchlets glandular; calyx -obes glandular-villous.................. ssp. **mariposa** (Dudley) Wells (*A. mariposa*)
      8. Fr. not very pulpy, 12-15 mm in diam., round-ovoid. **A. glauca** (incl. var. *puberula*)

Chimaphila

Somewhat shrubby perennial to about 15 cm tall with ± whorled lvs.; fls. white, turning pink, in ± umbel-like racemes; fruit. a capsule. About 5 species, 2 in CA. One species in Kern County: **C. menziesii** (D. Don) Spreng. (Near Tiger Flat, Greenhorn Mtns., Twisselmann,)

Hemitomes

Nongreen glabrous fleshy perennial with scale-like lvs.; infl. various; corolla pink to cream-colored, with hairy throats; fr. a berry about 1 cm wide. One species: **H. congestum** Gray (In dense fir stands, summit of Sunday Peak. Shevock 10522)

Pterospora (p. 301)

Stout simple purplish-brown herbaceous root parasite about 3-10 dm high; lvs. scale-like; fls. in long terminal raceme, nodding; corolla urn-shaped, 5-lobed; stamens 10, hypogynous; fr. a 5-loculed capsule. One species: **P. andromedea**

Pyrola (p. 301)

Low perennial herbs with a basal cluster of evergreen lvs.; fls. in simple raceme at end of leafless scape, 5-merous; petals concave; stamens 10; anthers opening by basal pores; fr. a 5-lobed capsule. About 15 species, 4 in CA. One species in Kern County: **P. picta** (Quite variable, forma *aphylla* and subsp. *dentata* included.)

Rhododendron (p. 302)

Loosely branched deciduous shrub with shredding bark; fls. showy, white with pink tinge, funnelform; stamens 5, exserted; fr. a capsule. About 1000 species, 2 in CA. One species in Kern County: **R. occidentale**

Sarcodes (p. 301)

Low fleshy red pubescent herbaceous perennial; saprophytic; lvs. scale-like; fls. many, in stout spicate raceme, red, subtended by red bracts; corolla bell-shaped, 5-lobed, red; anthers opening by terminal pores. One species: **S. sanguinea**

EUPHORBIACEAE

Herbs or shrubs usually with milky sap; lvs. simple, alternate or opposite; plants monoecious or dioecious; fls. with calyx and corolla present or absent; ovary superior, 3-celled; fr. a capsule.

1\. Plants woody .... **Ricinus**
1\. Plants herbaceous.
  2. Fls. with a calyx.

3. Plants armed with stiff stinging hairs... **Eremocarpus**
3. Plants not armed with stinging hairs.
 4. Herbage stellate-pubescent.. **Croton**
 4. Herbage glabrous... **Stillingia**
2. Fls. lacking a calyx.
 5. Glands of cyathium lacking petaloid appendages; lf.-bases symmetrical; annuals.. **Euphorbia**
 5. Glands of cyathium with petaloid appendages; lvs. with unequal bases; annuals to perennials.. **Chamaesyce**

Chamaesyce (p. 278)
(part of *Euphorbia*)

Monoecious herbs with milky juice; lvs. simple, opposite, with small stipules; fls. in a cyathium (invol. resembling a calyx or corolla with united lobes) with 1-5 nectariferous glands and with petaloid appendages; staminate fls. a single stamen subtended by a bractlet; pistillate fl. solitary in center of cyathium, becoming exserted; fr. a nodding 3-lobed capsule. About 250 species, 22 in CA.

1. Perennial.
 2. Stipules united into broad white scale; capsule glabrous. **C. albomarginata** (T&G) Small
 2. Stipules free; capsule hairy.. **C. vallis-mortae** Millsp.
1. Annual.
 3. Staminate fls. 40-60 per cyathium; lf.-margin entire; sandy soils. **C. ocellata** (Durra.d & Hilg.) Millsp.
  4. Lf. >10 mm long, lanceolate to ovate; Mojave Desert. ssp. **arenicola** (Parish) Thorne
  4. Lf. <10 mm long, ovate, sickle-shaped; Central Valley and foothills. ssp. **ocellata**
 3. Staminate fls. <20 per cyathium; lf.-margin finely toothed; various soils.
  5. Staminate fls. 4-5 per cyathium; lf.-blade often with dark central spot; weedy. **C. maculata** (L.) Small
  5. Staminate fls. 5-18 per cyathium; lf.-blade lacking dark central spot; dry places. **C. serpyllifolia** (Pers.) Small

Croton (p. 277)

Perennial stellate-scurfy herbs with alternate simple lvs.; dioecious; staminate fls. in terminal racemes, calyx 5-parted, petals present or absent; pistillate fls. below the staminate, petals absent; fr. globose 3-lobed capsule. About 950 species, 2 in CA. One species in Kern County: **C. californicus** (incl. var. *mohavensis*)

Eremocarpus (p. 278)

Low annual with dichotomously branching stems forming a dense rounded mass; lvs. alternate, ovate, entire, covered with gray stellate pubescence and longer stinging hairs; monoecious, petals absent.One species: **E. setigerus**

Euphorbia (p. 278)

Monoecious annual herbs with milky juice; lvs. simple, mostly alternate, lacking stipules; fls. in a cyathium (invol. resembling a calyx or corolla with united lobes) with 4 nectariferous glands lacking petaloid appendages; staminate fls. a single stamen subtended by a bractlet; pistillate fl. solitary in center of cyathium, becoming exserted; fr. a nodding 3-lobed capsule. About 1500 species, 14 in CA.

1. Glands elliptic or transversely oval; lvs. serrulate .... **E. spathulata**
1. Glands crescent-shaped or 2-horned; lvs. entire.
  2. Lvs. opposite .... **E. lathyris**
  2. Lvs. alternate .... **E. peplus**

Ricinus (p. 279)

Monoecious glabrous glaucous shrub or, in cold regions, annual with large alternate peltate 5-11-lobed lvs.; calyx 3-5-parted, petals none; ovary red, 3-celled; seeds 10-14 mm long, mottled. One variable species introduced from Asia: **R. communis**

Stillingia (p. 279)

Glabrous monoecious perennial herb with alternate linear lvs.; fls. apetalous, bracteolate in terminal spikes; capsule 3-lobed. About 30 species, 3 in CA. One species in Kern County: **S. paucidentata**

FABACEAE

(Leguminosae by some authors)

Herbs, shrubs or small trees usually with alternate compound lvs. and evident stipules; fls. regular or irregular and papilionaceous (large upright "banner," two lateral "wings" often fused at tips, two lower petals fused at tip to form a "keel"); stamens mostly 10, often with united filaments; pistil 1, simple and becoming a legume.

1. Corolla ± regular; stamens exserted; lvs. bipinnate .... **Prosopis**
1. Corolla irregular; stamens mostly included.
  2. Upper petal ("banner") internal in bud, enveloped by the lateral ones ("wings").
    3. Lvs. simple; fls. rose-purple .... **Cercis**
    3. Lvs. compound; fls. yellow to orange.
      4. Lvs. once-pinnate; low subshrubs leafless during most of year; fls. bright yellow .... **Senna**
      4. Lvs. twice-pinnate.

5. Tall herbs; corolla orange .... **Hoffmannseggia**
5. Large shrubs or small trees; fls. yellow.
6. Petiole short with rachis ending in a spine; rachillas long, flat with numerous small deciduous lfts.; stamens not much exserted. **Parkinsonia**
6. Petioles not becoming spines; stamens long-exserted, red.... **Caesalpinia**
2. Upper petal external in bud (papilionaceous).
7. Lvs. simple.
8. Corolla red; low stiff spiny shrub ,<1 m tall ....**Alhagi**
8. Corolla yellow; taller shrub, >1 m tall, not spiny.............................**Spartium**
7. Lvs. compound.
9. Lvs. palmate or appearing so.
10. Stamens 10, fused by filaments; pods dehiscent ....**Lupinus**
10. Stamens 9 fused by filaments ± 1 free or 5 and fused by filaments; pods mostly indehiscent.
11. Plants gland-dotted; lfts. 5-6 ....**Pediomelum**
11. Plants not gland-dotted; lfts. 3.
12. Fls. in ovoid to oblong heads ....**Trifolium**
12. Fls. not in heads; lvs. actually pinnately 3-foliate.
13. Fls. in axillary umbels .... **Lotus**
13. Fls. in spikes or racemes.
14. Pods curved or spirally coiled ....**Medicago**
14. Pods ovoid, straight .... **Melilotus**
9. Lvs. pinnate.
15. Lf.-rachis prolonged into a tendril.
16. Wings of corolla essentially free from keel; style bearded down inner face... **Lathyrus**
16. Wings adherent to keel.
17. Style bearded down one side ....**Pisum**
17. Style bearded in tuft at tip .... **Vicia**
15. Lf.-rachis not prolonged into a tendril.
18. Plants gland-dotted; pods indehiscent, 1-2-seeded.
19. Shrub; fls. dark blue; pod not prickly ....**Psorothamnus**
19. Herbaceous perennial; fls. light blue or yellow................**Glycyrrhiza**
18. Plants not gland-dotted; pods dehiscent.
20. Fls. in axillary umbels (quite reduced in some spp.).................. **Lotus**
20. Fls. in racemes.
21. Tree with large drooping racemes ....**Robinia**
21. Plants herbaceous.
22. Style incurved, longitudinally bearded................. **Sphaerophysa**
22. Style mostly straight, not bearded. ....**Astragalus**

Alhagi (p. 260)

Stiff spiny shrubs with small simple entire lvs.; fls. few, in axillary racemes; corolla reddish-purple. About 3 species, 1 introduced in CA. One species in Kern County: **A. pseudoalhagi** (Bieb.) Desv. (*A. camelorum*)

Astragalus (p. 260)

Annual or perennial herbs with alternate stipulate odd-pinnately compound lvs.; fls. in racemose or spikate infl.; papilionaceous; stamens 9 + 1; fr. variable in texture, shape and covering. About 2000 species, 96 in CA.

1. Pod strongly inflated.
  2. Pod 1-celled.
    3. Pod with stalk at base >2 mm long.
      4. Pod 2.5-4 cm long; calyx lacking black hairs; petals greenish-white, 11-15 mm long. .... **A. trichopodus** var. **phoxus** (Jones) Barn.
      4. Pods 2-2.4 cm long; calyx with black hairs; petals whitish, 10-12 mm long. **A. whitneyi**
    3. Pod sessile or with stalk <2 mm long.
      5. Fls. rose-lavender or purplish.
        6. Pod thinly villous-tomentulous .... **A. subvestitus**
        6. Pod shaggy, hairs >2 mm long ....**A. purshii** var. **tinctus** Jones
      5. Fls. whitish or yellowish-white.
        7. Pod conspicuously hairy, ovoid, with narrow beak-like tip; borne in dense clusters... **A. hornii**
        7. Pod glabrous or inconspicuously pubescent.
          8. Herbage strigulose.
            9. Plants with soft shaggy hairs; fr. 16-22 mm long. **A. ertterae** Barn. & Shevock (rare, sandy slopes, open pinyon woodland near Walker Pass)
            9. Plants minutely hairy; fr. 25-60 mm long .. **A. douglasii** var. **douglasii**
          8. Herbage and pod villosulous; banner red-veined................. **A. macrodon**
  2. Pod 2-celled.
    10. Pods shaggy-hairy, resembling a mass of wool; plant acaulescent; fls. rose-lavender... **A. purshii** var. **tinctus**
    10. Pods strigulose to hoary, not woolly .... **A. lentiginosus**
      11. Fls. small, keel <9 mm long.
        12. Beak of pod nearly linear, tubular, resembling a persistent style. var. **kernensis**
        12. Beak of pod deltoid or triangular, laterally flattened; petals mostly purple. var. **albifolius**
      11. Fls. larger, keel >8 mm long.
        13. Petals pale yellow; plant coarse.. var. **nigricalycis**
        13. Petals purple.
          14. Racemes compact, not >4 cm long .... var. **idriensis**
          14. Racemes loose, open, >4 cm long..................................... var. **variabilis**
1. Pod not strongly inflated.
  15. Annual.
    16. Pods 3-4 mm long, transversely wrinkled; petals pale.
      17. Pods deflexed, exserted from calyx, conspicuously gray-hairy. **A. gambelianus**
      17. Pods erect, little exserted, minutely hairy.......................... **A. didymocarpus**

18. Calyx hairs white; lflts. hairy on upper surface.................. var. **dispermus**
18. Calyx hairs black; lflts. ± glabrous on upper surface ... var. **didymocarpus**

16. Pods 12-20 mm long, not transversely wrinkled; petals purplish. **A. acutirostris**

15. Perennial.
19. Dorsal and ventral sutures of the pod prominent, externally forming thick ridge; petals white.. **A. pachypus** var. **pachypus**
19. Dorsal and ventral sutures not both externally prominent.
20. Pods sessile.
21. Pods 3-6.5 cm long, hirsute, mottled; petals yellowish-white, tipped with purple.. **A. layneae**
21. Pods 7.5-15 mm long, shaggy-villous; petals rose-lavender or purplish. **A. purshii** var. **tinctus**
20. Pods stipitate; petals yellowish-white.
22. Stipe 15-40 mm long; body of pod bladdery-inflated......**A. asymmetricus**
22. Stipe 3-11 mm long; body of pod turgid...............................**A. oxyphysus**

Caesalpinia (p. 270)
(*Poinciana*)

Straggling shrub to small tree without spines; lvs. bipinnately compound, secondary lfts, with black dots on lower side near margins; infl. densely glandular; fls. large, light yellow with 10 brilliant red stamens protruding 7-12 mm. About 200 species, 3in CA. One species introduced in Kern County: **C. gilliesii** (Hook.) Dietr.

Cercis (p. 263)

Rather large shrub with simple broad heart-shaped 7-9-veined lvs.; fls. red-purple in short clumps appearing before lvs.; corolla irregular; stamens distinct. About 7 species, 1 in CA. One species in Kern County: **C. occidentalis**

Glycyrrhiza (p. 263)

Perennial herbs with thick rootstocks and glandular odd-pinnately compound lvs.; fls. in axillary spikes; pods indehiscent. About 20 species, 2 in CA.

Pods bur-like with hooked prickles; fls. yellow-white................................ **G. lepidota**
Pods glabrous but glandular; fls. pale blue .... **G. glabra**

Hoffmannseggia (p. 263)

Herbaceous perennial with several scattered stems from underground rootstocks; lvs. with 5-11 pinnae, each with 6-10 pairs of lfts.; fls. in glandular racemes; corolla about 2 cm broad, ± regular, yellow to reddish; stamens 10, distinct. About 20 species, 2 in CA. One species in Kern County: **H. glauca** (Ortega) Eifert (*H. densiflora*)

Lathyrus (p. 264)

Perennial herbs with erect to twining angled to winged stems; lvs. pinnately-compound with terminal tendrils; fls. in axillary racemes; corolla showy, wings ± folded with keel but free; stamens with fused filaments. About 150 species, 24 in CA.

1. Lfts. 2; fls. purple or white, 20-25 mm long .... **L. latifolius**
1. Lfts. at least 4.
   2. Fls. tan to yellowish with darker veins, 10-15 mm long .... **L. sulphureus**
   2. Fls. white to violet, 16-22 mm long.
      3. Fls. white; calyx 10-14 mm long .... **L. vestitus** Nutt. (*L. laetiflorus*)
      3. Fls. blue-violet or orchid; calyx 7-10 mm long. **L. brownii** Eastw. (*L. pauciflorus*)

Lotus (p. 264)

Annual to perennial herbs, some ± shrubby; lvs. pinnately compound, 1-several-foliate; fls. axillary, solitary to umbellate; corolla white or yellow, often with some red, papilionaceous. About 150 species, 33 in CA.

1. Stipules expanded, not gland-like.
   2. Stipules green, resembling lft.; corolla bright yellow .... **L. corniculatus**
   2. Stipules membranaceous, not like lft.
      3. Lvs. glabrous; bract 1-5-pinnate, remote from 8-15-fld. umbel... **L. crassifolius**
      3. Lvs. pubescent; bract 1-3-foliate, closely subtending 1-5-fld. umbel. **L. oblongifolius**
1. Stipules reduced to dot-like glands.
   4. Annual.
      5. Fls. yellow, often aging red.
         6. Wings conspicuously longer than keel; style hairy .... **L. strigosus** (incl. var. *hirtellus* and *L. tomentellus*)
         6. Wings mostly as long as or shorter than keel; style glabrous.
            8. Fls. 1-5 in umbel; pods 1-2 mm wide .... **L. salsuginosus**
               9. Corolla 3-5 mm long; fr. 1-1.5 cm long .... var. **brevivexillus**
               9. Corolla 6-10 mm long; fr. 1.5-3 cm long .... var. **salsuginosus**
            8. Fls. solitary, subsessile; pods 3 mm wide.
               10. Plants subglabrous to strigose; calyx-teeth ca. as long as tube; pods 10-15 mm long. .... **L. wrangelianus** Fisch. & Meyer (*L. subpinnatus*)
               10. Plants densely villous; calyx-teeth ca. 2 x length of tube; pods 5-10 mm long. .... **L. humistratus**
      5. Fls. cream-white to pinkish.
         11. Calyx-teeth shorter than tube; pods not deflexed; lfts. mostly 3-10 mm long. **L. micranthus**
         11. Calyx-teeth longer than tube; pods deflexed; lfts. mostly 10-15 mm long. **L. purshianus**
   4. Perennial.
      12. Pod dehiscent, 3-4 cm long; corolla 15-24 mm long .... **L. grandiflorus**

12. Pod indehiscent, <1 cm long; corolla 3-9 mm long.
 13. Fls. on peduncles >10 mm long .... **L. nevadensis**
  14. Calyx-teeth ca. 1/2 length of tube; plant covered with fine hairs, scarcely silvery. .... var. **nevadensis**
  14. Calyx-teeth as long as tube; plant silvery.
   var. **davidsonii** (Greene) Isley (*L. davidsonii*)
 13. Fls. mostly sessile in lf.-axils.
  15. Plant subglabrous; pod well-exserted .... **L. scoparius**
   16. Keel prominent, >wings... var. **brevialatus**
   16. Keel not prominent, ± =wings.. var. **scoparius**
  15. Plant grayish with oppressed pubescence.
   17. Plant silvery-tomentose, prostrate to decumbent; pod scarcely exserted.
    L. argophyllus
   17. Plant silky-canescent, procumbent; pod well exceeding calyx.
    L. procumbens var. **jepsonii**

Lupinus (p. 266)

Annual to perennial herbs or shrubs; lvs. alternate, palmately compound; fls. in terminal racemes, showy; corolla papilionaceous; stamens 10, united by filaments. About 200 species, 71 in CA.

1. Plants annual.
 2. Fls. not at all whorled.
  3. Keel ciliate on lower edges near base.
   4. Corolla blue. .... **L. benthamii**
   4. Banner bright yellow, wings rose to purple .... **L. stiversii** Kell
    (Sandy granite roadcut, Sawmill Rd., Greenhorn Mtns. Shevock 9717)
  3. Keel not ciliate; corolla lilac to violet-purple, 6-10 mm long.
   5. Plants glabrous, ± succulent; corolla ca. 10 mm long.................. **L. odoratus**
   5. Plants densely hairy; corolla 6-8 mm long
    6. Plants acaulescent to 1 dm tall; fls. bluish-purple to pink.........**L. shockleyi**
    6. Plants much-branched to 2 dm tall; fls. lilac with yellow spot on banner.
     L. concinnus (incl. var. *orcuttii*)
 2. Fls. whorled.
  7. Keel ciliate on upper margins near apex.
   8. Pedicels 1-3 mm long; fls. 4-8 mm long ....**L. bicolor**
   8. Pedicels slender, 4-10 mm long; fls. 8-15 mm long.
    L. nanus (incl. *L. vallicola*)
  7. Keel ciliate on upper margins near claws (base).
   9. Keel ciliate on lower margins near claws.
    10. Plants 0.5-1.5 dm tall; peduncles 2-4 cm long; fls. 10-11 mm long.
     L. microcarpus Sims var. **horizontalis** (Heller) Jeps. (*L. horizontalis*)
    10. Plants 2-10 dm tall; peduncles 3-15 cm long; fls. 12-14 mm long.
     L. succulentus

9. Keel not ciliate on lower margins.................................. **L. microcarpus** Sims
  11. Calyx ± sparsely hairy; fr. on one side of infl. axis.
  var. **microcarpus** (*L. ruber* and *L. subvexus*)
  11. Calyx shaggy-hairy; fr. not on one side of infl. axis.
  var. **densiflorus** (Benth.) Jeps. (*L. densiflorus*)

1. Perennial.
  12. Plants shrubby.
    13. Keel narrowed towards base; banner pubescent on back................ **L. albifrons**
    13. Keel not narrowed toward base .... **L. excubitus**
  12. Plants herbaceous.
    14. Lvs. mostly glabrous on upper surface; wet places.
      15. Keel not ciliate on upper margins.
      L. polyphyllus var. **burkei** (Wats.) Hitch. (var. *surpurbus*)
      15. Keel ciliate on upper margins .... **L. latifolius**
        16. Fl. 13-18 mm long .... var. **parishii**
        16. Fl. 10-14 mm long .... var. **columbianus**
    14. Lvs. ± hairy above.
      17. Banner pubescent on back (sometimes concealed by calyx).
        18. Calyx spurred at base just above pedicel; lvs. not silvery-silky.
        L. arbustus
        18. Calyx not spurred; lvs. silvery-silky.
          19. Plants matted, <2 dm tall .... **L. breweri** var. **grandiflorus**
          19. Plants taller, 5-9 dm tall ....**L. elatus**
      17. Banner glabrous on back.
        20. Lvs. mostly in dense basal clusters; plants in dense clumps.
          21. Bracts 8-9 mm long, persistent; banner elliptic-ovate; damp places.
          L. lepidus Dougl. var. **confertus** (Kell.) Smith (*L. confertus*)
          21. Bracts 4-5 mm long, deciduous; banner round; dry places .... **L. grayi**
        20. Lvs. well distributed along stem.
          22. Keel ciliate on upper edges at least toward tip.
            23. Bracts of infl. ± persistent, conspicuous, 8-9 mm long.
            L. lepidus Dougl. var. **confertus** (Kell.) Smith (*L. confertus*)
            23. Bracts deciduous, inconspicuous, 4-7 mm long.
              24. Lvs. densely woolly with interwoven hairs; fls. 12-14 mm long.
              L. grayi
              24. Lvs. not woolly, but silky-strigose; fls. 14-18 mm long.
              L. exubiyus
                25. Shrubs mostly >7 dm tall ....var. **exubitus**
                25. Subshrubs mostly <7 dm tall...................... var. **austromontanus**
          22. Keel mostly glabrous on upper edges.
            26. Plants matted with ± prostrate stems; fls. 4-9 mm long ... **L. breweri**
              27. Leaflets >6 mm long..var. **breweri**
              27. Leaflets 3-5 mm long .... var. **bryoides**
            26. Plants not matted, fls. >9 mm long.

28. Banner narrow, acute; wings narrow; keel exposed. **L. albicaulis** (incl. var. *shastensis*)
28. Banner ovate to round, obtuse or rounded; wings wider, mostly covering keel.
  29. Lfts. thinly strigose above, greenish........................**L. andersonii**
  29. Lfts. silky-strigose above.
    30. Lfts. silvery-silky; fls. lavender to blue ....**L. elatus**
    30. Lfts. loosely silky, not silvery.
      31. Rhizomes absent , plant not resprouting below ground; corolla yellowish to lavender-violet. **L. adsurgens** (incl. var. *undulatus*)
      31. Rhizomes present, plant resprouting below ground; corolla purple...**L. formosus**
        32. Fl. 10-14 mm long; stem 3-4 mm diam....... var. **formosus**
        32. Fl. 16-18 mm long; stem 5-7 mm diam........ var. **robustus**

Medicago (p. 270)

Annual to perennial herbs with pinnately trifoliate lvs.; stipules evident; fls. in small racemes or spikes, papilionaceous; 9 filaments fused, 1 free; pods curved to spirally coiled. About 55 species, 7 introduced in CA.

1. Perennial; fls. blue .... **M. sativa**
1. Annual; fls. yellow.
  2. Pods ± kidney shaped, 1-seeded; fls. many in spikes ....**M. lupulina**
  2. Pods spirally coiled; fls. 2-5 in small racemes..........**M. polymorpha** (*M. hispida*)

Melilotus (p. 270)

Annual to biennial herbs with pinnately trifoliate lvs.; fls. small, white or yellow, in spike-like racemes, papilionaceous; 9 filaments fused, 1 free; pods ovoid. About 20 species, 3 introduced in CA.

1. Fls. white .... **M. alba**
1. Fls. yellow.
  2. Fls. 2-3 mm long; pedicels to 1 mm long ....**M. indica**
  2. Fls. 5-7 mm long; pedicels 1.5-2 mm long ....**M. officinalis**

Parkinsonia (p. 270)

Low trees with green branches and twigs; spines in pairs at each node and a spine at tip of primary leaf-rachis; lvs. bipinnately compound with 1-3 pairs of primary lfts. and many secondary lfts.; fls. in long racemes; banner yellow spotted with red; stamens 10; fr. 5-10 cm long, flattened between seeds. Two species, 1 in CA. One species in Kern County: **P. aculeata**

Pediomelum (p. 271)
(*Psoralea*)

Perennial herb with heavy-scented foliage; lvs. pinnately 5-6-foliate; stipules evident; fls. whitish-purple in dense pedunculate racemes; papilionaceous. About 22 species, 2 in CA. One species in Kern County: **P. californicum** (Wats.) Rydb.

Pisum (p. 270)

Glabrous annual with pinnately compound lvs. terminating in a pinnate tendril; stipules larger than lfts.; fls. 1-3 in lf.-axils, white, papilionaceous, the wings adhering to keel; 9 filaments fused, 1 free; style bearded down one side. About 2 species, 1 introduced in CA. One species in Kern County: **P. sativum**

Prosopis (p. 270)

Large shrub to low tree with paired spines at nodes; lvs. bipinnately compound with 1-2 pairs of primary lfts. and many secondary lfts.; fls. small, yellow, congested in axillary spikes; stamens 10, free and exserted; pods flat, constricted between seeds. About 44 species, 4 in CA. One species in Kern County: **P. glandulosa** Torr. var. **torreyana** (Benson) Johnston (*P. juliflora*)

Psorothamnus (p. 263)
(*Dalea*)

Shrub with odd-pinnately compound gland-dotted lvs.; fls. dark purple-blue in terminal racemes; pods ovoid with brown glands. About 9 species, 6 in CA.

Fr. with many small glands forming longitudinal lines; lvs mostly silvery hairy.
P. fremontii (Gray) Barneby var. **fremontii**
Fr. with few or several large, scattered glands; lvs. mostly glabrous.
P. arborescens (Gray) Barn. var. **minutifolius** (Parish) Barn.
(Red Rock Canyon, Faull & Charlton POM)

Robinia (p. 271)

Deciduous tree with dark brown bark and prickly branches; lfts. 9-19, 2-5 cm long; fls. white, showy, papilionaceous in drooping axillary racemes; pods 5-10 cm long, reddish-brown. About 4 species, 2 in CA. One species introduced in Kern County: **R. pseudoacacia**

Senna (p. 263)
(*Cassia*)

Much-branched rounded shrub with numerous yellowish-green striate stems leafless most of year; lf. with 1-4 pairs of roundish lfts.; racemes to 15 cm long with numerous bright yellow fls.; corolla nearly regular. About 260 species, 4 in CA. One species in Kern County: **S. armata** (Wats.) Irwin & Barneby

Spartium

Shrub to about 3 m tall with long nearly leafless green branchlets; lvs. alternate, simple; fls. yellow, 2-2.5 cm long, in loose terminal racemes; pods 5-10 cm long. One species. One species introduced in Kern County: **S. junceum** L. (Common to abundant in lower Kern Canyon along Hwy. 178)

Sphaerophysa (p. 271)

Perennial herb with tough rootstocks; lvs. pinnately compound with 13-21 ± rounded lfts.; fls. purplish-red, papilionaceous in axillary racemes; pod stalked, inflated. Two species, one in CA. One species introduced in Kern County: **S. salsula**

Trifolium (p. 271)

Herbs with mostly palmately trifoliate lvs.; stipules evident; infl. capitate, with or without an invol.; fls. small, papilionaceous; pods globose to elongate in persistent calyx. About 300 species, 45 in CA.

1. Heads without invol. at base of fls.
  2. Annual.
    3. Corolla inflated in age .... **T. depauperatum** var. **depauperatum**
    3. Corolla not inflated.
      4. Fls. sessile, not reflexed in age; calyx villous.
        5. Calyx 20-nerved; stipules with a long bristle-tip; heads 15-25 mm in dram. **T. hirtum**
        5. Calyx 10-nerved; stipules not bristle-tipped; heads smaller. **T. albopurpureum**
          6. Corolla about as long as calyx .... var. **albopurpureum**
          6. Corolla much shorter than calyx. var. **olivaceum** (Greene) Isley (*T. olivaceum*)
      4. Fls. pedicellate, reflexed in age; calyx not villous.
          7. Calyx ciliolate on margins; stipules 15-30 mm long. .... **T. ciliolatum**
          7. Calyx not ciliolate; stipules 8-10 mm long....................... **T. gracilentum**
  2. Perennial.
    8. Infls. axillary.. **T. repens**
    8. Infls. terminal.
      9. Peduncles 0, heads subtended by uppermost lvs.; fls. not reflexed; plant hairy...**T. pratense**

9. Peduncles present; fls. soon reflexed; plant glabrous.
T. macilentum Greene var. **dedeckerae** (Gillette) Barneby (uncommon, rocky slopes near summit of Spanish Needle Peak, Shevock 11211)
1. Heads with an invol. at base of fls.
10. Corolla inflated in age, 3-8 mm long, red-purple.
T. depauperatum var. **amplectens** (T&G) McDerm. (*T. stenophyllum*)
10. Corolla not inflated in age; invol.-bracts united.
11. Invol. bell-shaped to bowl-shaped.
12. Lobes of invol. entire; calyx pubescent, teeth subulate.... **T. microcephalum**
12. Lobes of invol. toothed; heads larger; calyx teeth 1-3 times trichotomously forked. .... **T. cyathiferum**
11. Invol. flat, wheel-shaped.
13. Perennial.
14. Corolla cream-color, keel purple-tipped; fls. 1-8 in a head.
T. monanthum var. **monanthum**
14. Corolla white to light purple; fls. many.
15. Invol. 12-15 mm broad; calyx not inflated in age......... **T. wormskioldii**
15. Invol. 5-6 mm broad; calyx inflated in age....................... **T. fragiferum**
13. Annual.
16. Plants sticky-pubescent; corolla 12-15 mm long, pale..... **T. obtusiflorum**
16. Plants glabrous.
17. Calyx-teeth dilated, mostly 3-toothed, lobes <tube.
T. willdenovii Spreng. (*T. tridentatum*)
17. Calyx-teeth bristle-tipped, lobes >tube............................ **T. variegatum**

Vicia (p. 273)

Vine-like herbs with pinnately compound lvs. terminating in tendrils; stipules evident; fls. in axillary racemes or spikes; wings adherent to keel; 9 filaments fused, 1 free; style-tip bearded. About 130 species, 15 in CA.

1. Fls. 1-4 in lf.-axils, subsessile, white with purple spot on wings..................... **V. faba**
1. Fls. borne near ends of evident peduncles.
2. Fls. 2-5 mm long .... **V. ludoviciana** Nutt. var. **ludoviciana** (*V. exigua*)
2. Fls. 10-18 mm long.
3. Lfts. 4-8 pairs; peduncles 3-9-fld., perennial.
V. americana var. **americana** (incl. *V. californica*)
3. Lfts. 8-12 pairs; peduncles many-flowered; annual .... **V. villosa**

FAGACEAE

Trees and shrubs with alternate simple petioled lvs.; plants monoecious, fls. apetalous, staminate in catkins, pistillate solitary and subtended by a ± woody invol. or "cup."

Fr. a spiny bur; lvs. persistent, rusty-tomentose beneath............................. **Chrysolepis**
Fr. an acorn with cup-like invol.. **Quercus**

Chrysolepis (p. 211)
(Castanopsis)

Low spreading rounded evergreen shrub with smooth bark; lvs. subentire 3-7 mm long, yellowish above, golden rusty-tomentose beneath; fr. with spiny invol. Two species, both in CA. One species in Kern County. **C. sempervirens** (Kell.) Hjelmq.

Quercus (p. 211)

Deciduous or evergreen trees or shrubs with ± contorted branches; staminate catkins slender and drooping; pistillate fls. solitary in cup-like invol.; fr. an acorn. About 300 species, 16 in CA.

1. Bark dark, not scaly; cups with thin, closely overlapping scales; acorn tomentose on inner surface (see no. 5). (Black Oaks)
  2. Deciduous; lvs. >5 cm long with large bristle-tipped lobes. .... **Q. kelloggii**
  2. Evergreen; lvs. mostly smaller.
    3. Lvs. thin, shallowly lobed with spinose tips .... **Q. Xmoreha** (not recognized in Jepson Manual)
    3. Lvs. thicker, tough, entire to spiny-toothed.
      4. Lvs. plane, glabrous beneath; cup 12-20 mm deep, 12-14 mm diameter with thin brownish-pubescent scales.............................. **Q. wislizenii** var. **frutescens**
      4. Lvs. convex above, stellate-pubescent especially in vein-axils underneath; cup 8-12 mm deep, 10-16 mm diameter, silky within. ...... **Q. agrifolia** var. **agrifolia**
1. Bark light, scaly; cups with tuberculate scales, sometimes hidden by felt-like hairs. (White Oaks)
  5. Acorn tomentose on inner surface; lvs. 2-5 cm long, yellowish-tomentose beneath; cup covered with felt-like hairs .... **Q. chrysolepis**
  5. Acorn glabrous on inner surface.
    6. Plants deciduous.
      7. Lvs. blue-green, toothed or wavy-margined; acorn 2-3 cm long, rounded. **Q. douglasii**
      7. Lvs. not blue-green, distinctly lobed.
        8. Stately tree; acorn 3-5 cm long, tapering .... **Q. lobata**
        8. Shrubby; acorn 2-3 cm long; twigs reddish .... **Q. garryana** var. **breweri**
    6. Plants evergreen and shrubby.
      9. Lvs. shining and subglabrous above, tough, minutely dentate to entire, 15-25 mm long. .... **Q. berberidifolia** Liebm. (*Q. dumosa*)
      9. Lvs. stellate-pubescent above, dull.
        10. Lvs. convex above, irregularly dentate; acorn abruptly pointed. **Q. Xalvordiana**
        10. Lvs. plane, spinose-dentate; acorn tapering, yellow-brown, 12-23 mm long. .... **Q. john-tuckeri** Nixon & Muller (*Q. turbinella*)

FRANKENIACEAE

Frankenia (p. 285)

Low herbaceous to suffrutescent herb with opposite entire subsessile lvs. with axillary fascicles; fls. regular, small, pinkish; ovary superior; fr. a capsule. About 90 species, 2 in CA. One species in Kern County: **F. salina** (Molina) Johnston (*F. grandifolia*)

GARRYACEAE

Garrya (p. 300)

Evergreen dioecious shrub with opposite leathery lvs.; fls. small, apetalous, in pendulous catkins; fr. ovoid densely silky drupe. About 14 species, 6 in CA. One species in Kern County: **G. flavescens** (incl. var. *pallida*)

GENTIANACEAE

Herbs with mostly opposite or whorled lvs.; fls. regular, perfect, axillary or terminal; calyx 4-12-lobed; petals connate, rotate to funnelform; stamens inserted alternate corolla-lobes; style 1; fr. a 2-celled capsule.

1. Corolla rotate, mostly whitish; fl. 4-merous .... **Swertia**
1. Corolla tubular, pink or blue.
    2. Anthers twisted; fls. pink .... **Centaurium**
    2. Anthers not twisted; fls. blue .... **Gentianopsis**

Centaurium (p. 303)

Glabrous erect branching annuals with opposite lvs.; fls. in forks at ends of branchlets; corolla trumpet-shaped with pale pink lobes; stigma-lobes fan-shaped. About 30 species, 7 in CA.

Corolla-lobes < half as long as tube, pale pink to white; transmontane..... **C. exaltatum**

Corolla-lobes > half as long as tube, rose with red spots in the white tube; cismontane. **C. venustum** (incl. var. *abramsii*)

Gentianopsis (p. 303)
(*Gentiana*)

Low herbs with opposite sessile lvs.; stem ending in a single fl. without subtending bract; corolla blue, funnelform with 4 lobes, often with intermediate folds; stigmas 2; caps. stalked. About 15 species, 2 in CA. One species in Kern County: **G. simplex** (Gray) Iltis

Swertia (p. 303)
(*Frasera*)

Perennial herbs with erect stems; lvs. opposite or whorled, entire, thickish; fls. 4-merous, in terminal panicle; corolla rotate, its 4 lobes bearing ± fringed glands; stigma 2-cleft. About 120 species, 7 in CA.

Stem-lvs. opposite; stems several .... **S. neglecta** (Hall) Jeps.
Stem-lvs. whorled; stems solitary .... **S. tubulosa** ((Cov.) St. John

GERANIACEAE

Annual to perennial herbs with stipulate palmate or pinnate lvs.; fls. 5-merous, hypogynous; ovary deeply lobed with elongate styles which separate and coil at maturity.

Lvs. pinnately-veined; outer stamens lacking anthers .... **Erodium**
Lvs. palmately-veined; outer stamens fertile .... **Geranium**

Erodium (p. 273)

Annual herb with pinnate lvs. in well-formed rosette; petals lavender; style-column very elongate, styles bearded on inside. About 75 species, 8 in CA.

1. Lvs. ± heart-shaped at base.
  2. Petals unequal, much exceeding sepals .... **E. texanum**
  2. Petals equal, rarely longer than sepals .... **E. macrophyllum**
1. Lvs. not heart-shaped at base.
  3. Lvs. simple, often divided; style-column 5-12 cm long.
    4. Style-column 5-9 cm long; sepals with short green tip.
      E. brachycarpum (Gord.) Thell. (*E. obtusiplicatum)*
    4. Style-column 9-12 cm long; sepals with prominent reddish tip, margins of one sepal marked with red. .... **E. botrys**
  3. Lvs. compound; style-column 2.5-4 cm long.
    5. Lfts. coarsely toothed; sepal-tips not setose .... **E. moschatum**
    5. Lfts. lobed or divided; sepal-tips setose .... **E. cicutarium**

Geranium (p. 275)

Herbs with forking stems and swollen nodes; lvs. palmately-lobed; fls. axillary, regular; corolla rose-pinkish to white; style-column beaked, styles mostly glabrous on inside. About 300 species, 13 in CA.

1. Petals 12-16 mm long; plant annual .... **G. californicum**
1. Petals 5-7 mm long; plant perennial.
  2. Lf.-lobes acute .... **G. dissectum**
  2. Lf.-lobes obtuse .... **G. carolinianum**

GROSSULARIACEAE
(part of Saxifragaceae)

Ribes (p. 254)

Shrubby plants with or without nodal spines; lvs. simple, alternate, palmate; fls. showy; ovary inferior; styles 2; fr. a pulpy several-seeded berry. About 120 species, 31 in CA.

1. Nodal spines lacking; berry spineless; fls. usually several in a raceme.
  2. Fls. greenish white to pink.
    3. Berry red, 5 mm diam.; lvs. 1-4 cm wide .... **R. cereum** var. **cereum**
    3. Berry black, 10-12 mm diam.; 3-8 mm wide .... **R. viscosissimum**
  2. Fls. rose to deep red.
    4. Sepals erect, 4-5 mm long; berry blue-black ....**R. nevadense**
    4. Sepals spreading, 3-4 mm long; berry purple-black.....................**R. malvaceum**
1. Nodal spines present; fls. 1-4, axillary.
  5. Ovary and berry without spines or prickles; fls. yellowish to whitish; berry dark, 4-8 mm diam.
    6. Ovary and berry glabrous; nodal spines mostly 1...................... **R. quercetorum**
    6. Ovary and berry velvety; nodal spines 1-3. **R. velutinum** (incl. var. *glanduliferum*)
  5. Ovary and berry with spines or bristles; fls. purplish; berry reddish to purplish, 9-25 mm diam.
    7. Fl.-tube above ovary 1-3 mm long.
      8. Young twigs bristly on internodes; lvs. glandular beneath; berry 15-25 mm diam.; fls. purplish. ....**R. menziesii** (incl. var. *ixoderme*)
      8. Young twigs not bristly on internodes; lvs. not glandular; berry 9-10 mm diam.; fls. greenish-purplish....................**R. californicum** var. **californicum**
    7. Fl.-tube above ovary 6 mm long; berry purple, spiny; fls. purplish-red. **R. roezlii**

HALORAGACEAE

Myriophyllum (p. 296)

Perennial aquatic with whorled pinnately dissected lvs.; fls. in spike above water; petals 4, small, greenish. About 40 species, 5 in CA. A single species: **M. hippuroides**

HIPPOTANACEAE

Aesculus (p. 280)

Large bush or tree to 17 m tall with broad rounded top; lvs. opposite, palmately 5-7-foliate; fls. white, showy, in large terminal clusters; fr. a leathery capsule with one large brown seed. About 15 species, one in CA. One species in Kern County: **A. californica**

HYDROCHARITACEAE
(incl. Najadaceae)

Submerged slender aquatic herbs with opposite to whorled slender lvs.; fls. unisexual, small, axillary; fr. an achene or capsule.

Fls. sessile; lvs. coarsely toothed .... **Najas**
Fls. peduncled; lvs. minutely toothed .... **Elodea**

Elodea (p. 175)

Submerged aquatic perennial herbs forming large tangles of branching elongate stems with occasional slender roots; lvs. whorled, thin, 1-veined; fls. staminate or pistillate, 1-3 in axillary sheathing bract. About 12 species, 2 in CA. One species in Kern County: **E. canadensis**

Najas (p. 174)

Slender branching submerged annuals with small linear, 1-5 cm long, 0.5-3 mm wide, mostly opposite lvs.; lv.-margins toothed; fls. very small, unisexual, solitary, axillary; staminate fl. with 1 stamen; pistillate fl. with 1 carpel. About 50 species, 5 in CA. One species in Kern County: **N. marina**

HYDROPHYLLACEAE

Herbs or shrubs with alternate lvs.; fls. 5-merous; calyx deeply lobed, with or without basal auricles between lobes; corolla wheel-shaped to bell-shaped to tubular, usually with small scale-like appendages at base of each filament; ovary 1-2-loculed; stigmas 2; fr. a capsule.

1. Styles separate to near base.
  2. Plants annual, prostrate .... **Nama**
  2. Plants perennial.
    3. Plants ill-scented perennials with crowded lvs .... **Turricula**
    3. Plant shrubby .... **Eriodictyon**
1. Styles ± united, not separate to near base.
  4. Plants perennial.
    5. Calyx-lobes similar.
      6. Lvs. entire; fls. solitary in lf.-axils; corolla hairy within .... **Hesperochiron**
      6. Lvs. deeply divided or lobed.
        7. Ovary 1-loculed; lvs. mostly basal .... **Hydrophyllum**
        7. Ovary appearing 2-loculed; lvs. mostly cauline .... **Phacelia**
    5. Calyx-lobes of 2 kinds, the outer 3 heart-shaped, the inner narrow ..... **Tricardia**
  4. Plants annual.
    8. Herbage sticky and scented; ovules borne on both surfaces of placentae. **Eucrypta**
    8. Herbage not sticky or scented; ovules borne only on front of placentae.

9. Ovary 1-loculed; infl. not a clearly marked cyme.
   10. Stems recurved-prickly; caps. prickly or bristly....................... **Pholistoma**
   10. Stems not recurved-prickly; caps. pubescent............................**Nemophila**
9. Ovary falsely 2-loculed by intrusion of placentae; infl. an evident coiled cyme.
   11. Stamens attached to corolla at different levels.................................**Nama**
   11. Stamens attached at same level near base of corolla.
      12. Corolla mostly deciduous, white to blue or purplish.................**Phacelia**
      12. Corolla persistent, pendulous, yellowish or pinkish......... **Emmenanthe**

Emmenanthe (p. 315)

Glandular-sticky annuals; fls. pendulous; corolla yellowish to pinkish, bell-shaped to tubular; stamens included; caps. 1-loculed, appearing 2-loculed by intrusion of placentae. One species: **E. penduliflora**

Key to varieties:
   Corolla yellowish .... var. **penduliflora**
   Corolla pink, drying white..var. **rosea**

Eriodictyon (p. 315)

Aromatic evergreen shrubs with shredding bark and open ± weedy growth from underground rootstocks; lvs. mostly crowded toward ends of branches; fls. in terminal branched scorpioid cymes; corolla deciduous, funnelform to bell-shaped; style divided to base; caps. falsely 2-loculed. About 9 species, 8 in CA.

1. Lvs.. densely hairy on upper surface.
   2. Calyx & corolla glandular; filaments glabrous; corolla 2-5 mm long. **E . tomentosum**
   2. Calyx & corolla not glandular; filaments hairy; corolla 5-16 mm long. **E. crassifolium** Benth. var. **nigrescens** Brand. (incl. var. *denudatum*)
1. Lvs. glabrous to sparsely hairy on upper surface.
   3. Twigs glabrous to sparsely hairy; corolla sparsely hairy................**E. californicum**
   3. Twigs densely hairy; corolla densely hairy...........**E. crassifolium** var. **nigrescens**

Eucrypta (p. 316)

Sticky scented hispid annual; lower lvs. opposite, pinnately divided; fls. in open cymes; corolla bell-shaped, deciduous; stamens included; style shortly bifid; caps. enveloped by stellate-spreading calyx. Two species, both in CA.

1. Lvs. 2-3-pinnatifid, mostly opposite; corolla yellowish-white.**E. chrysanthemifolia**
   2. Lvs. 7-9-lobed; corolla 2-3 mm broad......................................var. **bipinnatifida**
   2. Lvs. 9-13-lobed; corolla 6-8 mm broad........................... var. **chrysanthemifolia**
1. Lvs. 1-pinnate, alternate; corolla white to bluish..................................**E. micrantha**

Hesperochiron (p. 316)

Stemless perennial herb; lvs. in rosette, entire; fls. solitary in lf.-axils; corolla white to bluish, funnelform; stamens included; style shortly 2-cleft at apex. Two species, both in CA. One species in Kern County: **H. pumilus**

Hydrophyllum (p. 316)

Perennial herb; lvs. mostly basal, pinnatifid; fls. in subcaptiate cymes; corolla deciduous, bell-shaped, white to violet; stamens exserted; style shallowly bifid; caps. 1-loculed. Eight species, 4 in CA. One species in Kern County: **H. occidentale**

Nama (p. 316)
(incl. *Lemmonia*)

Low branching annual with alternate, entire lvs.; fls. solitary in axils and numerous at leafy tips; corolla tubular to bell-shaped; stamens attached at different levels; style ± bifid. About 55 species, 11 in CA.

Corolla 7-14 mm long, blue-purple to rose.
N. demissum var. **demissum** (incl. var. *deserti*)
Corolla 1-3 mm long, white to pale pink.
N. californicum (Gray) Bacon (*Lemmonia californicum*)

Nemophila (p. 316)

Diffusely branched weak stemmed annual; lvs. mostly opposite, toothed to pinnately divided; fls. solitary; calyx usually with auricles (little pointed lobes) between the sepals; corolla wheel-shaped to bell-shaped, deciduous; style ± deeply bifid; caps. 1-loculed. About 11 species, 8 in CA.

1. Corolla 1-4 cm broad; lvs. opposite.
  2. Fls. white with purple spot on each lobe .... **N. maculata**
  2. Fls. blue to white, without purple spot .... **N. menziesii**
    3. Upper lvs. like lower, divided into 5-13 lobes .... var. **menziesii**
    3. Upper lvs. not like lower, subentire to shallowly few-lobed ...... var. **integrifolia**
1. Corolla <1 cm broad.
  4. Lvs. all opposite.
    5. Auricles 1/3 as long as sepals; corolla white to pale blue, lobes mostly with dark blotch at base.
      6. Lvs. oblong or oval, deeply divided .... **N. pedunculata**
      6. Lvs. spatulate, shallowly lobed. .... **N. spatulata**
    5. Auricles minute or obsolete.
      7. Lvs. divided almost to midrib; corolla wheel-shaped, 5-12 mm broad. **N. pulchella**
        8. Corolla deep blue to violet .... var. **pulchella**

8. Corolla white.. var. **fremontii**
7. Lvs. not so deeply divided; corolla bell-shaped, 3-5 mm broad, white to bluish...**N. parviflora** var. **quercifolia**
4. Lvs. opposite below, alternate above.
9. Lvs. divided almost to midrib; corolla wheel-shaped, 5-12 mm broad, white. **N. pulchella** var. **gracilis**
9. Lvs. not so deeply divided; corolla bell-shaped, 2-4 mm broad, white to bluish. **N. parviflora** var. **parviflora**

Phacelia (p. 317)

Annual to perennial pubescent herbs; lvs. entire to bipinnate; fls. in dense to lax (in fr.) coiled cymes; corolla wheel-shaped to bell-shaped, scales attached in pairs at base of filaments; style bifid at apex to divided almost to base; caps. 1- to almost 2-loculed. About 175 species 94 in CA.

1. Plants perennial.
2. Lvs. pinnate into toothed or pinnatifid lobes; corolla dirty-white to bluish, deciduous, 6-8 mm long. **P. ramosissima** Lehm. var. **latifolia** (Tor.) Cronq. (var. *suffrutescens*)
2. Lvs. entire to pinnate with entire lobes or lfts.
3. Calyx-lobes broadly lanceolate to ovate, overlapping in fr.; corolla white, 4-7 mm long..**P. imbricata** var. **imbricata** (*P. stimulans*)
3. Calyx-lobes linear-lanceolate, not overlapping in fr.
4. Lvs. greenish, cauline mostly entire; corolla 4-6 mm long, lavender to whitish..**P. mutabilis**
4. Lvs. grayish, cauline mostly pinnatifid; corolla 7-9 mm long, white.. **P. egena**
1. Plants annual.
5. Ovules 2-8 (-16); seeds pitted, not corrugated.
6. Lvs. pinnately toothed to compound, their divisions further toothed or divided.
7. Seeds 10-40, small, <1 mm long; corolla 10-12 mm long, light blue to purplish.. **P. douglasii**
7. Seeds 1-8 (16), larger, 1.5-3 mm long; corolla mostly smaller.
8. Calyx much enlarged and papery in fr.; corolla 8-10 mm long, blue with pale center; stamens 9-13 mm long.. **P. ciliata**
8. Calyx not much enlarged or papery in fr.
9. Corolla blue or violet; calyx <10 mm long in fr.
10. Calyx-lobes oblanceolate to obovate, 4-5 mm long.
11. Stamens long-exserted, 10-14 mm long; corolla violet-purple. **P. crenulata**
11. Stamens not long-exserted; corolla blue to dirty white ....**P. distans**
10. Calyx-lobes linear to lance-linear, 6-8 mm long; corolla blue; stamens exserted..**P. tanacetifolia**
9. Corolla white to lavender; calyx 10 mm long in fr.

12. Fls. many, crowded, dirty-white; stems mostly not purple; cismontane .... **P. cicutaria**
12. Fls. few, remote, lavender; stems mostly purple; deserts. **P. vallis-mortae** (incl. var. *heliophila*)
6. Lvs. entire to pinnate, then with entire divisions.
13. Plants glandular-pubescent throughout.
14. Corolla tubular, white to pinkish, yellow below; lvs. roundish. **P. rotundifolia**
14. Corolla bell-shaped, long, lavender to violet; lvs. linear to ovate.
15. Lower lvs. entire, 1-5 cm long; corolla-tube >4.5 mm long.
16. Calyx-lobes obovate to narrow-spatulate, 5-15 mm long in fr. **P. purpusii**
16. Calyx-lobes linear to oblanceolate, 4-7 mm long in fr. **P. mohavensis**
15. Lower lvs. 1-6 pinnate, 2-7 cm long; corolla-tube <4.5 mm long. **P. novenmillensis** Munz
(Scattered along Pacific Crest Trail, Spanish Needle to Owens Peak. Shevock)
13. Plants not glandular except sometimes in infl.
17. Pedicels elongating in fr., spreading or recurving; corolla bluish to violet with white throat.
18. Corolla 5-8 mm long; lvs. mostly entire .... **P. curvipes**
18. Corolla 9-15 mm long; lower lvs. pinnate .... **P. davidsonii**
17. Pedicels not elongating, erect or ascending.
19. Corolla 6-15 mm long, lavender to violet .... **P. congdonii**
19. Corolla 3-6 mm long.
20. Style 1-2 mm long; stamens 2-4 mm long; fls. lavender. **P. austromontana**
20. Style 4-8 mm long; stamens 6-8 mm long; fls. violet. **P. humilis** var. **dudleyi**
5. Ovules many; seeds honeycombed to corrugate.
21. Corolla tubular, 5 mm long, white to pinkish; lvs. roundish, 5-20 mm wide. **P. rotundifolia**
21. Corolla bell-shaped.
22. Lvs. entire to crenate-dentate, not deeply lobed.
23. Corolla 5-7 mm long, violet; lvs. 20-25 mm wide .... **P. pachyphylla**
23. Corolla 15-22 mm long, bright blue .... **P. nashiana**
22. Lvs. pinnate to pinnatifid.
24. Corolla 3-5 mm long, lavender to whitish with pale yellow tube. **P. affinis**
24. Corolla 8-16 mm long.
25. Corolla 10-12 mm long, light blue without yellow tube...... **P. douglasii**
25. Corolla 9-16 mm long, with yellow tube.
26. Corolla bright blue to deep lavender; lvs. pinnate .... **P. fremontii**
26. Corolla purplish; lvs. bipinnate .... **P. bicolor**

Pholistoma (p. 320)

Succulent prostrate or reclining brittle-stemmed annual often with prickly stem-angles; lower lvs. opposite, upper alternate, pinnate, petioles often winged and clasping; calyx-lobes with or without auricles (little pointed lobes); corolla subrotate; stamens included; style cleft less than halfway; caps. globose. Three species, all in CA.

Calyx with auricles between lobes; stems with down-pointing prickles on angles; corolla 10-30 cm broad, lavender to purple with darker markings.
P. auritum var. **auritum**
Calyx lacking auricles; stems lacking down-pointing prickles; corolla 4-10 mm cm broad, white, usually with purple spot on each lobe........................**P. membranaceum**

Tricardia (p. 320)

Perennial herb branched from base; lvs. entire, in basal rosette; fls. purplish; calyx in 5 unequal lobes, 3 outer large and heart-shaped, 2 inner linear; stamens included; style 2-cleft. One species: **T. watsonii**

Turricula (p. 320)

Very glandular stout erect ill-scented somewhat shrubby perennial; lvs. crowded, alternate, sessile, lanceolate; fls. many in terminal branched coiled cymes; corolla purple, deciduous; stamens included; style deeply divided. (Plant causes severe dermatitis to some people.) One species: **T. parryi**

HYPERICACEAE
(Guttiferae)

Hypericum (p. 284)

Glabrous mostly perennial plants with opposite entire sessile lvs.; fls. regular 5-merous with numerous stamens; petals yellow to yellow-orange, deciduous; ovary superior with 3 styles; fr. a capsule. About 350 species, 6 in CA.

1. Petals not longer than sepals; caps. 1-celled......................................**H. anagalloides**
1. Petals much longer than sepals; caps. 3-celled.
   2. Sepals linear-deltoid, acuminate; stems with sterile shoots in lower lf.-axils.
   H. perforatum
   2. Sepals ovate to obovate, acute to obtuse; stems not as above.
   H. formosum Kunth var. **scouleri** (Hook.) Coulter (*H. scouleri*)

IRIDACEAE

Perennial herbs with flattened stems and mostly basal linear or sword- shaped lvs.; fls. terminal, showy, regular; ovary inferior.

Infl. an umbel; perianth-segments alike, <2 cm long ....**Sisyrinchium**
Infl. not an umbel; perianth with 3 erect and 3 spreading or drooping segments, >2 cm long .... **Iris**

Iris (p. 207)

Perennial herbs with erect stems with 1-several fls. at the top and large flattened basal lvs.; perianth with 3 outer broad, spreading or drooping sepals, the 3 inner narrower and erect; ovary 3-loculed; style with 3 petal-like branches that arch over stamens. About 150 species, 14 in CA.

1. Lvs. 2-9 mm wide; stem 2-5 dm tall.
   2. Fls. pale lilac to whitish with lilac veins; stigma 2-lobed; plants 2-5 dm tall. **I. missouriensis**
   2. Fls. yellow to cream to lavender; stigmas entire; plants 0.5-3 dm tall. **I. hartwegii** var. **hartwegii**
1. Lvs. 15-25 mm wide; stem 4-15 dm tall; fls. bright yellow with orange spot near base of sepals.. **I. pseudacorus**

Sisyrinchium (p. 207)

Tufted perennials with compressed stems and narrow grass-like lvs.; infl. an umbel subtended by 2 bracts; perianth-segments all alike, blue to purplish-red. About 70 species, 6 in CA.

1. Fls. yellow... **S. elmeri** Greene ("Yellow-eyed-grass", uncommon in meadow near Tiger Flat, Greenhorn Mtns. Shevock 10627)
1. Fls. blue to purple.
   2. Stem with 1 or more lf.-bearing nodes, each with 2 or more peduncles..**S. bellum**
   2. Stem leafless, simple, with single terminal spathe.
      3. Pedicels glabrous; perianth 10-15 mm long........... **S. idahoense** var. **idahoense**
      3. Pedicels hairy; perianth 9-10 mm long .... **S. halophilum**

ISOETACEAE

Isoetes (p. 168)

Small ± aquatic "onion-like" plants with corm crowned with many sedge-like elongate lvs. with expanded bases bearing sporangia. About 150 species, 6 in CA. One species in Kern County: **I. nuttallii**

JUGLANDACEAE

Juglans (p. 210)

Aromatic deciduous tree with alt. pinnately compound lvs.; fls. imperfect, staminate in catkins, pistillate with 2 plumose stigmas; fr. a walnut. About 21 species, 2 in CA. One species introduced in Kern County: **J. nigra**

JUNCACEAE

Perennial to annual herbs of moist places; stems mostly simple, tufted, terete (rounded) or compressed; lvs. alternate, sheathing and grass-like; infl. compound panicle; fls. small, regular, 6-parted into 3 outer and 3 inner papery segments; style with 3 long stigmas; fr. a capsule.

Lf.-sheaths open; lvs. stiff; stems mostly with spongy pith .... **Juncus**
Lf.-sheaths closed; lvs. soft, flat; stems hollow .... **Luzula**

Juncus (p. 199)

Glabrous perennials or annuals with open lf.-sheaths, some with auricles (small lobes) at summit; perianth greenish to brownish purple. About 255 species, 56 in CA.

1. Infl. appears lateral, the lowest bract terete and exactly like a continuation of the stem.
  2. Stamens 3; plants 6-13 dm tall; stems terete .... **J. effusus**
  2. Stamens 6.
    3. Upper lf.-sheaths usually with well developed blades; stems compressed, ± twisted, 2-6 dm high.. **J. mexicanus**
    3. Upper lf.-sheaths without blades; stems mostly terete, wiry, 3-8 dm high. **J. balticus**
1. Infl. appears terminal.
  4. Lf.-blades flat, the surface facing the stem without internal partitions.
    5. Fls. single, each subtended by 2 bractlets.
      6. Annuals; stems branched at base; infl. large................................. **J. bufonius**
        7. Caps. oblong, 3-5 mm long; perianth 4-6 mm long................ var. **bufonius**
        7. Caps. ± globose, 2-3 mm long; perianth 3-4 mm long. var. **occidentalis** Herm. (*J. sphaerocarpus*)
      6. Perennial; stems simple; infl. smaller ....**J. tenuis**
    5. Fls. in heads which are subtended by bracts.
      8. Annuals with bristly lf.-blades; stamens 3 ....**J. kelloggii**
      8. Perennials with flat lf.-blades; stamens 6.
        9. Plants 2-4 dm tall; auricles mostly lacking; heads 1-several, 6-10-fld. **J. orthophyllus**
        9. Plants 2-10 dm tall; auricles 1.5-3 mm long; heads 8-25, 3-5-fld. **J. macrophyllus**

4. Lf.-blades terete, or if flat, the edge toward the stem with ± evident internal partitions.
  10. Lf.-blades terete or ± flattened.
    11. Stem and lf.-surfaces roughened .... **J. rugulosus**
    11. Stem and lf.-surfaces smooth.
      12. Anthers shorter than filaments; stems 4-10 dm tall.................... **J. torreyi**
      12. Anthers longer than filaments.
        13. Caps. tapering into long beak, well exserted when mature; plants 4-10 dm tall. .... **J. dubius**
        13. Caps. lacking long beak, not well exserted; plants slender, 1.5-5 dm tall. .... **J. nevadensis**
  10. Lf.-blades flattened with edge facing stem.
    14. Anthers longer than filaments; styles mostly longer than ovary at anthesis.
      15. Caps. shorter than perianth; perianth purple-brown........... **J. macrandrus**
      15. Caps. longer than perianth; perianth greenish or with brown tinge. **J. oxymeris**
    14. Anthers shorter than filaments; style shorter.
      16. Stamens usually 6; lvs. 3-12 mm wide; plants 5-9 dm tall..... **J. xiphioides**
      16. Stamens usually 3; lvs. 2-5 mm wide; plants 2-5 dm tall........ **J. ensifolius**

Luzula (p. 200)

Perennial with slender hollow simple stems and flat grass-like soft lvs.; lf.-sheath closed; fls. clustered; stamens 6. About 80 species, 6 in CA. One species in Kern County: **L. comosa**

JUNCAGINACEAE
(incl. Lilaeaceae)

Emergent aquatic herbs with linear lvs.; infl. mostly terminal raceme; perianth segments 6 or 0, greenish.

Perianth parts 0 or 1; lvs. 1-5 mm wide, flat; some fls. unisexual; racemes < lvs. **Lilaea**

Perianth parts 3-6; lvs. ± 1 mm wide, subcylindric; fls. bisexual; racemes > lvs. **Triglochin**

Lilaea (p. 174)

Aquatic stemless annual herb with fibrous roots and basal spongy linear lvs. 5-30 cm long; infl. of 2 types, some basal and some in spikes on scapes 6-20 cm long; perianth purplish, reduced to bract subtending fl. One species: **L. scilloides**

Triglochin (p. 174)

Perennial marsh herb with linear basal lvs.; fls. small, in terminal raceme 5-15 cm long; perianth-segments concave, inconspicuous; stigmas 6. About 12 species, 4 in CA. One species in Kern County: **T. concinna** Davyvar. **debilis** (Jones) Howell

LAMIACEAE

(Labiatae by some authors)

Aromatic herbs or shrubs with opposite lvs. and (usually) 4-angled stems; fls. mostly irreg., clustered in bracteolate inflorescences; corolla usually 2-lipped; stamens mostly 4, sometimes 2 aborted; ovary 4-lobed, forming single nutlets.

1. Nutlets united; style not basal; stamens well-exserted.......................... **Trichostema**
1. Nutlets distinct; style basal (attached to base of nutlets).
    2. Calyx 2-lipped, the lips entire.
        3. Shrub; calyx inflated in fruit.......................... **Salazaria**
        3. Herb; calyx with crest on top.......................... **Scutellaria**
    2. Calyx regular or 2-lipped, the lips not entire.
        4. Corolla strongly 2-lipped.
            5. Calyx with spinescent hooked teeth.......................... **Marrubium**
            5. Calyx lacking hooked teeth.
                6. Annuals.
                    7. Fertile stamens 2.......................... **Salvia**
                    7. Fertile stamens 4.
                        8. Fls. in axillary and terminal clusters.......................... **Lamium**
                        8. Fls. in dense spikes, 1-3 cm broad.......................... **Pogogyne**
                6. Perennials.
                    9. Shrubs.......................... **Salvia**
                    9. Herbs.
                        10. Plants white-woolly; corolla white with dark veins.......................... **Stachys**
                        10. Plants subglabrous; corolla rose-violet.......................... **Agastache**
        4. Corolla regular or nearly so.
            11. Fls. in terminal heads; dry places.......................... **Monardella**
            11. Fls. in axillary whorles; wet places.
                12. Fls. white; fertile stamens 2.......................... **Lycopus**
                12. Fls. lavender to purple; fertile stamens 4.......................... **Mentha**

Agastache (p. 327)

Tall perennial herbs with serrate ovate lvs.; fls. in dense terminal spikes; corolla rose-violet, 10-15 mm long; stamens 4. About 22 species, 2 in CA. One species in Kern County: **A. urticifolia**

Lamium (p. 328)

Sparsely pubescent annual with broadly ovate to roundish crennate lvs.; corolla purple-red, 12-16 mm long with very slender tube. About 40 species, 2 introduced in CA. One species in Kern County: **L. amplexicaule**

Lycopus (p. 328)

Stoloniferous glabrous perennial herb with lobed or pinnatifid lvs.; fls. congested in lf.-axils; corolla 2 mm long, white; stamens 2. About 14 species, 3 in CA. One species in Kern County: **L. americanus**

Marrubium (p 328)

Perennial herb with "wrinkled" toothed woolly lvs.; fls. small, in dense whorles; calyx cylindric, with 5-10 recurved bristle-tips (that get caught in your socks); stamens 4. About 30 species, 1 introduced in CA. One species in Kern County: **M. vulgare**

Mentha (p. 328)

Aromatic perennial herbs with toothed punctate lvs.; fls. in axillary clusters or terminal spikes; corolla 2-lipped, lower lip 3-lobed; stamens 4. About 25 species, 5 introduced in CA.

1. Fl.-whorles exceeded by subtending lvs .... **M. arvensis**
1. Fl.-whorles in terminal spikes.
  2. Plants villous-woolly.................................. **M. suaveolens** Ehrh. (*M. rotundifolia*)
  2. Plants mostly glabrous.
    3. Lvs. subsessile; spikes narrow, interrupted.................... **M. spicata** var. **spicata**
    3. Lvs. petioled; spikes dense, thick........................... **M. Xpiperita** L. (*M. citrata*)

Monardella (p. 328)

Annual or perennial herbs with smallish entire to serrate lvs. on rounded stems; fls. in terminal heads subtended by broad invol.-bracts; corolla-lobes linear to oblong; stamens 4. About 20 species, all in CA.

1. Perennials, mostly woody at base,
  2. Stems and lvs. mostly glabrous.. **M. glauca** Greene
  2. Stems and lvs. silvery with dense minute hairs.
    3. Lf.-margin wavy, ± entire, <10 mm long.
       M. beneolens Shevock, Ertter & Jokerst
    3. Lf.-margin not wavy, entire, 10-30 mm long.
      4. Outer bracts of infl. absent; middle bracts papery .... **M. linoides**
        5. Lvs. 15-25 mm long... ssp. **linoides**
        5. Lvs. 10-15 mm long...ssp. **oblonga**

4. Outer bracts spreading, narrowly ovate; middle bracts papery.
M. odoratissima ssp. **odoratissima** (incl. ssp. *parvifolia*)

1. Annuals.
   6. Bracts white-margined; calyx-teeth white-tipped.
      7. Bracts acute, narrowly white-margined; calyx 13-veined ....**M. candicans**
      7. Bracts abruptly acuminate with white tips; calyx 15-veined..................**M. exilis**
   6. Bracts not at all white; calyx-teeth not white-tipped.
      8. Bracts acute, with prominent secondary veins.............................. **M. lanceolata**
      8. Bracts abruptly acuminate without conspicuous secondary veins..... **M. breweri**

Pogogyne (p. 329)

Small aromatic annual with elliptic to oblong entire to serrate glabrous lvs.; fls. congested in axils of linear, bristly bracts; corolla 1-2 cm long, lavender with yellow marks in throat; stamens 4. About 7 species, all in CA. One species in Kern County: **P. douglasii**

Salazaria (p. 329)

Low dense divaricately branched shrub with spinescent branchlets; calyx equally 2-lobed, lobed entire, becoming inflated and globular in fruit. One species: **S. mexicana**

Salvia (p. 329)

Herbs or shrubs with aromatic opposite lvs.; fls. in whorles in interrupted spikes; corolla strongly 2-lipped, the upper concave, the lower spreading and 3-lobed; stamens 2, anthers widely separate on long thin connective. About 900 species, 19 in CA.

1. Annuals; both connective-branches bearing fertile anthers.
   2. Lvs. spiny, tomentose; corolla lavender, 25-35 mm long .... **S. carduacea**
   2. Lvs. not spiny or tomentose; corolla blue, 12-16 mm long.............**S. columbariae**
1. Shrubs; lower connective branch obsolete.
   3. Stamens not exserted; corolla-tube whitish; erect shrub.......................**S. mellifera**
   3. Stamens exserted; corolla-tube not white; erect or prostrate shrubs.
      4. Lf.-base ± cordate; hairs branched.. **S. leucophylla**
      4. Lf.-base not cordate; hairs simple.
         5. Corolla blue-violet to rose, tube 13-23 mm long; bract 10-20 mm long, papery...**S. pachyphylla** Munz (Spanish Needle, Owens Peak, Skinner Peak, Scodie Mtns. Shevock 9810)
         5. Corolla blue (rarely purple-rose), tube 6-13 mm long; bract 5-12 mm long, soft-hairy.............. **S. dorrii** (Kell.) Abrams var. **pilosa** (Gray) Strach. & Rev.

Scutellaria (p. 330)

Low herbaceous perennials; fls. 1-3 in axils; calyx swollen, 2-lipped, lips entire, top of calyx-tube with crest-like projection; stamens 4, in 2-pairs. About 300 species, 8 in CA.

1. Mid and upper petioles 1-3 mm long; fls. white............**S. bolanderi** ssp. **bolanderi**
1. Petioles 5-30 mm long; fls. blue-violet.
   2. Lvs. dentate; corolla 15-20 mm long...**S. tuberosa**
   2. Lvs. entire; corolla 24-30 mm long........**S. siphocampyloides** Vatke (*S. austinae*)

Stachys (p. 330)

White-woolly perennial herbs with cordate serrate lvs.; fls. in terminal spikes; corolla white with purple veins; stamens 4. About 300 species, 8 in CA. One species in Kern County: **S. albens**

Trichostema (p. 330)

Strong-scented annuals; corolla blue or lavender, 5-lobed with slender tube; stamens 4, exserted. About 17 species, 10 in CA.

1. Stamens 10-20 mm long.
   2. Lvs. lanceolate, at least 3 times longer than wide; corolla-tube 5-10 mm long. **T. lanceolatum**
   2. Lvs. ovate, not more than 2 times as long as wide; corolla-tube 2-6 mm long. **T. ovatum**
1. Stamens 2-7 mm long.
   3. Corolla 3 mm long; stamens 5-6 mm long; lvs. 15-25 mm long. **T. austromontanum**
   3. Corolla 5-6 mm long; stamens 3-5 mm long; lvs. 20-30 mm long .... **T. oblongum**

LAURACEAE

Umbellularia (p. 239)

Pungently aromatic evergreen shrub or small tree with alternate, entire lvs.; fls. in simple umbels in upper axils; sepals 6, petaloid; petals none; fr. a drupe. One species : **U. californica**

LEMNACEAE

Very small floating plants with leaf-like stems (fronds) and few rootlets; fls. rare, separate staminate and pistillate, lacking perianth.

1. Rootlets absent; fronds strap-shaped .... **Wolffiella**

1. Rootlets present.
   2. Fronds 5-15-nerved, with several rootlets ....**Spirodela**
   2. Fronds 1-5-nerved, with one rootlet .... **Lemna**

Lemna (p. 198)

Minute plants with flattened fronds, each with 1 rootlet and 2 vegetative pouches. About 13 species, 6 in CA.

1. Outline of frond oblique, faintly 1-nerved, smooth, 1-1.5 mm wide.....**L. valdiviana**
1. Outline of frond symmetrical; minutely bumpy along midnerve.
   2. Fronds oblong-elliptical, 1-nerved, 1-1.5 mm wide.
   L. minuscula Herter (*L. minima*)
   2. Fronds rounded, 3-5-nerved, 2-4 mm wide .... **L. gibba**

Spirodela (p. 198)

Fronds flattened, disc-shaped, 5-15-nerved with 2-several rootlets; vegetative pouches 2, triangular; plants purplish beneath. About 3 species, 2 in CA. One species in Kern County **S. polyrrhiza**

Wolffiella (p. 198)

Minute plants with thin tongue-shaped frond, 2-6 mm long, 1-3 mm wide, rootless. About 9 species, 2 in CA. One species in Kern County: **W. lingulata**

LENNOACEAE

Pholisma (p. 301)

Fleshy root-parasitic herb lacking chlorophyll; part above ground 1-2 dm tall; fls. in dense terminal spike; corolla tubular, purplish with white border. Three species, 2 in CA. One species in Kern County: **P. arenarium**

LILIACEAE

Herbaceous to ± woody plants with alternate, opposite, whorled or basal parallel-veined lvs.; fls. mostly showy; perianth of six mostly petal-like parts; ovary superior; fr. a 3-celled capsule or berry.

1. Plants woody with stiff sword-like lvs. (Agavaceae by some authors)
   2. Perianth 3-13 cm long; fls. all perfect ....**Yucca**
   2. Perianth <1 cm long; some fls. imperfect ....**Nolina**
1. Plants herbaceous.
   3. Fls. in umbel at end of long leafless stem (Amaryllidaceae by some authors).
      4. Perianth-segments mostly distinct.
         5. Each filament with cup-shaped winged basal appendage............... **Bloomeria**

5. Filaments not appendaged.
6. Pedicels subtended by bractlets above main bract; onion-like odor absent. **Muilla**
6. Pedicels lacking bractlets; strong onion-like odor present. .... **Allium**
4. Perianth-segments united into definite tube.
7. Anthers erect, pressed against style; filaments attached near base; stigma ± 3-lobed.
8. Lvs. narrow, rounded on back; stigma obviously 3-lobed; stamens 3. **Brodiaea**
8. Lvs. flat, keeled on back; stigma with 3 small lobes........... **Dichelostemma**
7. Anthers not pressed to style; filaments attached near middle of anther; stamens 6 in number .... **Triteleia**
3. Fls. not in terminal umbel, inflorescence various.
9. Plants with a bulb or corm.
10. Stems leafy, sometimes branched above.
11. Styles 3, separate to base; fls. erect .... **Zigadenus**
11. Style 1, ± lobed at tip.
12. Perianth segments unlike, the outer sepal-like and the inner petal-like. **Calochortus**
12. Perianth segments alike, petal-like.
13. Filaments attached near middle of anthers; perianth curved backwards... **Lilium**
13. Filaments attached near base of anthers; perianth not curved backwards.. **Fritillaria**
10. Stems with lvs. mostly at base; inflorescence a widely branching panicle. **Chlorogalum**
9. Plants lacking bulbs or corms.
14. Stems "fern-like," the needle-like branchlets borne in axils of scale-like lvs. **Asparagus**
14. Stems with true lvs.
15. Panicles 2-6 dm long; lvs. 2-5 dm long .... **Veratrum**
15. Panicle or raceme <2 dm long; lvs. <2 dm long........................ **Smilacina**

Allium (p. 200)

Perennial herbs with long linear basal lvs. from an underground bulb; strong onion-like odor; stems unbranched with terminal umbel subtended by sheath-like bract; fr. a capsule. About 500 species, 51 in CA.

1. Lvs. solitary, round in cross-section; ovary with 6 terminal crests or lobes.
2. Stamens equal to or longer than perianth; perianth pale violet to deep rose with dark midveins.. **A. howellii** var. **howellii**
2. Stamens included, <perianth.
3. Inner perianth parts dentate to jagged. **A.denticulatum** (Traub.) McNeal (*A. fimbriatum* var. *denticulatum*)
3. Perianth parts ± entire.

4. Ovary crests notched to irregular (not dentate); perianth with maroon recurved tips..**A. shevockii** McNeal
(Endemic to Spanish Needle Peak. Shevock 11219)
4. Ovary crests dentate to deeply cut... **A.fimbriatum**
5. Perianth deep red-purple, tips recurved-spreading........... var. **fimbriatum**
5. Perianth pale pink to whitish with pink midveins; tips erect.
var. **mohavense**
1. Lvs. >1, if solitary then strongly flattened.
6. Lvs. flattened.
7. Ovary-crests 6, minutely dentate; perianth parts ± curved inward.
A. campanulatum
7. Ovary-crests 0-3, smooth; perianth parts not curved inward.
8. Stems 50-100 cm tall; perianth rose to white....................... **A. validum** Wats.
(Wet meadows near Tiger Flat. Shevock 10534)
8. Stems <50 cm tall; perianth various.
9. Perianth white........**A. obtusum** Lemmon var. **obtusum** (*A. tribracteatum*)
9. Perianth rose-purple to pink.
10. Stamens well-included; perianth ± pink............... **A. cratericola** Eastw.
(Piute Mtn. Road above Bodfish Gap. McNeal)
10. Stamens exserted or ± = to perianth; perianth dull purple with dark midveins... **A. burlewii**
6. Lvs. round in X-section.
11. Perianth parts white to pale pink, erect to spreading, 5-10 mm long.
12. Perianth becoming translucent & folded over ovary................. **A. hyalinum**
12. Perianth not becoming translucent or folded over ovary.
A. lacunosum S. Watson
13. Stems 10-25 cm long; infl. dense.......... var. **kernensis** McNeal & Ownby
13. Stems 15-35 cm long; infl. open.
var **davisae** (M.E. Jones) McNeal & Ownby (*A. davisae*)
11. Perianth parts rose to purple, often recurved, 8-15 mm long.
14. Inner perianth parts dentate, curled.. **A. crispum**
14. Inner perianth parts entire, not curled......... **A. peninsulare** var. **peninsulare**

Asparagus (p. 201)

Perennial herb with much-branched stems with small needle-like branchlets borne in axils of scale-like lvs.; fls. small, greenish yellow; perianth-segments alike; fr. a red berry. About 300 species, 2 introduced in CA. One species in Kern County: **A. officinalis**

Bloomeria (p. 201)

Perennial herbs with linear keeled basal lvs.; Fls. several in loose umbels subtended by membranous bracts; perianth segments yellow with dark median line, distinct; fr. a caps. Two species, both in CA. One species in Kern County: **B. crocea** (incl. vars. *aurea & montana*)

Brodiaea (p. 202)

Perennial herbs with long grass-like lvs. from an underground bulb-like corm; fls. in terminal umbel subtended by papery bracts; perianth-segments petal-like, fused at base, violet; fr. a capsule. About 14species, all in CA.

Scape 5-25 cm tall; sterile stamens bent inwards, sometimes covering stamens.
B. coronaria

Scape 3-7 cm tall; sterile stamens erect.
B. terrestris Kell. ssp. **kernensis** (Hoover) Niehaus (*B. coronaria* var. *kernensis*)

Calochortus (p. 202)

Glabrous perennial herbs from underground bulbs; lvs. usually linear, the basal solitary and large, the cauline reduced; fls. conspicuous; perianth-segments well-differentiated into 3 petals and 3 sepals; fr. a 3-angled caps. About 65 species, 42 in CA.

1. Fls. closed, nodding; petals deep rose .... **C. amoenus**
1. Fls. open, bowl-shaped, erect.
  2. Surface of gland on petal naked but bearded above; petals fringed, white to cream with some rose............................. **C. westonii** Eastw. (*C. coeruleus* var. *westonii*)
  2. Surface of gland hairy.
    3. Gland not depressed, not surrounded by a membrane
      4. Infl. with distinct axis; petals white to lavender with brownish spot above gland.. **C. palmeri** var. **palmeri**
      4. Infl. subumbellate with 1-4 fls.
        5. Fls. lavender with purple veins .... **C. striatus**
        5. Fls. various, lacking purple veins.
          6. Gland square or shaped like inverted V, not crescent-shaped; petals white to yellowish or lavender with dark median blotch.
            7. Gland square; petals sometimes also dark red, often with second blotch above the first... **C. venustus**
            7. Gland shaped like inverted V; blotch on petals surrounded by zone of bright yellow... **C. supurbus**
          6. Gland crescent-shaped.
            8. Petals white to purplish with median red-brown blotch surrounded by pale yellow; gland double crescent... **C. vestae**
            8. Petals deep yellow, often with median blotch; gland a single crescent. **C. luteus**
    3. Gland depressed, surrounded by membrane.
      9. Fls. white to purplish.
        10. Sepals with purplish spot; petals narrowly obovate.
        C. panamintensis (Ownby) Rev. (Near summit of Owens Peak. Shevock)
        10. Sepals lacking spot; petals wedge-shaped............................ **C. invenustus**
      9. Fls. red-orange to yellow.

11. Fls. brilliant red-orange .... **C. kennedyi** var. **kennedyi**
11. Fls. lemon yellow, often with reddish markings on sepals and transverse line on petals above the gland.
C. clavatus Wats. ssp. **pallidus** (Hoover) Munz

Chlorogalum (p. 204)

Perennial herb with several linear wavy basal lvs. and a tall almost leafless paniculate infl.; fls. bloom in evening; perianth white with greenish to purplish midveins. About 5 species, all in CA. One species in Kern County: **C. pomeridianum** var. **pomeridianum**

Dichelostemma (p. 202)
(part of *Brodiaea*)

Perennial herbs with long narrow lvs. from an underground bulb-like corm; lvs. flat, keeled on lower side; fls. in terminal umbel, each subtended by a bract; perianth-segments petal-like, fused at base. About 12 species, 6 in CA.

Anthers 6; stems erect 30-80 cm tall; fls. violet.
D. congestum (Sm.) Kunth (*B. pulchella*)
Anthers 3; stems twining 50-150 cm long; fls. rose-pink ...... **D. volubile** (Kell.) Heller

Fritillaria (p. 204)

Perennial herb with simple, erect stem; lvs. sessile, linear to ± ovate; fls. on terminal raceme; perianth of 6 distinct petal-like segments with gland at base. About 100 species, 18 in CA.

1. Style not or barely divided; perianth pink to purplish.
   2. Fls. 25-35 mm long; often with red stripes .... **F. striata**
   2. Fls. 12-17 mm long; "rice-like" bulblets present .... **F. brandegei**
1. Style obviously 3-cleft.
   3. Lower lvs. near the ground; fls. greenish-yellow outside, purplish inside, with strong odor. .... **F. agrestis**
   3. Lower lvs. well above ground; fls. mottled greenish-yellow on purple.
   F. pinetorum

Lilium (p. 204)

Perennial herb with simple, tall leafy stems; lvs. narrow, sessile, upper ones whorled; fls. fragrant, nodding, orange to yellow with maroon dots; segments 3-5 cm long, recurved. About 90 species, 12 in CA. One species in Kern County: **L. kelleyanum**

Muilla (p. 205)

Perennial herb form underground bulb-like corm; lvs. long, somewhat rounded in cross section; fls. in terminal umbel subtended by papery bracts; perianth spreading, with 6 slightly united whitish to greenish segments with dark 2-nerved midveins. About 6 species, 4 in CA.

Filaments slender .... **M. maritima**
Filaments flattened and nearly overlapping to form a crown .... **M. coronata**
(Also common on Kern Plateau and Scodie Mtns. Shevock 10796)

Nolina (p. 205)

Lvs. many in a dense ± rosette, narrowly linear, 6-10 dm long, fls. in dense terminal compound panicle with long white bracts subtending main branches. About 25 species, 3 in CA. One species in Kern County: **N. parryi** (incl. ssp. *wolfii*)

Smilacina (p. 205)

Perennial herb with creeping rootstock and simple stems with alternate sessile lvs.; fls. small, white, in terminal racemes; fr. a red-purple berry. About 25 species, 2 in CA.

Infl. a well-developed panicle; stamens longer than perianth...... **S. racemosa** (L.) Link
(Shirley Meadows; Greenhorm Mtns. Shevock 9722)
Infl. a raceme; stamens shorter than perianth.. **S. stellata**

Triteleia (p. 202)
(part of *Brodiaea*)

Perennial herbs with long linear lvs. from underground bulb-like corm; lvs. flat, keeled on lower side; fls. in terminal umbel subtended by distinct bracts; perianth-segments petal-like, fused at base; filaments attached near middle of stamens. About 14 species, 12 in CA.

1. Perianth blue to white, tube 12-25 mm long... **T. laxa** Benth.
1. Perianth cream to golden yellow with dark midvein, tube 4-12 mm long.
   2. Perianth-tube = lobes.
      T. dudleyi Hoover (Near Inspiration Point, Piute Mtns. Shevock)
   2. Perianth-tube shorter than lobes.........**T. ixioides** (Wats.) Greene (*Brodiaea lutea*)
      3. Fls. golden yellow; fl.-tube >6 mm deep .... ssp. **ixioides**
      3. Fls. cream to deep yellow; fl.-tube <6 mm deep.
         4. fls. straw colored to deep yellow; short filaments >4 mm long; anthers yellowish..ssp. **scabra** (Greene) Lenz
         4. Fls. dull yellow; short filaments <4 mm long, anthers blue.
            ssp. **analina** (Greene) Lenz

Veratrum (p. 205)

Tall stout leafy perennial herb with broad, clasping, strongly veined lvs.; fls. greenish-white in large terminal panicle, segments with green gland near base. About 25 species, 4 in CA. One species in Kern County: **V. californicum** var. **californicum**

Yucca (p. 205)

Tree or ± shrubs with narrow stiff linear lvs. with expanded bases and terminal spine; fls. large, pendent, with ± fleshy perianth. About 40 species, 4 in CA.

Tree-like with definite trunk; style with 3 stigmas. **Y. brevifolia** (incl. vars. *brevifolia* & *herbertii*)
Shrub-like, stemless with long terminal infl.; style with capitate stigma. **Y. whipplei** (incl. ssp. *caespitosa*)

Zigadenus (p. 207)

Perennial herb with bulb; stems simple with linear lvs.; fls. greenish to yellowish in terminal panicles; perianth with gland at base. About 18 species, 7 in CA.

1. Stamens longer than perianth; perianth-segments 3-4 mm long, clawed. .**Z. venenosus** var. **venenosus**
1. Stamens not longer than perianth; perianth-segments 6-8 mm long.
   2. Lvs. 8-30 mm wide, 50-70 cm long .... **Z. exaltatus**
   2. Lvs. 6-8 mm wide, 15-20 cm long ....**Z. brevibracteatus**

LIMNANTHACEAE

Limnanthes (p. 275)

Low tender glabrous annual herbs of moist places; lvs. pinnately dissected; fls. 5-merous, regular, white; carpels separating as nutlets. Nine species, 8 in CA. One species in Kern County: **L. montana**

LINACEAE

Annual or perennial herbs with simple entire alternate or opposite lvs.; fl. 5-merous; ovary superior, 2-5 celled; fr. a capsule.

Petals blue, 10-20 mm long; styles 5...**Linum**
Petals pink to white, 2-3 mm long; styles 3... **Hesperolinon**

Hesperolinon (p. 276)
(part of *Linum*)

Slender-stemmed annual; basal lvs. whorled, becoming alternate above; fls. in open cymes on slender pedicels. About 13 species, all in CA. One species in Kern County: **H. micranthum** (Gray) Small

Linum (p. 275)

Glabrous perennial with several densely lfy. stems from a branched caudex; fls. blue, petals early deciduous. About 200 species, 5 in CA. One species in Kern County: **L. lewisii** var. **lewisii**

LOASACEAE

Herbs or ± shrubby plants with rough often stinging hairs; lvs. alternate; fls. regular, 5-merous with numerous stamens; ovary inferior, fr. a capsule.

1. Stamens 5; style entire.. **Petalonyx**
1. Stamens many; style cleft.
    2. Stamens inserted below the petals; style 3-cleft to entire.................... **Mentzelia**
    2. Stamens inserted on the petals; style 5-cleft.. **Eucnide**

Eucnide (p. 287)

Rounded bush 3-6 dm high, very hispid with barbed hairs; lvs. ovate, toothed; petals cream, 2-4 cm long. Eight species, two in CA. One species in Kern County: **E. urens**

Mentzelia (p. 287)

Annual to perennial herbs covered with tenacious barbed hairs; lvs. alternate, brittle in age; stamens numerous, inserted below petals; fr. a turbinate or cylindric capsule. About 50 species, 28 in CA.

1. Outer stamens with narrow filaments.
    2. Perennial; petals 5-8 cm long.. **M. laevicaulis**
    2. Annuals; petals <3 cm long.
        3. Seeds angular with grooves along 3 sides; caps. not narrowed at base.
            4. Petals ovate; lvs. deeply lobed; below 5000 feet ....**M. affinis**
            4. Petals broadly oval; lvs. entire or dentate; above 4500 feet. **M. dispersa** (incl. var. *obtusa*)
        3. Seeds not regularly grooved, irregularly angled or round; caps, narrowed toward base (Note: the following species form a polyploid complex which can only be distinguished with difficulty; up to 5 species may grow together.).
            5. Floral bracts lobed; all capsules erect.
                6. Bracts without white area at base; petals yellow..................... **M. albicaulis**

6. Bracts with white area at base; petals orange.
  7. Petals 6-20 mm long; styles 6-11 mm long.
    8. Stems greenish; lvs. entire to pinnatifid; petals 12-20 mm long. **M. gracilenta**
    8. Stems whitish; lvs. mostly pectinately pinnatifid; petals 6-10 mm long. **M. pectinata**
  7. Petals 2-7 mm long; styles 2-7 mm long.
    9. Fls. terminal, crowded, ± concealed by bracts.................... **M. congesta**
    9. Fls. not as above.
      10. First caps. 9-17 mm long; cauline lvs. mostly entire, linear; above 4500 ft.. **M. montana**
      10. First caps. 14-30 mm long; lvs. never entire; below 5000 ft. **M. veatchiana**
5. Floral bracts usually entire; oldest caps. recurved.
  11. Oldest caps. recurved >90 degrees; seeds rounded.
    12. Sepals <8mm long; petals 4-15 mm long................................ **M. nitens**
    12. Sepals >8 mm long; petals 12-24 mm long. **M. eremophila** (Jeps.) Thomp. & Roberts (Red Rock Canyon, Twisselmann)
  11. No caps. recurved >90 degrees; seeds angled........................ **M. albicaulis**
1. Outer stamens with broad filaments.
  13. Bracts scarious with conspicuous white area and narrow green margin. **M. involucrata**
  13. Bracts green without white area.
    14. Filament-teeth shorter than anther stalk. **M. tridentata** (Davidson) Thom. & Roberts (*M. tricuspis* var. *brevicornuta*)
    14. Filament-teeth longer than anther stalk................................**M. tricuspis** Gray (Red Rock Canyon, Twisselmann T2408, T15408)

Petalonyx (p. 288)

Somewhat shrubby herbs, scabrous with short barbed hairs; fls. small, whitish, terminal; stamens 5; fr. a small oblong caps. Five species, 3in CA. One species in Kern County: **P. thurberi** ssp. **thurberi**

LYTHRACEAE

Herbs with alternate or opposite lvs., no stipules; fl.-tube enclosing ovary, petals 4-6; fr. a capsule.

Fl.-tube short, bell-shaped; lvs. opposite; petals 4....................................... **Ammannia**
Fl.-tube cylindric; lvs. alternate; petals 4-6..**Lythrum**

Ammannia (p. 290)

Annual with angled stems, 1-4 dm tall; lvs. linear-lanceolate, heart-shaped; fls. purple, axillary. About 25 species, 2 in CA. One species in Kern County: **A. coccinea**

Lythrum (p. 290)

Slender glabrous herbs with alternate or opposite lvs.; fls. purple, regular. About 35 species, 5 in CA.

1. Infl.-bracts ovate; petals 8-14 mm long; stem ± hairy .... **L. salicaria** L.
(Becoming common in wet pastures, South Fork Valley. Shevock 11966)
1. Infl.-bracts linear; petals 2-8 mm long; stem glabrous.
   2. Perennials; stem erect; branches spreading to ascending; pedicels 1-2 mm long. **L. californicum**
   2. Annuals; stems mostly prostrate; pedicels 0 (fls. sessile) .... **L. hyssopifolia**

MALVACEAE

Herbs or soft-woody shrubs with stellate or branched hairs; lvs. alternate, palmately lobed; fls. regular, 5-merous; stamens numerous, filaments united into a tube surrounding style; fr. various.

1. Fr. a capsule; invol. of large heart-shaped bracts; seeds cottony............. **Gossypium**
1. Fr. of several carpels, usually separating from each other and receptacle when mature.
   2. Style-branches terminating in capitate stigmas.
      3. Carpels sharply differentiated into reticulate basal and wider, winged, smooth apical portion... **Sphaeralcea**
      3. Carpels not so differentiated.
         4. Fls. yellowish; pubescence whitish-stellate .... **Malvella**
         4. Fls. white, pink or purplish; pubescence not very whitish.
            5. Annual; carpels indehiscent... **Eremalche**
            5. Perennial; carpels splitting into halves at maturity............. **Malacothamnus**
   2. Style-branches filiform, longitudinally stigmatic.
      6. Involucel of 3-9 connate bractlets.. **Alcea**
      6. Involucel of 0-3 distinct bractlets.
         7. Bractlets of involucel 3; weedy annual .... **Malva**
         7. Bractlets 0 or 1; native annuals or perennials..................................... **Sidalcea**

Alcea (p. 282)
(*Althaea*)

Biennial, 2-3 m tall; lvs. round-cordate, 5-7-lobed; fls. 6-10 cm wide in elongate terminal spicate raceme, of many colors. About 60 species, 1 introduced in CA. One species in Kern County: **A. rosea** L.

Eremalche (p. 282)

Low annual herbs with orbicular or palmately-parted, stellate-pubescent lvs.; fls. in upper lf.-axils, white to rose-purple; stigmas capitate, carpels 10-40, indehiscent, reticulate or ridged on back. Three species, all in Kern County.

1. Lvs. scalloped; petals with dark spot .... **E. rotundifolia**
1. Lvs. 5-4-lobed; petals not spotted.
   2. Petals 5-6 mm long; calyx 3-5 mm long .... **E. exilis**
   2. Petals 10-30 mm long; calyx 7-12 mm long .... **E. parryi**
      3. Petals white to pale lavender, 10-13 mm long.
      ssp. **kernensis** (Wolf) Bates (*E. kernensis*)
      3. Petals purplish, 12-30 mm long .... ssp. **parryi**

Gossypium (p. 283)

Stout branched annual with broad 3-lobed lvs. 7-15 cm wide; fls. white or yellowish turning pinkish; fr. a dehiscent capsule ("boll") with cottony seeds. Widely cultivated with several species, 1 introduced in CA. One species in Kern County: **G. hirsutum**

Malacothamnus (p. 283)

Shrubs with slender flexuous branches, densely stellate-pubescent; lvs. rounded, lobed; fls. pink or lavender in axillary clusters; carpels several, dehiscent, not rugose. About 20 species, 18 in CA.

Lvs. cordate; petals 12-16 mm long; calyx-lobes 3-7 mm long.
M. fremontii Gray (*M. orbiculatus*)
Lvs. truncate; petals 16-19 mm long; calyx-lobes 6-8 mm long......... **M. marrubioides**

Malva (p. 283)

Annual or biennial sparsely pubescent herbs with roundish lvs.; fls. in axillary clusters, involucels present; petals white to pinkish; fr. disk-like, separating into many compressed reniform carpels. About 11 species, 4 introduced in CA.

Petals 2 times length of calyx .... **M. neglecta**
Petals ca. same length as calyx .... **M. parviflora**

Malvella (p. 283)
(*Sida*)

Decumbent or prostrate whitish-stellate perennial herb; lvs. roundish to deltoid, dentate; fls. axillary; involucel usually absent; petals yellowish; fr. of 6-10 carpels, indehiscent. About 4 species, 1 in CA. One species in Kern County: **M. leprosa** (Ortega.) Krapov. (*Sida hederacea*)

Sidalcea (p. 283)

Perennial coarsely stellate-pubescent herb with rounded palmately-parted mostly basal lvs.; fls. in terminal racemes; petals rose-pink, white-veined. About 25 species, 18 in CA.

Calyx stellate-pubescent and with some slender soft hairs to 2 mm long; carpels glabrous to finely stellate. .... **S. ranunculacea**
Calyx mostly stellate, no long soft hairs; carpels sparsely glandular-puberulent. .**S. malvaeflora** ssp. **sparsifolia**

Sphaeralcea (p. 284)

Perennial somewhat shrubby herb with yellowish-canescent stems; lvs. thickish, suborbicular; fls. in open panicles with few to many fls.; petals grenadine; fr. of 12-16 half-dehiscent rugose carpels. About 50 species, 8 in CA. One species in Kern County: **S. ambigua** var. **ambigua** (incl. var. *monticola*)

MARSILIACEAE

Marsilea (p. 167)

Small perennial ± aquatic clover-like ferns with 4 leaflets; sporocarps bean-shaped with 2 teeth. About 70 species, 2 in CA. One species in Kern County: **M. vestita**

MARTYNIACEAE

Proboscidea (p. 345)

Coarse sticky-pubescent annual with opposite, petioled lvs.; lf.-blades rounded, 1-3 dm wide; fls. large, showy, dull white-yellowish with purple markings; caps. 6-10 cm long with 2 long (8-20 cm) curved horns. About 8 species, 4 in CA. One species in Kern County: **P. louisianica**

MELIACEAE

Melia (p. 279)

Deciduous round-topped tree with alternate bipinnate lvs. 3-8 dm long; fls. purplish, fragrant, in open panicle 1-2 dm long; fr. a drupe 10-15 mm diam. About 10 species, one introduced in CA. One species in Kern County. **M. azedarach**

MOLLUGINACEAE
(part of Aizoaceae)

Prostrate annuals with alt. to whorled simple lvs.; infl. clustered; sepals 5; petals 0; stamens 3-20; fr. a caps.

Plants hairy; sepals 4-7 mm long..**Glinus**
Plants glabrous; sepals 1-2 mm long...**Mollugo**

Glinus (p. 232)

Low stellate-pubescent much branched prostrate annuals; lvs. entire, ± whorled; fls. densely clustered in upper nodes; fl.-tube bell-shaped; sepals 5; petals 0; stamens 3 or 5. About 12 species, one introduced in CA. One species in Kern County: **G. lotoides**

Mollugo (p. 232)

Low glabrous much branched annuals with linear whorled lvs.; fls. axillary on slender pedicels; sepals 5, white; petals 0; stamens 3 or 5; fr. a capsule. About 20 species, 2 introduced in CA. One species in Kern County: **M verticillata**

MORACEAE

Trees, often with milky juice, with alternate simple lvs.; fls. small, inconspicuous, imperfect, usually aggregated into heads or spikes.

1. Fls. external in dense spike or head.
   2. Lvs. thin, light green, may be 3-5-lobed; tree to 15 m tall..............................**Morus**
   2. Lvs. entire, long-pointed; axillary spines present ....**Maclura**
1. Fls. inside a closed receptacle; lvs. with 3-5 palmate lobes.................................. **Ficus**

Ficus (p. 210)

Small much-branched deciduous tree with thick palmately 3-5-lobed lvs.; fr. a fig. About 500 species, one occasionally adventive in CA. One species in Kern County: **F. carica**

Maclura (p. 211)

Deciduous tree or large shrub with axillary spines and entire, 5-12 cm long lvs.; thick fleshy fr. yellow, looks much like an orange. About 12 species, 1 introduced in CA. One species in Kern County: **M. pomifera**

Morus (p. 211)

Wide-spreading rounded tree to 15 m tall; lvs. broad ovate, scallop-toothed, often 3-5-lobed; fr. 2.5-5 cm long, fleshy, white to dark violet. About 12 species, one adventive in CA. One species in Kern County: **M. alba**

NYCTAGINACEAE

Herbs or subshrubs with swollen nodes and fragile stems; lvs. mostly opposite; fls. perfect, regular, subtended by bracts which are frequently united into a calyx-like invol.; sepals bell-shaped to trumpet-shaped, 4-5-lobed, constricted above the ovary; petals none; ovary superior (though appearing inferior due to constriction of perianth-tube); fr. an achene.

Plants annual; stigma linear; perianth trumpet-shaped .... **Abronia**
Plants perennial; stigma subglobose; perianth funnelform or bell-shaped, subtended by calyx-like invol .... **Mirabilis**

Abronia (p. 231)

Viscid, branching, frequently prostrate or decumbent annuals with opposite, unequal, ± fleshy lvs.; infl. capitate, subtended by invol. of 5-8 dry bracts; fls. showy, trumpet-shaped with slender tube. About 25 species, 9 in CA.

1. Fr. with 2-5 conspicuous wings.
    2. Fr. mostly with 2 wings; bracts ovate; fls. white to rose-pink....... **A. pogonantha**
    2. Fr. mostly with 3-5 wings; bracts lanceolate; fl. purplish-rose. **A. villosa** var. **villosa**
1. Fr. wingless; fl. white or pinkish.. **A. turbinata**

Mirabilis (p. 231)

Perennial somewhat shrubby herbs with repeatedly dichotomous stems; lvs. opposite, entire; fls. 1-several in a 5-lobed bell-shaped calyx-like invol.; calyx surpassing invol., funnelform to bell-shaped, petaloid. About 40 species, 12 in CA.

1. Invol. 3-10-fld.; 3-5 cm long; perianth funnelform. **M. multiflora** (Torr.) Gray var. **pubescens** Wats. (*M. froebelii*)
1. Invol. 1-fld., <1.5 cm long; perianth bell-shaped.
    2. Fls. rose to red; lvs. subglabrous to rough-pubescent. **M. californica** Gray (*M. laevis*)
    2. Fls. white to pale pink; lvs. viscid-villous... **M. bigelovii**
        3. Stems viscid-villous, fr. with 10 pale vertical lines. var. **bigelovii** (incl. var. *aspera*)
        3. Stems subglabrous below, short backwards-bending hairs above; fr. sometimes striate. .... var. **retrorsa**

OLEACEAE

Tree or shrub with opposite deciduous simple or pinnately compound lvs.; fls. regular, unisexual, inconspicuous, in clusters; calyx usually 4-lobed; corolla, if present, 2-4-lobed; stamens 2, hypogynous; fr. a samara or drupe.

Lvs. compound; fr. a samara; tree .... **Fraxinus**
Lvs. simple; fr. a blue-black drupe; desert shrub .... **Forestiera**

Forestiera (p. 303)

Deciduous shrub with simple opposite often clumped lvs. and spiny branchlets; fls. small, in sessile clumps; corolla usually absent; fr. a blue-black ellipsoid drupe. About 20 species, one in CA. One species in Kern County: **F. pubescens** Nutt. (*F. neomexicana)*

Fraxinus (p. 303)

Deciduous tree 10-25 m high with opposite pinnately compound lvs.; dioecious; fls. small, in crowded panicles; calyx ca. 1 mm long; petals none; stamens 2; samara winged ca. half its length. About 65 species, 4 in CA. One species in Kern County: **F. latifolia**

ONAGRACEAE

Herbs or shrubs with alternate or opposite lvs.; fls. regular, mostly 4-merous; ovary inferior; fr. a capsule.

1. Sepals persistent; wet places .... **Ludwigia**
1. Sepals deciduous after blooming.
  2. Fls. 2-merous .... **Circaea**
  2. Fls. 4-merous.
    3. Seeds with tuft of hairs (coma) at one end .... **Epilobium**
    3. Seeds without a tuft of hairs.
      4. Fr. a dehiscent capsule.
        5. Ovary 4-loculed; fl.-tube prolonged beyond ovary.
          6. Stigma 4-lobed; petals white to lavender.
            7. Petals 2-lobed; sepals erect .... **Epilobium**
            7. Petals entire; sepals reflexed.
              8. Anthers erect, attached near base; fls. usually not white .... **Clarkia**
              8. Anthers attached near middle; fls. white .... **Oenothera**
          6. Stigma capitate, entire .... **Camissonia**
        5. Ovary 2-loculed; fls. minute on capillary branches .... **Gayophytum**
      4. Fr. indehiscent, nut-like and hard .... **Clarkia**

Camissonia (p. 291)

Annuals with taproot; lvs. basal or cauline, alternate; petals 4, yellow, rose or white; stamens 8; stigma capitate. About 62 species, 43 in CA.

1. Ovary with long slender sterile projection below fl.-tube; plants without leafy stem.
  2. Petals 5-18 mm long, yellow aging red; plants with soft hairs .... **C. graciliflora**
  2. Petals 2-4 mm long, yellow; plants with stiff straight hairs.................. **C. palmeri**
1. Ovary lacking sterile projection; plants mostly with obvious leafy stem.
  3. Caps. on well-defined pedicel, not coiled or contorted; corolla white to yellow.
    4. Lvs. mostly in basal rosette; fls. crowded; petals white aging reddish, 4-6 mm long.. **C. claviformis** ssp. **claviformis**
    4. Lvs. well-distributed; fls. few; petals bright yellow often with red spot, 10-14 mm long.. **C. kernensis** ssp. **kernensis**
  3. Caps. sessile, often coiled or contorted.
    5. Petals white; fls. open in evening.
      6. Caps. not thickened near base, not beaked; fls. absent in lower nodes. **C. refracta**
      6. Caps. thickened near base, attenuate into a slender beak................ **C. boothii**
        7. Caps. with simple curve, its tip spreading away from stem axis. ssp. **decorticans**
        7. Caps. contorted so its tip points down; stem epidermis white. ssp. **desertorum**
    5. Petals yellow; fls. open in day.
      8. Lvs. pinnately lobed; petals usually with minute maroon flecks near base; caps. often reflexed at maturity... **C. californica**
      8. Lvs. subentire; petals without flecks; caps. not reflexed.
        9. Caps. quadrangular at maturity; basal nodes mostly with fls.; lvs. mostly lanceolate to ovate.
          10. Plants conspicuously gray-strigose; branches usually decumbent. **C. pallida** ssp. **pallida**
          10. Plants not gray-strigose.
            11. Caps. <1 mm thick; upper lvs. ± ovate; plants erect. ....**C. hirtella**
            11. Caps. 1-2 mm thick; upper lvs. lance-ovate; plants decumbent. **C. micrantha**
        9. Caps. subterete; basal nodes lacking fls.; lvs. mostly linear.
          12. Sepals reflexed singly; plants with spreading pubescence.
            13. Stigma elevated well above anthers at blooming; petals 8-16 mm long.. **C. kernensis**
              14. Lvs. clustered near base; plant compact, densely hairy. ssp. **kernensis**
              14. Lvs. not clustered near base; plant ± open, sparsely hairy. ssp. **gilmanii** (Munz) Raven (Red Rock Canyon, Penalosa 473)
            13. Stigma surrounded by anthers at blooming; petals 2-4 mm long. **C. parvula**
          12. Sepals usually reflexed in pairs.

15. Stigma held above anthers at blooming; petals 4-15 mm long. **C. campestris** ssp. **campestris**
15. Stigma surrounded by anthers at blooming; petals 2-5 mm long.
  16. Infl. glandular-pubescent; petals 2-4 mm long; plant strigulose. **C. strigulosa** (Fisch. & Meyer) Raven (*C. dentata*)
  16. Infl. very little glandular-pubescent; petals 3-5 mm long. **C. contorta**

Circaea (p. 293)

Low slender perennial herb with opposite lvs.; fls. small, white to pinkish; fr. nut-like with hooked hairs. About 8 species, one in CA. One species in Kern County: **C. alpina** ssp. **pacifica**

Clarkia (p. 293)

Annuals with exfoliating epidermis; lvs. simple, mostly entire; fls. in spikes or racemes, showy; petals sometimes with broad tips (limb) and narrow bases (claw); stigma 4-lobed; fr. a caps. About 41 species, 40 in CA.

1. Fr. 2-3 mm long, indehiscent, nut-like. **C. heterandra** (Torr.) Lewis & Raven (*Heterogaura heterandra*)
1. Fr. >5 mm long, dehiscent.
  2. Petals conspicuously lobed, lavender.
    3. Limb 2-lobed with subulate tooth in sinus .... **C. xantiana**
    3. Limb obscurely 3-lobed, claw expanded into a pair of lateral lobes. **C. rhomboidea**
  2. Petals not lobed.
    4. Buds, at least older ones, pendulous or deflexed.
      5. Claw of petals slender, equal to or longer than limb.
        6. Some long hairs on ovary and calyx; lf.-width 1/4-2/3 length; petal-width 1/3-3/4 length, petals lavender to dark red-purple. .... **C. unguiculata**
        6. Long hairs absent; lf.-width less than 1/3 length; petal-width less than 1/2 length.
          7. Lvs. bright green; petals pink, with or without purple spot; style equalling anthers. .... **C. exilis**
          7. gray-green; petals like C. exilis or wrinkled and unexpanded; style longer than anthers. **C. tembloriensis** ssp. **calientensis** (Vasek) Holsinger
      5. Claw lacking or very short; limb lavender, flecked with red-purple.
        8. Fl.-tube 1-3 mm long with ring of hairs near summit............. **C. dudleyana**
        8. Fl.-tube 3-5 mm long with ring of hairs near middle............... **C. cylindrica**
    4. Buds erect; petals lavender to purple-red, mostly with dark central spot.
      9. Ovary densely spreading-pubescent .... **C. purpurea** ssp. **viminea**
      9. Ovary not spreading-pubescent .... **C. speciosa** ssp. **polyantha**

Epilobium (p. 294)

Shrubs to herbs, some with turions (tuber-like buds which persist at base of stem); lvs. opposite or alternate, ± sessile; petals notched, white to reddish-purple. About 170 species, 30 in CA.

1. Annual; lvs. opposite only near base; seed hair-tuft none.
  2. Petals 6-12 mm long; fls. in leafy spikes.
  E. densiflorum (Lindl.) Hoch & Raven (*Boisduvalia densiflorum*)
  2. Petals 1-4 mm long; fls. axillary.
    3. Fl.-bracts ovate to oblong; petals 2-4 mm long; caps. 6-8 mm long, straight.
    E. pygmaeum (Speg.) Hoch & Raven (*Boisduvalia glabella*)
    3. Fl.-bracts linear; petals 1-2 mm long; caps. 8-10 mm long, curved.
    E. torreyi (Wats.) Hoch & Raven (*Boisduvalia stricta*)
1. Perennial to subshrub; lvs. opposite above; seed hair-tuft present.
  4. Hypanthium 17-34 mm long; corolla red........................**E. canum** (Greene) Raven
    5. Subshrubs 2-9 dm tall; lvs. linear to lanceolate.
    ssp. **canum** (*Zauschneria californica*)
    5. Mostly herbs 1-5 dm tall; lvs. wider to ovate.......ssp. **latifolium** (Hook.) Raven (*Zauschneria latifolia* and *Z. viscosa*)
  4. Hypanthium <17 mm long; corolla white to purple.
    6. Stems mostly erect, with offset rosettes...**E. ciliatum**
      7. Petals 2-6 mm long, white to pink; lvs. reduced in open infl.
      var. **ciliatum** (*E. adenocaulon*)
      7. Petals 4-15 mm long, pink to rose-purple; lvs. little reduced in crowded infl.
      var. **glandulosum** (*E. brevistylum*)
    6. Stems ± ascending, clumped or tufted.
      8. Plants clumped; pedicels 5-25 mm long.................................**E. glaberrimum**
      8. Plants matted; pedicels 20-65 mm long .... **E. oregonense**

Gayophytum (p. 294)

Slender-stemmed annuals; fls. in upper axils, sepals reflexed, petals white drying pink; fr. a capsule. About 9 species, 8 in CA.

1. Petals 3-8 mm long; stigma surpassing stamens.............................. **G. eriospermum**
1. Petals 0.5-3 mm long; anthers surpassing stigma.
  2. Caps. very irregular, lumpy; many seeds aborted ....**G. heterozygum**
  2. Caps. not irregular, only slightly lumpy; most seeds mature.
    3. Caps. flat; plant mostly branched at base.
      4. Two cap.-valves remain attached; petals <1.5 mm long.................. **G. humile**
      4. All valves fall off; petals 1.3-1.8 mm long............................... **G. racemosum**
    3. Caps. ± not flat; plant branched above.
      5. Branches ± throughout, generally 2-8 nodes between branches. **G. decipiens**
      5. Branches at base or not, many above, generally 0-1 node between branches.
      G. diffusum ssp. **parviflorum**

Ludwigia (p. 295)

Herbs often floating in water; fls. axillary; fr. a capsule. About 82 species, 4 in CA.

Petals 5, yellow, 10-14 mm long .... **L. peploides**
Petals none .... **L. palustris**

Oenothera (p. 295)

Ours coarse annuals to biennials with basal rosettes; fls. aging red, tube well developed; petals 2-4 cm long; stigma with 4 linear lobes. About 120 species, 17 in CA.

1. Fls. yellow.
   2. Annual, 0-1 dm tall .... **O. primiveris** ssp. **primaveris**
   2. Biennial, 4-12 dm tall; stems red.
      O. elata Kunth ssp. **hirsutissima** (Wats.) Dietr. (*O. hookeri* ssp. *deltoides*)
1. Fls. white.......... **O. deltoides**
   3. Hairs on buds straight; desert .... ssp. **deltoides**
   3. Hairs on buds curly; valley .... ssp. **cognata**

OPHIOGLOSSACEAE

Botrychium (p. 165)

Fern with single vegetative pinnately compound frond with a fertile sporebearing spike arising from its base. About 45 species, 8 in CA.

Sterile lf.-blade 10-40 cm long, several times pinnately compound.......... **B. multifidum**
Sterile lf.-blade 1-5 cm long, lobed to two-pinnate .... **B. simplex**

ORCHIDACEAE

Perennial herbs with sheathing lvs., sometimes reduced to scales; fls. irregular, solitary or in spikes or racemes; sepals 3; petals 3, 2 like sepals, the third saccate and often spurred; filaments 1 or 2, united to style to form a column; pollen in little sacs; fr. a capsule.

1. Lvs. reduced and scale-like; plants brown or yellow to purplish, not green. **Corallorhiza**
1. Lvs. larger, green.
   2. Corolla-lip clearly spurred at base.
      3. Sepals 1-nerved; lvs. near base of stem withering before flowering; spur 1-5 mm long.......... **Pipera**
      3. Sepals 3-nerved; lvs. usually scattered and green when flowering; spur >5 mm long.......... **Platanthera**
   2. Corolla-lip not spurred.

4. Lvs. 2, opposite, ± sessile; fls. greenish .... **Listera**
4. Lvs. >2, alternate; fls. purplish .... **Epipactis**

Corallorhiza (p. 207)

Saprophytic herb with brownish or yellow to purple stems and reduced lvs., about 20-70 cm tall; fls. crimson-purple to greenish, in loose terminal raceme. About 10 species, 4 in CA. One species in Kern County: **C. maculata**

Listera

Slender perennial to 2 dm tall with two opposite, sessile, ovate lvs.; fls. in terminal raceme, greenish-yellow; spur absent. About 25 species, 3 in CA. One species in Kern County: **L. convallarioides** (Sw.) Nutt. (Rare, near Tiger Flat, Greenhorn Mts., Shevock 10621)

Pipera (p. 208)
(part of *Habenaria*)

Perennial herb with 2-5 basal lvs. and several bract-like cauline lvs.; infl. spike-like, terminal; fls. whitish green; sepals 1-veined; corolla spurred. About 9 species, all in CA. One species in Kern County: **P. unalascensis** (Spreng.) Rydb.

Platanthera (p. 208)
(part of *Habenaria*)

Perennial herb with simple erect stems with lvs. near base; fls. small, spurred, greenish white in terminal spike or raceme. About 85 species, 4 in CA.

Fls. on spike dense, white; spur 10-20 mm long .... **P. leucostachys** Lindl.
Fls. on spike loosely arranged, greenish; spur 6-8 mm long.
P. sparsiflora (Wats.) Schltr.

OROBANCHACEAE

Fleshy root-parasitic herbs lacking chlorophyll; lvs. scale-like; corolla tubular, curved, 2-lipped, purplish to yellowish; stamens 4; fr. a capsule.

Plants glabrous; filaments hairy at base; capsule 4-valved .... **Boschniakia**
Plants sticky-pubescent; filaments not hairy; capsule 2-valved .... **Orobanche**

Boschniakia

Reddish-brown, glabrous plants with scale-like overlapping lvs.; fls. sessile in dense spike; corollas bent at middle. Three species, two in CA. One species in Kern County: **B. strobilacea** Gray (N-facing slopes near summit of Spring Canyon, Scodie Mtns. Shevock 10804 CAS)

Orobanche (p. 345)

Sticky-pubescent root parasites; calyx 5-cleft into acute lobes; corolla curved, 2-lipped, yellowish to purplish; stamens included. About 140 species, 11 in CA.

1. Pedicels 3-10 mm long; bractlets lacking.
 2. Pedicels 1-3 times stem length; calyx-lobes longer than tube............... **O. uniflora** (incl. ssp. *occidentalis*)
 2. Pedicels not exceeding stem length; calyx-lobes not longer than tube. **O. fasciculata**
1. Pedicels shorter and subtended by bractlets.
 3. Lower lip of corolla spreading, 7-15 mm long; corolla 25-30 mm long. **O. californica** Cham. & Schldl. ssp. **feudgei** (Munz) Heck. (*O. grayana* var. *feudgei*)
 3. Lower lip of corolla erect, 4-8 mm long; corolla 20-25 mm long.
  4. Corolla pale, yellowish with darker veins; anthers subglabrous. **O. parishii** (Jeps.) Heck. ssp. **parishii** (*O. californica* var. *parishii*)
  4. Corolla purplish; anthers woolly; transmontane........................ **O. corymbosa**

OXALIDACEAE

Oxalis (p. 275)

Herbaceous perennials from underground bulblets or rooting at nodes; lfts. 3, folding at night; fls. in umbel-like cymes on axillary peduncles; regular, 5-merous; petals yellow; fr. a capsule. About 480 species, 12 in CA.

Plants mostly stemless; petals ca. 2 cm long ....**O. pes-caprae**
Plants with leafy stems; petals 4-8 mm long ....**O. corniculata**

PAPAVERACEAE
(incl. Fumariaceae)

Herbs or shrubs, some with milky juice; lvs. opposite or alternate without stipules; fls. mostly regular, mostly solitary; sepals 2, 3 or 4, early deciduous; petals 4-6; stamens mostly numerous; ovary superior; fr. a capsule.

1. Fls. bilateral; petals 4, outer ones pouched at base....................................... **Dicentra**
1. Fls. radial; petals 4+, outer ones not pouched at base.
 2. Lvs. mostly opposite or whorled.
  3. Basal lvs. broadly linear without petiole; plant pubescent.............. **Platystemon**
  3. Basal lvs. spatulate, narrowed at base; plants mostly glabrous...........**Meconella**
 2. Lvs. mostly alternate or basal.
  4. Plants shrubby... **Dendromecon**
  4. Plants herbaceous.
   5. Plants prickly-spiny; fls. large, white.. **Argemone**

5. Plants not prickly-spiny.
 6. Lvs. much divided into linear segments .... **Eschscholzia**
 6. Lvs. not much divided.
  7. Plants <4 cm tall; stamens 6-9 .... **Canbya**
  7. Plants >4 cm tall; stamens many.
   8. Style slender; stigma capitate .... **Stylomecon**
   8. Style lacking; stigma discoid .... **Papaver**

Argemone (p. 240)

Annual or perennial spinescent-prickly perennial with yellow latex; lvs. lobed, clasping, prickly; fls. 5-13 cm in diameter; petals white; stamens numerous; capsule spinescent. About 30 species, 2 in CA. One species in Kern County: **A. munita** (incl. ssp. *rotundata*)

Canbya (p. 240)

Minute glabrous tufted annual 2-3 cm high; sepals 3; petals 3 mm long, white; capsule ovoid. Two species, 1 in CA. One species in Kern County: **C. candida**

Dendromecon (p. 240)

Glabrous openly branched evergreen shrubs with alternate simple entire tough grayish green lvs.; fls. solitary; sepals 2; petals 4, yellow, 2-3 cm long; capsule linear. Two species, both in CA. One species in Kern County: **D. rigida**

Dicentra (p. 244)

Perennial herbs with basal or basal and cauline dissected lvs.; fls. solitary or in terminal panicle, flattened and heart-shaped, irregular; outer two petals saccate at base; fr. a capsule. About 16 species, 6 in CA.

Fls. golden yellow; tall plant ca. 5-15 dm high .... **D. chrysantha**
Fls. pinkish; small stemless plant <1 dm high .... **D. pauciflora**

Eschscholzia (p. 240)

Annual to perennial herbs with colorless juice; lvs. alternate, glabrous, finely dissected; fls. yellow to red-orange; torus (receptacle) expanded to form funnel-shaped base for pistil; sepals 2, united into a cap which falls early (calyptra); capsule elongate. About 12 species, 10 in CA.

1. Torus with 2 rims, the inner erect and translucent, the outer spreading; plants annual to perennial .... **E. californica** (incl. *E. procera*)
1. Torus with 1-rim, erect and translucent; annuals.
 2. Buds nodding, usually with some hairs; lvs. ± covered with grayish fine hairs.

3. Sepals 11-21 mm long; petals 15-32 mm long, orange or deep yellow. **E. lemmonii** ssp. **lemmonii**
3. Sepals 5-13 mm long; petals 5-16 mm long, yellow (some with orange base). **E. hypecoides**

2. Buds erect, usually glabrous; lvs. glabrous.
   4. Stems scapose, lvs. mostly crowded at base.
      5. Lvs. usually twice ternate (in threes) into relatively few rather long ultimate lobes; seeds bur-like .... **E. lobbii**
      5. Lvs. usually 3-4 times ternate; seeds reticulate or pitted.
         6. Lf.-blades mostly 4 times ternate into short pointed lobes.......... **E. parishii**
         6. Lf.-blades mostly 3 times ternate into blunt lobes.
            7. Calyptra abruptly long-pointed.
               8. Petals bright yellow, 10-25 mm long........ **E. caespitosa**
               8. Petals deep orange, 25-40 mm long. **E. lemmonii** Greene ssp. **kernensis** (Munz) Clark (*E caespitosa* ssp. *kernensis*)
            7. Calyptra tapering to sharp point .... **E. glyptosperma**
   4. Stems leafy.
      9. Petals 3-6 mm long.........**E. minutiflora** (incl. var. *rutaefolia* and *E. parishii*)
      9. Petals 10-30 mm long........ **E. caespitosa**

Meconella (p. 243)
(incl. *Hesperomecon*)

Glabrous slender-stemmed annuals with mostly basal lvs.; sepals 3, often reddish; petals 6 white to yellow. Four species, 3 in CA.

Lvs. basal; petals cream with yellow base.............**M. linearis** (Benth.) Nels. & Macbr. (*Hesperomecon linearis*)
Lvs. basal and cauline; petals white........ **M. californica**

Papaver (p. 243)

Annual with yellowish juice and pinnately-lobed lvs.; fls. solitary on long peduncles, drooping in bud; petals 4, rose; capsule ovoid, dehiscent by pores under the edge of apical disk. About 50 species, 4 in CA. One species introduced in Kern County: **P. rhoeas** L. (*P. hybridum*)

Platystemon (p. 243)

Low villous annuals with opposite entire lvs.; fls. scapose on long peduncles; petals 6, cream, 8-20 mm long; carpels 6-25, united at first, separate in fr. One variable species: **P. californicus**

Stylomecon (p. 243)

Erect annual with yellow juice and pinnatifid lvs.; fls. axillary on elongate peduncles, 5-20 cm long; petals 4, orange-red with basal spot, 1-2 cm long; caps. rounded. One species: **S. heterophylla**

PINACEAE

Resinous evergreen trees with linear needle-like solitary or clustered (fascicled) lvs.; monoecious; staminate cones drying and deciduous after producing pollen; pistillate cones with ± woody scales each bearing 2 seeds.

1. Cones erect, the scales falling separately; lvs. solitary .... **Abies**
1. Cones reflexed or pendulous; scales persistent
   2. Cone-bracts inconspicuous; needle-lvs. in fascicles .... **Pinus**
   2. Cone-bracts conspicuous; needle-lvs. solitary .... **Pseudotsuga**

Abies (p. 168)

Tall conical trees with flattened linear sessile lvs. which are 2-ranked (lower branches) or curved upward (upper branches); pistillate cones on topmost branches erect with numerous deciduous overlapping scales.About 40 species, 7 in CA.

Lvs. flat, with slight basal twist; old bark gray .... **A. concolor**
Lvs. ± 4-sided, lacking basal twist; old bark dark red ......**A. magnifica** var. **magnifica**

Pinus (p. 168)

Evergreen trees with long needle-like lvs. borne in clusters (fascicles) of 1-5 in axils of small, deciduous scale-like lvs.; pistillate cones becoming woody, scales with ± thickened terminal exposed part with a protuberance (umbo). About 80 species, 20 in CA.

1. Fascicles with single thick needle, 2.5-3.5 cm long .... **P. monophylla**
1. Fascicles with 2-5 needles.
   2. Fascicles with 2 needles, 3-6 cm long; cone 3-5 cm long.
      P. contorta Loudon ssp. **murrayana** (Grev. & Balf.) Critch. (*P. murrayana*)
   2. Fascicles with >2 needles; needles and cones larger.
      3. Fascicles with 3 needles, 12-30 cm long.
         4. Cone very large, heavy; needles 20-30 cm long, gray-green; trunk mostly branched... **P. sabiniana**
         4. Cone smaller; needles 12-30 cm long, green.
            5. Cone 15-25 cm long, umbo with deflexed prickle .... **P. jeffreyi**
            5. Cone 7-15 cm long, umbo with protruding prickle................. **P. ponderosa**

3. Fascicles with 5 needles, 3-10 cm long.
   6. Cones 8-25 cm long, with short stalks .... **P. flexilis**
      (Also at Owens Peak and Inspiration Point, Piute Mtns. Shevock)
   6. Cones 25-40 cm long, with conspicuous stalks........................ **P. lambertiana**

Pseudotsuga (p. 170)

Tree 12-20 m tall with slender pendulous branchlets; lvs. flat, solitary, 2-3 cm long, pointed; cones 10-15 cm long with conspicuous 3-toothed bract. Four species, 2 in CA. One species in Kern County: **P. macrocarpa**

PLANTAGINACEAE

Plantago (p. 345)

Annual or perennial herbs with basal longitudinally ribbed lvs. and scapose bracteate spikes; fls. regular, 4-merous; sepals with scarious margin; corolla trumpet-shaped with 4 spreading or reflexed scarious persistent lobes; ovary superior; fr. a capsule, dehiscing like a "lid." About 270 species, 15 in CA.

1. Corolla-lobes spreading or reflexed; stamens 4.
   2. Anterior sepals connate; stamens conspicuously exserted; perennial. **P. lanceolata**
   2. Anterior sepals free; stamens not conspicuous.
      3. Heavy-rooted perennial; bracts not longer than calyx .... **P. major**
      3. Annual.
         4. Seeds dull, brown; bracts and sepals not similar.. **P. erecta** Morris (*P. hookeriana*)
         4. Seeds shining, reddish-yellow; bracts very much like sepals. **P. ovata** Forsskal (*P. insularis*)
1. Corolla-lobes erect and forming a beak over capsule; stamens 2. **P. elongata** (incl. *P. bigelovii*)

PLATANACEAE

Platanus (p. 255)

Deciduous tree with thin exfoliating bark and large alternate palmately lobed lvs.; petioles dilated at base and covering buds; infl. spherical, becoming hard in fruit, 2-3 cm in diameter. Nine species, one in CA. One species in Kern County: **P. racemosa**

POACEAE

(Gramineae by some authors)

Herbaceous to ± woody with hollow stems (culms) solid at the swollen nodes; lvs. parallel-veined with an open sheath and expanded blade with membranous appendage (ligule) at its junction with the sheath; fls. usually perfect, lacking a perianth, arranged in groups (spikelets) subtended by a pair of bracts (glumes), each floret consisting of two bracts, the outer (palea) ± overlapping the inner (lemma) and both enclosing the single fl.; stamens 3 on delicate filaments; styles 2, plumose.

Key to Tribes of Poaceae

1. Florets falling separately, glumes persistent (dry infl. breaks apart leaving empty glumes on plant).
  2. Spikelets in 2 rows, mostly sessile.
    3. Spikelets on one side of axis forming 1-sided spikes; spikes mostly more than one. .... **CHLORIDEAE**
    3. Spikelets on two sides of axis; spikes terminal, single .... **HORDEAE**
  2. Spikelets in an open or spike-like raceme or panicle, mostly on distinct pedicels.
    4. Spikelets with one fl.
      5. Spikelets with 2 sterile awnless scale-like lemmas below the fertile lemma; panicles dense and spike-like .... **Phalaris**
      5. Spikelets lacking sterile lemmas .... **AGROSTIDEAE**
    4. Spikelets with 2 to many fls.
      6. Glumes larger than lemmas; awn attached to back of lemma, usually bent. **AVENEAE**
      6. Glumes smaller than lemmas; awn attached to tip of lemma (or absent), straight. .... **FESTUCEAE**
1. Florets falling with glumes (spikelets shed as a unit, some with reduced or not obvious glumes).
  7. Spikelets strongly compressed, very flat, in panicles; glumes reduced. .... **Oryza**
  7. Spikelets not or slightly flattened; glumes larger.
    8. Lemma and palea very thin, translucent; glumes firmer than lemma.
      9. Spikelets in pairs, one sessile, one with pedicel .... **ANDROPOGONEAE**
      9. Spikelets not in pairs .... **AGROSTIDEAE**
    8. Lemma and palea papery to leathery, not thin .... **PANICEAE**

AGROSTIDEAE

1. Florets falling separately, glumes persistent.
  2. Lemma hardened and tightly enclosing grain; callus (thickening at base of lemma) bearded.
    3. Awn 3-branched .... **Aristida**
    3. Awn simple.
      4. Awn twisted and bent, much longer than fr.

5. Panicle narrow, slender; lemma <4 times length of palea; awn length variable. **Achnatherum**
5. Panicle broad, open; lemma >4 times length of palea; twice-bent awn 6-11 cm long. **Nassella**
4. Awn not twisted, rarely much longer than plump fr.
6. Lemmas smooth; blades flat; glumes about 3 mm long **Pipatherum**
6. Lemmas hairy; blades inrolled; glumes 6-7 mm long **Achnatherum**
2. Lemma thin, membranous; callus not well developed.
7. Glumes longer than lemma (but not much in some Agrostis).
8. Glumes compressed, with sharp keel; panicle dense and cylindric **Phleum**
8. Glumes not compressed, not hairy on margin **Agrostis**
7. Glumes mostly shorter than lemma.
9. Lemma awned from the tip **Muhlenbergia**
9. Lemma awnless or awned from the back.
10. Fr. permanently enclosed in lemma and palea.
11. Panicle short, partly enclosed in sheath; ligule absent **Crypsis**
11. Panicle elongate, more exposed; ligule present. **Muhlenbergia**
10. Fr. falling from lemma and palea **Sporobolus**
1. Florets falling with the glumes
12. Glumes long-awned **Polypogon**
12. Glumes awnless **Alopecurus**

ANDROPOGONEAE

Spikelets all alike, fertile, surrounded by copious soft hairs; infl. a narrow panicle. **Imperata**

Spikelets unlike, the sessile perfect, the pedicellate staminate or sterile; infl. an open panicle, purplish **Sorghum**

AVENEAE

1. Florets 2, lower perfect and awnless, upper staminate and awned **Holcus**
1. Florets 2 or more, all alike except the reduced upper ones.
2. Lemmas cleft at tip with very small awn between the lobes; glumes white-margined. **Schismus**
2. Lemmas toothed but not deeply cleft; awns, when present, on back of lemma.
3. Glumes 2-3.5 cm long, 7-9-veined; spikelets 2-fld., pendulous; annuals.. **Avena**
3. Glumes <2 cm long, 1-5-veined; spikelets not pendulous.
4. Lemmas keeled (like the bottom of a boat); the awn, when present, from above the middle.
5. Stems in spikelets short, mostly glabrous **Koeleria**
5. Stems in spikelets slender, with long soft hairs **Trisetum**
4. Lemmas rounded on back, awned from below the middle **Deschampsia**

CHLORIDEAE

1. Spikelets with more than 1 perfect floret.
   2. Spikes many, slender, raceme-like on long axis.................................. **Leptochloa**
   2. Spikes few, arising from one point ....**Eleusine**
1. Spikelets with only 1 perfect floret; spikes arising from one point.
   3. Spikelets with 1 or more reduced florets above the perfect one..................**Chloris**
   3. Spikelets without additional modified florets ....**Cynodon**

FESTUCEAE

1. Plants tall stout reeds with large plume-like panicles.
   2. Lemmas lacking hairs; stems in spikelets hairy ....**Phragmites**
   2. Lemmas hairy; stems in spikelets not hairy ....**Arundo**
1. Plants not reed-like, <1.5 m tall.
   3. Plants dioecious, spreading by stolons or creeping rhizomes; in salty or alkaline soils.. **Distichlis**
   3. Plants not dioecious.
      4. Spikelets of 2 forms, fertile with 1 perfect floret long-awned; sterile spikelets with many sterile lemmas; infl. one-sided.. **Lamarckia**
      4. Spikelets all alike in same infl.
         5. Lemmas with 3 prominent nerves; glumes shorter than first floret. **Eragrostis**
         5. Lemmas 5-many-nerved.
            6. Lemmas keeled (like the bottom of a boat).
               7. Spikelets strongly compressed, crowded in 1-sided clusters at ends of branches.. **Dactylis**
               7. Spikelets not strongly compressed, not crowded in 1-sided clusters.
                  8. Lemmas awned from minutely 2-toothed tip, or nearly awnless; spikelets large...**Bromus**
                  8. Lemmas without awns; spikelets small ....**Poa**
            6. Lemmas rounded on back.
               9. Glumes papery; lemmas strongly veined, firm with dry margins; spikelets purplish...**Melica**
               9. Glumes not papery.
                  10. Veins of lemma parallel, not converging at tip.
                     11. Lf.-sheaths fused; veins of upper empty glumes single......**Glyceria**
                     11. Lf.-sheaths open; veins of upper empty glumes 3.
                        12. Lemmas faintly 5-veined; rhizomes 0; salty soils.......**Puccinellia**
                        12. Lemmas clearly 7-9-veined; rhizomes present; freshwatersoils. **Torreochloa**
                  10. Veins of lemma converging toward tip.
                     13. Lemmas awned or awn-tipped from minutely cleft tip ....**Bromus**
                     13. Lemmas entire, pointed, awnless or awned from the tip.
                        14. Spikelets awned.
                           15. Perennials; florets opening; anthers 3.......................... **Festuca**

15. Annuals; florets mostly don't open; anthers 1-3 .... **Vulpia**
14. Spikelets awnless .... **Poa**

HORDEAE

1. Spikelets solitary at each node of rachis.
  2. Spikelets 1-fld., sunken in hollows in the rachis; spikes slender, cylindric; low annuals. .... **Scribneria**
  2. Spikelets 2-several-fld., not sunken in rachis.
    3. Spikelets placed edgewise to the rachis, the lateral ones with a single glume. **Lolium**
    3. Spikelets placed flatwise to rachis.
      4. Plants perennial.
        5. Spikelets strongly overlapping; infl. internodes mostly <3 mm long. **Agropyron**
        5. Spikelets not much overlapping; infl. internodes >4 mm long.
          6. Awns on lemmas and glumes absent.. **Elytrigia**
          6. Awns on lemmas >5 mm long... **Elymus**
      4. Plants annual
        7. Glumes broad, 3-veined .... **Triticum**
        7. Glumes awl-shaped, 1-veined .... **Secale**
1. Spikelets mostly >1 at each node, sometimes reduced to awns.
  8. Spikelets 3 at each node, 1-fld., the lateral pair stalked, reduced to long awns. **Hordeum**
  8. Spikelets 2 or more at each node, alike, 2-6-fld.
    9. Lemmas with awns 16-30 mm long.. **Elymus**
    9. Lemmas with awns <5 mm long to awnless.. **Leymus**

PANICEAE

1. Spikelets subtended or surrounded by 1-many distinct or ± fused bristles forming an invol.
  2. Bristles persistent, spikelets deciduous .... **Setaria**
  2. Bristles falling with spikelets at maturity.
    3. Bristles united at base, barbed .... **Cenchrus**
    3. Bristles not united at base, slender .... **Pennisetum**
1. Spikelets not subtended by bristles.
  4. Glumes awned or short-pointed .... **Echinochloa**
  4. Glumes awnless; tip of palea enclosed by lemma.
    5. Spikelets in panicles .... **Panicum**
    5. Spikelets in 1-sided spike-like racemes.
      6. Racemes several along main axis.. **Eriochloa**
      6. Racemes aggregated at summit of culm.
        7. Racemes slender, 3-12 .... **Digitaria**
        7. Racemes stout, in pairs .... **Paspalum**

Achnatherum (p. 187, 188)
(part of *Oryzopsis* and *Stipa*)

Tufted perennial bunchgrasses; ligule membranous; infl. panicle-like; floret 1, glumes > lemma; lemma hard, broad, awned from tip. About 75 species, 17 in CA.

1. Awn hairy at base, some hairs >1 mm long.
  2. Lowest hairs on awn longer than upper hairs.
    3. Awn with 1 bend; ligules ciliate. .... **A. speciosum** (Trin. & Rupr.) Barkw. (*Stipa speciosa*)
    3. Awn with 2 bends; ligules glabrous...................... **A. latiglumis** (Swall.) Barkw. (*Stipa latiglumis*)
  2. Lowest hairs on awn shorter than upper hairs.
    4. Palea >half lemma length........**A. nevadensis** (John.) Barkw. (*Stipa nevadensis*)
    4. Palea less than or about half lemma length......... **A. occidentalis** (John.) Barkw.
      5. Hairs of lemma-tip = hairs at awn base. ssp. **californicum** (Mer. & Davy) Barkw. (*Stipa californicum*)
      5. Hairs of lemma-tip >hairs at awn base............ssp. **pubescens** (Vasey) Barkw. (*Stipa elmeri*)
1. Awn scabrous at base, not hairy.
  6. Lf.-blades <2 mm wide; awn straight.
    7. Awns 3-6 mm long, deciduous. **A. hymenoides** (Roem. & Schultes) Barkw. (*Oryzopsis hymenoides*)
    7. Awns 16-30 mm long, persistent.......................... **A. lemmonii** ((Vasey) Barkw. (*Stipa lemmonii*)
  6. Lf.-blades 2-7 mm wide; awn bent twice. ....**A. coronatum** (Thurb.) Barkw. (*Stipa coronata*)

Agropyron (p. 182)

Perennial bunchgrasses; auricles present; spikes erect, green to purplish; spikelets solitary at nodes, strongly overlapping, several-fld., sessile, placed flatwise at each joint; glumes 2, equal, several-veined; lemmas rounded on back. About 12 species, one introduced in CA. One species in Kern County: **A. desertorum**

Agrostis (p. 186)

Perennial grasses with ligules and open to contracted panicles of small spikelets; spikelets 1-fld., falling separate from glumes; lemmas usually shorter than glumes, with or without dorsal awn. About 200 species, 21 in CA.

1. Palea evident, 2-veined, at least half as long as lemma.
  2. Panicle contracted, lobed, the short branches densely whorled. **A. viridis** Gouan (*A. semiverticillata*)
  2. Panicle open, not lobed or densely whorled; turf grass .... **A. stolonifera** (incl. *A. alba*)
1. Palea very small, not evident.

3. Rhizomes present.................... **A. pallens** Trin. (incl. *A. lepida* and *A. diegoensis*)
3. Rhizomes absent; tufted bunchgrasses.
   4. Panicle narrow, interrupted, 5-30 cm long; ligule 4-6 mm long. **A. exarata** (incl. var. *pacifica*)
   4. Panicle open, diffuse with slender branches.
      5. Panicle 5-10 cm long; plants 1-3 dm tall ....**A. idahoensis**
      5. Panicle 15-25 cm long; plants 3-9 dm tall ....**A. scabra**

Alopecurus (p. 187)

Annual or perennial with flat blades and soft dense spike-like panicles; spikelets 1-fld., falling with glumes, strongly laterally compressed; glumes hairy on keel; lemmas with slender awn from below the middle. About 35 species, 6 in CA.

Annual with loose panicle 2-4 cm long; upper sheaths inflated .... **A. saccatus**
Perennial with slender panicle 2-7 cm long .... **A. aequalis**

Aristida (p. 187)

Erect or prostrate-spreading perennial with narrow blades to 3 mm wide; panicle large, diffuse; spikelets 1-fld., falling separate from glumes; lemmas with 3-branched awns 1-1.5 cm long. About 300 species, 6 in CA. One species in Kern County: **A. divaricata**

Arundo (p. 175)

Very tall stout perennial reed to 6 or 7 m tall with large plume-like terminal panicles; spikelets several-fld., falling separate from glumes. Three species, one introduced in CA. One species in Kern County: **A. donax**

Avena (p. 185)

Annual grasses, rather tall with open relatively few-fld. panicles of large spikelets; spikelets 2-3-fld., falling separate from the large papery glumes; lemmas with a bent and twisted awn. About 15 species, 5 introduced in CA.

1. Lemma-teeth with thin extended tips; pedicels very slender; lemmas with long reddish hair. ....**A. barbata**
1. Lemma-teeth sharp pointed; pedicels stouter.
   2. Lemmas with long brownish hairs; spikelets mostly 3-fld........................ **A. fatua**
   2. Lemmas mostly glabrous; spikelets mostly 2-fld...................................... **A. sativa**

Bromus (p. 175)

Low to tall annuals or perennials with closed sheaths, flat blades and mostly open panicles of large spikelets; spikelets falling apart from glumes; glumes unequal, pointed; lemmas 2-toothed, awned from between the teeth or awnless. About 150 species, 22 in CA.

1. Plants annual, mostly weedy.
  2. Spikelets strongly flattened; teeth of lemmas not >0.5 mm long.
    3. Lemmas awnless or with awn <3 mm long.
      B. catharticus Vahl (*B. wildenovii* and *B. unioloides*)
    3. Lemmas with awns 7-15 mm long .... **B. arizonicus**
  2. Spikelets only somewhat flattened or not flattened at all; lemma-teeth >0.5 mm long.
    4. Awn bent, twisted at base .... **B. trinii**
    4. Awn straight or spreading, not twisted at base.
      5. Lemmas broad, rounded apically, the teeth <1 mm long; 1st glume 3-5-veined.
        6. Panicles compact, spikelets longer than pedicels; lemmas prominently-veined, hairy...**B. hordaceus** L. (*B. mollis*)
        6. Panicles more open, spikelets not longer than pedicels; lemmas not prominently-veined.
          7. Glumes mostly equal, hairy; awn 10-16 mm long................ **B. arenarius**
          7. Glumes unequal, the lower much longer than the upper, glabrous; awn 4-8 mm long.......................................**B. japonicus** Murr. (*B. commutatus*)
      5. Lemmas narrow, elongate, tapering at tip, the teeth 2-5 mm long; 1st glume 1-veined.
        8. Panicle erect, not spreading; awn 1-2 cm long......................**B. madritensis**
          9. Stems hairy below panicle; sheaths hairy...........ssp. **rubens** (L.) Husnot (*B. rubens*)
          9. Stems and sheaths glabrous .... ssp. **madritensis**
        8. Panicle open, branches spreading or drooping.
          10. Awns 12-14 mm long; lemmas about 1 cm long................. **B. tectorum**
          10. Awns 30-50 mm long; lemmas >2 cm long .... **B. diandrus**
1. Plants perennial, mostly native.
  11. Spikelets strongly flattened, the lemmas strongly keeled.
    B. carinatus var. **carinatus** (*B. breviaristatus* and *B. marginatus*)
  11. Spikelets cylindrical to ± flattened but lemmas not strongly keeled
    12. Second glume 5-veined; awns 3-5 mm long...................... **B. laevipes** Shear (*B. pseudolaevipes*)
    12. Second glume 3-veined; awns 5-8 mm long .... **B. orcuttianus**

Cenchrus (p. 191)

Low branching annual forming clumps; sheaths hairy on margins and at throat; racemes of spiny burs; spikelets 1-fld., subtended by globular involucre of rigid spines 3.5-7 mm long. About 20 species, 4 introduced in CA. One species in Kern County: **C. longispinus**

Chloris (p. 189)

Tufted perennials or annuals with 2-several feathery spikes ± arising from common point; spikelets with 1 perfect floret, sessile, in 2 rows along one side of rachis; glumes ± unequal, the 1st shorter; lemma with short awn. About 50 species, 3 introduced in CA.

Plants with long stout stolons; awns 1-5 mm long.. **C. gayana**
Plants lacking stolons; awns 5-10 mm long... **C. virgata**

Crypsis (p. 187)
(*Heleochloa*)

Low spreading tufted annuals with dense oblong spike-like panicles, the subtending lvs. with inflated sheaths and reduced blades; spikelets 1-fld.; glumes mostly persistent. About 8 species, 3 introduced in CA. One species in Kern County: **C. schoenoides** (L.) Lam.

Cynodon (p. 190)

Creeping perennial with stolons and rhizomes; lvs. short to about 3 cm long, ligule a ring of white hairs; spikes several, slender, arising from common point at ends of erect stems; spikelets 1-fld., in 2 rows. About 10 species, 3 introduced in CA. One species in Kern County: **C. dactylon**

Dactylis (p. 177)

Coarse tufted perennial with panicles densely compacted in clusters; spikelets 2-6-fld., compressed, strongly overlapping. One species, introduced in Kern County: **D. glomerata**

Deschampsia (p. 185)

Annual to perennial grasses with flat or inrolled lvs.; spikelets 2-fld., breaking apart above glumes; rachilla extending beyond florets, hairy; lemmas hairy at base, with slender straight or bent awn from back. About 40 species, 4 in CA.

1. Annual; panicle open; spikelets 4-8 mm long................................ **D. danthonioides**
1. Perennial bunchgrasses.
   2. Panicle narrow, elongate, 1-3 mm long; lvs. 1-1.5 mm wide................ **D. elongata**
   2. Panicle open, nodding, 1-2 dm long; lvs. 1.5-4 mm wide.................. **D. caespitosa**

Digitaria (p. 191)

Much-branched purplish annual with few to several slender racemes branching from common point; spikelets in 2's or 3's, in 2 rows; first glume minute, second 1.5 mm long, fringed with hairs. About 200 species, 2 introduced in CA. One species in Kern County: **D. sanguinalis**

Distichlis (p. 177)

Low dioecious salt-tolerant perennial with wiry stems; ligule short, toothed; lvs. 2-ranked; spikelets few to many-fld.; glumes unequal, broad, 3-7-veined. About 5 species, 1 in CA. One species in Kern County: **D. spicata**

Echinochloa (p. 191)

Coarse annual grasses with compressed sheaths, long flat blades and terminal panicles of stout short densely fld. 1-sided racemes; spikelets 1-fld; glumes unequal, with stiff bristly hairs. About 15 species, 2 introduced in CA.

Awns mostly longer than spikelets; lf.-blades 5-15 mm wide..................... **E. crus-galli**
Awns shorter than spikelets, or absent; lf.-blades 3-6 mm wide...................... **E. colona**

Eleusine (p. 190)

Annual grass with 2-6 stout spikes arising mostly from common point at tip of branch; spikelets compressed and closely overlapping in 2 rows; glumes broad, 1-veined, shorter than lemma. About 9 species, 2 introduced in CA. One species in Kern County: **E. indica**

Elymus (p. 182, 184)
(incl. some *Agropyron* and all *Sitanion*)

Tufted perennials with membranous ligules; infl. spike-like, open to dense; spikelets 1-4 per node; glumes lanceolate to awl-shaped; lemmas rounded on back, usually awned from the tip. About 150 species, 11 in CA.

1. Glumes awl- to awn-shaped, base <1.5 mm wide; awn 25-100 mm long.
   2. Glumes divided into 3+ outward-curving awns.
      E. multisetus (Smith) Davy (*Sitanion jubatum*)

2. Glumes divided into 2 outward-curving awns.
E. elymoides (Raf.) Swezy ssp. **elymoides** (*Sitanion hystrix*)
1. Glumes flat, base >1.5 mm wide; awn <25 mm long.
3. Spikelets 2-4 per node .... **E. glaucus**
3. Spikelets 1 per node.
4. Infl.-internodes mostly >15 mm.......... **E. stebbinsii** Gould (*Agropyron parishii*)
4. Infl.-internodes mostly <15 mm long.................. **E. trachycaulis** (Link) Shinn.
5. Lemma-awn <7 mm long............ ssp. **trachycaulis** (*Agropyron trachycaulis*)
5. Lemma-awn 8-30 mm long.
ssp. **subsecundus** (Link) Gould (*Agropyron subsecundus*)

Elytrigia (p. 182)
(part of *Agropyron*)

Erect to tufted perennials; blades 2-8 mm wide, sheath glabrous to ciliate; infl. spike-like; spikelets on 2 sides of axis, strongly overlapping, 1 per node; glumes and lemmas awnless. About 25 species, 5 introduced in CA.

Rhizomes absent .... **E. elongata** (Host) Nevski
Rhizomes present.................................. **E. intermedia** (Host) Nevski ssp. **intermedia**
(*Agropyron intermedia & A. trichophorum*)

Eragrostis (p. 178)

Annuals with loose to dense terminal panicles; spikelets strongly compressed, few to many-fld.; glumes and lemmas fall with the seed; glumes keeled, shorter than spikelets; lemmas 3-veined, broad, keeled. About 300 species, 10 in CA.

1. Plants glandular or warty along lf.-margins and/or lemma keel; panicle open; spikelets 10-40-fld.. **E. cilianensis**
1. Plants not glandular on lemmas or lf.-margins.
2. Fr.-tip truncate, surface checkered, with evident groove on side opposite embryo.
E. mexicana
3. Spikelet ovoid, 1.5+ mm wide..ssp. **mexicana**
3. Spikelet linear, ± 1 mm wide...............ssp. **virescens** (Presl.) Koch. & Sanchez
(*E. orcuttiana*)
2. Fr. pear-shaped, tip rounded, surface smooth, not grooved.
E. pectinacea (Mich.) Nees var. **pectinacea** (*E. diffusa*)

Eriochloa (p. 191)

Annuas with several racemes along axis; spikelets solitary or in 2's in two rows along one side of a narrow rachis; lower rachilla thickened, forming a ring-like callus below 2nd glume. 1st glume minute and sheathing this. About 30 species, 3 in CA. One species in Kern County: **E. acuminata** (Presl.) Hitchc. var. **acuminata** (*E. gracilis*)

Festuca (p. 178)
(see also *Vulpia*)

Cespitose perennials with spikelets in narrow or open panicles; florets breaking off above glumes, uppermost floret reduced; glumes narrow, thin, acute, unequal; lemmas rounded on back, 3-veined, awned from tip, occasionally awnless. About 80 species, 15 in CA.

1. Awns scarcely if at all evident, <1 mm long.
   2. Panicle 1-2 dm long; lemmas 5-7 mm long .... **F. pratensis**
   2. Panicle 1.5-3.5 dm long; lemmas 7-10 mm long........................... **F. arundinacea**
1. Awns well developed, >2 mm long.
   3. Awn of lemma longer than lemma body; lf.-sheaths green..............**F. occidentalis**
   3. Awn of lemma shorter than lemma body; lf.-sheaths red-brown with conspicuous veins...**F. rubra**

Glyceria (p. 179)

Perennial of wet places, with creeping rhizomes; lf.-sheaths fused; panicle diffuse, nodding; spikelets 6-8-fld.; glumes 1 to 1.5 mm long, remaining after florets fall off. About 40 species, 6 in CA. One species in Kern County: **G. elata** (*G. erecta* and *G. pauciflora* are in *Torreyochloa)*

Holcus (p. 183)

Velvety-pubescent perennial with flat blades and contracted panicles with purple-tinged spikelets; spikelets 2-fld., falling with the glumes; 1st lemma awnless, 2nd lemma with short hooked awn. About 8 species, 2 introduced in CA. One species in Kern County: **H. lanatus**

Hordeum (p. 183)

Annual and perennial grasses with flat blades and dense bristly spikes; spikelets 1-fld., 3 at a node, one or all spikelets fertile; glumes narrow, long-awned; lemmas narrow, rounded on back, awned. About 32 species, 7 in CA.

1. Perennial.
   2. Glumes and awns 1.8-8 cm long .... **H. jubatum**
   2. Glumes and awns <1.8 cm long................................ **H. brachyantherum** Nevski
      3. Blades pubescent, 1.5-5 mm wide; glumes of central spikelet 1.5-2.5 times length of palea......ssp. **californicum** (Covas. & Stebb.) Both. (*H. californicum*)
      3. Blades glabrous, 3-9 mm wide; glumes of central spikelet scarcely longer than palea... ssp. **brachyantherum** (*H. brachyantherum*)
1. Annual.
   4. Glumes of central spikelet with hairy margins .... **H. murinum** L.
      5. Spike dense, 6-8 spikelets per 1 cm of rachis.
         ssp. **glaucum** (Steud.) Tzvel. (*H. glaucum*)

5. Spike less dense, 3-5 spikelets pere 1 cm............ ssp. **leporinum**)Link) Arcang. (*H. leporinum*)

4. Glumes not with hairy margins.

6. Auricles (lobes at base of lf.-blade) conspicuous; all 3 spikelets sessile and fertile. .... **H. vulgare**

6. Auricles mostly absent; lateral spikelets with pedicels, mostly sterile.

7. Spike somewhat oblong, most <5 cm long; glumes and awns strongly spreading........................ **H. marinum** Huds. ssp. **gussoneanum** (Parl.) Thell. (*H. geniculatum*)

7. Spike more linear, mostly over 5 cm long; awns and glumes suberect. **H. depressum**

Imperata (p. 191)

Erect rather coarse perennial with narrow silky terminal panicles; spikelets paired, awnless, surrounded by long silky hairs; glumes and lemmas thin. About 8 species, 1 introduced in CA. One species in Kern County: **I. brevifolia**

Koeleria (p. 185)

Tufted grasses with narrow blades and shining spike-like panicles; spikelets 2-4-fld.; glumes persistent, 1st narrow and 1-veined, 2nd broader and 3-5-veined; lemmas may be awned from tip. About 30 species, 2 in CA.

Perennial, 3-6 dm tall; stems somewhat hairy below panicle. **K. macrantha** (Ledeb.) Schultes (*K. cristata*)

Annual, 1-3 dm tall; stems glabrous below panicle .... **K. phleoides**

Lamarckia (p. 179)

Low erect annual with flat blades and prominent ligules; panicles compact, 1-sided, with crowded clustered spikelets; shining, golden to purplish; florets slender, awned. One species: **L. aurea**

Leptochloa (p. 190)

Annual grasses with simple elongate spikes on a common axis forming a long panicle; spikelets 2-several-fld. along one side of a slender rachis; glumes persistent, unequal, awnless. About 40 species, 4 in CA.

1. Sheaths rough to the touch; panicle <10 cm long, tinged purple................**L. viscida**

1. Sheaths smooth; panicle mostly >10 cm long.

2. Lemmas awned; spikelets 7-12 mm long .... **L. fascicularis**

2. Lemmas awnless; spikelets 5-7 mm long .... **L. uninervia**

Leymus (p. 182)
(part of *Elymus*)

Perennials with lf.-blades to 12 mm wide; ligules present; infl. spike-like with 2-several sessile spikelets per node; glumes awl-like, rigid; lemmas rounded on back, usually awned from tip. About 31 species, 8 in CA.

1. Plants tall, >1 m tall; lf.-blade >10 mm wide
   2. Infl. unbranched; spikelets sessile on central axis. **L. cinereus** (Scribn. & Mer.) Love
   2. Infl. branched at lower nodes; some spikelets stalked. **L. xmultiflorus** (Gould) Barkw. & Dewey (*E. triticoides* ssp. *multiflorus*)
1. Plants shorter, <1 m tall; lf.-blade <10 mm wide........ **L. triticoides** (Buckley) Pilger

Lolium (p. 184)

Perennials with flat blades and terminal flat spikes; spikelets solitary, sessile, placed edgewise in little depressions in rachis; inner glume absent. About 8 species, 3 introduced in CA.

Lemmas awnless; auricles inconspicuous; spikelets 6-10-fld........................**L. perenne**
Lemmas awned; auricles conspicuous; spikelets 10-20-fld...................**L. multiflorum**

Melica (p. 180)

Perennial with closed sheaths, ligules and flat narrow lvs.; panicle mostly narrow; spikelets 1-6-fld., terminal floret sterile; glumes with papery margins and tips, pointed, 3-5-veined. About 80 species, 12 in CA.

1. Glumes falling with florets; spikelets bent downwards ....**M. stricta**
1. Glumes persistent, florets breaking off above them; spikelets not bent downwards.
   2. Lemmas awned, awn 5-12 mm long ....**M. aristata**
   2. Lemmas not awned.
      3. Fertile florets 1-2 ....**M. imperfecta**
      3. Fertile florets >2; plants densely tufted ....**M. californica** (incl. var. *nevadensis*)

Muhlenbergia (p. 187)

Perennial or annual grass with ligules and flat or inrolled lvs.; infl. mostly narrow; spikelets 1-fld. falling off above glumes; lemmas mostly longer than glumes, 3-veined. About 160 species, 18 in CA.

1. Annual; usually purplish.
   2. Lemma awned; lf.-blades 1 mm wide ....**M. microsperma**
   2. Lemma awnless ....**M. filiformis**
1. Perennial.
   3. Rhizomes present; panicle 2-15 cm long.

4. Panicles open, pedicels very thin, 2-15 mm long .... **M. asperifolia**
4. Panicles narrow, pedicels stouter, short .... **M. richardsonis**
3. Rhizomes absent, plants tufted bunchgrass; panicle 25-50 cm long, slender. **M. rigens**

Nassella (p. 188)
(part of *Stipa*)

Perennial with panicle-like infl.; glumes tapered from base to tip, tip reddish; floret 1, rounded; awn 45-110 mm long, distal segment wavy. About 80 species, 4 in CA. One species in Kern County: **N. cernua** (Stebb. & Love) Barkw.

Oryza (p. 191)

Erect annual to 2 m tall with dense drooping panicle 15-40 cm long; spikelets 1-fld., strongly flattened, with 2 short glumes falling with spikelet; lemmas awned. About 20 species, 1 introduced in CA. One species in Kern County: **O. sativa**

Panicum (p. 192)

Annual to perennial grasses with ligules and ± flattened spikelets in open panicles; glumes green-veined, the 1st often much reduced; 1st lemma sterile, resembling a 3rd glume; fertile lemma papery hard, rounded with inrolled margins. About 450 species, 9 in CA.

1. Perennial; lf.-blades hairy .. **P. acuminatum** Sw. var. **acuminatum** (*P. occidentale*)
1. Annuals.
2. First glume very short; ring of white hairs at ligule................ **P. dichotomiflorum**
2. First glume as much as half as long as spikelet.
3. Panicle erect, large; spikelets 2.5-4 mm long .... **P. capillare**
3. Panicle drooping; spikelets 4.5-5 mm long .... **P. miliaceum**

Paspalum (p. 192)

Perennial with 2-several long slender racemes spreading from tip of stems; spikelets 1-fld. oval, solitary or in pairs; lemma and glume papery hard with inrolled margins. About 300 species, 4 in CA.

Racemes 2; spikelets solitary .... **P. distichum**
Racemes 4-10; spikelets in pairs.. **P. dilatatum**

Pennisetum

Tufted perennial to 1 m tall with dense purplish panicles; spikelets in groups of 1-3, surrounded by bristles which are plumose at base; spikelets deciduous with glumes. About 80 species, 3 introduced in CA. One species in Kern County: **P. setaceum** (Forsk.) Chiov. (garden escape. Monroe 59)

Phalaris (p. 190)

Annual to perennial with many flat blades and dense spike-like panicles; spikelets 1-fld., flattened, with 1-2 sterile lemmas below fertile floret; glumes equal, boat-shaped, strongly keeled. About 20 species, 10 in CA.

1. Spikelets in groups of 7, 1 fertile surrounded by 6 sterile, the group falling together. **P. paradoxa**
1. Spikelets all alike.
  2. Perennials; panicle 5-18 cm long ....**P. aquatica** L. (*P. stenoptera*)
  2. Annuals.
    3. Glumes broadly winged; panicles ovate.
      4. Sterile lemma solitary; fertile lemma 3 mm long .... **P. minor**
      4. Sterile lemmas 2; fertile lemma 4-6 mm long............................**P. canariensis**
    3. Glumes mostly wingless; panicles subcylindric, often purplish.......**P. lemmonii**

Phleum (p. 188)

Annual or perennial with tall erect stems, flat blades and dense cylindrical panicles; spikelets 1-fld., flattened; glumes equal, ciliate on keels. About 15 species, 2 in CA. One species introduced in Kern County: **P. pratense**

Phragmites (p. 180)

Perennial reed to 4 m tall with large terminal panicles; lvs. 2-6 dm long, 1-5 cm wide; panicle brownish yellow, 1-3 dm long; spikelets 12-15 mm long, 3-7-fld., the rachilla covered with long silky hairs. About 3 species, one in CA. One species in Kern County: **P. australis** (Cav.) Steudlel (*P. communis* var. *berlandieri*)

Piptatherum (p. 187)
(part of *Oryzopsis*)

Slender tufted perennial with ligules and inrolled lvs.; panicles open; spikelets 1-fld.; glumes persistent, lemmas hairy, awned. About 30 species, 2 in CA. One species in Kern County: **P. miliaceum** (L.) Cosson

Poa (p. 180)

Annual or perennial grasses with ligules and narrow, flat or inrolled lvs. ending in a boat-shaped tip; panicles open to contracted; spikelets 2-8-fld.; glumes ± unequal, the lower 1-veined, upper 3-veined, mostly persistent; lemmas without awns. About 500 species, 33 in CA.

1. Annuals.
  2. Lemmas hairy but without cobwebby base .... **P. annua**
  2. Lemmas glabrous to somewhat hairy, with a tuft of cobwebby hairs at base.
    3. Panicle open, broad; plants not in desert.

4. Lemmas glabrous except for web, veins faint .... **P. bolanderi**
4. Lemmas ± hairy, the 5 veins distinct .... **P. howellii**
3. Panicle narrow; tufted desert winter annual...................................... **P. bigelovii**
1. Perennials.
5. Plants with creeping rhizomes.
6. Panicle open, pyramid-shaped; lemmas with tuft of cobwebby hairs at base. **P. pratensis**
6. Panicle narrow, elongated; lemmas without obvious basal tuft; stems flattened. **P. compressa**
5. Plants lacking creeping rhizomes.
7. Florets mostly of dark purple bulblets .... **P. bulbosa**
7. Florets normal; bunchgrasses.
8. Lemmas glabrous; ligule obvious.
P. secunda Presl ssp. **juncifolia** (Scribn.) Soreng (*P. nevadensis*)
8. Lemmas hairy.
9. Spikelets flattened; lemmas with keel, ± hairy.
P. fendleriana (Steudel) Vasey ssp. **longiligula** (Scribn. & Williams) Soreng (*P. longiligula*)
9. Spikelets more rounded; lemmas rounded on back, crisp-hairy towards base........... **P. secunda** Presl. ssp **secunda** (incl. *P. scabrella* & *P. incurva*)

Polypogon (p. 188)

Annual or perennial grasses with dense bristly panicles; spikelets 1-fld., breaking off below the glumes; glumes with slender straight awns; lemmas with shorter awns. About 18 species, 5 introduced into CA.

Annuals; glume-awns 6-10 mm long... **P. monspeliensis**
Tufted perennials; glume-awns 3-5 mm long.. **P. interruptus**

Puccinellia (p. 181)

Low mostly pale tufted annuals or perennials of alkaline habitats; panicle narrow to open; spikelets several-fld., somewhat rounded; lf.-sheath open. About 50 species, 8 in CA.

1. Annual to 20 cm tall .... **P. simplex**
1. Perennials, mostly larger.............**P. nutkaensis** (Presl) Fern. & Weath. (*P. grandis*)

Schismus (p. 185)

Low tufted annuals with narrow lvs. and small panicles; spikelets several-fld., florets breaking off above glumes; lemmas rounded on back, several-veined, 2-toothed at tips. About 5 species, 2 introduced in CA.

Glumes 4-5 mm long; lemmas about 2 mm long, with small notch at tip.... **S. barbatus**
Glumes 5-6 mm long; lemmas 2.5-3 mm long, with 2 pointed lobes at tip.. **S. arabicus**

Scribneria (p. 184)

Low Annual with short narrow lvs. and cylindrical spikes; spikelets 1-fld., solitary, flattened and pressed flatwise against rachis; glumes equal, 1st 2-veined, 2nd 4-veined. One species: **S. bolanderi**

Secale (p. 184)

Annual to 1.5 m tall, blue-green; blades to 12 mm broad; spike slender, nodding; spikelets with 2 fertile fls., alternating on zigzag rachis; glumes awn-like. Five species, one escaped in CA. One species in Kern County: **S. cereale**

Setaria (p. 192)

Annual to perennial grasses with flat lvs. and cylindrical spike-like or looser panicles; spikelets surrounded by few to many awn-like bristles. About 65 species, 6 in CA.

1. Bristles below spikelets numerous, >5.
  2. Annual; spikelets 3 mm long ... **S. pumila** (Poir.) Roem. & Schultes (*S. lutescens*)
  2. Perennial; spikelets 2-3 mm long.
    3. Panicle 4-9 cm long; bristles yellow to purple, 3-18 mm long. **S. gracilis** Kunth (*S. geniculata*)
    3. Panicle 8-15 cm long; bristles orange to purple, 3-6 mm long..... **S. sphaceolata**
1. Bristles below spikelets 1-3; annuals.
  4. Bristles with short stiff downward pointing hairs .... **S. verticillata**
  4. Bristles lacking short stiff hairs .... **S. viridis**

Sorghum (p. 193)

Tall annuals or perennials with flat blades often >2 cm wide and a broad white midrib; ligule a short membrane fringed with hairs; large open terminal panicle with little racemes scattered along branchlets; spikelets in pairs, the sessile one fertile, the pedicelled one developed but sterile. About 20 species, 2 introduced in CA.

1. Perennial with rhizomes .... **S. halepense**
1. Annuals .... **S. bicolor** (*S. sudanense*)

Sporobolus (p. 188)

Perennial bunchgrass 3-10 dm tall with soft straight hairs on throat of lf.-sheath and ligule; lf.-blades rolled inward to flat, narrow; panicle diffuse with 1-fld. spikelets on ends of branches. About 150 species, 6 in CA. One species in Kern County: **S. airoides**

Stipa

(see *Achnatherum* and *Nassella*)

Torreyochloa (p. 179)
(*Glyceria*)

Perennials to 1.4 m tall; ligules membranous; infl. panicle-like; glumes smaller than lowest floret; florets 3-8; lemma rounded on back; awns absent. About 5 species, 2 in CA.

Infl. ± linear, length 5-15 times width, <1 cm wide.............. **T. erecta** (Hitchc.) Church
Infl. ± ovate, length 1-6 times width, 1-12 cm wide.
T. pallida (Presl) Church var. **pauciflora** (Presl) Davis

Trisetum (p. 186)

Tufted perennial with open drooping shining panicles with whorled thin branches terminated by 3-fld. spikelets; lemmas short-hairy at base, 2-toothed at tip, awned from back with curved awn 5-10 mm long. About 65 species, 5 in CA. One species in Kern County: **T. cernuum**

Triticum (p. 184)

Annuals with flat blades and thick spikes; spikelets 2-5-fld., solitary, placed flatwise to rachis; glumes rigid, keeled, sharp-pointed; lemma broad, keeled, terminating in short awn. About 15 species, one in CA. One species introduced in Kern County: **T. aestivum**

Vulpia (p. 178)
(part of *Festuca*)

Annuals with spikelets in narrow or open panicles; florets 2-10, breaking off above glumes; glumes unequal, upper 3-veined; lemmas rounded on back, 3-5-veined, awned from tip. About 20 species, 4 in CA.

1. Lower glume less than half upper glume length.................... **V. myuros** (L.) Gmelin
   2. Lemma margin ciliate near tip.....................var. **hirsuta** (Hackel) Asch & Graebn. (*F. megalura*)
   2. Lemma margin glabrous.. var. **myuros**
1. Lower glume more than half upper glume length.
   3. Florets 5-10; lemma-awn <4 mm long.
   V. octoflora (Walter) Rydb. var. **octoflora**
   3. Florets 1-5; lemma-awn 4-22 mm long.
      4. Lowest infl.-branches appressed to erect........................ **V. bromoides** (L.) Gray (*F. dertonensis*)
      4. Lowest infl.-branches spreading to reflexed....... **V. microstachys** (Nutt.) Benth.
         5. Spikelets not obviously hairy.
         var. **pauciflora** (Beal) Leonard & Gould (var. *simulans* & *F. reflexa*)
         5. Spikelets obviously hairy.

6. Glumes & lemmas hairy.
var. **ciliata** (Beal) Leonard & Gould (*F. grayi* & *F. eastwoodae*)
6. Either glumes or lemmas hairy, not both.
7. Glumes hairy, lemmas glabrous.
var. **confusa** (Piper) Leonard & Gould (*F. confusa* & *F. tracyi*)
7. Glumes glabrous, lemmas hairy.................................. var. **microstachys**

POLEMONIACEAE

Annual or perennial herbs or shrubs; lvs. alternate or opposite, simple to dissected to compound; fls. regular, mostly 5-merous; petals connate, bell-shaped to funnelform; stamens equally or unequally inserted; style simple, stigma-lobes usually 3; fr. a 3-loculed capsule.

1. Calyx growing with capsule, not ruptured.
2. Lvs. compound; corolla blueish. ....**Polemonium**
2. Lvs. simple to dissected; corolla pink, white or yellowish........................ **Collomia**
1. Calyx not growing with capsule, sinuses ruptured.
3. Lvs. mostly opposite.
4. Lvs. entire; stamens unequally inserted.. **Phlox**
4. Lvs. palmately cleft; stamens equally inserted.
5. Plants shrubby; lvs. spine-tipped...**Leptodactylon**
5. Plants not shrubby, mostly annuals; lvs. not spine-tipped...............**Linanthus**
3. Lvs. alternate.
6. Calyx-lobes subequal; infl. variable.
7. Corolla regular; lvs. mostly not spine-tipped.
8. Upper cauline lvs. reduced ....**Gilia**
8. Upper cauline lvs. well-developed.
9. Lf.-lobes not pointed; upper lvs. palmate ....**Allophyllum**
9. Lf.-lobes pointed; upper lvs. pinnate...**Ipomopsis**
7. Corolla mostly irregular; lvs. spine-tipped................................**Loeseliastrum**
6. Calyx-lobes unequal; fls. in dense bracteate heads.
10. Plants cobwebby-pubescent; lvs. not spine-tipped....................... **Eriastrum**
10. Plants not cobwebby-pubescent; lvs. spine-tipped.......................**Navarretia**

Allophyllum (p. 307)

Erect annual herbs 1-6 dm high with leafy well-branched stems; lvs. alternate, the uppermost digitately 3-5-lobed with linear-lanceolate segments; fls. in loose to ± congested 2-8-fld. glomerules; calyx glandular pubescent; corolla funnelform; caps. 3-loculed. Four species, all in CA.

1. Lf.-segments 3-15 mm wide, terminal segment 3-8 times longer than broad.
2. Corolla-tube 6-16 mm long, red-violet to purple with pink lobes.. **A. divaricatum**
2. Corolla-tube 5-8 mm long, white with white lobes......................**A. integrifolium**
1. Lf.-segments not over 3 mm wide, terminal segment 8-14 times longer than broad; fls. dark blue-violet... **A. gilioides**

3. Infl. glomerate, 4-8 fls. subtended by a lf.; many lvs. pinnately lobed. ssp. **gilioides**
3. Infl. loose, 2-3 fls. subtended by a lf.; few lvs. pinnately lobed. ssp. **violaceum** (Heller) Day (*A. violaceum*)

Collomia (p. 307)

Erect annual herbs with alternate lvs.; fls. congested; calyx herbaceous, bell-shaped, the sinuses (clefts between lobes) in age distended at base, growing with capsule; corolla narrow-funnelform to trumpet-shaped. About 15 species, 8 in CA.

1. Lower lvs. divided into narrow lobes to incised, petiolate; fls. pink. **C. heterophylla**
1. Lower lvs. mostly entire, sessile.

2. Corolla 5-15 mm long, the tube little exceeding calyx, the limb pink or white. **C. linearis**
2. Corolla 15-30 mm long, the tube much longer than calyx, the limb salmon-yellow.......... **C. grandiflora**

Eriastrum (p. 307)

Annual or perennial mostly herbaceous puberulent to cobwebby woolly; lvs. alternate, entire and linear or pinnately divided into linear divisions; fls. mostly sessile in bracteate heads; calyx-segments subulate; corolla funnelform to ± salverform. About 13 species, all in CA.

1. Plants perennial; corolla pale- to deep-blue.......... **E. densifolium**
2. Lf.-lobes about as long as width of rachis; lvs. commonly recurved; corolla pale blue to lavender.......... ssp. **mohavense**
2. Lf.-lobes 2-10 times as long as width of rachis; lvs. mostly ascending; fls. deep blue.
3. Bracts mostly 1-5-lobed; fl.-clusters axillary, mostly <15-fld..... ssp. **elongatum**
3. Bracts mostly 5-9-lobed; fl.-clusters mostly terminal, 15-20-fld. ssp. **austromontanum**
1. Plants annual.
4. Stamens inserted in sinus (cleft between lobes) of corolla. **E. pluriflorum** (incl. ssp. *sherman-hoytae*)
4. Stamens inserted well below sinuses.
5. Corolla 15-16 mm long, violet, ± irregular; bracts 5-9-lobed......... **E. eremicum**
5. Corolla <15 mm long, regular; bracts 3-lobed.
6. Corolla-lobes about as long as tube.......... **E. sapphirinum**
6. Corolla-lobes much shorter than tube.
7. Anthers ± exserted.......... **E. sparsiflorum**
7. Anthers included.......... **E. hooveri**

Gilia (p. 308)

Annuals with alternate entire to pinnately dissected lvs., well-developed below, reduced above; fls. solitary or paniculately branched to crowded; corolla tubular-funnelform to trumpet-shaped; stamens equally inserted in corolla-tube or throat; fr. a capsule. About 70 species, 47 in CA.

1. Pollen white to yellow.
  2. Corolla bright red-pink, funnelform.. **G. latifolia**
  2. Corolla white to yellow to violet.
    3. Corolla white to violet.
      4. Corolla-tube longer than calyx, 7-11 mm long, violet to white....... **G. sinuata**
      4. Corolla-tube shorter than calyx, 3-6 mm long, white with yellow throat. **G. inyoensis**
    3. Corolla yellow, 5-6 mm long.. **G. filiformis**
1. Pollen blue.
  5. Plants cobweb-woolly on lower stems and lvs.
    6. Stems glabrous at base.
      7. Corolla purple throughout, 12-16 mm long .... **G. tenuiflora**
        8. Corolla-lobes 2-4 mm wide; infl. diffuse .... ssp. **tenuiflora**
        8. Corolla-lobes 4-6 mm broad, infl. ± congested.............. ssp. **amplifaucalis**
      7. Corolla with some yellow.
        9. Corolla 4-6 mm long....................................... **G. ochroleuca** ssp. **bizonata**
        9. Corolla 8-25 mm long.. **G. latiflora**
          10. Stems glabrous at base.
            11. Corolla-throat mostly yellow and white.
              12. Calyx 4-5 mm long; corolla 15-20 mm long .... ssp. **latiflora**
              12. Calyx 2-3 mm long; corolla 10-15 mm long ....ssp. **cuyamensis**
            11. Corolla-throat purple in lower half, 18-24 mm long...........ssp. **davyi**
          10. Stems cobwebby near base.
            13. Corolla-tube 2-3 times as long as lobes; stout. ssp. **davyi** (incl. ssp. *excellens*)
            13. Corolla-tube 3-6 times longer than lobes, slender...........ssp. **elongata**
    6. Stems cobwebby-pubescent at base.
      14. Upper corolla-throat violet or white, sometimes yellow, lacking purple.
        15. Fls. 4-13 mm long; calyx glabrous to lightly pubescent with small glandular dots or hairs.
          16. Corolla 4-7 mm long.
            17. Corolla violet to white; calyx gland-dotted.
              18. Corolla-lobes pointed; corolla-tube violet to white. **G. transmontana**
              18. Corolla-lobes rounded; corolla-tube deep purple, pollen bright blue.. **G. malior**
            17. Corolla yellowish, pale violet in throat. **G. ochroleuca** ssp. **ochroleuca**

16. Corolla 6-12 mm long, bright violet with some yellow in throat; stamens long-exserted..**G. aliquanta**

15. Fls. 10-32 mm long; calyx pubescent with large glandular dots or cobwebby hairs.

19. Fls. 5-9 mm long; corolla 20-32 mm long with purple tube, yellow throat and pinkish lobes..**G. cana** ssp. **speciosa**

19. Caps. 3-4 mm long.

20. Corolla-tube yellow, 2-4 times as long as calyx; stamens long-exserted, corolla-lobes pink-violet............ **G. leptantha** ssp. **purpusii**

20. Corolla-tube 1-1.5 times as long as calyx.

21. Stamens equal to or exceeding corolla-lobes, corolla-tube pale violet...**G. leptantha** ssp. **pinetorum**

21. Stamens shorter than corolla-lobes; corolla-tube violet, 3-5 mm long; corolla-lobes deep pink-violet.
G. ochroleuca ssp. **vivida** (Grant & Grant) Grant & Grant
(*G. leptantha* ssp. *vivida*)

14. Upper corolla-throat yellow or at least partly purple.

22. Infl. ± congested to glomerate; calyx glandular and at maturity equal to or longer than caps.

23. Stamens and style long-exserted......................................**G. brecciarum**

24. Corolla-throat broad, white to violet; corolla 9-20 mm long.
ssp. **neglecta** (incl. ssp. *argusana*)

24. Corolla-throat deep violet with purple tube, 5-7 mm long.
ssp. **jacens** (Grant & Grant) Day (*G. jacens*)

23. Stamens and style appearing at orifice of corolla or slightly exserted.

25. Corolla-veins beneath sinuses violet, lobes 2-3.5 mm long, violet to white...**G. brecciarum** ssp. **brecciarum**

25. Corolla-veins beneath sinuses white, upper throat with yellow spots.
G. modocensis Eastw. (*G. tetrabreccia*)

22. Infl. diffuse; calyx moderately glandular; corolla throat yellow to purple.

26. Stamens short, maturing at orifice; calyx shorter than caps......**G. minor**

26. Stamens unequal, longest ones well-exserted; calyx equal to or longer than caps..**G. interior**

5. Plants not cobweb-woolly on lower stems or lvs.

27. Infl. with loose to head-like glomerules of 3-many fls. subtended by a single lf.; upper cauline lvs. well-developed and pinnately dissected.

28. Fls. in dense-spherical heads, light blue; stamens exserted.
G. capitata ssp. **abrotanifolia**

28. Fls. not in dense heads; stamens not exserted.

29. Corolla 10-20 mm long; style exserted.

30. Corolla funnelform, blue-violet. ....**G. achilleaefolia** ssp. **multicaulis**

30. Corolla bell-shaped, purple with yellow throat with purple spots.
G. tricolor ssp. **diffusa**

29. Corolla 6-8 mm long with yellow tube and 5 pairs of purple spots in throat..**G. clivorum**

27. Infl. loose with 1-8 fls. subtended by a single lf.; upper cauline lvs. reduced or simple.
  31. Basal rosette absent; corolla 3-6 mm long, violet to pink or white, tube sometimes yellow, throat sometimes with purple spots; mountains. **G. capillaris**
  31. Basal rosette well-developed; deserts.
    32. Basal lvs. 1-3 times pinnate with narrow rachis; stem lvs. pinnately lobed.
      33. Corolla rose-lavender, 10-14 mm long with yellowish tube well exserted. .... **G. scopulorum**
      33. Corolla pale blue to lavender, 6-10 mm long with purple spots in throat, tube included in the calyx. .... **G. stellata**
    32. Basal lvs. strap-shaped, sinuate to bipinnatifid; stem lvs. narrowly linear and entire.
      34. Corolla 2-9 mm long; basal lvs. pinnately toothed or lobed; lf.-rachis or-blade 2-10 mm wide.
        35. Corolla-throat narrow, the tube and throat thread-like; calyx-sinuses U-shaped in fr. .... **G. leptomeria**
        35. Corolla-throat full, not thread-like, mostly included in calyx; calyx-sinuses V-shaped in fr. .... **G. micromeria**
      34. Corolla 8-14 mm long with violet tube, yellowish throat and white with violet-streaked lobes; lf.-rachis 1-2 mm wide.......... **G. hutchinsifolia**

Ipomopsis (p. 312)

Perennial herb with basal rosette of pinnately dissected lvs.; infl. an elongate panicle, the fls. short-pedicelled, each subtended by a bract; corolla tubular-funnelform, 2.5-3.5 cm long, magenta; caps. ovoid. About 30 species, 8 in CA. One species in Kern County: **I. aggregata** ssp. **bridgesii**.

Langloisia

(see *Loseliastrum*)

Leptodactylon (p. 312)

Tufted glandular-villous subshrub with alternate palmately-cleft lvs., their lobes pungent (prickle-tipped); fls. mostly sessile in upper axils and in terminal clusters; calyx-lobes pungent; corolla 1.5-2.5 cm long, narrow-funnelform, white to pink; caps. cylindric. About 7 species, 3 in CA. One species in Kern County: **L. pungens** (incl. ssp. *pulchriflorum* and ssp. *hallii*)

Linanthus (p. 212)

Annuals or perennials with simple or divaricately branched stems; lvs. mostly opposite, palmately parted into linear segments; infl. open to subcapitate; corolla funnelform or trumpet-shaped; stigma mostly 3-lobed. About 40 species, 35in CA.

1. Plant obscurely woody perennial; corolla with yellow tube, white lobes. **L. nuttallii**
1. Plant annual.
  2. Calyx with conspicuous hyaline membrane in sinuses (cleft between lobes).
    3. Pedicels >5 mm long.
      4. Corolla barely exserted, pale blue .... **L. pygmaeus**
      4. Corolla 2-5 times exceeding calyx.
        5. Corolla yellow .... **L. aureus**
        5. Corolla pink-lilac.
          6. Corolla 10-30 mm long .... **L. liniflorus** (incl. ssp. *pharnaceoides*)
          6. Corolla 4-6 mm long .... **L. filipes**
    3. Pedicels <5 mm long.
      7. Corolla with kidney-shaped arches in throat below each lobe, blue, white or yellow .... **L. parryae**
      7. Corolla without kidney-shaped arches.
        8. Corolla-tube included in calyx; plants bloom in evening; fls. whitish.
          9. Corolla 10-15 mm long; lvs. mostly entire .... **L. bigelovii**
          9. Corolla 15-30 mm long; lvs. mostly 3-5-parted .... **L. dichotomus**
        8. Corolla-tube well exserted from calyx.
          10. Calyx ca. 5 mm long .... **L. nudatus**
          10. Calyx 7-10 mm long.
            11. Corolla 12-25 mm long. lobes 2-4 mm long .... **L. ciliatus**
            11. Corolla 25-30 mm long, lobes 5-8 mm long .... **L. montanus**
  2. Calyx not membranous in sinuses.

    12. Corolla-lobes 5-8 mm long, rose to lilac or yellow; throat dark.
      13. Corolla 20-35 mm long .... **L. parviflorus**
      13. Corolla 10-15 mm long .... **L. serrulatus**
    12. Corolla-lobes 3-5 mm long, pink to white; throat and tube yellow.... **L. bicolor**

Loeseliastrum (p. 312)
(part of *Langloisia*)

Diffuse low rigid annuals; lvs. pinnately-toothed, the lobes bristle-tipped; calyx-lobes bristle-tipped; corolla tubular-funnelform, mostly 2-lipped; caps. 3-sided. Two species, both in CA.

Calyx 5-6 mm long; corolla-lobes rounded or truncate at apex, whitish to pink, with reddish pattern on upper lobes .... **L. matthewsii** (Gray) Timb.
Calyx 3-4 mm long; corolla-lobes pointed at apex, pale lavender to white, yellowish or pinkish with purple spots .... **L. schottii** (Torr.) Timb. (incl. *L. flaviflora*)

Navarretia (p. 314)

Annuals with rigid stems; lvs. alternate, entire to pinnately lobed, acerose or spine-tipped; fls. in spiny, densely bracted heads; corolla small, trumpet-shaped to funnelform; stigma entire to 2-3-lobed; caps. ovoid. About 30 species, 27 in CA.

1. Main stems with fine hairs bent downwards.
  2. Corolla pale blue, lobes 4-6 mm long .... **N. intertexta**
  2. Corolla yellow, lobes 1-2 mm long .... **N. nigellaeformis**
1. Main stems with spreading hairs to nearly glabrous.
  3. Corolla 4-7 mm long, pink to white or purple.
    4. Bracts palmate; branches from beneath terminal head. .... **N. divaricata**
    4. Bracts pinnate; branches not from beneath terminal head .... **N. penninsularis**
  3. Corolla 7-18 mm long, pink to purple or blue.
    5. Bracts constricted in middle .... **N. setiloba**
    5. Bracts not constricted, laciniately-lobed; corolla blue .... **N. pubescens** (incl. *N. mitracarpa*)

Phlox (p. 315)

Annuals or cussion-like perennials with opposite entire linear-subulate lvs.; fls. showy, pink to lilac or white; corolla trumpet-shaped, tube 9-13 mm long, lobes 6-7 mm long. About 45 species, 12 in CA.

Perennial .... **P. austromontana** Cov. (*P. diffusa* ssp. *subcarinata*)
Annual .... **P. gracilis** Greene

Polemonium (p. 315)

Leafy annual with pinnately-compound lvs.; infl. cymose; calyx bell-shaped; corolla open bell-shaped, white to bluish, 2-6 mm long. About 20 species, 7 in CA. One species in Kern County: **P. micranthum**.

POLYGONACEAE

Herbs or shrubs with mostly alternate lvs.; lvs. simple with stipules sometimes forming a membranaceous sheath (ocrea) above swollen node; fls. small; calyx 2-6-parted, in 1 or 2 series, often resembling petals; petals 0; stamens mostly 6-9; ovary superior; fr. an achene.

1. Lvs. without stipules.
  2. Lvs. cauline, opposite, notched; invol.-bracts 2-winged .... **Pterostegia**
  2. Lvs. mostly basal or alternate, not notched; invol. ± tubular to absent.
    3. Invol.-bracts free.
      4. Lf.-blade rounded, petiole obvious; calyx yellow .... **Goodmania**
      4. Lf.-blade linear to oblanceolate, petiole often indistinct; calyx greenish.
        5. Calyx woolly; stamens 6 .... **Hollisteria**

5. Calyx sparsely hairy; stamens 3 .... **Lastarriaea**
3. Invol.-bracts fused, funnel- to bell-shaped.
6. Invol.-lobes rounded to acute .... **Eriogonum**
6. Invol.-lobes bristle-tipped (awned).
7. Invol. funnel-shaped, awns 4 .... **Oxytheca**
7. Invol. bell-shaped; awns 3-6.
8. Bracts 2-3 per infl.-node .... **Chorizanthe**
8. Bracts 1 per infl.-node, 3-lobed.
9. Invol. awned at base and tip; bracts 1-10 mm long, on one side of node.. **Centrostegia**
9. Invol. awned only at tip; bracts broad, 5-25 mm long, encircling node. **Mucronea**
1. Lvs. with evident stipular sheaths (ocrea).
10. Calyx 6-parted, the inner 3 larger than outer 3.. **Rumex**
10. Calyx 5-parted, mostly equal.. **Polygonum**

Centrostegia (p. 216)
(part of *Chorizanthe*)

Glandular-hairy annual to 30 cm tall with basal glabrous lvs.; infl. to 50 cm broad; bract 1 per node, 3-lobed, awned; invol. 1 per node, sessile with swollen 3-awned base and 5 upper awns. One species: **C. thurberi** Benth.

Chorizanthe (p. 215)

Low dichotomously or trichotomously branched erect to prostrate annuals; lvs. basal or along stem; infl. capitate or cymose; invols. sessile, cylindric urnshaped or funnelform, mostly 1-fld., 3-6-toothed; achenes 3-angled, glabrous. About 50 species, 33 in CA.

1. Invols. 6-toothed and 6-ribbed.
2. Invol. teeth with marginal papery membrane.. **C. membranacea**
2. Invol. teeth without marginal membrane.
3. Invol. teeth equal or outer 3 subequal.
4. Outer and inner calyx lobes alike or nearly so. **C. brevicornu** ssp. **brevicornu**
4. Inner calyx lobes much smaller than outer.. **C. xanti**
3. Invol. teeth unequal, anterior one usually longer than the tube.
5. Elongate anterior tooth straight; outer calyx lobes linear-oblong. **C. uniaristata**
5. Elongate tooth hooked; outer calyx lobes ovate.. **C. clevelandii**
1. Invols. not 6-toothed and 6-ribbed.
6. Invol. teeth 4-5.
7. Invol. teeth and ribs 4-5, one tooth much exceeding others; plants prostrate. **C. spinosa**
7. Invol. teeth 5, one leaf-like and enlarged; plants erect. .... **C. watsonii**
6. Invol. teeth 3, ± equal.. **C. rigida**

Eriogonum (p. 217)

Annual or perennial herbs or shrubs with basal or cauline (on stem) alternate simple entire lvs.; fls. borne in involucres which are 4-10-lobed or -toothed and few- to many-flowered; pedicels ± exserted, jointed at summit with base of perianth or with a slender stalk-like base; perianth a "calyx", 6-parted, petaloid in 2 series; stamens 9; styles 3. About 250 species, 113 in CA.

(Some of the following marked with an asterisk (*) are probable for Kern County as reported in Reveal (1989) but need confirmation.)

1. Calyx stalked at base; plants ± shrubby; fls. yellow, fading pinkish.
  2. Invols. solitary at nodes.......... **E. saxatile**
  2. Invols. clustered, subtended by bracts.......... **E. umbellatum**
    3. Infl. a simple umbel, rays unbranched.......... * var. **nevadense** Gand. (1500-3200 m, Jeffrey pine forests, east side of Sierra)
    3. Infl. compound, some rays branched.
      4. Lf.-blade ± glabrous on both sides.......... * var. **chlorothamnus** Rev. (1600-2900 m, Sagebrush shrubland, east side of Sierra)
      4. Lf.-blade hairy, at least below.
        5. Lf.-blade sparsely hairy.......... * var. **subaridum** Stokes (1300-2800 m Sagebrush shrubland, Tehachapis to Transverse Range)
        5. Lf.-blade densely tomentose, at least below.
          6. Lf.-blade remaining tomentose above.......... var. **bahiiforme**
          6. Lf.-blade shedding above, becoming ± glabrous.
            7. Infl. subglabrous.......... * var. **furcosum** Rev. (1200-3000 m, sand or gravel soils, Sierra)
            7. Infl. densely tomentose.......... * var. **munzii** Rev. (1500-2900 m, rocky areas, Transverse Range)
1. Calyx not stalked at base; bracts not leafy.
  8. Invols. bell-shaped to turbinate, not angled or ribbed, 4- or 5-toothed or -lobed; mostly peduncled; annuals (*E. inflatum* a perennial).
    9. Lvs. basal and on lower nodes.
      10. Basal lvs. linear, hispid-pilose, with margins rolled backwards. **E. spergulinum**
        11. Internodes with capitate glands.......... var. **spergulinum**
        11. Internodes with stipitate glands.......... var. **reddingianum**
      10. Basal lvs. oblanceolate, tomentose or with tufts of woolly hairs.
        12. Invols. ± glandular-puberulent to pubescent externally.
          13. Calyx-segments dissimilar, outer broader than inner.
            14. Outer segments obovate to elliptic, not obviously inflated; inner segments spatulate; stamens conspicuously exserted, fls. rose, tipped with white.......... **E angulosum**
            14. Outer segments roundish or obovate, conspicuously inflated, inner segments narrowly lanceolate; stamens included; fls. pink to yellowish with reddish spot on outer segments.

15. Outer segments inflated at base, the sides incurved; inner segments obtuse to acute.......... **E. maculatum**
15. Outer segments inflated above middle, the apex incurved; inner segments acute to acuminate.......... **E. viridescens**
13. Calyx segments similar.
16. Fls. concealed by cotton-like tomentum inside invols.; invols. 3 mm long; fls. white.......... **E. gossypinum**
16. Fls. not concealed; invols. 2 mm long; fls. rose, tipped with white. **E. gracillimum**
12. Invols. glabrous or densely tomentose, not glandular.
17. Invols. glabrous; fls. white, tinged with pink or yellow .... **E. ordii**
17. Invols. with soft white hairs; fls. white with green midribs. **E. temblorense**
9. Lvs. strictly basal, with soft white hairs at least below.
18. Invols. 4-lobed or -toothed.
19. Calyx pubescent with hooked hairs externally, reddish.
20. Invols. 2-fld.; calyx ca. 1 mm long.......... **E. hirtiflorum**
20. Invols. 4-6-fld.; calyx ca. 1.5 mm long. **E. inerme** (incl. var. *hispidulim*)
19. Calyx puberulent or with small stiff hairs externally, the hairs not hooked; calyx yellow or reddish.
21. Calyx ± 0.5 mm long, pink to red.......... * **E. parishii** Wats. (1200-3200 m, granitic sandy soils, eastern Sierra to Transverse Range)
21. Calyx >0.5 mm long, yellowish to reddish.
22. Lvs. short-hirsute, suborbicular, 1-2.5 cm long. **E. trichopes** var. **trichopes**
22. Lvs. with tufts of soft hairs or glabrous, ovate to oblanceolate, 2-8 cm long.......... **E. ordii**
18. Invols. 5-lobed or toothed.
23. Calyx glabrous externally, whitish.
24. Stems glabrous or glandular, not woolly; peduncles deflexed.
25. Stems glandular, to 4 dm tall, not inflated.......... **E. brachypodum**
25. Stems glabrous, to 10 dm tall, often inflated. **E. deflexum** Torr.var. **baratum** (Elmer) Rev. (*E. baratum*)
24. Stems woolly.......... **E. temblorense**
23. Calyx pubescent externally, yellowish.
26. Invols. glandular-puberulent externally; calyx yellow with reddish midribs.......... **E. pusillum**
26. Invols. glabrous externally; calyx yellow.
27. Outer calyx-segments not saccate-dilated.
28. Lf.-blades short-hirsute; invols. turbinate.
29. Plants strictly annual; branchlets numerous and whorled at each node; lower stems sometimes weakly inflated. **E. trichopes** var. **trichopes**
29. Plants perennial but flowering the first year; branchlets few at each node, not whorled; lower stems conspicuously inflated. **E. inflatum** var. **inflatum**

28. Lf.-blades woolly; invols. broadly bell-shaped.......... **E. reniforme**
27. Outer calyx-segments saccate-dilated at each side of the heart-shaped base, yellow maturing reddish.. **E. thomasii**
8. Invols. cylindric or cylindric-turbinate, often 5-6 nerved, angled, or ribbed, mostly sessile, solitary or in heads; annuals or perennials.
30. Plants annual; lvs. mostly in basal rosettes.
31. Flowering branches with short branchlets, usually of a single internode; invols. axillary and terminal.
32. Calyx 1-1.5 mm long, yellow... **E. mohavense**
32. Calyx 1.5-2.5 mm long, white to rose or red.
33. Stems with soft white hairs .... **E. temblorense**
33. Stems glabrous... **E. covilleanum**
31. Flowering branches elongate, wand-shaped, bearing invols. at nodes.
34. Invols. 2-5 mm long.
35. Lvs. oblong-obovate to oblanceolate; stems with soft white hairs, calyx yellow to pinkish
36. Invol. 4-5 mm long with minute teeth .... **E. roseum**
36. Invol 2-3 mm long with prominent teeth.
37. Infl.-axes ± upcurved; petiole winged.... * **E. cithariforme** Wats. (500-2400 m, sandy soil, extreme SW Kern County)
37. Infl.-axes straight; petiole slender...**E. gracile** Benth var. **gracile** (*E. roseum* var. *leucocladon*)
35. Lvs. rounded; stems usually glabrous except sparsely woolly at base, calyx white or tinged pink.
E. davidsonii (*E. molestum* var. *davidsonii*)
34. Invols. 1-1.5 mm long.
38. Outer calyx-segments fan-shaped, their sides incurved below the broad truncate tips, yellowish to reddish; branches incurved at summit in age... **E. nidularium**
38. Outer calyx-segments not as above.
39. Calyx-segments 1.5-2 mm long, white to pink.
E. baileyi var. **baileyi**
39. Calyx-segments 1-1.5 mm long, yellow............. **E. brachyanthum**
30. Plants perennial; lvs. often on stem as well as basal.
40. Invols. solitary at nodes, lateral ones appressed.
41. Calyx stipitate with long, winged attenuated bases, 5-7 mm long, pinkish to yellowish.. **E. saxatile**
41. Calyx not stipitate, 2-5 mm long.
42. Invols. 6-7 mm long; herbs 8-18 dm tall; calyx pinkish.
E. elongatum
42. Invols. 2-6 mm long.
43. Infl. with invols. placed racemosely along the branches; invols. with soft white hairs; plants ± shrubby; calyx white to pink.
E. wrightii
44. Low woody subshrubs, 2-4 dm high; lvs. 1.5-3 cm long.
var. **trachygonum**

44. Low loosely matted, 1-2.5 dm high; lvs. 5-12 mm long. var. **subscaposum**

43. Infl. with invols. in cymes or panicles.

45. Lvs. >2 cm long .... **E. nudum** Benth. (*E. latifolium*)

46. Lvs. scattered on stem.

47. Stems inflated, robust; calyx yellow to white. var. **indictum** (Jeps.) Rev. (*E. indictum*)

47. Stems not inflated; calyx cream to pink. var. **auriculatum** (Benth.) Jeps.

46. Lvs. in basal rosettes; calyx whitish to pinkish.

48. Invols. 2-6 in a cluster.

49. Fls. glabrous externally, white to pink to yellow. var. **nudum**

49. Fls. pubescent externally, yellow to white.

50. Fls. white.................................. var. **pubiflorum** Benth. (*E. latifolium* ssp. *saxicola*)

50. Fls. bright yellow............ var. **westonii** (Stokes) Howell (*E. latifolium* ssp. *westonii*)

48. Invols. solitary, rarely in pairs..........var. **pauciflorum** Wats.

45. Lvs. <2 cm long; calyx yellow or white.

51. Infl. a compact terminal cyme; invols. tomentose or glabrous. **E. microthecum**

52. Lf. mostly 8-12 mm long, barely rolled under. var. **laxiflorum**

52. Lf. mostly 5-12 mm long, strongly rolled under. var. **panamintensis** Stokes (Limestones, Erskine Creek Canyon. Shevock)

51. Infl. a divaricately branched panicle; invols. glabrous.

53. Outer calyx-segments yellow, round, ± heart-shaped at base; branches green, dichotomous, ascending. **E. heermannii** var. **heermannii**

53. Outer calyx-segments white, obovate, narrowed at base; branches grayish, mostly horizontal................ **E. plumatella**

40. Invols. mostly clustered or in heads.

54. Plants cespitose (matted), herbaceous with woody caudex; flowering stems scape-like.

55. Calyx hairy.. **E. breedlovei** (Howell) Rev.

56. Infl. ± glabrous; invol.-teeth rounded, obviously papery margined. var. **shevockii** Howell (granite rock outcrops, Kern Plateau, crest of southern Sierra to Scodie Mtns., Shevock)

56. Infl. densely glandular; invol.-teeth acute, inconspicuously papery margined. .... var. **breedlovei** (Endemic to metamorphics, Piute Mtns. Shevock)

55. Calyx glabrous.. **E. kennedyi**

57. Lf. oblong, 2-6 mm long (Sierra Nevada)

58. Lf. gray to rusty; invols. 2-3.5 mm long, tomentose.
* var. **pinicola** Rev.
(1400-1800 m, dry ridges, western flank of Sierra)
58. Lf. white; invols. 1-2 mm long, subglabrous.........* var. **purpusii** (Brand.) Rev. (1200-2500 m, sandy granite flats, Indian Wells Canyon)
57. Lf. oblanceolate, 2-12 mm long (Transverse Range).
59. Lf. mostly 6-12 mm long; infl.-axis 8-15 cm long.
var. **austromontanum**
59. Lf. mostly 2-5 mm long; infl.-axis 4-12 cm long.
* var. **kennedyi**
(1400-1800 m, dry gravelly areas, probable on Mt. Pinos)
54. Plants shrubby or, if herbaceous, not cespitose.
60. Plants herbaceous, only base is woody............. **E. elatum** var. **elatum**
60. Plants definitely shrubby; calyx white to pinkish....... **E. fasciculatum**
61. Lvs., calyx and invols. pubescent.
var. **foliolosum** (Nutt.) Abrams (ssp. *foliosum*)
61. Lvs. densely covered with white hairs above.
var. **polifolium** (DC) T & G (ssp. *polifolium*)

Goodmania (p. 222)
(part of *Oxytheca*)

Spreading annual to 8 cm tall with basal, rounded, tomentose lvs.; infl.-bracts leaf-like to linear and awned; invol.-bracts 5, free, awned; calyx yellow. One species: **G. luteola** (Parry) Rev. & Ertter

Hollisteria (p. 222)

Slender-stemmed diffuse white-woolly annual; stipules lanceolate, spine-tipped; invol. solitary, sessile in axils, consisting of three 2-flowered bracts; calyx 6-cleft, greenish yellow, densely woolly. One species: **H. lanata**

Lastarriaea (p. 215)
(part of *Chorizanthe*)

Hairy annual to 15 cm tall with basal, linear lvs.; infl.-bracts 5, awns hooked; true invol. absent, calyx invol.-like, greenish with hooked awns; Three species, one in CA. One species in Kern County: **L. coriacea** (Goodm.) Hoov.

Mucronea (p. 215)
(part of *Chorizanthe*)

Glandular-hairy annual to 30 cm tall with basal oblanceolate lvs.; infl.-bracts 1 per node, 3-lobed, encircling stem; invol. ± bilateral, 4-lobed, white to pink, awned. Two species, both in CA. One species in Kern County: **M. perfoliata** (Gray) Heller

Oxytheca (p. 222)

Slender-stemmed dichotomously branched annuals with small stalked glands at nodes; bracts 3, encircling stem; invol.-lobes 4, each terminated by a bristle or awn; perianth white to yellow. About 8 species, 7 in CA. One species in Kern County: **O. perfoliata**

Polygonum (p. 222)

Annual or perennial herbs often with swollen joints; lvs. entire, alternate, stipules usually papery, sheathing, forming ocrea; calyx mostly 5-parted, petaloid; styles 2-3; achenes 3-angled. About 300 species, 32 in CA.

1. Fls. in axillary clusters or in spikes with leafy bracts; ocreae hyaline and frequently lacerate; annuals.
  2. Stems ± rounded in cross section, often spreading to prostrate.
    3. Achenes dull, scarcely shining; fls. borne in axils of scarcely reduced lvs.; calyx pinkish. .... **P. arenastrum**
    3. Achenes smooth and shining; upper lvs. bracteate .... **P. argyrocoleon**
  2. Stems and branches strongly angled, usually suberect .... **P. douglasii**
    4. Pedicels recurved, at least in fr.; calyx reddish. ssp. **douglasii** (incl. var. *latifolium*)
    4. Pedicels erect.
      5. Fls. in small axillary clusters, not spicate.......... ssp. **johnstonii** (Munz) Hickm. (*P. sawatchense*)
      5. Fls. in terminal leafy-bracteate spikes; calyx green; stamens 3. **P. polygaloides** Meis ssp. **kelloggii** (Greene) Hickm. (*P. kelloggii*)
1. Fls. clustered in terminal spikes w/o leafy bracts; ocreae rarely hyaline or lacerate.
  6. Stems simple; infl. a single terminal spike; calyx white; perennial of wet meadows. .... **P. bistortoides**
  6. Stems branched; spikes on main stem and branches.
    7. Mature ocreae not ciliate; calyx rose to purplish.
      8. Perennial with creeping rhizomes; ± aquatic. **P. amphibium** L. var. **emersum** Mich. (*P. coccineum*)
      8. Annuals.
        9. Peduncles with stalked glands; spikes erect; bracts of infl. ciliate. **P. mexicanum**
        9. Peduncles with or w/o sessile glands; spikes arching.......... **P. lapathifolium**
    7. Mature ocreae fringed with cilia.
      10. Calyx punctate with glands, white; perennial.......... **P. punctatum**
      10. Calyx not glandular-punctate, pink or purplish.......... **P. persicaria**

Pterostegia

Dichotomously diffusely branched slender stemmed annual; lvs. opposite, elliptical, entire to 2-lobed; bracts foliaceous, reddish, subtending a fl., enlarged in fr.; calyx reddish, 6-parted. One species: **P. drymarioides** Fisch. & Meyer (Shaded areas, lower Kern Canyon, L. (Cooley 229; common throughout lower elevations).

Rumex (p. 224)

Annual to perennial herbs; fls. numerous, greenish, crowded and commonly whorled in panicled racemes; lvs. mostly basal, cauline alternate; petioles sheathing at base; calyx in 2 series of 3 each, the inner larger, ± colored and appressed to the 3-angled achene (called valves in fr.). About 140 species, 16 in CA.

1. Plants dioecious, low, slender, acid to the taste; valves small, without callosities (hardened thickenings) .... **R. acetosella**
1. Plants monoecious, taller, not acid to the taste.
    2. Stems ascending to decumbent; basal lvs. absent .... **R. salicifolius**
        3. Valves without callosities........ var. **denticulatus** Torr. (*R. californicus*)
        3. One or all valves with callosity.
            4. Valves 2-3 mm long, only one with callosity........ var. **salicifolius**
            4. Valves 3-4 mm long, yellowish, mostly all with a callosity. var. **transitorius** (Rech.) Hickm. (*R. transitorius*)
    2. Stems, erect; basal lvs. present when young.
        5. Valves without callosities........ **R. hymenosepalus**
        5. At least one valve with distinct callosity.
            6. Valves denticulate; plants annual or perennial.
                7. Perennial; lvs. to >70 cm long; valves 3-5 mm wide .... **R. obtusifolius** L. (Irrigation sump, west of Bakersfield. M. Pelz 28)
                7. Annual; lvs. <20 cm long; valves 2-3 mm wide........ **R. violacens**
            6. Valves entire; plants perennial.
                8. Lvs. small, flat; valves small, scarcely larger than callosity. **R. conglomeratus**
                8. Lvs. large, somewhat crisped or undulate; valves larger than callosities; plant dirty brick-red when dry........ **R. crispus**

POLYPODIACEAE

Polypodium (p. 165)

Epiphytic ferns with deciduous fronds 1-3.5 dm high; fronds pinnately divided nearly to rachis, veins dark; sori round, naked. About 125 species, 6 in CA. One species in Kern County: **P. calirhiza** Whitmore & Smith (*P. californicum*)

PONTEDERIACEAE

Eichornia (p. 198)

Perennial floating aquatic herb rooting at nodes; lvs. rounded with inflated petioles; fls. showy, to 5 cm long, violet. About 7 species, one in CA. One species introduced in Kern County: **E. crassipes**

PORTULACACEAE

Annual to perennial ± succulent herbs; lvs. alternate or opposite or basal; fls. regular; sepals 2; petals mostly 5, sometimes 0; ovary superior; styles 2-9; fr. a 1-locular capsule, sometimes circumscissile (top opening as a lid).

1. Calyx fused to lower part of ovary, its lobes coming off summit of circumscissile caps. .... **Portulaca**
1. Calyx and ovary free.
   2. Caps. circumscissile; sepals 2-8 .... **Lewisia**
   2. Caps. opening by 2-3 valves; sepals 2.
      3. Style 1; caps. 2-valved; infl. with fls. on one side .... **Calyptridium**
      3. Styles 3; caps. 3-valved.
         4. Cauline lvs. alternate; petals red.......... **Calandrinia**
         4. Cauline lvs. opposite; petals white to pink.
            5. Lvs. mostly basal; cauline lvs. mostly 2 .... **Claytonia**
            5. Lvs. ± gradually reduced upward; cauline lvs. >2.......... **Montia**

Calandrinia (p. 232)

Simple subglabrous annual with several branches from base; lvs. alternate, entire; fls. red (rarely white), in leafy racemes; sepals 2; petals 5; caps. ovoid, pointed. About 150 species, 4 in CA. One species in Kern County: **C. ciliata** (incl. var. *menziesii*)

Calyptridium (p. 232)

Annual or perennial herbs with alternate or basal spatulate lvs.; fls. small, in scorpioid spikes; sepals 2, rounded with papery margin; petals 2-4; stamens 1-3; style simple with 2 stigmas. About 8 species, all in CA.

1. Style short; petals in age folding as a cap over the elongate caps.
   2. Caps. 2-4 times as long as sepals; sepals 1-2 mm long .... **C. monandrum**
   2. Caps. <2 times as long as sepals; sepals 2-4 mm long.......... **C. parryi**
1. Style long-filiform; petals in age twisting around the style; caps. ± round. **C. umbellatum**

Claytonia (p. 232)
(part of *Montia*)

Glabrous ± fleshy annuals with mostly basal and 2 cauline, opposite lvs.; infl. terminal. 1-sided; pedicels reflexed; petals pinkish to white. About 28 species, 14 in CA.

1. Stem lvs. perfoliate, forming flattish disk.......... **C. perfoliata** Willd.
1. Stem lvs. not perfoliate, united only on part of one side.
   2. Petals 2.5-4 mm long; infl. mostly 3-6 fld.

3. Cauline lvs. 10-20 mm long, not exceeding infl...**C. exigua** T. & G. var. **exigua** (*M. spathulata* var. *exigua*)

3. Cauline lvs. 2-6 cm long, exceeding infl. **C. parviflora** Hook. var. **viridis** (Davids.) Miller & Chamb. (*M. spathulata* var. *viridis*)

2. Petals 5-7 mm long; infl. mostly 8-15 fld........ **C. gypsophiloides** Fisch. & Meyer

Lewisia (p. 232)

Fleshy perennials from thick fleshy base; lvs. mostly in basal rosette, entire; sepals 2-6(8); petals 4-18, rose to white, styles 3-8, united at base; caps. circumscissile, globose or ovoid. About 20 species, 16 in CA.

Sepals 2; bracts ovate.. **L. disepala** Rydb.
(Sandy ridgetops, Kern Plateau & Scodie Mtns. Flowers very early. Shevock 10799)
Sepals 6-8; bracts awl-like.. **L. rediviva** (incl. var. *minor*)

Montia (p. 232)

Annual or perennial glabrous, often glaucous, herbs; lvs. ± fleshy, basal and opposite; pedicels enlarged at base of fr., often recurved; sepals 2; petals 2-5, pink or white; style-branches 3; caps. globose to ovoid with 1-3 seeds. About 12 species, 7 in CA.

Delicate annual; petals 1-3 mm long... **M. fontana**
Perennial, ± fleshy; petals 5-8 mm long.. **M. chamissoi**

Portulaca (p. 233)

Fleshy herb with scattered alternate or ± opposite lvs.; fls. sessile, yellow, opening mostly in sunshine; sepals 2; petals 5; stamens 7-20; style deeply 5-6-parted; caps. 1-loculed, many-seeded. About 100 species, 2 in CA. One species introduced in Kern County: **P. oleracea**

POTAMOGETONACEAE
(incl. Ruppiaceae)

Perennial aquatic herbs rooting at nodes; lvs. all submerged and thin or some also floating and thicker, all with sheathing bases; fls. small, clustered in axillary spikes; perianth of 4 rounded segments or none; stamens 2-4; carpels 4, free, becoming hard and 1-seeded.

Perianth parts 4; stamens 4; fr. sessile.. **Potamogeton**
Perianth parts none; stamens 2; fr. long-stalked .... **Ruppia**

Potamogeton (p. 173)

Aquatic herb with submerged linear lvs. and some species with floating elliptic lvs.; perianth of 4 segments; stamens 4. About 90 species, 17 in CA.

1. Lvs. all submerged and similar.
   2. Lvs. oblong, 5-8 mm wide..**P. crispus**
   2. Lvs. linear, <5 mm wide.
      3. Stipules fused with base of lf. forming a sheath at least 1 cm.long; lvs. 3-15 cm long.. **P. pectinatus**
      3. Stipules mostly free from lvs.
         4. Lvs. 2-6 cm long, with a pair of basal glands; nerves mostly 3.......**P. pusillus**
         4. Lvs. 3-10 cm long, lacking basal glands, 3-5-nerved......................**P. foliosus**
1. Lvs. submerged and floating, dissimilar.
   5. Submerged lvs. lanceolate to ovate, 1-4 cm wide.................................**P. nodosus**
   5. Submerged lvs. linear, <1 cm wide.
      6. Stipules fused to lf.-base; infls. both submerged and emergent.
         P. diversifolius Raf. (Vernal pools, Lynn's Valley near Glennville. Shevock)
      6. Stipules free from lf.-base; infls. all emergent, cylindric ....**P. nodosus**

Ruppia (p. 174)

Aquatic herbs with branching thread-like stems with internodes 2-8 cm long; lvs. thread-like, 2-10 cm long; fls. few, small, in terminal spikes; perianth absent; stamens 2. Two species, both in CA. One species in Kern County: **R. maritima**

PRIMULACEAE

Scapose or leafy-stemmed annual to perennial herbs; lvs. simple, opposite, whorled or alternate; fls. perfect, regular, 5-merous; petals fused; stamens 5, inserted opposite corolla-lobes; ovary 1-celled, superior; fr. a capsule.

1. Plants with leafy stems; fls. not in umbels, salmon-colored........................**Anagallis**
1. Plants scapose; fls. in bracteate umbels on long stalk.
   2. Corolla-lobes spreading, whitish..**Androsace**
   2. Corolla-lobes reflexed, acute, magenta to white ....**Dodecatheon**

Anagallis (p. 302)

Low spreading diffusely branched glabrous annual; lvs. ovate to oval, sessile; fls. small, solitary in axils; corolla rotate, salmon, rarely white; fr. a capsule. About 20 species, one in CA. One species introduced in Kern County: **A. arvensis**

Androsace (p. 303)

Small puberulent annual; lvs. in basal rosette; scapes 1-6, with involucrate umbels of small whitish fls.; corolla trumpet-shaped with short tube; fr. globose. About 100 species, 4 in CA. One species in Kern County: **A. elongata** ssp. **acuta**

Dodecatheon (p. 303)

Low perennial scapose herbs with basal lvs.; scape ending in an umbel; fls. 4-5-merous, nodding; calyx and corolla reflexed; filaments short, broad, often united; style filiform, exserted. About 14 species, 8 in CA.

Plants glabrous; scape 0.8-2.5 dm high; stamen connective maroon to black. **D. hendersonii** Gray (*D. hansenii*)
Plants glandular-pubescent; scape 2-4 dm high; stamen connective yellow. **D. clevelandii** ssp. **patulum**

PTERIDACEAE
(Polypodiaceae)

Terrestrial ferns with pinnately compound fronds with marginal or submarginal sori protected by incurled lf.-margin or sporangia covering backs of fronds.

1. Sporangia scattered on veins on backs of fronds .... **Pentagramma**
1. Sporangia near margins of fronds.
   2. Lower surface of lf. with scales, hairs or glands .... **Cheilanthes**
   2. Lower surface of lf. glabrous.
      3. Sterile lf.-segments ± sessile... **Aspidotis**
      3. Sterile lf.-segments stalked.
         4. Sporangia along outer third to two thirds of vein; lf.-margin only slightly curled over sporangia... **Argyrochosma**
         4. Sporangia only at tip of veins; lf.-margin curled over sporangia.......... **Pellaea**

Argyrochosma (p. 166)
(*Notholaena*)

Small ferns of rock crevices; lvs. 2-3 pinnate with roundish blue- to gray-green thickish segments; petioles dark brown, glabrous; sporangia at least on outer third of lf.-segments. About 20 species, 2 in CA. One species in Kern County: **A. jonesii** (Maxon) Windham

Aspidotis (p. 165)
(part of *Cheilanthes*)

Small ferns with several 0.5-3.5 dm long fronds with dark brown wiry petioles; frond finely dissected with narrow acute segments; sori small in sinuses of the teeth. Four species, three in CA.

Sporangia continuous along both sides of midvein; lvs. leathery.
A. densa (Brackenr.) Lellinger (*Cheilanthes siliquosa*)
Sporangia not continuous along both sides of midvein; lvs. not leathery.
A. californica (Hook.) Copel.

Cheilanthes (p. 165)

Small drought resistant ferns with pinnate to compound hairy to scaly fronds; sori marginal covered by inrolled indusium. About 150 species, 10 in CA.

1. Lf.-blade without scales, clear sticky exudate present................................**C. viscida**
1. Lf.-blade with scales, exudate absent.
   2. Lf.-segments with dissected scales on upper surface; lower surface with pale to reddish scales...**C. intertexta**
   2. Lfts. glabrous above, scales beneath large, whitish................................**C. covillei**

Pellaea (p. 166)

Small ferns with erect glabrous persistent 2-4 times pinnately compound fronds; lfts. round-oval to linear; sori concealed by strongly inrolled lf.-margins. About 35 species, 6 in CA.

Petioles 4-20 cm long, purple-brown; lfts. gray-green.
P. mucronata var. **californica** (Lemm.) Munz & Johnston
Petioles 5-40 cm long, straw-colored; lfts. reddish.........................**P. andromedaefolia**

Pentagramma (p. 166)
(*Pityrogramma*)

Small ferns with crowded 1-3 times pinnately compound fronds; fronds densely wax-powdered beneath; sori along veins; indusium absent. Two species, both in CA.

Upper surface of frond glabrous, under surface yellow; petiole reddish-brown.
P. triangularis ssp. **triangularis**
Upper surface of frond glandular, under surface white; petiole blackish.
P. pallida (Weath.) Yatsk. (*P. triangularis* var. *pallida*)

RANUNCULACEAE

Herbs or woody climbers with simple or compound lvs. without stipules; fl.- parts free, stamens numerous, hypogynous; pistils few to many, separate; fr. a follicle or achene.

1. Pistils several-ovuled; fr. a follicle.
  2. Fls. irregular.
    3. Upper sepal spurred; petals 4 .... **Delphinium**
    3. Upper sepal hooded; petals 2 .... **Aconitum**
  2. Fls. regular.
    4. Petals absent; sepals petaloid .... **Isopyrum**
    4. Petals spurred, red-orange .... **Aquilegia**
1. Pistils 1-ovuled; fr. an achene.
  5. Lvs. opposite; petals absent; woody vines .... **Clematis**
  5. Lvs. alternate.
    6. Petals present.
      7. Sepals spurred; small annuals with basal lvs .... **Myosurus**
      7. Sepals not spurred; lvs. cauline and basal .... **Ranunculus**
    6. Petals absent .... **Thalictrum**

Aconitum (p. 236)

Perennial 5-20 dm tall; lvs. 3-5-cleft; fls. purple-blue, 3-4 cm long; fr. a follicle. About 100 species, 1 in CA. One species in Kern County: **A. columbianum**

Aquilegia (p. 237)

Perennial 1-3 dm tall; lvs. mostly basal, compound; fls. pendent; sepals red-orange, laminae yellow; petals red-orange, spurred; fr. a follicle. About 70 species, 3 in CA. One species in Kern County: **A. formosa** (incl. vars. *pauciflora* and *truncata*)

Clematis (p. 237)

Half-woody vines which climb by twining petioles; lvs. opposite, pinnately 5-7-foliate; petals none; sepals white, pubescent; fr. an achene with long feathery style. Over 250 species, 3 in CA. One species in Kern County: **C. ligusticifolia**

Delphinium (p. 237)

Erect perennial herb with palmately lobed or divided lvs.; fls. irregular in showy spikes or racemes; Sepals 5, one prolonged into a spur containing spurs of 2 upper petals; fr. a follicle. About 250 species, 21 in CA.

1. Fls. pink to rose.
  2. Plants 7-10 dm tall; sepals acuminate .... **D. purpusii**
  2. Plants 2-4 dm tall; sepals rounded .... **D. gracilentum**

1. Fls. blue to purple.
  3. Stem very slender at base, easily separated from roots.
    4. Infl. glandular-pubescent; lower elevations (Foothill woodland). **D. gracilentum** (*D. paten s* ssp. *greenii*)
    4. Infl. glabrous; higher (Yellow Pine Forest)................**D. patens** ssp. **montanum**
  3. Stem not attenuate at base.
    5. Fls. not strongly bicolored.
      6. Fls. deep blue to purple.
        7. Stems glabrous...**D. parishii** ssp. **parishii**
        7. Stems pubescent...............................**D. parryi** ssp. **parryi** (ssp. *seditiosum*)
      6. Fls. pale blue to whitish.
        8. Stems glabrous.
          9. Fls. white to pinkish.
            10. Follicles ca. 20 mm long; sepals 15-20 mm long...... **D. gypsophilum**
            10. Follicles 7-9 mm long; sepals 10-14 mm long. ....**D. inopinum**
          9. Fls. light blue; follicles 10-14 mm long. **D. parryi** ssp. **purpureum** (Lewis & Epling) Warn. (*D. parishii* ssp. *purpureum*)
        8. Stems pubescent.
          11. Lvs. cauline; fls. white to pink................. **D. hesperium** ssp. **pallescens**
          11. Lvs. mostly basal; fls. greenish to bluish.............................. **D. hanseni**
            12. Sepals violet-purple to maroon (foothills <600 m). ssp. **ewanianum** Warn.
            12. Sepals dark blue-purple to whitish-pink (above 600 m). ssp. **kernense** (Davidson) Ewan
    5. Fls. strongly bicolored, sepals blue, petals white......................... **D. recurvatum**

Isopyrum (p. 238)

Low slender glabrous perennial herbs; lvs. 2-3-ternate; sepals white to pinkish; petals none; fr. a follicle. About 30 species, 2 in CA. One species in Kern County: **I. occidentale**

Myosurus (p. 238)

Small tufted annuals with basal entire linear lvs.; sepals spurred; petals greenish-yellow; fr. an achene. About 15 species, 4 in CA. One species in Kern County: **M. minimus** (incl. var. *filiformis*)

Ranunculus (p. 238)

Annual or perennial herbs with basal and alternate entire to compound lvs.; fls. yellow or white mostly 5-merous; stamens and pistils many; fr. an achene. About 250 species, 30 in CA.

1. Petals white; submersed aquatic ....**R. aquatilus**
1. Petals yellow; terrestrial.

2. Plants annual.
   3. Plants scapose; carpel with long beak (>2 mm) and 2 basal lobes. **R. testiculatus** Crantz (Alder Creek Campground, Greenhorn Mtns. Moe 9147)
   3. Plants not scapose, cauline lvs. present; carpel spheric with short beak (<1 mm) **R. hebecarpus**
2. Perennials.
   4. Plants glabrous, stoloniferous....................... **R. cymbalaria** var. **saximontanus**
   4. Plants pubescent.
      5. Achenes mostly 4-5 mm long..**R. canus**
      5. Achenes mostly 2-3 mm long.
         6. Petals mostly 5-6, 1-2 times as long as broad....................... **R. occidentalis**
         6. Petals mostly 8-15, 2-3 times as long as broad.................... **R. californicus**

Thalictrum (p. 239)

Erect perennial herbs with alternate, ternately (in threes) decompound lvs. with basally dilated petioles; fls. unisexual, paniculate; stamens and pistils numerous; fr. an achene. About 80 species, 4 in CA. One species in Kern County: **T. fendleri** Gray var. **polycarpum** Torr. (*T. polycarpum*)

RESEDACEAE

Oligomeris (p. 284)

Low branching annual with linear entire often fascicled lvs.; fls. small, whitish, in terminal spikes; sepals 4; petals 2; fr. a capsule. About 3 species, 1 in CA. One species in Kern County: **O. linifolia**

RHAMNACEAE

Shrubs with simple alternate or opposite lvs.; fls. small, regular in terminal or axillary panicles; petals 0 or 4-5, clawed; fr. hard.

Fr. fleshy, drupe-like ....**Rhamnus**
Fr. dry, capsular; petals clawed and hooded ....**Ceanothus**

Ceanothus (p. 280)

Shrubs, often with divaricate twigs; lvs. mostly obviously 3-nerved on back due to prominent veins; fls. small, showy. About 45 species, 44 in CA.

1. Stipules thin and early deciduous; lvs. alternate; caps. without horns.
   2. Ultimate branchlets flexible.
      3. Lf.-margin entire to minutely toothed at tip; erect shrub to about 3 m; deciduous.. **C. integerrimus** (incl. all vars.)
      3. Lf.-margin serrate; prostrate shrub <30 cm tall; evergreen .... **C. diversifolius**

2. Ultimate branchlets rigidly divaricate.
   4. Shrubs 2-4 m tall; fls. whitish to blue, the clusters 3-8 cm long; lvs. 1-2.5 cm long.......... **C. leucodermis**
   4. Shrubs <2 m tall; fls. white, the clusters 1.5-3 cm long; lvs. 1-2 cm long. **C. cordulatus**

1. Stipules with thick corky persistent bases; lvs. opposite, .5-1.5 cm long; caps. with 3 horns; fls. white.
   5. Lvs. mostly concave on upper surface; horns of caps. lateral and spreading, small. **C. greggii** var. **vestitus** (Greene) McMinn (*C. vestitus*)
   5. Lvs. not concave; horns of caps. apical or subapical, conspicuous. **C. cuneatus** var. **cuneatus**

Rhamnus (p. 281)

Shrubs with alternate pinnately veined persistent lvs.; berry green, black or red. About 125 species, 9 in CA.

1. Petals none; lvs. 1-4 cm long; winter buds with bud scales.
   2. Branchlets rigid, divaricate, ending in weak spines; lf.-blades 10-17 mm long, ± clustered.......... **R. crocea**
   2. Branchlets not divaricate, seldom ending in weak spines; lf.-blades 18-40 mm long. .... **R. ilicifolia**
1. Petals present; lvs. 3-8 cm long; winter buds without bud scales. **R. tomentella** Benth. (*R. californica*)
   3. Lf-margins entire or with blunt teeth .... ssp. **tomentella**
   3. Lf.-margins dentate with sharp teeth .... ssp. **cuspidata** (Greene) Sawyer

ROSACEAE

Herbs, shrubs or small trees with alternate stipulate simple or compound lvs.; fls. regular, perigynous, solitary or clustered; pistils 1 to many; fr. an achene, drupe, pome or cluster of drupelets.

1. Ovary inferior; fr. a pome; large shrubs.
   2. Lvs. deciduous; fls. racemose; petals 8-12 mm long.......... **Amelanchier**
   2. Lvs. persistent; fls. in large corymbose panicles; petals to 4 mm long. **Heteromeles**
1. Ovary or ovaries superior; fr. an achene, drupe or drupelets.
   3. Fruits drupes or drupelets.
      4. Trailing prickly vines; lvs. lobed to compound.......... **Rubus**
      4. Shrub or small tree w/o prickles; lvs. serrate.......... **Prunus**
   3. Fruit a dry achene.
      5. Lvs. simple; plants woody.
         6. Petals present.
            7. Fls. solitary; lvs. cuneate, 3-lobed.......... **Purshia**
            7. Fls. paniculate; lvs. not lobed.
               8. Lvs. sublinear, fascicled.......... **Adenostoma**

8. Lvs. ovate to rounded..**Holodiscus**
6. Petals absent.
9. Fl.-tube slender, the limb deciduous and pushed off by the elongating feathery style.. **Cercocarpus**
9. Fl.-tube campanulate; achene not long-styled............................ **Coleogyne**
5. Lvs. compound or pinnately dissected.
10. Plants shrubby.
11. Pistils usually i.
12. Lvs. 3-lobed; fls. solitary...**Purshia**
12. Lvs. pinnately dissected, with strong odor........................ **Chamaebatia**
11. Pistils 4-many; carpels enclosed in fleshy fl.-tube (hip); plants prickly. **Rosa**
10. Plants herbaceous.
13. Styles easily deciduous.
14. Filament dilated at base; petals white-pinkish.
15. Stamens 10; leaflets 7-15 per side.. **Horkelia**
15. Stamens 20; leaflets 15-30 per side.................................. **Horkeliella**
14. Filaments filiform; petals white to yellow.
16. Plant hanging in clumps from rock walls; petals yellow............ **Ivesia**
16. Plants not hanging in clumps; petals white to yellow.
17. Lf. cylindric with >100 leaflets; petals white......................... **Ivesia**
17. Lf. flat with <50 leaflets; petals yellow ....**Potentilla**
13. Styles persistent.
18. Pistils many; petals 4-10 mm long...**Geum**
18. Pistil 1; petals none .... **Aphanes**

Adenostoma (p. 255)

Evergreen shrub with numerous small fascicled linear resinous lvs.; fls. small, white, crowded in terminal panicles turning rusty red in age; pistil 1 becoming an achene. Two species, both in CA. One species in Kern County: **A. fasciculatum**

Amelanchier (p. 255)

Deciduous shrub with rigid erect to spreading branches; lvs. oval to broadly obovate, thickish, pubescent, lower side paler than upper, entire to toothed to middle; racemes 4-6-fld.; petals white, 8-11 mm long; fr. a purplish-black pome. About 10 species, 2 in CA. One species in Kern County: **A. utahensis** Koehne (*A. pallida*)

Aphanes (p. 255)
(*Alchemilla*)

Small slender-branched annual with alt. palmately lobed lvs. with sheathing stipules; fls. small, in axillary clusters; petals absent; stamen 1; pistils 1-2. About 20 species, one in CA. One species in Kern County: **A. occidentalis** (Nutt.) Rydb.

Cercocarpus (p. 255)

Evergreen shrub with alt. simple straight-veined lvs. on short branchlets; fls. small, with persistent lower fl.-tube and deciduous bowl-shaped upper part (which is often pushed off by elongating style); petals 0; pistil 1 with elongating silky style. About 13 species, 5 in CA.

1. Lvs. toothed, not revolute (margins rolled backward); anthers hairy. **C. betuloides** var. **betuloides**
1. Lvs. entire, strongly revolute; anthers glabrous.
   2. Lvs. mostly 12-30 mm long, margins partly revolute. **C. ledifolius** var. **intermontanus** Holmg.
   2. Lvs. mostly <10 mm long, margins revolute nearly to midrib............**C. intricatus**

Chamaebatia (p. 256)

Glandular-pubescent aromatic evergreen subshrub with exfoliating bark; lvs. thrice pinnate, fern-like; fls. white, about 2 cm diameter; stamens many; pistil becoming an achene. Two species, both in CA. One species in Kern County: **C. foliolosa**

Coleogyne (p. 256)

Intricately branched deciduous shrub with opposite spinescent branches; lvs. in opposite fascicles, linear 5-15 mm long; fls. solitary; sepals 4, petals 0, stamens numerous; pistil becoming an achene. A single species: **C. ramosissima**

Geum (p. 256)

Perennial herb with pinnately compound stipulate lvs.; fls. few in terminal open cymes; petals 4-8 mm long, yellow; stamens numerous; pistils numerous with conspicuously jointed and kinked style on achenes. About 50 species, 3 in CA. One species in Kern County: **G. macrophyllum**

Heteromeles (p. 256)
(*Photinia*)

Large evergreen shrub with simple toothed lvs.; fls. small, white, many in large terminal panicles; stamens 10; fr. a red berry-like pome. A single species: **H. arbutifolia** (Lindl.) Roem.

Holodiscus (p. 256)

Small compact shrubs with pubescent young twigs; lvs. alternate, simple toothed, without stipules; fls. whitish, small; petals 5; stamens 20; pistils 5 becoming achenes. About 5 species, 2 in CA. One species in Kern County: **H. discolor** (Pursh) Maxim (*H. boursieri*)

Horkelia (p.256)
(part of *Potentilla*)

Perennial herbs with thick caudex; lvs. pinnately compound with numerous toothed or divided lfts.; fls. white, in cymes; hypanthium well developed; stamens 10 with dilated filaments; pistils many becoming achenes. About 17 species, 15 in CA. One species in Kern County: **H. rydbergii** Elmer (*P bolanderi* var. *parryi*)

Horkeliella (p. 257)
(part of *Potentilla*)

Tufted perennial with odd-pinnately compound lvs.; hypanthium subtended with 5 bractlets; petals whitish; stamens 20; pistils many. Two species, both in CA. One species in Kern County: **H. purpurascens** (Wats.) Rydb.

Ivesia (p. 257)
(part of *Potentilla*)

Perennial herbs with thick caudex; lvs. mostly basal, pinnately compound with many deeply lobed or divided lfts.; fls. white to tinged with purple; stamens 15-20 with filiform filaments; fr. an achene. About 30 species, 22 in CA.

Lvs. ± flat; pistils 3-20; petals yellow .... **I. saxosa** (Greene) Ertter
Lvs. culindric; pistil 1; petals white .... **I. santolinoides** Gray

Potentilla (p. 256)

Perennial herbs with pinnately or palmately compound lvs.; fls. yellow to cream, without evident hypanthium; stamens few to many; filaments flattened; pistils numerous, becoming achenes. About 250 species, 24 in CA.

1. Styles lateral; basal lvs. pinnate .... **P. glandulosa**
1. Styles terminal (on tip of ovary).
  2. Styles to about 1 mm long; petals shorter than sepals; annuals.
    3. Some stem hairs glandular; leaflets 3 .... **P. biennis**
    3. Stem hairs not glandular; leaflets 3-5.
       P. rivalis.var. **millegrana** (*P. millegrana*)
  2. Styles >1 mm long; petals longer than sepals; perennials.
    4. Plants of sandy soils, mostly <25 cm tall .... **P. wheeleri** Wats.
       (Mack Meadow, Piute mtns. Shevock 8715)
    4. Plants of dry meadows, mostly >20 cm tall .... **P. gracilis**
      5. Leaflets lobed >3/4 to midvein; petals 5-8 mm long.
         var. **elmeri** (Rydb.) Jeps. (*P. pectinisecta*)
      5. Leaflets toothed mostly <1/2 to midvein; petals 4-7 mm long.
         var. **fastigiata** (Nutt.) Wats. (ssp. *nuttallii*)

Prunus (p. 257)

Trees and shrubs with simple usually serrate lvs.; stipules small, early deciduous; petals 5, white to rose; stamens numerous; pistil 1 becoming a drupe. About 400 species, 15 in CA.

1. Evergreen; lvs. ± leathery, coarsely spinose-toothed; fr. red..................... **P. ilicifolia**
1. Deciduous; lvs. thin.
  2. Shrubby with many spinose branchlets.
    3. Lvs. broadly ovate to rounded, serrulate, not fascicled; fr. red-purple, glabrous. **P. subcordata**
    3. Lvs. oblanceolate, obscurely serrulate, fascicled.
      4. Petals 5-6 mm long, rose-pink; pedicels 6-8 mm long................ **P. andersonii**
      4. Petals 2-3 mm long, white; pedicels about 1 mm long. .... **P. fasciculata**
  2. Shrubby to tree-like; branchlets not spinose.
    5. Fls. 1-10, not in well-developed racemes; twigs red and shiny; fr. red. **P. emarginata**
    5. Fls. many in terminal racemes; twigs brownish, pubescent; fr. dark red. **P. virginiana** L. var. **demissa** (Nutt.) Torr. (*P. demissa*)

Purshia (p. 258)

Shrubs with crowded ± fascicled 3-cleft lvs.; fls. solitary; petals 5, cream to yellow; stamens about 25; fr. an achene with short persistent style. Five species, two in CA. One species in Kern County: **P. tridentata**

Key to varieties:

Lvs. pubescent above, not punctate with sunken glands; twigs glandular and hairy; lvs. to 3 cm long.. var. **tridentata**

Lvs. glabrous above, punctate; twigs glandular but not hairy; lvs. to 1 cm long. var. **glandulosa** (Curran) Jones (*P. glandulosa*)

Rosa (p. 259)

Erect or sprawling thorny shrubs with alt. deciduous pinnately compound lvs.; stipules evident; fls. large, rose-pink; stamens numerous; fl.-tube becomes fleshy (hip) and encloses the hairy achenes. About 100 species, 11 in CA.

1. Sepals pectinate-glandular, laciniately lobed; fr. orange-scarlet........... **R. eglanteria**
1. Sepals entire.
  2. Sepals glandular-hispid on back; lfts. mostly 5................................ .**R. pinetorum**
  2. Sepals not glandular-hispid on back; lfts. 5-7.
    3. Stems with stout flattened recurved prickles; pedicels villous. **R. californica** (incl. *R. aldersonii*)
    3. Stems with straight or ascending weak slender prickles; pedicels glabrous. **R. woodsii** var. **ultramontana**

Rubus (p. 259)

Shrubs to trailing vines, unarmed or armed with prickles; lvs. 3-5-lobed or foliate; fls. rather large, white to pink; stamens numerous; carpels many, crowded on receptacle, becoming drupelets of an aggregate fruit. About 500 species, 11 in CA.

1. Lvs. simple, palmately 3-lobed; stems unarmed with peeling bark; fr. scarlet. **R. parviflorus**
1. Lvs. compound, mostly 3-5-foliate; stems prickly.
   2. Drupelets adhering to fleshy receptacle and falling separately or with it; fr. oblong, black (blackberries).
      3. Fls. borne in large terminal panicles; prickles stout. **R. discolor** Weihe & Nees (*R. procerus*)
      3. Fls. borne in small clusters; prickles slender, bristle-like. **R. ursinus** (incl. *R. vitifolius* var. *titanus*)
   2. Drupelets forming hollow cone which separates from the dry receptacle as single aggregate fruit (raspberries).
      4. Sepals deflexed; pedicels and sepals mostly glandless................**R. leucodermis**
      4. Sepals not deflexed; pedicels and sepals with stalked glands.......**R. glaucifolius**

RUBIACEAE

Shrub or small herb with opposite or whorled entire lvs.; fls. perfect or imperfect, mostly 4-merous; corolla regular; ovary inferior.

Lvs. in whorls of 4 or more; plants herbs or basally woody .... **Galium**
Lvs. opposite; plant shrubby; fls. in heads .... **Cephalanthus**

Cephalanthus (p. 346)

Shrub 2-9 m tall, young twigs reddish; lvs. opposite, petioled; fl.-heads 10-25 mm in diam., corolla white. About 17 species, one in CA. One species in Kern County: **C. occidentalis** var. **californicus**

Galium (p. 346)

Annual or perennial herbs, sometimes ± shrubby, with 4-angled slender stems; lvs. in apparent whorles due to large leaf-like stipules; fls. small, sepals obsolete, corolla rotate; fr. twin-like, of 2 indehiscent carpels. About 400 species, 47 in CA.

1. Lvs. 5-8 in a whorl.
   2. Ovary and fr. glabrous, not bristly-hairy; stems with some rough hairs bent backwards; perennial....................................... **G. trifidum** (incl. var. *subiflorum*)

   2. Ovary and fr. bristly-hairy; annual.

3. Fls. several on each side-branch, the upper part of plant forming an open panicle; fr. 1 mm wide .... **G. parisiense**
3. Fls. mostly 2-3 in axillary cymules; stems easily clinging to other vegetation; fr. 4-5 mm wide. .... **G. aparine**

1. Lvs. 2-4 in a whorl.
4. Annual; fr. with hooked hairs .... **G. bifolium**
4. Perennial, woody at base of plant.
5. Mature fr. mostly glabrous.
6. Lvs. linear-subulate; plants forming low dense mats; stems densely leafy. **G. andrewsii**
6. Lvs. lance-linear or broader; plants taller, more open.
7. Stems with rough hairs bent backwards on the angles, woody; lvs. 2-6 (10) mm long. .... **G. nuttallii**
7. Stems lacking such hairs, scarcely woody; lvs. 6-20 mm long; fls. reddish-purple. .... **G. bolanderi**
5. Mature fr. densely hispid.
8. Lf. linear, widest above middle........ **G. angustifolium**
9. Corolla hairy........ ssp. **onycense**
9. Corolla glabrous........ ssp. **angustifolium**
8. Lf. ovate, widest below middle.
10. Fl.-clusters drooping; mature fruits 4-5 mm in diam. (including hairs). **G. hallii**
10. Fl.-clusters not drooping; mature fruits 3-4 mm in diam..... **G. matthewsii**

SALICACEAE

Dioecious deciduous trees or shrubs with simple alternate lvs.; fls. in early spring, in catkins; perianth none; stigmas 2-4; fr. a capsule; seeds with a tuft of hairs.

Lvs. ovate to deltoid; catkin-scales laciniate; mostly trees .... **Populus**
Lvs. narrow; catkin-scales entire; mostly shrubby .... **Salix**

Populus (p. 208)

Trees with pale furrowed bark and rather broad simple lvs.; winter buds with many scales. About 40 species, 5 in CA.

1. Lvs. not much longer than broad, blade serrate to tip.
2. Petioles rounded; lvs. dark green above, silvery beneath. **P. balsamifera** L. ssp. **trichocarpa** (T. & G.) Braysh. (*P. trichocarpa*)
2. Petioles flattened laterally; lvs. yellowish-green on both sides. **P. fremontii** ssp. **fremontii**

1. Lvs. much longer than broad, blade not serrate near tip. **P. angustifolia** James (*P. accuminata*)

Salix (p. 209)

Shrubs or small trees with mostly narrow lvs.; winter buds with a single scale; fls. in catkins; each fl. subtended by small bract; fr. a capsule. About 400 species, 29 in CA.

1. Lf.-blades entire.
  2. Mature lvs. glabrous beneath.
    3. Lf.-blades 1.5-4 cm wide, mostly 3-4 times longer.
      4. Filaments and capsule mostly glabrous ....**S. lutea**
      4. Filaments and capsule hairy; catkin scales dark............................**S. lemmonii**
    3. Lf.-blades 1-2 cm wide, mostly about 5 times as long.
      5. Catkin-bracts yellow; stamens 3-9; trees.
        6. Petioles glandular above.
          S. lucida Muhlenb. ssp. **lasiandra** (Benth.) Murray (*S. lasiandra*)
        6. Petioles not glandular ....**S. gooddingii**
      5. Catkin-bracts dark; stamens 2; shrub or small tree....................... **S. lasiolepis**
  2. Mature lvs. hairy beneath.
    7. Capsule glabrous .... **S. lasiolepis** (incl. var. *bracelinae*)
    7. Capsule hairy to slightly hairy.
      8. Catkin-bracts dark to nearly black .... **S. scouleriana**
      8. Catkin-bracts yellow with reddish tips .... **S. geyeriana** var. **argentea**
1. Lf.-blades distinctly serrulate; catkin-bracts yellow.
  9. Mature lvs. glabrous beneath.
    10. Capsule silky with soft hairs ....**S. gooddingii**
    10. Capsule essentially glabrous.
      11. Lf.-blades 6-15 mm wide .... **S. melanopsis**
      11. Lf.-blades 15-40 mm wide ....**S. laevigata**
  9. Mature lvs. pubescent beneath...............................**S. exigua** (incl. *S. hindsiana*)

SANTALACEAE

Comandra (p. 215)

Smooth perennial herb 1-3 dm high with alternate entire subsessile lvs.; stems with longitudinal lines; fls. purplish-green, in clusters; sepals 5, petals 0; fr. drupe-like. Four species, one in CA. One species in Kern County: **C. umbellata** (L.) Nutt. ssp. **californica** (Rydb.) Piehl. (*C. pallida*)

SAURURACEAE

Anemopsis (p. 215)

Perennial stoloniferous herbs with swollen nodes; lvs. mostly basal; spike conical with basal petaloid white invol.; each fl. subtended by small white bract; fr. a capsule. One species: **A. californica**

SAXIFRAGACEAE

Herbs with mostly basal palmately veined lvs.; stipules absent; fls. perfect, perigynous; carpels 1-2; styles 2; fr. a capsule or follicle.

1. Fertile stamens 10.
   2. Styles 3; petals lobed, clawed .... **Lithophragma**
   2. Styles 2; petals broadly elliptic, not clawed ....**Saxifraga**
1. Fertile stamens 5 ....**Heuchera**

Heuchera (p. 253)

Perennial herbs with long-petioled mostly basal roundish-lobed lvs.; fls. white, small; stamens 5. About 50 species, 13 in CA.

Stamens long exserted, longer than sepals; styles long-exserted; petals very slender.
H. rubescens var. **albicola**
Stamens shorter than sepals; styles scarcely exserted; petals wider than filaments.
H. cespitosa

Lithophragma (p. 253)

Perennial herb with long slender simple scapose flowering shoots; petals white, narrow at base (clawed) and much broader at tip, lobed to toothed. About 12 species, 9 in CA.

1. Basal lvs. deeply lobed, nearly to base; petals with 3-7 oblong lobes.
L. parviflorum
1. Basal lvs. not lobed near to their bases; petals nearly entire.
   2. Pedicels 1-3 mm long; stem-lvs. alternate... **L. bolanderi**
   2. Pedicels 5-10 mm long; stem-lvs. mostly opposite.......................... **L. cymbalaria**

Saxifraga (p. 255)

Perennial herb with basal lvs. 1-5 cm long on ± shorter petioles; fl.-shoots 1-3 dm tall, loosely branched; fls. purplish, small. About 400 species, 17 in CA. One species in Kern County: **S. californica**

SCROPHULARIACEAE

Annual to perennial herbs or shrubs; lvs. simple, mostly alternate; fls. perfect, in racemes or panicles; calyx and corolla mostly 2-lipped; stamens 5, sometimes 4 or 2; ovary superior, 2-celled; fr. a capsule.

1. Anthers 5; corolla wheel-shaped, nearly regular....................................... **Verbascum**
1. Anthers 4 or 2; corolla mostly irregular.
   2. Anthers 2.
   3. Corolla-lobes 5; caps. ovoid.
      4. Corolla with some yellow.
         5. Corolla lemon yellow, 15-20 mm long.. **Mohavea**
         5. Corolla white with yellow tube, 5-7 mm long................................ **Gratiola**
      4. Corolla blue-violet, 7-9 mm long.. **Lindernia**
   3. Corolla-lobes 4; caps. flattened.. **Veronica**
   2. Anthers 4; corolla lobes 5.
   6. Plants lacking obvious leafy stem; corolla nearly wheel-shaped, 1.5 mm long. **Limosella**
   6. Plants with obvious leafy stem.
      7. Stigmas distinct, flattened.
         8. Corolla bell-shaped; white; sepals ± distinct; plant ± aquatic. .... **Bacopa**
         8. Corolla bilabiate; sepals united.. **Mimulus**
      7. Stigmas united, capitate or punctiform.
         9. Corolla spurred or saccate on lower side near base.
            10. Corolla swollen or saccate at base...................................... **Antirrhinum**
            10. Corolla with long spur at base, yellow.. **Linaria**
         9. Corolla not spurred or saccate on lower side near base.
            11. Corolla with upper lip open, not forming a galea (beak-like upper lip).
               12. Plants annual; corolla swollen on upper side of base............ **Collinsia**
               12. Plants perennial; corolla not swollen on upper side.
                  13. Corolla brownish, <10 mm long, inflated; stem square. **Scrophularia**
                  13. Corolla not brownish, 10-14 mm long, tubular or funnelform; stem round.
                     14. Fertile filaments densely hairy at base.......................... **Keckiella**
                     14. Fertile filaments glabrous.. **Penstemon**
            11. Corolla with upper lip forming a galea which encloses stamens.
               15. Anther-cells equal in size and position; perennial with well-developed basal rosette.. **Pedicularis**
               15. Anther-cells unequally placed, upper attached by its middle, lower attached or abortive.
                  16. Calyx-tube split ventrally, dorsal portion bract-like; stigma dot-like, not expanded.. **Cordylanthus**
                  16. Calyx-tube surrounding lower portion of corolla-tube, its lobes placed laterally; stigma expanded.

17. Anther-sacs 1 per stamen; lower corolla-lip strongly 3 pouched. **Triphysaria**
17. Anther-sacs 2 per stamen; lower corolla-lip reduced to ± 3 pouches.. **Castilleja**

Antirrhinum (p. 334)

Erect or diffuse annuals; lower lvs. opposite, upper alternate; fls. in terminal racemes; corolla 2-lipped, swollen or saccate at base; fertile stamens 4. About 36 species, 15 in CA.

1. Pedicels 3-10 cm long, twining; fls. bright yellow .... **A. filipes** Gray (Short Canyon, E. slope of Sierra, M. A. Henry - needs confirmation)
1. Pedicels <3 cm long, not twining; fl.-color various.
   2. Corolla-throat open; upper lip pink, lower whitish. **A. ovatum** Eastw. (*Howelliella ovata*)
   2. Corolla-throat closed.
      3. Fls. 10-12 mm long, bluish with yellow palate. **A. leptaleum** Gray (*A. cornutum* var. *leptaleum)*
      3. Fls. 16-19 mm long, reddish with cream palate ....**A. multiflorum** Pennell (Zenda mine, Caliente Canyon. Moe 1813)

Bacopa (p. 334)

Perennial herb with opposite lvs. and axillary fls.; corolla bell-shaped, white, upper lip 2-lobed, lower 3-lobed; stamens 4. About 100 species, 4 in CA. One species in Kern County: **B. eisenii**

Castilleja (p. 334)
(incl. many *Orthocarpus*)

Herbs or somewhat shrubby partially parasitic plants; lvs. alternate, sessile, becoming colored bracts in infl.; calyx tubular; corolla tubular, ± greenish, upper lip (galea) elongate, mostly reddish; lower lip small, 3-lobed; stamens 4, inside galea. About 200 species, 37 in CA.

1. Annual; lvs. and bracts entire.
   2. Infl. red; lower lip of corolla 3 small greenish lobes. **C. minor** (Gray) Grat ssp. **spiralis** (Jeps.) Chuang & Heckard (*C. stenantha*)
   2. Infl. not red, greenish to yellow to purplish; lower lip of corolla 3-pouched.
      3. Bracts green throughout.......................... **C. tenuis** (Heller) Chuang & Heckard (*Orthocarpus hispidus*)
      3. Bracts tipped purplish, yellowish, or whitish.
         4. Corolla-beak hooked, densely hairy. **C. exserta** (Heller) Chuang & Heckard (*Orthocarpus purpurascens*)
            5. Corolla white to yellow or rose to purple-red............................ssp. **exserta**

5. Corolla bright rose-red with orange-tipped lower lip.
ssp. **venusta** (Heller) Chuang & Heckard
(*Orthocarpus purpurascens* var. *ornatus*)
4. Corolla-beak straight, puberulent.
6. Bract-lobes 3, tipped white or tellow.
C. attenuata (Gray) Chuang & Heckard (*Orthocarpus attenuata*)
6. Bract-lobes 3-5, tipped pink.
C. lineariloba (Benth.) Chuang & Heckard (*Orthocarpus lineariloba*)
1. Perennial; lvs. and bracts divided.
7. Calyx-lobes united much farther dorsally than ventrally, their tips curved up.
8. Floral bracts entire or somewhat 3-lobed; lvs. pubescent.
C. subinclusa Greene (*C. jepsoni*)
8. Floral bracts deeply 3-cleft; lvs. glabrous...................................... **C. linearifolia**
7. Calyx-lobes not united as above; their tips straight.
9. Lvs. with simple hairs, green to grayish-green.
10. Plants glandular-pubescent below infl............................. **C. appelgatei** Fern.
11. Galea 6-10 mm long; lower lip about 1/3 as long.
ssp. **pallida** (Eastw.) Chuang & Heckard (*C. breweri*)
11. Galea 10-25 mm long; lower lip to 1/5 as long.
12. Galea 15-25 mm long; calyx-lobes rounded.
ssp. **martinii** (Abrams) Chuang & Heckard (*C. roseana*)
12. Galea 10-17 mm long; calyx-lobes sharp-pointed.
13. Lvs. lanceolate; lower corolla-lip not exserted.............ssp. **appelgatei**
13. Lvs. linear; lower corolla-lip exserted.
ssp. **disticha** (Eastw.) Chuang & Heckard (*C. disticha*)
10. Plants not glandular-pubescent below infl.
14. Main cauline lvs. entire; wet places... **C. miniata**
14. Main cauline lvs. with 1-2 pairs of lobes; desert.
C. angustifolia (Nutt.) Don (*C. chromosa*)
9. Lvs. with branched hairs, grayish to white-woolly.
15. Calyx 10-15 mm long; infl. yellowish .... **C. plagiotoma**
15. Calyx 15-20 mm long; infl. red.. **C. foliolosa**

Collinsia (p. 335)

Annuals with opposite lvs.; the upper sessile and clasping, becoming bracts in infl.; corolla 5-lobed, saccate at base on upper side, 2-lipped, the upper 2-lobed, the lower 3-lobed, the middle lobe keel-shaped and enclosing the 4 stamens and style. About 18 species, 16 in CA.

1. Fls. congested in whorles; pedicels shorter than calyx.
2. Lateral lower corolla-lobes bearded; lvs. pubescent beneath................ **C. tinctoria**
2. Lateral lower corolla-lobes glabrous; lvs. glabrous.
3. Upper pair of filaments with distinct basal appendages 1-2 mm long.
C. heterophylla
3. Upper pair of filaments without appendages....... **C. bartsiifolia** var. **davidsonii**
1. Fls. not whorled; pedicels as long as or longer than calyx.

4. Infl. not glandular-pubescent; calyx-lobes exceeding caps.; corolla 4-7 mm long. **C. parviflora**
4. Infl. glandular-pubescent; calyx-lobes not exceeding caps.
  5. Fruiting pedicels ascending; upper bracts at least 2 mm long.
    6. Calyx 3 mm wide at base in fr. .... **C. childii**
    6. Calyx 5 mm wide at base in fr. .... **C. callosa**
  5. Fruiting pedicels defexed-spreading; upper bracts <2 mm long .... **C. torreyi**
    7. Corolla 7-9 mm long .... var. **torreyi**
    7. Corolla 4-6 mm long .... var. **wrightii** (Wats.) Johnson (*C. wrightii*)

Cordylanthus (p. 336)

Branched annuals with yellow roots; fls. yellowish to purple, in spikes and subtended by bracts which are seldom colored; inner bract foliose or modified into calyx-like structure; calyx ± tubular, split to near base ventrally, extended dorsally; corolla 2-lipped, the upper a galea; stamens 4 or 2. About 18 species, 16 in CA.

1. Lvs. entire, oblong; infl. an elongate spike; galea with spreading hairs; stamens 2 **C. mollis** Gray ssp. **hispidus** (Penn) Chuang & Heckard (*C. hispidus*)
1. Lvs. or lf.-divisions linear; infl. a 1-6-fld. head; galea glabrous to minutely pubescent; stamens 4.
  2. Corolla >3 times longer than wide; corolla pinkish, throat with maroon blotches. **C. eremicus** (Cov. & Mort) Munz ssp. **kernensis** Chuang & Heckard (Spanish Needle Peak, Shevock 11306 CAS)
  2. Corolla to 3 times longer than wide; corolla yellowish white.
    3. Fls. 1-3 in clusters; anther sacs 1 per stamen; corolla white with yellow tips and purple-veined pouch .... **C. nevinii**
    3. Fls. 5-15 in dense clusters; anther sacs 2 per stamen; corolla yellowish with white pouch. .... **C. rigidus**
      4. Corolla whitish with some purple .... ssp. **rigidus** (*C. ferrisianus*)
      4. Corolla light yellow .... ssp. **brevibracteatus** (*C. compactus*)

Gratiola (p. 337)

Erect or diffuse glabrous herb; lvs. opposite, slightly clasping at base; fls. axillary; corolla white, small; fertile stamens 2; stigmas 2. About 20 species, 3 in CA. One species in Kern County: **G. ebracteata**

Keckiella (p. 342)
(part of *Penstemon*)

Shrub with ± opposite (or in 3's) drought-deciduous lvs.; infl. panicle- or spike-like; corolla whitish to reddish, upper lip hooded; filaments densely hairy at base. About 7 species, all in CA.

1. Corolla pink to reddish, tube 16-25 mm long. **K. ternata** (Torr.) Straw var. **septentrionalis** (Munz) Holm.

1. Corolla white, yellowish to brownish, tube <15 mm long.
   2. Corolla 12-18 mm long; stems glabrous .... **K. breviflora** (Lindl.) Straw
   2. Corolla 11-16 mm long; stems densely short-hairy......**K. rothrockii** (Gray) Straw

Limosella (p. 337)

Tufted stoloniferous glabrous herb; lvs. palmately veined, in tufts; fls. solitary on long pedicels from the tufts; corolla small, whitish; stamens 4; stigma capitate. About 15 species, 3 in CA. One species in Kern County: **L. aquatica**

Linaria (p. 337)

Perennial herb with opposite clasping lvs.; fls. in terminal raceme; corolla tubular with long spur at base, yellow, 3-4 cm long; fertile stamens 4. About 100 species, 8 introduced in CA. One species in Kern County: **L. genistifolia** (L.) Miller ssp. **dalmatica** (L.) Maire & Petitm. (*L. dalmatica*)

Lindernia (p. 337)

Small slender-stemmed glabrous annual; lvs. opposite; fls. axillary; corolla blue-violet, 2-lipped; fertile stamens 2; stigma 2-lobed. About 50 species, 1 in CA. One species in Kern County: **L. dubia** (L.) Penn. var. **anagallidea** (Mich.) Cooperr. (*L. anagallidea*)

Mimulus (p. 338)

Annual or perennial herbs or low shrubs, often sticky or glandular-pubescent; lvs. opposite; fls. axillary or in racemes; sepals united into a tubular ± saccate calyx; corolla 2-lipped with a pair of ridges running down lower side of throat; stamens 4; stigmas 2. About 100 species, 63 in CA.

1. Plants perennial.
   2. Pedicels longer than calyx; plants herbaceous; wet places.
      3. Corolla scarlet..**M. cardinalis**
      3. Corolla yellow.
         4. Mature calyx not much inflated, the lobes mostly straight; corolla-throat open.
            5. Lvs. pinnately veined; stems 1-3 dm long............................ **M. moschatus**
            5. Lvs. palmately veined; stems to 1 dm long. **M. primuloides** (incl. var. *pilosellus*)
         4. Mature calyx strongly inflated, the lower lobes curved up against the others; throat ± closed... **M. guttatus**
   2. Pedicels shorter than calyx; plants shrubby; dry places. **M. aurantiacus** Curt. (*Diplacus lcalycinus*)
1. Plants annual.
   6. Pedicels longer than calyx; corolla deciduous.

7. Mature calyx strongly inflated, lower lobes curved up against the others.
M. guttatus

7. Mature calyx not inflated, lower lobes straight.
 8. Fls. mostly yellow.
  9. Calyx not ridged along sepal-midribs; caps. glandular-puberulent.
M. pilosus
  9. Calyx ridged along sepal-midribs; caps. glabrous.
   10. Lvs. ovate to lance-ovate, dentate; calyx-lobes triangular.
M. floribundus (incl. *M. dudleyi*)
   10. Lvs. linear to oblong, entire; calyx-lobes rounded.
M. montioides (*M. barbatus*)
 8. Fls. mostly purple or red.
  11. Anthers pubescent.
   12. Corolla 13-20 mm long plant glandular-pubescent............... **M. palmeri**
   12. Corolla 8-10 mm long; plant mostly glabrous........... **M. inconspicuous**
  11. Anthers glabrous.
   13. Corolla 10-15 mm long.
    14. Calyx 10-12 mm long in fr.; lvs. denticulate..................... **M. parishii**
    14. Calyx 5-6 mm long; lvs. entire...............**M. montoides** (*M. barbatus*)
   13. Corolla 6-10 mm long.
    15. Pedicels 5-20 mm long; to 2 times length of bract; anthers included.
M. breweri
    15. Pedicels 15-25 mm long; 3-5 times length of bract; anthers exserted.
M. androsaceus

6. Pedicels shorter than calyx; corolla persistent on developing caps.
 16. Fls. yellow.
  17. Pedicel 2-7 mm long; calyx 5-11 mm long; corolla-throat with 3 maroon lines... **M. mephiticus** Greene
(Sandy places, Owens Peak, Spanish Needle. Shevock)
  17. Pedicel 1-3 mm long; calyx 4-8 mm long; corolla-throat with 5 dark lines.
M. whitneyi
 16. Fls. purple to magenta.
  18. Caps. symmetrical, membranous, soon dehiscent.
   19. Corolla lobes subequal; stigma lobes subequal.
    20. Mature calyx little or not inflated, nearly plane; alpine.......**M. whitneyi**
    20. Mature calyx inflated, ridges raised, glandular-pubescent.
     21. Inner face of corolla-lobes glabrous; cismontane.
      22. Anthers glabrous... **M. fremontii**
      22. Anthers ciliate.
       23. Glandless hairs absent; fls. scattered..........................**M. viscidus**
       23. Glandless hairs present; fls.crowded near branch-tips.
M. constrictus
     21. Inner face of corolla-lobes pilose; deserts.
      24. Lowest pair of calyx-lobes <1/2 length of upper; Frazier Park area.
M. johnstonii
      24. Lowest pair of calyx-lobes >1/2 length of upper; deserts.
M. bigelovii

19. Corolla-lobes unequal; stigma-lobes unequal.
  25. Lower corolla-lip magenta; plants to 90 cm tall, densely hairy. **M. bolanderi**
  25. Lower corolla-lip yellowish with maroon dots; plants to 12 cm. tall, puberulent. .... **M. shevockii** Heckard & Bacigal. (Kelso Creek & Cyrus Canyon areas. Shevock)

18. Caps. asymmetrical, swollen or oblique at base, cartilaginous, tardily dehiscent.
  26. Fls. cleistogamous, not or barely opening, 2-8 mm long........ **M. douglasii**
  26. Fls. opening normally.
    27. Corolla 6-12 mm long, nearly regular, white with rose veins.. **M. pictus**
    27. Corolla 13-50 mm long, irregular.
      28. Lower lip of corolla <1/3 length of upper .... **M. douglasii**
      28. Lower lip of corolla >1/3 length of upper.
        29. Anthers glabrous; corolla 15-25 mm long; calyx 9-12 mm long. **M. congdonii**
        29. Anthers pubescent; corolla 30-45 mm long; calyx 12-15 mm long.
          30. Pedicels 2-3 mm long; lower lip of corolla at least as long as upper. .... **M. tricolor**
          30. Pedicels 3-5 mm long lower lip of corolla shorter than upper. **M. kelloggii**

Mohavea (p. 341)

Erect annual herb with lanceolate alternate lvs.; fls. in dense leafy spikes; corolla with short tube, bilabiate with hairy lower lip, lemon yellow; fertile stamens 2; caps. ovoid. Two species, both in CA. One species in Kern County: **M. breviflora**

Pedicularis (p. 342)

Perennial herbs with mostly basal pinnatifid lvs.; fls. in spike-like raceme; corolla strongly 2-lipped, upper galeate, arched; lower 3-lobed; stamens 4. About 500 species, 11 in CA.

Corolla pale yellow; stems mostly underground.................................... **P. semibarbata**
Corolla purple-red; stems exceeding lvs...**P. densiflora**

Penstemon (p. 342)

Perennial herbs or shrubs; lvs. opposite, the lower petioled, the upper sessile; fls. showy, in racemose panicles; corolla tubular, throat often inflated, bilabiate; fertile stamens 4 with arched filaments; the 5th stamen a sterile staminode often bearded. About 250 species, 53 in CA.

1. Anthers densely comose; corolla rose-red .... **P. newberryi**
1. Anthers glabrous or sparsely hairy.

2. Anther-sacs completely opening, usually extending away from each other after opening.
 3. Corolla scarlet.
  4. Corolla 25-33 mm long; plant glaucous; low elevations.... **P. centranthifolius**
  4. Corolla 32-40 mm long; plant not glaucous; high elevations......... **P. labrosus**
 3. Corolla not scarlet.
  5. Lvs. entire or obscurely denticulate, not auriculate.
   6. Infl. glandular-puberulent, not secund; staminode densely bearded. **P. incertus**
   6. Infl. not glandular-puberulent, secund; staminode mostly glabrous. **P. speciosus** (incl. var. *piliferus*)
  5. Lvs. definitely toothed, auriculate (basal lobes extending around stem). **P. grinnellii** var. **scrophularioides** (Jones) Holmg. (ssp. *scrophularioides*)
2. Anther-sacs opening across apex only, lobes remaining saccate, parallel.
 7. Corolla scarlet, tubular; upper lip erect, lower reflexed. **P. rostriflorus** Kell. (*P. bridgesii*)
 7. Corolla bluish; both lips spreading.
  8. Lvs. basal; corolla-throat bearded.. **P. caesius**
  8. Lvs. well distributed; corolla-throat glabrous.
   9. Infl. glandular-pubescent.. **P. laetus**
   9. Infl. glabrous... **P. heterophyllus**

Scrophularia (p. 344)

Coarse perennial herbs with 4-angled stems; lvs. opposite; fls. in loose terminal panicles; calyx 5-parted; corolla brownish maroon, the tube open, short, bilabiate; fertile stamens 4, sterile filament developed, brownish. About 150 species, 5 in CA.

Main lf.-blades dark green, 6-12 cm long; infl. open, 3-6 dm long; cismontane. **S. californica** ssp. **floribunda** (Greene) Shaw (var. *floribunda*)
Main lf.-blades yellow-green, 3-6 cm long; infl. compact, 1.5-3 dm long; transmontane................ **S. desertorum** (Munz) Shaw (*S. californica* var. *desertorum*)

Triphysaria (p. 341)
(part of *Orthocarpus*)

Annuals with sessile ± linear, pinnately divided lvs.; infl. a spike with ± leaf-like bracts; calyx 4-lobed; corolla long-tubular, lower lip 3-pouched, yellow; anther sacs 1 per stamen. About 5 species, all in CA.

Corolla 8-15 mm long, ± = bract; pouches 1-2 mm deep. **T. micrantha** (Gray) Chuang & Heckard (*Orthocarpus erianthus*)
Corolla 10-25 mm long, >bract; pouches 3-4 mm deep. **T. eriantha** (Benth.) Chuang & Heckard

Verbascum (p. 344)

Biennial to perennial herbs, erect to 1.5 m tall; basal lvs. in rosette, simple; fls. in racemes or crowded spikes; corolla rotate (wheel-shaped), ± irregular, yellow or white; stamens 5, the filaments with long soft hairs; stigma capitate. About 360 species, 4 introduced into CA.

Plants very woolly; stem lvs. decurrent (extending below node) .... **V. thapsus**
Plants not woolly; cauline lvs. not decurrent .... **V. virgatum**

Veronica (p. 344)

Annual or perennial erect or prostrate herbs; lvs. opposite; corolla small, subrotate, 4-lobed; stamens 2; stigma entire; caps. flattened. About 250 species, 16 in CA.

1. Stems with lateral racemes below tip.
  2. Lvs. short-petioled; corolla 7-10 mm wide, violet-blue.................... **V. americana**
  2. Lvs. sessile, clasping; corolla 3-7 mm wide, pale with violet lines.
     V. anagallis-aquatica L. (Kern River. Stenger 106, S. Cooley J-107)
1. Stems ending in a simple raceme.
  3. Perennial from underground rootstocks; fls. only in upper axils; corolla pale blue.
     V. serpyllifolia ssp. **humifusa** (Dics.) Syme (var. *humifusa*)
  3. Annual; fls. from several axils.
    4. Corolla 2-3 mm wide; pedicels 1-2 mm long.
      5. Corolla white; lvs. linear-oblong.
         V. peregrina ssp. **xalapensis** (Kunth) Pennell (var. *xalapensis*)
      5. Corolla blue; lvs. rounded to oval... **V. arvensis**
    4. Corolla 6-11 mm wide; deep blue; pedicels 10-25 mm long............... **V. persica**

SELAGINELLACEAE

Selaginella (p. 167)

Low leafy terrestrial freely branched plants with numerous small overlapping lvs.; sporangia inconspicuous. About 700 species, 10 in CA.

1. Stems erect to ascending, 5-20 cm long, rooting near base......................**S. bigelovii**
1. Stems prostrate to pendent, 3-6 cm long; lvs. white-margined.
  2. Bristles on distal lvs. ± soft, forming conspicuous tufts......................... **S. asprella**
  2. Bristles on distal lvs. ± rigid, not forming conspicuous tufts.
     S. watsonii Underw. (Pinyon Peak, Scodie Mtns. Shevock 10800 CAS)

SIMARUBACEAE

Ailanthus (p. 277)

Tree 5-20 m tall spreading freely underground; lvs. 3-6 dm long, pinnately 11-25-foliate with strong odor; fr. a ± reddish samara. About 15 species, one introduced in CA. One species in Kern County. **A. altissima**

SOLANACEAE

Herbs or shrubs with alternate lvs.; fls. regular; corolla rotate or tubular; stamens 5, alternate corolla lobes; ovary superior, mostly 2-celled; fr. a berry or capsule.

1. Desert shrubs with spiny branches; corolla funnelform.................................. **Lycium**
1. Shrubs without spines, or herbs.
   2. Corolla rotate (wheel-shaped).
      3. Anthers connivent (together but not united), longer than filaments; calyx small in fruit.
         4. Anthers opening lengthwise; corolla yellow............................. **Lycopersicon**
         4. Anthers opening by terminal pores.. **Solanum**
      3. Anthers not connivent, shorter than filaments; calyx enlarged and bladder-like in fruit.. **Physalis**
   2. Corolla tubular to funnelform.
      5. Corolla 6-18 cm long, white; caps. prickly, 2-5 cm long.......................... **Datura**
      5. Corolla 2-7 cm long, yellow; caps. not prickly, smaller....................... **Nicotiana**

Datura (p. 331)

Erect or spreading coarse rank-smelling herbs; fls. solitary, large, white to light violet, opening in evening; fr. a spiny caps. About 13 species, 4 in CA.

Corolla 6-8 cm long, 3-5 cm broad; annual.............. **D. stramonium** (incl. var. *tatula*)
Corolla 15-18 cm long, 10-20 cm broad; perennial.. **D. wrightii** Regel. (*D. meteloides*)

Lycium (p. 332)

Spiny shrub with lvs. in clusters; fls. solitary or few in axils; corolla whitish, funnelform; fr. dry or fleshy. About 100 species, 10 in CA.

Lvs. oblanceolate to spatulate, 5-10 mm broad; corolla 8-12 mm long; fr. dry, constricted in middle.. **L. cooperi**
Lvs. subterete, pearshaped, 1-3 mm broad; corolla 10-16 mm long; fr. fleshy, red. **L. andersonii** (incl. var. *deserticola*)

Lycopersicon (p. 332)

Annual or short-lived perennial strong-smelling glandular herb; lvs. odd-pinnate; fls. yellow; anthers opening lengthwise; fr. a red or yellow pulpy berry. The cultivated tomato. About 6 species, 2 introduced in CA. One species to Kern County: **L. esculentum**

Nicotiana (p. 332)

Narcotic-poisonous heavy-scented annual or perennial herbs or shrubs; fls. in terminal panicles or racemes; corolla funnelform to nearly tubular; fr. a capsule. About 60 species, 7 in CA.

1. Plant shrubby; fls. yellow.. **N. glauca**
1. Plant herbaceous; fls. whitish.
   2. Upper cauline lvs. sessile; corolla 4-7 cm long.
      N. quadrivalvis Pursh (*N. bigelovii*)
   2. Upper cauline lvs. petioled; corolla 2-3 cm long................................ **N. attenuata**

Physalis (p. 333)

Annual or perennial herbs; fls. solitary in lf.-axils; calyx bell-shaped, 5-toothed, enlarged and bladdery-inflated in fr.; corolla whitish to yellow; fr. a berry. About 85 species, 9 in CA.

Corolla whitish with yellow center, 10-20 mm broad.................................**P. acutifolia**
Corolla yellow, 3-8 mm broad.. **P. lanceifolia**

Solanum (p. 333)

Herbs or shrubs with simple lvs.; fls. in umbels or cymes; calyx 5-toothed; corolla rotate, 5-lobed; stamens 5; anthers grouped around style, dehiscent by terminal pores; ovary 2-celled; fr. a many-seeded berry. About 1500 species, 19 in CA.

1. Plants prickly; berries yellow..**S. eleagnifolium**
1. Plants not prickly; berries mostly not yellow.
   2. Plants annuals to herbaceous perennials.
      3. Corolla white to bluish, 25-30 mm broad; large underground tubers produced (cultivated potato)..**S. tuberosum**
      3. Corolla white, 3-10 mm broad.
         4. Lvs. entire or shallowly toothed; berry yellowish or black.
            5. Stems conspicuously villous or hirsute; berry yellowish..... **S. sarrachoides**
            5. Stems nearly glabrous; berry glossy black.
               S. americanum Miller (*S. nodiflorum*)
         4. Lvs. pinnatifid or deeply lobed; berry green................................ **S. triflorum**
   2. Plants somewhat shrubby perennials; corolla violet to blue.

6. Hairs of upper stem branched; berry white................. **S. umbelliferum** Eschsch. (*S. californicum*)
6. Hairs unbranched; berry green.. **S. xanti** (incl. vars. *montanum* and *intermedium*)

STERCULIACEAE

Fremontodendron (p. 284)

Open spreading evergreen shrub with simple, alternate, stellate-pubescent lvs.; fls. large; petals absent; calyx yellow, stellate-pubescent, subtended by about 3 bractlets; fr. a bristly-hairy capsule. Two species, both in CA. One species in Kern County: **F. californicum**

TAMARICACEAE

Tamarix (p. 285)

Shrubs or small trees with small thickish scale-like lvs. on slender branchlets; fls. small in panicled racemes; fr. a capsule. About 70 species; 5 introduced in CA.

1. Fls. 5-merous, pinkish or white.
  2. Tree 6-10 m tall; lvs. sheathing.. **T. aphylla**
  2. Loosely branched shrub; lvs. not sheathing........................ **T. ramosissima** Ledeb. (*T. pentandra*)
1. Fls. 4-merous, pink.. **T. parviflora** DC (*T. tentandra*)

TYPHACEAE
(incl. Sparganiaceae)

Perennial emergent herbs with cylindric, solid stems and linear lvs.; infl. head- or spike-like; plants monoecious; fr. an achene.

Infl. head-like, spheric.. **Sparganium**
Infl. spike-like, cylindric.. **Typha**

Sparganium (p. 173)

Perennial emergent aquatic herbs with grass-like, sheathing lvs.; fls. unisexual in globose heads, the staminate above the pistillate; perianth of few papery scales. About 15 species, 4 in CA. One species in Kern County: **S. eurycarpum** ssp. **eurycarpum**

Typha (p. 173)

Tall (to 3 meters) perennial herbs with simple stems submerged at base; lvs. long, linear; fls. unisexual, crowded into compact "saussage-like" spikes, the staminate above the pistillate; perianth of many slender threads. About 10 species, 3 in CA.

Interval between staminate and pistillate spikes 1-4 cm long; lf.-blades rounded on back.......... **T. domingensis**
Interval between staminate and pistillate spikes none; lf.-blades flat........... **T. latifolia**

ULMACEAE

Trees or shrubs with rounded branchlets and alternate simple pinnate deciduous lvs.; fls. perfect or imperfect; corolla absent; styles 2.

Fr. a dry samara; fls. perfect; trees .... **Ulmus**
Fr. an orange-brown drupe; fls. imperfect; large shrub .... **Celtis**

Celtis (p. 214)

Deciduous shrub to 6 m tall; lvs. serrate, 3-6 cm long, 3-nerved at base; fls. greenish. About 60 species, one in CA. One species in Kern County: **C. reticulata** Torr. (*C. douglasii)*

Ulmus (p. 214)

Trees with simple or double-serrate lvs. and linear stipules; calyx persistent; fr. winged nearly all the way around, notched at apex. About 20 species, 3 in CA.

1. Lvs. doubly-serrate, unequal at base; tree to 35 m tall.......... **U. minor** Miller (*U. procera*)

1. Lvs. mostly simple-serrate, subequal at base.
   2. Branchlets soon glabrous; fls. in spring; tree to 25 m tall.......... **U. pumila**
   2. Branchlets pubescent; fls. in late summer; tree to 15 m tall.......... **U. parvifolia**

URTICACEAE

Herbs, some with stinging hairs; lvs. opposite or alternate; dioecious or monoecious; fls. clustered, 4-merous, small and greenish; fr. an achene.

1. Stinging hairs present; lvs. opposite.
   2. Pistillate calyx 4-parted, segment ± distinct .... **Urtica**
   2. Pistillate calyx saccate, 2-4-toothed at apex .... **Hesperocnide**
1. Stinging hairs absent; lvs. alternate .... **Parietaria**

Hesperocnide (p. 213)

Annual, 2-5 dm tall; lvs. opposite, ovate, scalloped, 5-15 mm long; fls. in axillary heads. Two species, one in CA. One species in Kern County: **H. tenella**

Parietaria (p. 214)

Loosely branched annual; stems slender, procumbent; lvs. ovate, entire; fls. in axillary clusters. About 25 species, 3 in CA. One species in Kern County: **P. hespera** Hinton var. **californica** Hinton (*P. floridana*)

Urtica (p. 214)

Annual or perennial with stinging hairs; lvs. opposite, stipulate; fls. clustered in paired racemes. About 50 species, 2 in CA.

Annual; staminate & pistillate fls. mixed .... **U. urens**
Perennial; staminate & pistillate fls. in separate clusters.
U. dioica L. ssp. **holosericea** (Nutt.) Thorne (*U. holosericea*)

VALERIANACEAE

Plectritis (p. 349)

Subglabrous annuals with slender stems with tufts of hairs at nodes; lvs. simple, opposite, upper sessile; infl. capitate; calyx obsolete, corolla mostly bilabiate, funnelform, spurred at base; stamens 3; ovary inferior, fr. a winged achene. About 11 species, all in CA.

1. Corolla nearly regular, white or light pink, usually lacking red spots, stout, 2-4 mm long; fr. without brush-like rows of hairs .... **P. macrocera**
1. Corolla strongly bilabiate, pink to light red, usually with 2 red spots at base of lower lip, slender; fr. with 2 rows of brush-like hairs .... **P. ciliosa**
   2. Corolla 5-9 mm long .... ssp. **ciliosa**
   2. Corolla 1.5-3.5 mm long.
   ssp. **insignis** (Suskd.) Morey (*P. macroptera* var. *patelliformis*)

VERBENACEAE

Herbs or shrubs with opposite or whorled lvs.; fls. irregular, in terminal or axillary infl.; corolla-tube cylindrical, 4-5-lobed; stamens 4; ovary 2-4-celled, forming single-seeded nutlets, or a drupe in Vitex.

1. Fls. sessile, in spikes; lvs. simple; herbs.
   2. Nutlets 4; fls. in terminal spikes; plants erect.. **Verbena**
   2. Nutlets 2; fls. in dense head-like spikes; plant matted.................................. **Phyla**
1. Fls. pedicelled; lvs. palmately compound; shrubs... **Vitex**

Phyla (p. 326)
(*Lippia*)

Matted herbs with opposite lvs.; fls. in compact head-like spikes; corolla 2-lipped, rose to white; nutlets 2. About 15 species, 2 in CA.

Lvs. with short petiole; fl.-heads rounded to 1 cm long.
P. nodiflora (L.) Greene var. **nodiflora**
Lvs. mostly sessile; fl.-heads elongate to 2 cm long.
P. nodiflora (L.) Greene var. **incisa** (Small) Mold. (Lokern, Yinger 642)

Verbena (p. 327)

Annual or perennial herbs with mostly opposite lvs.; fls. bracteate in terminal spikes; corolla 2-lipped. About 100 species, 11 in CA.

1. Fls. mostly in head-like clusters; corolla-limb 10 mm broad, rose-violet. **V. tenuisecta**
1. Fls. in slender spikes; corolla-limb 2-4 mm broad.
  2. Bracts shorter than fls., inconspicuous; corolla purple.................... **V. lasiostachys**
    3. Spikes 1-3 dm long in fr., lvs. canescent; corolla-limb 3-4 mm broad. var. **lasiostachys** (incl. *V. abramsi*)
    3. Spikes to 1 dm long in fr.; lvs. bright green; corolla-limb 2-3 mm broad. var. **scabrida** Mold (*V. robusta*)
  2. Bracts longer or as long as fls., conspicuous; corolla blue................. **V. bracteata**

Vitex (p. 327)

Deciduous shrub to about 3 m high; lvs. digitately 5-7-foliate, gray-pubescent; fls. pale violet; corolla tubular; fr. a small drupe. About 250 species, one introduced in CA. One species in Kern County: **V. agnus-castus**

VIOLACEAE

Viola.(p. 286)

Annual to perennial herbs; early fls. showy, late fls. cleistogamous, apetalous; corolla irregular, lower petal spurred; stamens 5; ovary superior, 1-celled; fr. a capsule. About 300 species, 24 in CA.

1. Petals white; plant lacking leafy stems .... **V. macloskeyi**
1. Petals with some yellow.
  2. Lvs. not cut or lobed.
    3. Plants erect; lvs. crowded; caps. acute..**V. glabella**
    3. Plants erect to prostrate; lvs. well distributed; caps. obtuse.
      4. Basal blade 1.5-6 times longer than wide, 15-80 mm long, ± linear.
      V. pinetorum Greene ssp. **grisea** (Jeps.) Little (*V. purpurea* ssp. *xerophyta*)

4. Basal blade 1-1.5 times longer than wide, 10-50 mm long, ovate to round. **V. purpurea**
5. Cauline lf. deeply toothed...... ssp. **mohavensis** (Baker & Clausen) Clausen (*V. aurea* ssp. *mohavensis*)
5. Cauline lf. entire, jagged, not deeply toothed.
6. Lf. green, tinged purple, glabrous above............................ssp. **purpurea**
6. Lf. gray-green, puberulent.
ssp. **quercetorum** (Baker & Clausen) Little (*V. quercetorum*)
2. Lvs. deeply cut or lobed.
7. Lvs. 2 times palmately divided; petals pale yellow...........................**V. sheltonii**
7. Lvs. 2-3 times pinnatifid; petals yellow, brownish on back............ **V. douglasii**

VISCACEAE
(Loranthaceae)

Parasitic on woody plants; branches dichotomous, swollen at nodes; lvs. opposite, simple, entire; dioecious.

Berry on recurved pedicels; pistillate sepals 2; on Pinus .... **Arceuthobium**
Berry sessile; pistillate sepals 3 .... **Phoradendron**

Arceuthobium (p. 214)

Yellowish-brown to olive-green much-branched to tufted fragile stems; lvs. reduced to scales; anthers 1-celled; berry fleshy. Genus recently revised. About 45 species, 12 in CA.

(All *A. campylopodum* in Twisselmann)
1. On White Fir ....**A. abietinum** (Engelm.) Hawksworth & Wiens
1. On Pines.
2. On Pinyon Pine..**A. divaricatum** Engelm.
2. On other pines.
3. On pines with 5 needles per bundle.
4. On Limber Pine..**A. cyanocarpum** Coulter & Nels.
4. On Sugar Pine....................................**A. californicum** Hawksworth & Wiens
3. On pines with 203 needles per bundle.
5. On Ponderosa or Jeffrey Pine..**A. campylopodum**
5. On Digger Pine..**A. occidentale** Engelm.

Phoradendron (p. 214)

Greenish much-branched rather brittle stems; lvs. broad or reduced to scales; anthers 2-celled; berry fleshy. About 190 species, 4 in CA.

1. Pistillate spikes with 2 fls at each joint.
2. Lvs. reduced to connate scales; plant glabrous; berry white; on *Calocedrus*. **P. juniperinum** var. **libocedri**

2. Lvs. green, oblanceolate; berry straw-colored.................................... **P. bolleanum**
3. Much-branched, dense, 1-2 dm high; lvs. sessile, 10-15 mm long; on *Juniperus*. var. **densum**
3. More open, 2-3 dm high; lvs. subpetiolate, 15-30 mm long; on *Abies*. var. **pauciflorum**

1. Pistillate spikes with 6+ fls. at each joint; lvs. green; berry white to pink-tinged. **P. flavescens**
4. Internodes canescent; on *Populus, Platanus, Fremontodendron*, etc. var. **macrophyllum**
4. Internodes densely short-villous; on Quercus .... var. **villosum**

VITACEAE

Woody vines climbing by tendrils; lvs. opposite, palmate; fls. small, petals 4-5; fr. a berry.

Lvs. simple, palmately lobed .... **Vitis**
Lvs. palmately compound .... **Parthenocissus**

Parthenocissus (p. 281)

Tendrils disk-like at tips, branched; berries nearly black. About 15 species, one introduced in CA. One species in Kern County: **P. vitacea** (Knerr.) Hitchc. (*P. insertus*)

Vitis (p. 281)

Lvs. palmately lobed or dentate; calyx minute; berry ovoid, pulpy. About 65 species, 3 in CA.

1. Skin of fr. adhering to pulp; young growth mostly glabrous .... **V. vinifera**
1. Skin of fr. loose; young growth tomentose.
2. Young growth arachnoid-pubescent, greenish; berry purple, glaucous. **V. californica**
2. Young growth densely white-tomentose; berry black .... **V. girdiana**

ZANNICHELLIACEAE

Zannichellia (p. 174)

Submerged aquatic monoecious herbs with long slender opposite lvs. 2-8 cm long, with sheathing bases; fls. very small, axillary in spathe-like invol.; perianth absent. About 2 species, one in CA. One species in Kern County: **Z. palustris**

ZYGOPHYLLACEAE

Herbs or shrubs, often with branches jointed at nodes; lvs. opposite, compound; fls. regular, 5-merous, axillary.

1. Lfts. 1 pair; stamens with scale-like appendage.
   2. Large shrub; lvs. resinous; fr. hairy .... **Larrea**
   2. Suffrutescent; fr. glabrous .... **Zygophyllum**
1. Lfts. 4-7 pairs; low trailing herb .... **Tribulus**

Larrea (p. 276)

Evergreen strong-scented resinous shrub; lvs. opposite, compound with 2 lfts.; fls. solitary, yellow; fr. hairy. About 5 species, one in CA. One species in Kern County: **L. tridentata** (*L. divaricata*)

Tribulus (p. 277)

Weak prostrate pubescent annual with opposite pinnately-compound lvs.; petals yellow; carpels crested and armed with 2-4 spreading spines which separate as spinose nutlets. About 12 species, one introduced in CA. One species in Kern County: **T. terrestris**

Zygophyllum (p. 277)

Much branched or suffrutescent glabrous herb; lvs. opposite, compound with 2 lfts.; fls. copper and yellow; pedicel reflexed in fr.; capsule 5-angled. About 70 species, one introduced in CA. One species in Kern County: **Z. fabago** (incl. var. *brachycarpum*)

GLOSSARY

acuminate. Gradually tapering to a short point.
alternate. A single leaf per node. Between, as stamens alternate with petals.
anthesis. Opening of flower.
arcuate. Somewhat curved, like a bow.
auricle. An ear-shaped appendage.
axillary. In an axil, the upper angle formed by petiole or leaf with stem.
beaked. With a long firm tip or appendage.
bifid. Divided in two.
bract. A reduced leaf subtending a flower.
calyx. Collective term for the sepals.
canescent. Covered with grayish white hairs.
capitate. Head-shaped; aggregated in tight clusters.
capsule. A dry dehiscent fruit composed of two or more carpels.
cauline. On the stem.
cismontane. West of the Cascade-Sierran crest; not in the deserts.
claw. Narrow petiole-like base of some petals and sepals.
comose. With a tuft of hairs.
connate. Members of same whorl fused together.
cordate. Heart-shaped with notch at base.
coriaceous. Leathery, tough in texture.
corymb. A flat-topped cluster of flowers.
crenulate. Margin finely scalloped; with small rounded teeth.
deciduous. Falling off all at once.
decurrent. Extending down the stem below point of insertion.
dehiscent. Opening when dry.
deltate. Triangular.
dioecious. Taxa with separate male and female plants.
divaricate. Widely divergent, spreading apart.
exserted. Protruding, as beyond corolla.
falcate. Sickle-shaped.
fascicled. Clustered together, usually leaves or stamens.
filiform. Thread-like.
gibbous. Swollen on one side.
glabrous. Without hairs (not necessarily smooth).
glandular. Bearing glands, usually sticky depressions, protuberances or hairs.
glomerate. Densely compacted in clusters or heads.
hastate. Shaped like an arrowhead with basal tips turned outward.
head. A dense cluster of sessile flowers.
hispid. Rough with stiff or bristly hairs.
hyaline. Translucent.
hypanthium. Cup- or flask-shaped extension of the receptacle to which calyx, corolla, and stamens are attached.
hypogynous. Inserted below ovary.
included. Not protruding beyond throat of corolla.

inferior. Ovary fused with hypanthium and below calyx-lobes.
inflorescence. Cluster or arrangement of flowers.
involucel. Secondary bracts within an inflorescence.
involucre. A whorl or whorls of bracts subtending flowers.
irregular. Bilaterally symmetric or asymmetric, petals not all alike in size or shape.
lanceolate. Longer than broad.
limb. Expanded part of corolla or single petal.
linear. Long and narrow, of uniform width.
lobulate. With rounded lobes.
lyrate. Lyre-shaped; divided with terminal lobe larger than others.
margin. Edge of leaf blade.
-merous. Having parts of a certain number, e.g. 5-merous = parts in 5's.
monoecious. Taxa with separate male and female flowers on the same plant.
nut. A hard-shelled indehiscent single-seeded fruit.
ob-. Inversely (e.g., oblanceolate is inversely lanceolate).
opposite. Two leaves per node. One in front of another (as stamens opposite petals).
palmate. Radiately lobed, as a hand with fingers spread.
panicle. A compound racemose inflorescence.
papilionaceous. See description of Fabaceae.
pedicel. The stalk of a flower.
peduncle. The stalk of an inflorescence.
peltate. Like a thumbtack with long stalk.
perfoliate. Leaf bas or bases entirely surrounding stem.
perianth. Collective term for calyx and corolla.
pericarp. Ripened ovary wall.
perigynous. Borne around an ovary (as opposed to below it - hypogynous).
petaloid. Having the aspect of petals.
petiole. Stalk of a leaf.
pinnate. Veins or leaflets arranged along a central axis.
pinnatifid. Pinnately cleft into narrow lobes.
placenta. Tissue bearing ovules (seeds) in the ovary.
pleated. With accordion-like folds.
plumose. Feathery.
pome. A fruit like an apple or pear.
puberulent. Covered with very short soft hairs.
punctate. Dotted with glands, dots or little holes.
pustulate. With irregular blisters or pustules, usually at base of hairs.
raceme. A simple inflorescence with each flower having a pedicel.
rachis. Axis of an inflorescence or compound leaf.
recurved. Bent or curved backwards.
reflexed. Abruptly bent or turned downwards or backwards.
regular. Radially symmetric, petals alike in size.
retrorse. Pointing downward or backwards.
rosette. A crowded cluster of leaves, usually at ground level.
rotate. Wheel-shaped, corolla with wide rim and shallow bowl or tube.
saccate. With a sac or pouch.
sagittate. Shaped like an arrowhead.

salverform. Corolla with narrow tube and flat limb.
samara. A winged indehiscent fruit.
scapose. With basal rosette and leafless flowering stem.
scarious. Thin, dry and papery, not green.
scorpioid. Coiled like a scorpion tail.
secund. Arranged on one side only.
sessile. Attached directly, not stalked.
setaceous. Bristle-like or bearing bristles.
spatulate. Shaped like a spatula.
spike. A simple inflorescence with each flower being sessile.
spur. A slender sac-like process from a sepal or petal, usually containing nectar.
squarrose. Spreading rigidly at right angles (such as tips of sepals or bracts).
stellate. Star-shaped.
stipe. Stalk, of fern leaf or ovary.
strigose. With stiff, appressed, straight hairs.
subulate. Awl-shaped.
suffrutescent. Woody at base but not shrubby; not necessarily low.
superior. Ovary attached above and not fused with stamens and petals.
terete. Cylindrical, round in cross section.
truncate. Cut off squarely at tip or base.
utricle. Small bladdery, one-seeded, indehiscent fruit.
viscid. Sticky, glutinous.
whorled. Ring of leaves (or other appendages) at a node or within a flower.
woolly. Having long soft entangled hairs.

A FLORA OF KERN COUNTY, CALIFORNIA

by Ernest C. Twisselmann

Illustrated by

Eben and Gladys McMillan

Kern Joshua trees (*Yucca brevifolia* var. *Herbertii*) on the western Mojave Desert; Soledad Mountain in the distance.

THE WASMANN JOURNAL OF BIOLOGY

VOL. 25, NOS. 1 AND 2 SPRING AND FALL, 1967

A FLORA OF KERN COUNTY, CALIFORNIA

ERNEST C. TWISSELMANN
Cholame, California
Illustrated by EBEN and GLADYS McMILLAN

THE NATURAL HISTORY OF PLANTS IN KERN COUNTY

Kern County, California, is an unusual coincidence of a political unit that is also a natural region of botanical merger and transition. Here the floras of the mountains of southern and northern California merge, as do those of the coastal regions, the great Central Valley, and the broad American deserts.

The county, California's third largest and the nineteenth largest in the United States, is rectangular except for the series of right-angled set-backs on the western border. From east to west it is 130 miles wide; from north to south, 67 miles. It has a land area of 8172 square miles or 5,160,960 acres; this is larger than the land area of Massachusetts, New Jersey, or Hawaii, or of Connecticut, Delaware, and Rhode Island combined.

The central portion of the county is occupied by the broad head of the San Joaquin Valley. Encompassing it is a horseshoe of mountains. To the west is the Temblor Range, the southernmost and most arid of the low south inner Coast Ranges and bisected by the San Luis Obispo County line. To the east and south are the county's complex mountain masses. The Greenhorn Range and Breckenridge Mountain border the valley on the northeast; on the southeast, the Tehachapi Mountains. To the south, the high level summit of the Mt. Pinos region that rises more than 7000 feet from the valley floor furnishes a dramatic backdrop for the head of the valley.

The county is a land of contrasts. Elevations vary from a low of 205 feet in the Buena Vista-Tulare Lake drainage system to a high of 8750 feet just northwest of the summit of Mt. Pinos.

Geographic and floristic transitions are rapid; one can easily drive from the desolate alkali plains at Buena Vista Lake at an elevation of 300 feet to the Jeffrey pine forest at the 8,286-foot elevation of the summit of Mt. Abel in less than an hour, passing through nine of the county's seventeen plant associations en route. Temperatures, too, can vary greatly in a short distance; in May the temperature at Bakersfield may be well over 100° F. while less than fifty miles away at the summit of the Greenhorn Range there may be snow flurries. Precipitation is equally variable; a low of less than 0.5 inch in a year has been recorded at Armitage on the desert; more than sixty inches may fall on Sunday Peak in the Greenhorn Range. Such a region, with its complex and varied mountain ranges, its broad inland valley, and its picturesque desert, seems certain to yield floristic records that will also show sharp contrasts. These this paper reports.

A list of the flowering plants of the county has long been desired, not only by the professional botanists but also by those who are concerned with the plants in their daily life: ranchers, forest rangers, wild life specialists, range technicians, and others who are either directly or indirectly concerned with the flora and the vegetation. Further, the county and the state have a large and increasing group of informed lay people with a curiosity about the land around them and an awareness of both the importance and the beauty of the natural scene. The goal of this work is to try to supply a floral inventory useful to all these groups.

Mary Austin once wrote of arid inner California, "You will do well to avoid that range uncomforted by singing floods. You will find it forsaken of most things but beauty and madness and death and God. Many such ranges quicken the imagination with a sense of purpose not revealed, but the ordinary traveller brings nothing away from them but an intolerable thirst." This flora has been especially prepared as a botanical guide for the growing number who are not ordinary travellers.

Geography

Kern County's 8,172 square miles consist of four main geographical regions. These are the upper San Joaquin Valley, the mountain chain that extends from the Tulare County line to Mt.

Pinos at the head of the valley, the Temblor Range that bounds the county on the west, and the transmontane Mojave Desert. As an understanding of the geography is essential to an intelligent comprehension of the flora, a brief and highly generalized discussion of each of these regions follows.

THE SAN JOAQUIN VALLEY. The upper San Joaquin Valley is both geographically and economically the heartland of Kern County. Originally it was a region of broad arid plains, often with an extensive cover of salt bush (*Atriplex polycarpa*), sloping gradually to the south. These plains surrounded the wet center of the valley that was the drainage basin for Kern River and the lesser streams that drained the neighboring mountains.

Although the valley was discovered in 1772 by the Spanish explorer, Don Pedro Fages, it wasn't settled until long after the coastal regions. As late as the mid-1860's, Colonel Baker's field at Kern River was the only fenced pasture between Firebaugh's Ferry near Fresno and Los Angeles where travelers could graze their horses untethered. Not until 1866 was Kern County formed from portions of Los Angeles and Tulare counties, 16 years after California had been admitted to the union. It was 1869 before a post office was established in Bakersfield; the county seat wasn't moved there from the mountain mining town of Havilah until 1876. In the 1880's the valley finally began to be well settled.

Very early, however, the broad plains were used for grazing sheep. Ownership of the lands was of little concern to the migratory herdsmen. Some would spent almost an entire year away from their base in the Los Banos region in the central San Joaquin Valley, feeding south in late winter and spring, then north through the Sierra Nevada in summer and fall, returning home the following winter. But by the 1880's most of the valley grazing lands in Kern County were owned or controlled by resident sheepmen. The warm San Joaquin Valley with its early growth in good years was ideal for winter sheep range and for lambing the ewes. Summers were spent grazing the surrounding mountains. One of the notable trails that enabled the sheep to graze the San Joaquin Valley in winter and early spring, the Mojave Desert in late spring, and the Sierra Nevada in summer and fall was the "Long Trail" that led from Bakersfield through

FIGURE 1. Sunrise from the mesas west of Red Rock Canyon.

the Tehachapi Mountains, thence up the desert to distant Mono Lake and the high eastern Sierra Nevada. (One of the valley forks of this trail, now the unimproved road from Arvin to Tehachapi, is still known as "the Sheep Trail.") In times of drouth, the valley was severely grazed. What permanent effect this had on the flora and vegetation is a matter of conjecture and debate. However, the very large herds of tule elk and San Joaquin antelope that roamed the valley before the coming of the white man must have had much the same effect on the plant cover in dry years as the sheep had.

Irrigation developed rapidly in the region from Bakersfield to Buena Vista Lake in the late 1880's and in the central valley regions around Shafter and Wasco in the early 1900's. In the 1920's large areas in the Buttonwillow region were reclaimed for farmlands. Finally, the development of economically practical deep well pumping after World War II and the construction of the Kern-Friant Canal that brought water from the central Sierra Nevada saw most of the remaining east-side rapidly

FIGURE 2. The southern Kern Plateau across Kern Valley from the east slope of the Greenhorn Range; in the foreground is a digger pine (*Pinus Sabiniana*).

FIGURE 3. The desert-like eastern South Fork Valley; in the middle distance are extensions of the Kern Plateau and the Piute Mountains; the Greenhorn Range is in the far distance.

converted to farm lands. Projects now being constructed that will bring water from northern California's Feather River Dam will make possible cultivation of much of the arid west-side plains in the next few years. Thus, the seemingly imperishable vegetation of the upper San Joaquin Valley's semi-desert plains will soon almost entirely be gone, to be replaced by intensive agriculture.

Currently the principal crops are cotton, potatoes, oranges, alfalfa, and grapes; others of importance are rice, plums, peas, sugar beets, rose bushes, almonds, cantaloupes, and water melons. In addition, many small acreages are devoted to specialty crops. The croplands have a rich flora of weeds and much of interest can be learned from the native species that persist in waste places, vacant lots, and other oddments of land the onrush of civilization has passed by.

Oil was discovered early in Kern County's history; the first oil company in California was incorporated in 1863 on the borders of the San Joaquin Valley at the Temblor Ranch. The shallow very heavy crude oil was mined, distilled for kerosene, and

FIGURE 4. The arid southern Temblor Range. Brush Mountain and Mt. Abel are in the distance.

transported by wagon across the Coast Ranges to Port Hartford (now Avila), thence by boat to San Francisco. Much local oil activity of this sort continued into the early 20th century. Then the great discoveries in the Rio Bravo region near Bakersfield in 1900 and the spectacular Lakeview gusher near Maricopa in 1910 led to rapid development and expansion of the oil fields that made the county a great producing center for the new petroleum demands of the automobile age. Both the sudden wealth of the oil fields and the rich agriculture gave Bakersfield a lusty, raw frontier tradition that was not to disappear until World War II. Since then, its rapid growth has been orderly and the city of some 65,000 population (the metropolitan area has a population of more than 180,000) is now culturally essentially a suburb of Los Angeles. In the opinion of one who has known both Bakersfields, the old was infinitely more interesting than the new. The important valley farming centers are Arvin, Greenfield, Buttonwillow, Shafter, Wasco, and, on the Tulare County line, Delano. The only west-side city is the oil community of Taft.

THE KERN COUNTY MOUNTAINS. Kern County's complex of mountains includes the southern Sierra Nevada (including Breckenridge Mountain and the Piute Mountains), the Tehachapi Mountains, and the Mt. Pinos region, including the San Emigdio Range.

The Kern County portion of the Sierra Nevada has four features: the Greenhorn Range, which is the southern end of the Great Western Divide, Breckenridge Mountain, the Piute Mountains, and the Kern Plateau. Of these, the Greenhorn Range and Breckenridge Mountain border the San Joaquin Valley on the east. The Kern Plateau lies to the east of the Greenhorn Range and the Piute Mountains to the east of Breckenridge Mountain.

Unlike the other mountains, the Greenhorn Range has a broad belt of foothills, mostly of gentle relief. It is a region of relatively high rainfall and has good grassland at the lower elevations, Douglas oak woodland at the middle elevations, and a well developed yellow pine forest at the higher elevations.

The principal drainage is Poso Creek, which heads to the northeast at Tobias Peak and Portuguese Pass in Tulare County. The creek follows zones of weakness far to the southwest before it abruptly swings northwest just before it debouches on the

FIGURE 5. The foothills of the San Emigdio Range.

plains north of Bakersfield, to drain eventually into Tulare Lake. Poso Creek and its tributaries are a Kern County rarity: these are creeks that usually have at least some water all year. The principal tributaries are Sandy and McFarland creeks that head in the Sunday Peak region, Alder and Cedar creeks that drain the Greenhorn Pass region, Lumreau Creek on the north slope of Basket and Hatchet peaks, and Little Poso Creek that heads in the southwest portion of the summit region.

An interesting feature of the range is Lynns Valley, a broad valley with valley white oaks that form a parkland. It has moist subsoil that supports an often luxurious growth of wet soil plants and has a scattering of wet meadows and vernal pools. A number of Sierra Nevada plants reach their southern limits in this pleasant valley; several poolbed plants have their only Kern County occurrence here.

The Greenhorn Range was the first region in the county to be permanently settled; it thrived in the 1850's and '60s as a service and farming center for the Kern River gold fields. Since that time cattle ranching has been the principal occupation. Most of the ranches are relatively small (although often greatly supplemented by summer grazing on U. S. Forest Service lands at the higher elevations). Many are in the third or fourth generation of ownership by the same family. Thus, the ranch people have an unusually high degree of respect for their land and an enviable sophistication in range management practices. Such a combination of attitudes and practices is reflected in the general well being of both the native flora and fauna. Regrettably, this is not always true elsewhere in the county.

The only town is Glennville, a small but active hamlet that is the trading center for the region. Woody, a store and post office in the lower foothills, was once a community of local importance. Granite Station remains only as a landmark on the old stage route from Bakersfield and the San Joaquin Valley to Glennville and the gold regions.

The crest of the range is of moderate relief. At the north end, it is dominated by 8,320-foot Sunday Peak. Just south of Greenhorn Pass, Shirley, Cooks, and Greenhorn peaks form a cluster

FIGURE 6. The plains south of Point of Rocks in the western San Joaquin Valley; this region will soon be converted to irrigated farmlands.

of relatively gentle topography. In a range where granite is the common rock, these three are an unusual crest of metamorphic strata. From this group, the trend of the range swings sharply southwest, running approximately twenty-five miles to its terminus at the mouth of Kern Canyon on the border of the San Joaquin Valley. During and following the mining days until early in this century, the forest was intermittently timbered. For

FIGURE 7. Cement plant at Monolith in Tehachapi Valley. Air pollution from many sources has become a serious problem throughout the county.

many decades this had ceased, but currently it is being aggressively logged under the direction of the U. S. Forest Service.

The steep east slope of rather bold relief has been of little economic use other than for the old gold mines, mostly productive only in the last century, and for the usually scant grazing for cattle.

Recreation is an important use of the forest zone, and one that threatens to be devastating to the flora and fauna. Summer camping, fishing in the permanent streams, deer hunting, and the very limited winter sports afforded at the recent development at Shirley Meadows have all proved popular. The impact of recreation seekers, most of whom are indifferent to natural values, is certain to have deleterious effects, especially in a region of limited extent that is very poorly policed.

The Greenhorn Range is separated from Breckenridge Mountain by lower Kern Canyon. This steep gorge with its high canyonsides is the water course for Kern River from Kern Valley to its mouth at the plains east of Bakersfield. The canyon was considered impassable for many years; the route to the mountains was by way of Granite Station, Glennville, and Greenhorn Pass. In what must have been a monumental construction job at the time, a road was completed up the canyon to give more direct access to Kernville, the mining towns, and the mountain country early in this century. Hydroelectric plants took advantage of the water's rapid flow; the canyon has been an important source of electric power since 1897. The hot springs at Delonegha, Democrat, and Hobo were developed as therapeutic resorts about the turn of the century; although the national fad for mineral springs has long since passed, these are still visited by large numbers of health and pleasure seekers. The river is popular for fishing and picknicking. The slow, winding but picturesque highway along the south bank is soon to be replaced by a high speed freeway on the north side of the canyon. Although increasingly heavy traffic has made the new highway a necessity, any naturalist familiar with recent California freeway planning can only view the proposed road with apprehension.

Steep Breckenridge Mountain, directly south of the Greenhorn Range, has much the same geography compressed into a smaller area. Because of the rough terrain and the large part in the National Forest, permanent inhabitants are few. Cattle grazing, and in the past, lumbering, are the chief economic uses. Summer residences of Bakersfield people at the higher elevations were once common and fashionable. However, with the development of modern high speed highways that make other regions easily accessible, these have largely been abandoned.

Kern and South Fork valleys separate the Greenhorn Range, the Kern Plateau, and the Piute Mountains. Kern River flows

FIGURE 8. The county is changing rapidly; no place seems too remote or unpromising for development or exploitation.

south through Kern Valley into Lake Isabella; the river has its sources in upper Kern Canyon and is the main drainage for the western and central part of the Sierra Nevada north to Mt. Whitney. South Fork Valley, which converges with Kern Valley at Lake Isabella, extends about twenty-five miles east to the southern Sierra crest in the Walker Pass and Mt. Owen region. The oddly misnamed South Fork of the Kern River flows west through this valley after its course turns abruptly from the north where it drains the Sierra Nevada's Kern Plateau, which extends to Siberian Pass just southwest of the Mt. Whitney region in Tulare County. The rivers were dammed in 1952 for the purpose of controlling floods along Kern River in the San Joaquin Valley. However, the lake impounded behind the dam has become an unexpectedly popular recreation center; this use probably now overshadows the flood control purpose.

The confluence of Kern River and its South Fork was the site of old Kernville, now under the lake (the new Kernville, a thriving recreation center, is located upstream seven miles). Originally a typical crude bustling mining town, for many decades before the construction of the dam Kernville was a quiet trading center for ranchers of the mountain regions. Although Lake Isabella is an unattractive body of water in an arid setting and has a hot summer climate, its surprising popularity for recreation has caused rapid but poorly planned growth at Isabella and Bodfish just below the dam, and at Wofford, on the west side. Kernville, a few miles above the lake, has developed as a resort village of charm and distinction.

South Fork Valley, with its broad floor of alluvial soils, has long been used for irrigated farming, mostly for hay for the cattle ranches of the region. The two hamlets in the valley are Weldon and Onyx.

The southern Kern Plateau is the least known part of Kern County. This inaccessible region, abruptly bounded by the east wall of upper Kern Canyon on the west and by the plateau's

FIGURE 9. Shearing time on a mesa east of Tejon Pass. The county's livestock economy is in large part dependent on the Upper Sonoran grassland.

steep southern terminus north of South Fork Valley, is a region of granite domes and ridges, with a scattering of dry meadows. Forest Service roads began to penetrate the plateau in 1959 from a route along Brush Creek up the rugged east side of Kern Canyon near Roads End in Tulare County. Since the completion of roads, the plateau has been subjected to highly controversial selective logging under the direction of the U. S. Forest Service. The main existing roads terminate north of Kern County; a fork of one leads to Pine Flat at the Kern County line on the southwest corner of the plateau, another to the headwaters of Bartolas Creek. An arid picturesque region of granite domes and pinyon

FIGURE 10. Low foothills east of Famosa. In a good year following prolonged drouths nitrate poisoning may be a severe livestock hazard in the Lower Sonoran grassland.

and Jeffrey pine forests in the southeastern portion, largely in Tulare County but with a significant part extending south to Kern County, has been set aside as the Domelands Wilderness; it is presumably inviolate for roads or economic development. Eleva-

FIGURE 11. The trough of the San Andreas Fault has been filled with sediments to form pleasant Cuddy Valley in the Mt. Pinos region. In the right background is Frazier Mountain; in the foreground are young Jeffrey pines.

tions on the plateau in Kern County range from 6,000 to 7,800 feet. Prior to the current logging, the only economic use has been for summer grazing of the meadows.

The Piute Mountains lie south of South Fork Valley. This long relatively narrow range is dominated by Piute Mountain, the central block that has an elevation of 8,400 feet. The relatively steep lower slopes are covered by chaparral or shin oak brush; the summit crest by an open Jeffrey pine forest. In the past, there was considerable gold mining in the range, especially near the now vanished town of Claraville. Currently the range has the same controversial logging program as the Kern Plateau. Most of the meadows on the summit plateau east of the main crest are privately owned; these have been subjected to heavy and deleterious grazing for many years.

The north central portion of the range is drained by Erskine Creek, which usually flows until summer and apparently in years past flowed all year. Cottonwood Creek, a stream that usually has flowing water only following storms, drains the south end; it empties into Jawbone Canyon and ultimately into the desert Koehn Lake near Cantil. Short streams on the east slope are tributaries of Kelso Creek, which rises near Landers Meadow on the east central part of the range. This creek usually has water until late summer and sometimes all year; it flows into the South Fork of Kern River near Weldon.

Forest Service roads are maintained in the summer. There are scattered summer places and cattle camps on the summit plateau. The range has only a few year-round inhabitants.

The main crest of the Sierra Nevada extends south from the Kern Plateau past historic Walker Pass to Kiahvah (Scodie) Mountain. It ceases to be a highly elevated crest in the Mount Owen region north of Walker Pass. South of this region, its summit rarely exceeds 6,000 feet. The crest terminates with the end of the Sierra Nevada Fault at Jawbone Canyon. The southern crest, essentially a desert range south of Walker Pass, is sepa-

FIGURE 12. Severely eroded Quaternary sediments cover the basic rocks of the low west end of the Rand Mountains.

rated from the Piute Mountains by Kelso Canyon and the desert-like floor of Kelso Valley.

The complex Tehachapi Mountains reach their northeast limits at Jawbone Canyon on the east and the Caliente watershed on the west. In this northeastern region, the Tehachapis consist of a group of northeasterly trending ridges that begin at Keene near the summit of the west slope of the central Tehachapi Mountains and culminate in the wild terrain of upthrust volcanic rocks in the Cache Peak region. These ridges are largely covered with a highly specialized xeric chaparral but there are occasional open slopes of Douglas oak woodland and Upper Sonoran grassland. Soils tend to be poor and porous; rainfall is unpredictable. The only economic use of importance is cattle grazing and browsing; even for this purpose, large areas of the region are marginal.

The central Tehachapi Mountains, long considered an extension of the Sierra Nevada, are now thought to be a horst, a plateau uplifted by a complex of faults. The northwest side is dominated by Bear Mountain; the southeastern by the complex of Cummings Mountain, Tehachapi Peak, and Double Mountain. These summits are forested and floristically are essentially a southern extension of the Sierra Nevada flora. Between them lie Cummings, Bear, Brites, and Tehachapi valleys. These montane valleys are approximately 4,000 feet elevation; their soils are rich deep alluviums. Early in this century, they were planted extensively to orchards, primarily of pears, but drouth and adverse markets resulted in these being abandoned in the late 1920's and early 1930's. The hill country is devoted to cattle and sheep.

Of the four valleys, Brites and Cummings have long been farmed. Tehachapi Valley, the area of greatest settlement, is highly developed for farming and for suburban residences. Only Bear Valley is relatively unspoiled; owned by the same family for many decades and served by no public road, its present vegetation with extensive vernal pools on the valley floor has probably not changed substantially in the past 100 years.

In recent years, the character of the Tehachapi region has greatly changed. Once again the orchards have been planted; this time dependent on irrigation rather than rainfall. New specialty crops, such as grass seed and lawn turf, have become important in the mountain valleys. Almost all of the old ranches have been

FIGURE 13. An almost vertical tilting of Tertiary formations in San Emigdio Canyon.

sold, mostly to Los Angeles people of means, either for country estates or speculative investments. The old cement plant at Monolith has been expanded; a large new one is in operation in Oak Creek Canyon. Finally, the state is building a large prison in Cummings Valley.

At Tejon Pass south of Double Mountain, the range swings sharply to the southwest and has a high rounded crest that is steep on the east slope; the west slope consists of a long series of westerly trending ridges that extend to the San Joaquin Valley. The low west slopes are covered with Upper Sonoran grassland; the middle slopes, Douglas oak woodland (often well mixed with out-of-range valley white oaks). Broad areas of the higher slopes have dense impenetrable stands of shin oak brush. Highly local groves of ponderosa pines occur in Tejon Canyon and Jeffrey pines grow on the ridges along upper Catskin Creek, just south of Tejon Pass.

FIGURE 14. Low granite outcrops mark the boundary of the Sierran Cretaceous granite and the uplifted San Joaquin Valley Tertiary sedimentary formations in the Granite Station region.

The region southwest of Tejon Pass is almost entirely owned by the Tejon Ranch Company; its ownership of 288,000 acres in fee makes the company one of the largest land owners in California. The core of the ranch was originally a group of Mexican grants purchased by General Edward Fitzgerald Beale. Later, the General added many homesteads to his holdings, especially in the Rose Station region at the head of the San Joaquin Valley. The mountain portion of the ranch is used for grazing sheep and cattle; sheep were dominant in the last century, cattle in this. No public roads intrude on the property and private police protect it from vandals, promiscuous shooting, and the other nuisances that afflict much of the county.

Grapevine Canyon and Castac Valley mark the southwest terminus of the Tehachapi Mountains. Castac Valley, once a parkland well covered with valley white oaks, is the site of old Ft. Tejon, established in 1854 and abandoned in 1864. The region is served by the small town of Lebec, on the west side of the valley which is now bisected by a broad freeway. Castac Lake, a small land-locked body of highly mineralized water, lies on the southeast side of the valley at the mouth of long, straight Bear Trap Canyon, which is the trough-like western end of the Garlock Fault.

FIGURE 15. Normal erosion along a small fault in the San Emigdio Range has been greatly accelerated by drouth, heavy grazing, and thunderstorms. The sharp angle of the tributary ditches suggests the fault below the central wash predates the present uplift.

The west slope of the range is drained by a series of parallel creeks that are usually short and steep above the valley floor; they flow into Buena Vista Lake. The northern portion is drained by Caliente Creek, which has numerous tributaries, several of which rise to the north in the Piute Mountains. Cottonwood Creek is the western drainage for Bear Mountain; Chanac, Tejon, El Paso, Pastoria, and Tunis creeks drain the southwestern portion of the range. The east slope is drained by a similar series of relatively steep short creeks in the usually dry desert-like canyons. However, Jawbone and Cache creeks are

FIGURE 16. Boulders deposited at the mouth of San Emigdio Canyon during the Pleistocene erosion cycles.

lengthy and drain extensive areas from Tehachapi Valley to the Piute Mountains. Cottonwood (Alamo) Creek drains the east slope of the Tejon Pass region. Others, many without names, flow down the eastern canyons only after storms.

Tehachapi is the only town of consequence; it is important as a farming, railroad, and highway service center.

West of Castac Valley, the Kern County mountains terminate with high Mt. Pinos and Sawmill Mountain, which lie on the Ventura County line, and Mt. Abel, which is entirely within Kern County. These three summits are part of the same crest; together they form the Mt. Pinos complex. The steep north slope and the high broad summit of Mt. Pinos forms a dominating backdrop for the entire southern San Joaquin Valley. Exposed on all four sides, with no ranges of similar altitude anywhere nearby, the broad summit plateau is severely battered by storms and winds and is an unusual, arid, subalpine region with a highly specialized flora. The summit, at 8,831 feet elevation, lies in Ventura

FIGURE 17. The ponderosa pine forest at Shirley Meadows in winter.

County; the highest point in Kern County is on the north slope of the summit outcrops at 8,750 feet elevation. The two-mile long plateau is accented with picturesque lightning-distorted and wind-battered limber and Jeffrey pines. Historically, the mountain has been used for scant cattle grazing and marginal timber cutting. Now an excellent paved highway reaches nearly to the summit. Its proximity to Los Angeles results in heavy recreational use for the mountain, both for winter sports in the usually modest snow and for many summer camps, particularly those of youth groups. The region is heavily infested with deer hunters each fall (although deer are scarce on the public lands). It seems incredible that agencies of the state and federal governments not only allow but even encourage such excessive hunter concentration.

The Mt. Pinos complex is aligned against the San Andreas Fault. At the east base of Mt. Pinos its trough forms sunny Cuddy Valley. At the steep north base the sediments in the fault trough form the group of mostly small meadows and cienegas that collectively are the Mil Potreros. Since the earliest settlement of the county these have been grazed. (The old Mexican

San Francisco-Los Angeles ox road, El Camino Viejo, passed through the lower end before beginning its tortuous drop through San Emigdio Canyon to the San Joaquin Valley.) In the past few decades, this prolonged heavy grazing has greatly accelerated natural erosion processes and the meadows are being rapidly cut through with ditches that are draining the subsurface moisture and changing the wet cienegas to dry meadows (and, unless checked, will eventually reach the final result of the process: a steep dry canyon bottom).

The only towns are on the eastern border of the region; these are the resort and retirement community of Frazier Park and its neighboring development, Lake of the Woods.

The San Emigdio Range lies north of the San Andreas Fault. Although geographically a part of the Mt. Pinos region, geologically it is a distinct range. It is severely eroded, fractured, and contorted by tremendous recent and continuing geological forces. The higher elevations of San Emigdio Canyon and the long easterly trending Tecuya Ridge along the San Andreas Fault northwest of Frazier Park were formerly the scene of many mostly unsuccessful mining prospects, chiefly for antimony. Currently the mines are either inactive or have been abandoned. Once most of the lower foothills of the San Emigdio Range were homesteaded. These small units were not economically feasible in such an arid region and have been added to the few large cattle ranches. (The historic Kern County Land Company owns the eastern two-thirds of the range; a few private ranches the western one-third.) Except for the oil fields at Wheeler Ridge on the northeast end, the range has no public roads; in common with most private property in Kern County, visitors are discouraged, so much of the range remains, as on the Tejon Ranch in the Tehachapi Mountains, unspoiled by excessive hunting and other aggressive recreation.

The drainage is almost entirely to the north to the San Joaquin Valley's Buena Vista Lake. Most of the creeks have water only following rains, but San Emigdio Creek which is the drainage for the Mt. Pinos region often flows until late summer. Apparently in early times this creek usually had at least some water all year. The southwest slope of the long northwest flank of Mt. Abel drains into the Cuyama River in Cuyama Valley; here the dry creeks in Ballinger and Quatal canyons have water briefly following storms. This small region is the only part of Kern County that drains into the distant Pacific Ocean.

THE TEMBLOR RANGE. The Temblor Range, northwest of the San Emigdio Range, is southeast trending and is largely of gentle relief. It has not been considered a part of the Kern County mountain complex as geologically, geographically, and biologically its affinities are much closer to the San Joaquin Valley than to the other mountain ranges. The northern portion is a region of rounded hills; the central crest is somewhat more severely eroded, resembling a desert range.

The range was settled late; much of it was not homesteaded until the first quarter of this century. It became well populated in the homesteading days but these settlers found that their 320 acres were far, far too few for economic survival in an arid region. These homesteads were rather rapidly consolidated into comparatively large holdings. Except for dry-farm wheat and barley farming, largely in the northern portion, the land is used for grazing, primarily of cattle although sheep are also important.

The east slope drains into the San Joaquin Valley; the arid steep short canyons have water only following rains; the range is without permanent or even winter and spring streams. The west slope, largely in San Luis Obispo County, has the land-locked Soda Lake for drainage for the southern part, the northern two-thirds of the range drains into the Salinas Valley and ultimately Monterey Bay.

The base of the range along the San Joaquin Valley is about 1,000 feet elevation; the highest point, McKittrick Summit, is 4,322 feet elevation. The Temblor Range occupies less than ten percent of the county's area. It has no cities or towns.

THE MOJAVE DESERT. The northeastern portion of the Mojave Desert lies in Kern County, occupying about thirty percent of the county's area. Geographically, it is divided in three parts: the broad plains and sinks with their picturesque buttes in the southern two-thirds; the lonely Rand and El Paso mountains in the middle portion; and Indian Wells Valley in the northwest. Although Indian Wells Valley and the El Paso Mountains are geologically part of the Basin Ranges province, for geographic and other purposes they are always considered part of the Mojave Desert.

The Mojave plains reach from the base of the Tehachapi Mountains across Kern County and on far to the east and southeast. A distinctive feature of these desert plains is the numerous buttes that are cores of Tertiary volcanoes. One, Soledad Mountain, a few miles south of Mojave, is a notable landmark on the western border of the desert. Others, northeast of Mojave, are Desert and Castle buttes and Galileo Hill and, on the San Bernardino County line, Saddleback Mountain. Low ridges (such as

FIGURE 18. Fog fills the San Joaquin Valley and infiltrates Adobe Canyon in the Greenhorn Range; on the distant sky line above the fog-filled valley is the Mt. Pinos region.

Nine Mile Hill) east and northeast of Mojave are the surviving surfaces of the ancient mountain mass of Mohavea, now almost completely eroded away. The Muroc Sink occupies the southeastern corner of the county; it is the ancient drainage basin for the mountains to the west. This sink is made up of a series of

FIGURE 19. Soils modified by wind during dry years are characteristic of the upper San Joaquin Valley.

dry lakes (Rosamond, Buckhorn, and Rogers) with packed hard surfaces highly impregnated with saline minerals. All of these are now within the bounds of Edwards Air Force Base, which uses Rogers Lake, the largest, for the landing field of supersonic test aircraft. In the Boron region to the northeast, the old lake bed is covered with wind deposited sand or drifting low dunes. Here America's largest deposit of boron has been mined intermittently for decades; since 1958 it has been extensively developed to become one of California's leading mines (it is the state's largest open-pit mine). Mojave is the principal town of these desert plains. It was formerly primarily a center of railroad transportation (and once was important as the southern terminus of the mule-team freight lines from the borax mines in the Death Valley region). Although still a railroad center, currently the business generated by transcontinental Highway 46 that intersects the Los Angeles-Reno highway in the town is much more important. Rosamond, near the Los Angeles County line, is a community center for irrigated farming, and of lesser importance, for the

FIGURE 20. A desert dust storm filters into lower Last Chance Canyon in the El Paso Range.

nearby air-force base. Boron and North Edwards are mining and air-force towns on the eastern border of the county. California City, a recently established highly speculative real estate promotion, lies at the head of Fremont Valley.

The picturesque and severely eroded Rand and El Paso mountains mark the geological limits of the Mojave Desert. Historically, the ranges were most important for mining. The Rand region was active from the 1890's until the 1930's, yielding large quantities of gold and lesser amounts of silver and tungsten. One mine, the great Yellow Aster, produced more than twelve million dollars worth of gold. The mining towns were Randsburg and neighboring Johannesburg, and just across the San Bernardino County line, bawdy Red Mountain; these are now only remnants of their former size and importance.

The Rand Mountains have moderate relief on the southeast side and are steep and abrupt on the northwest slope. The El Paso Range has steeper slopes on both sides, with abrupt deep canyons that drain to the south. The westernmost is colorful severely eroded Red Rock Canyon. The central portion is drained by the equally interesting but less well known Last Chance Canyon. Finally, the east end is drained by the less severely eroded Iron Canyon and Goler Gulch. The El Paso Range was once a region with numerous placer mines; unlike the Rand Mountains, the lode for the El Paso placer deposits has never been discovered. The range is now virtually uninhabited.

North of the El Paso Range lies Indian Wells Valley. This north and south trending valley, nearly twenty miles wide, separates the southern Sierra Nevada and the Argus Mountains, a desert basin range. China Lake lies at the east base of the valley. This is an ancient lake that was part of the interconnecting series that received the Pleistocene drainage of the Sierra Nevada and extended from Mono Lake on the north to prehistoric Lake Manly in what is now Death Valley. China Lake, most of the Argus Range in San Bernardino and Inyo counties, and other large sections of the desert are now part of the China Lake Naval

FIGURE 21. This long-fallen Jeffrey pine at Basque Encina in the Tehachapi Mountains is several miles below the nearest living one. Old fallen out-of-range Jeffrey pines are occasional throughout the Tehachapi Mountains, suggesting a recent increase in the aridity of the region.

Figure 22. **The alkali sink association south of Buena Vista Lake; the shrubs belong to *Suaeda fruticosa*.**

Ordinance Test Station, effectively closing a great acreage of the desert to the public. In this day of rampant vandalism and abuse on the desert, the military at Edwards and China Lake are unintentionally perserving sanctuaries for the flora and the fauna.

Apart from mining, the principal historical use of the desert has been for usually brief spring grazing for sheep. In recent years there has been considerable irrigated farming in the southern portion near Rosamond and Willow Springs and at Cantil in Fremont Valley. The prevailing harsh winds and the fertile but very light soils that are difficult to stabilize have presented serious problems for these farming ventures.

One of the great changes in the recent few years is the greatly increased recreational use of the desert, especially of the holiday and weekend sort. The current vogue for motorcycles and four-wheel drive vehicles has had especially damaging effects. The desert floor and slopes are fragile; this sort of use has greatly increased both wind and rain erosion. The promiscuous shooting that is rapidly eliminating the native fauna in many regions is certain to have a strong but poorly understood effect on the flora (all the more deplorable as most of the animals are victims of target rather than game shooting). Such activities are subject to virtually no police control; indeed, public officials to date seem to see no reason to be concerned.

SUMMARY. Kern County consists of a complex of mountains, a broad expanse of desert, and a great inland valley. While it is a region of contrasts, almost all of the county's geography, econ-

FIGURE 23. The creosote bush association below Last Chance Canyon; in the distance it merges with the alkali sink association at Koehn Dry Lake.

omy, and biology is that of a land of little rain. Thus, the original flora and fauna were highly adapted for survival in a dry climate with drouths that were frequent and often prolonged. To be successful, man and his economy had to acquire the same specialized ability to cope with the challenging climate.

Figure 24. The Lower Sonoran grassland near Belridge in a good year.

Figure 25. The same region in time of drouth. The shrubs are common salt bush (*Atriplex polycarpa*).

Man's impact on the natural scene, even though recent, is difficult to measure. Certainly the enormous number of sheep that wintered in the San Joaquin Valley must have had an effect on the native flora. However, as their food preferences and habits are largely the same as those of the great herds of tule elk and San Joaquin antelope that roamed the valley before the white man came, it is doubtful the sheep actually eliminated many species or even greatly altered the character of the vegetation. The impact on the mountains and desert is not clear. Although these have been grazed, often heavily, since at least 1860, changes are difficult to assess. But it is of great importance in understanding the floristics of the county to remember that the agricultural economy was largely one of seasonal grazing well into the 20th century and this remains true even now in more than half of the county.

There can be no doubt that occurrences of moist soil have been greatly reduced, even apart from the elimination of the valley

FIGURE 26. The Upper Sonoran grassland in the hills southwest of Woody.

FIGURE 27. In the foothills north of Granite Station there is a broad area of transition between the Upper Sonoran grassland and the Douglas oak woodland.

marshlands and sloughs. Development of springs for livestock and domestic water eliminated many seeps, springs, and cienegas in the mountains. Deep-well pumping in the montane valleys lowered the water table sufficiently to eliminate meadow floras in some regions. Beyond this, it would seem that the activities of man have lowered the general available soil moisture generally in poorly understood fashion; the much more xeric vegetation of the last forty years can only in part be attributed to the prolonged drouths that have occurred.

Irrigated farming that began in the last century increased rapidly since about 1920, with even greater impetus following World War II. This has destroyed the natural flora of great stretches of arid plains that were converted to rich cropland. Meanwhile, the broad sinks of the San Joaquin Valley that received the southern Sierra Nevada drainage were dried as their water was diverted or impounded for irrigation. With the develment of deep-well pumping and the increasing use of soil-modifying agricultural chemicals, the process of converting arid and even alkaline lands to farming will proceed even more rapidly. During the decade of field work for this study, thousands of acres have gone under cultivation, not only in the valley, but in the montane Tehachapi and South Fork valleys and even on the Mojave Desert. Thus, this work, intended as a botanical guide, even before its completion has become in part only a historical record.

The increasing impact of human activities can easily be measured by population growth. In 1940, the county had fewer than 140,000 inhabitants; in 1950 the population had grown to 228,-309; in 1966, 338,600. Such dramatic growth has inevitably had great impact on the natural scene. However, the population is largely concentrated in the central valley, particularly in the Bakersfield region, and broad areas of both the desert and the mountains are uninhabited. Probably more than half the county has a population density of less than one person to a square mile. So, in the great arid expanses of Kern County, there will always be places, often only a few minutes from the heavily populated areas, where the discriminating can find the natural scene unspoiled.

Geology and Soils

Just as Kern County is a region of merger and transition floristically, so it is geologically. Five of California's eleven geologic provinces occur in the county; these are the Sierra Nevada, the Great Valley, the Coast Ranges, the Mojave Desert, and the Basin Ranges provinces (Troxel and Morton, 1962:19). These geological provinces that meet in the county form six local geological regions. The most clearly limited of these is the broad head of the San Joaquin Valley. The horseshoe of mountains enclosing it compose four geological regions; these are the southern Sierra Nevada and Tehachapi Mountains on the east, the Mt. Pinos-San Emigdio complex on the south, and, on the west, the Temblor Range. In addition, the northwestern portion of the Mojave Desert lies within the county.

In Paleozoic and early Mesozoic time, more than 100 million years ago, most if not all of Kern County was submerged under a widespread open sea in which sedimentary deposits accumu-

FIGURE 28. Subshrubs, mostly gray California buckwheat (*Eriogonum fasciculatum* ssp. *polifolium*), densely cover the rocky slopes in upper Chico Martinez Canyon in the Temblor Range. In the foreground coastal live oaks (*Quercus agrifolia*) grow at a spring.

lated to great thickness. During Jurassic and early Cretaceous time, 120 million to 90 million years ago, the county became part of a regional uplift and the sedimentary rocks became intensely deformed and metamorphosed to schist, quartzite, and marble. These were extensively intruded by the granite magmas that form the present granitic batholiths of the Sierra Nevada and Mojave Desert regions, and the whole region was elevated to mountainous terrain. This was followed by a long period during which the region was eroded to low relief.

During Cretaceous time the San Joaquin Valley formed by subsidence as the area to the east rose to form the Sierra Nevada. From that time to the present, the valley sank almost continuously, became submerged under the sea, and received great quantities of debris eroded from the continually rising Sierra Nevada. Also during this time the Coast Ranges formed as local islands uplifted by great compressive forces, probably along the San Andreas Fault, which may have been active since late Cretaceous time. During Pliocene time, the San Joaquin Valley became elevated above sea level as the Sierra Nevada was rejuvenated, sharply uplifted, and tilted westward.

During the present uplift, probably in the late Pleistocene, the upper San Joaquin Valley became covered with a broad fresh water lake when sedimentary ridges from the Coast Ranges on the west and the Sierra Nevada on the east joined to form a barrier across the valley that impeded and finally blocked the drainage to the sea. The drying of this lake has been rapid; considerable evidence suggests the present arid climate developed in the past 1200 years. The aridity of the region is probably still increasing.

On the east side of the valley, from the mesas east of Bakersfield north through the Poso Hills to the Tulare County line, a series of low hills at the base of the Sierra Nevada was formed as small generally westward-tilted fault blocks uplifted the Tertiary non-marine strata. In some regions these hills, with their nearly

FIGURE 29. The northwest slope of Breckenridge Mountain is well covered with the digger pine phase of the Douglas oak woodland; in the foreground is the upper part of the steep wall of Kern River Gorge.

FIGURE 30. Chaparral and Upper Sonoran grassland alternate near the mouth of Lecheria Canyon on the southeastern slope of the Tehachapi Mountains.

level crests, form a belt as much as twelve miles wide. The low granite outcrops that are a distinctive feature in the Granite Station region and are evident to the north and south mark the boundary between the Sierra Nevada and the uplifted Tertiary sediments of the San Joaquin Valley.

Several other groups of low hills in the valley, such as Lost Hills and Semitropic Ridge, are the result of simple anticlinal folding of the valley's underlying sedimentary strata. A structural feature of interest is the bleak Elk Hills that are prominent on the west side of the valley. This small but impressive range, that runs into the valley from the McKittrick region, is an excellent example of the beginning of mountain building with the simplest form of uplift, a simple arch fold in the marine sedimentary strata.

FIGURE 31. Shin oak brush at Tejon Pass; white firs (*Abies concolor*) grow in the moister places.

The soils of the valley are almost entirely sedimentary and are unusually deep. Those of the west side are erosion products of the Coast Range sedimentary rocks; these are especially fine. Those on the east side, mostly of granitic origin, tend to be coarser. Great areas of the low central part have alkaline soils; these have become much more extensive since the drainage of the old marshes and sloughs.

The northeast side of the San Joaquin Valley in Kern County is bounded by the complex southern limits of the Sierra Nevada. South of the Mt. Whitney region, the Sierra Nevada is divided into two great blocks by the Kern River Canyon that follows, at least for the most part, the zone of the Kern Canyon Fault that predates the present and probably the previous uplifts of that great mountain range. The Kern Plateau forms the block on the east side of the canyon; on the west is the Great Western Divide that culminates in the magnificent Kaweah Peaks. Both of these mountain masses have their southern terminus in Kern County.

FIGURE 32. The ponderosa pine phase of the yellow pine forest in the Greenhorn Range.

The Great Western Divide terminates with the Greenhorn Range and associated Breckenridge Mountain; the Kern Plateau at South Fork Valley. The eastern boundary of the Sierra Nevada is the abrupt main crest that extends south from the Kern Plateau to Jawbone Canyon. This eastern crest is uplifted along the Sierra Nevada Fault zone about fifteen miles east of the Kern Canyon Fault and parallel to it. This fault zone marks the eastern limits of the Tertiary and Quaternary westward tilting of the great block that forms the range.

The Greenhorn Range trends sharply southwest from the main axis of the Great Western Divide. Although locally often considered to end at Greenhorn Pass, for the purposes of this study the northeastern limit is at Portuguese Pass, just north of the Kern County line in Tulare County. The crest of the range is dominated by the 8,300 foot summit of Sunday Peak near the northern limit of the range. Other summits along the central crest include the group composed of Greenhorn, Shirley, and

Cooks peaks, all under 7,000 feet. From this group, located just south of Greenhorn Pass, the range veers sharply southwest to its termination at the vertical scarp of the Kern River Fault at the mouth of Kern Canyon.

The west slope of the Greenhorn Range is characteristic of the west slope of the Sierra Nevada in general: a relatively gentle slope dissected by sometimes deep canyons. The lower foothills, however, are distorted by complex relatively recent faulting. These do not follow the normal pattern of the foothills of the Sierra Nevada to the north, but have many varied alignments accented by such local features as Blue Mountain, a steep peak that dominates the landscape west of Glennville.

FIGURE 33. The arid Jeffrey pine forest on the southwest rim of the Kern Plateau.

The erosional processes in the Greenhorn Range are only moderately advanced. On the west slope the rock structures are usually well covered with soil, mostly heavy loam. While the underlying rocks for the most part are the granite that forms the core of the Sierran system, a complex of metamorphic rocks also occurs. These are most extensive from Greenhorn Pass to Shirley and Cooks peaks and from Glennville and Fulton southwest to the northeast slope of Basket Peak. Iron Mountain, southwest of Woody, is a highly local occurrence of metamorphic strata.

The eastern slope of the range is rather abruptly bounded by the Kern Canyon Fault that in Kern County lies along Kern Valley and the southern end of upper Kern Canyon. The escarpment that forms the east face of the Greenhorn Range, while steep, is not in alignment and is much broken and altered, par-

FIGURE 34. Well developed Jeffrey pine forest near Pine Flat on the Kern Plateau. In the foreground is a white fir (*Abies concolor*).

tially by erosion and partially by local subsidence. Black Mountain and Cane Peak, that form an easterly spur from the main crest, dominate the landscape northwest of Kern Valley; these are probably the result of their hard granite offering more re-

FIGURE 35. The arid shrub association on the mesa above Jawbone Canyon; in the foreground is a Kern Joshua tree (*Yucca brevifolia* var. *Herbertii*); Chuckwalla Mountain is in the background.

sistance to erosion than the coarse granitoid rocks originally around them.

Directly south of the Greenhorn Range and a part of the same tectonic block is Breckenridge Mountain. These are separated by the deep Kern River Gorge[1]. This is a young canyon, the result of vigorous erosion in Pleistocene and Recent time, that cuts deeply in its plunge from South Fork Valley to its mouth on the plains northeast of Bakersfield. In the central part of the canyon, the south canyonside rises fully 5,900 feet to the summit of Breckenridge Mountain, 1,500 feet of which form the canyonside itself. On the north side, the summit of the Greenhorn Range at the head of Delonegha Creek is 3,900 feet above the canyon bottom. The river is entrenched in a gorge that is antecedent

1. Common usage applies the name "Kern Canyon" to both the gorge between Isabella and Bakersfield and the long canyon between Kernville and the Mt. Whitney region; general adoption of the geologist's term for the lower canyon would not only be more descriptive but would eliminate much confusion caused by the current ambiguous usage.

FIGURE 36. The creosote bush association extends for scores of miles across the Mojave Desert. Fremont Valley is in the foreground; in the middle distance are the desert buttes, the San Bernardino Mountains are on the far horizon.

across the west block of the Sierra Nevada uplift. South Fork and Kern Valley were occupied by a large lake into which waters of the retreating glaciers of the central Sierra Nevada flowed; the overflow from this lake cut the Kern River Gorge through resistant rocks of the west block of the Sierra Nevada as it rose.

The east face of Breckenridge Mountain is the result of vertical movement on the Breckenridge Fault (closely related to and possibly an extension of the Kern Canyon Fault). This fault is much more distinct than the Greenhorn Range's eastern alignment against the Kern Canyon Fault; its steep east slope reaches from Havilah nearly to the summit of the mountain.

The west slope of Breckenridge Mountain, while steeper than typical in the Sierran system, is not nearly so abrupt as the east slope, and gradually merges into the low foothills and mesas east of Bakersfield. A local feature of interest is the vertical scarp of the Pleistocene Kern River Fault which is clearly visible for about two miles at the northwest base of the mountain, crossing Kern River at the mouth of the canyon and forming the southwest base of the Greenhorn Range for about five miles. This sort of vertical uplift is rare on the western slope of the Sierra Nevada although common on the east slope. Its steep vertical thrust can easily be seen from the Kern Canyon highway west of the mouth of Kern Canyon, where it rises from the Quaternary terrace deposits.

Most of Breckenridge Mountain is composed of Sierran granite. However, there are exposures of Pampa schist and other Paleozoic metamorphic rocks as well as some Cretaceous basic rocks on the west slope. A local lava flow in Tertiary stream-laid sediments is exposed near Bena, at the extreme southwest base of the mountain. Most of the rocks are well covered with soil, even at high elevations; often these are loam of high quality.

Directly to the east of Breckenridge Mountain, separated by Bodfish Gap, Havilah, and Walker Basin, is the massive fault

FIGURE 37. The shadscale scrub association on the desert plains north of Boron.

block that forms Piute Mountain, a fitting southern end to the central Sierra Nevada. Although separated from the Kern Plateau by South Fork Valley, it is essentially a southern extension of that great mountain mass. Piute Mountain, some twelve miles long, has a relatively level crest. Unlike most Sierran blocks, the mountain is moderately steep on all sides. Although the principal exposed rocks are granitic, the Piute Mountains also have a complicated and involved series of metamorphic rocks. The largest exposure of these forms a huge curve around the north end of the mountain, extending from Kelso Canyon northwest to Erskine Creek and Bodfish, then south in a long narrow exposure along the crest and west side to Caliente Canyon. In the Mount Laura region of Erskine Creek, exposures of limestone are common. The crest reaches an elevation of 8,432 feet at Piute Peak on the south end and 8,014 feet at Liebel Peak on the north end. East of the crest there is a summit plateau of gentle topography, tilting to the southeast.

The east base is abrupt and well eroded; it is bounded by Kelso Canyon on the north and Kelso Valley on the south. Here there are broad slopes of the coarse deep sandy talus that is common in the desert ranges. To the south, the divide between the head of Caliente and Jawbone canyons separates the Sierra Nevada and the northeastern Tehachapi Mountains.

An interesting feature of the Piute Mountains is the view from Lookout Point (a local name for the site of the old fire lookout). The point, a massive metamorphic outcrop of 8,312-foot elevation, is the end of a northeast spur from the south end of Piute Mountain. On the now uncommon clear days, the point affords an exceptional view of inner central California's mountain complexes. It is an excellent vantage point for examining such structural features as South Fork Valley, Kern Canyon from Lake Isabella to its head in the Mt. Whitney region and the Kaweah Peaks. Much nearer at hand, one is directly above South Fork Valley and, beyond it, there is an excellent view of the granite domes of the Kern Plateau. To the east are the barren southern crest of the Sierra Nevada, and the desert Argus, Inyo, Slate, and El Paso mountains, and in the far distance, the deep desert Panamint Range. To the southeast are the Muroc Dry Lakes of the Mojave Desert and faintly visible above the desert skyline, the distant San Bernardino Mountains. To the south there is a panoramic vista of the Tehachapi Mountain complex, Mt. Pinos, and a portion of the Liebre Range. Even a small part of the very distant Coast Ranges can be seen over the northwest end of the Piute main crest. Apart from the magnificence of the view, it is an excellent vantage point for study of an impressive grouping of geological features. (The point can be reached by a primitive but passable forest road from the French Meadow-Piute Mountain road.)

Piute Mountain is separated from the Kern Plateau of the south central Sierra Nevada by the long relatively narrow South Fork Valley, probably a depressed block that once connected the plateau and the Piute Mountains. The Kern Plateau (an inappropriate name for a region of ridges and granite domes) is a broad triangular area that has its apex near the Mt. Whitney region far to the north. It is bounded on the west and northwest by the steep high walls of upper Kern Canyon and on the east by the main Sierran crest. Only its steep southern rim is in Kern County; here the massive block forms the north wall of South

FIGURE 38. A remnant of the vanished marshlands: a pool in the Buena Vista Slough near Buttonwillow.

Fork Valley. It is the terminus of a region of picturesque granite domes, flats covered with local alluvium, and often rugged canyons. Not only is the Kern County portion, even at as much as 7,500 feet elevation, an arid region but it undoubtedly has been for a considerable time. This is shown by the broad slopes of granite sand, a disintegration product of the common granitoid rocks of the region. Erosion has not proceeded rapidly enough to remove this sand and the climate has been too dry to support the vegetation that converts it to loam. Thus, much of the relatively coarse sand remains in place. Although seventy per cent of the Kern Plateau in Kern County is composed of granite, northeast of Lake Isabella there is a large area of Paleozoic metasedimentary rocks. Northwest of these, east of Kern River almost to Kernville, there are extensive exposures of Cretaceous basic rocks. A large area from Spanish Needle Creek to Mt. Owen is also composed of Cretaceous basic rocks, these also occur locally along the east base of the Sierra Nevada above the desert. Northwest of Pilot Knob in

FIGURE 39. A drainage canal near Tupman in the old Buena Vista-Tulare lakes basin.

the Bartolas Country of the south central Plateau there are five exposures of Tertiary volcanic rocks.

Rugged Mt. Owen, the southernmost of the series of prominent peaks that extend far to the north, marks the southern limit of the Kern Plateau. From Mt. Owen the main Sierran crest continues south with much lower summits, seldom exceeding 6,000 feet elevation south of Walker Pass. This southernmost portion is a typical desert range that little resembles the main crest to the north.

Colorful Jawbone Canyon marks the southern end of the main structural feature of the Sierra Nevada, the great series of vertical scarps that form the easterly crest of the mountains along the Sierra Nevada Fault zone. This fault zone is much more pronounced north of our region; it reaches its most spectacular development in the sheer east wall of the Sierra Nevada from Olancha Peak to Mt. Whitney. Nonetheless, it is clearly marked in Kern County as far south as Kiahvah Mountain, just south of Walker Pass. South of Kiahvah Mountain, the Sierra Nevada Fault is less distinct and obvious and is notable for its sharp curve to the west. Here, where arid Gold, Dove, and Butterbread peaks are landmarks, it veers several miles westerly from the front that ends at Kiahvah Mountain. Finally, nearing its southern end, the zone veers sharply east again just west of Red Rock Canyon, where it terminates against the Garlock Fault, one of the prime structural features of the Tehachapi Mountains. The severely eroded stark badlands with their markedly deformed beds that occur in this southern region would be recognized at once by enthusiasts of western motion pictures and television programs.

The Tehachapi Mountains lie southwest of the Sierran limits at Caliente and Jawbone canyons. The range has been eroded from the granitic igneous rocks that formed part of the Cretaceous Sierra Nevada batholith and from more recent marine metasedimentary rocks now exposed in the western portion. The Tehachapi Mountains were uplifted as a result of compression against the Garlock Fault on the east and along the recently active White Wolf Fault on the northwest. Both slopes of the Tehachapi Mountains are steep; thus, it is a range bounded by steep slopes with deep severely eroded canyons but with summits that are

FIGURE 40. Plains at the mouth of San Emigdio Canyon. The flora is essentially the same as it was when it was collected and described by Alice Eastwood in 1893.

relatively gentle and round and that have a good soil cover. An excellent view of the Garlock Fault with offset alluvial fans that show recent and even current movement can be seen anywhere along the highway from Mojave to Cantil, especially along Pinosa Ridge northeast of Cache Canyon. The White Wolf Fault forms the northwest base of Bear Mountain; it can be observed from the Bakersfield-Tehachapi highway.

From the west, the dominant topographic feature is 6,898 foot Bear Mountain, sharply uplifted on the White Wolf Fault. Its rounded summit dominates the southeastern end of the San Joaquin Valley and is clearly visible even from the distant south inner Coast Ranges. To the southeast of this mountain, separated by park-like Cummings, Bear, and Brites valleys, are the crests of round-topped Cummings Mountain, Tehachapi Peak, and the eroded jagged summit of Double Mountain. Considerable evidence suggests this north central portion of the range was once a broad plateau. Bear Mountain on the northwestern border and Cummings Mountain, Tehachapi Peak, and Double Mountain on the southeast are the surviving surfaces of the old plateau. Brites,

Cummings, Bear, and Tehachapi valleys are now elevated to 4,000 feet and abandoned by whatever streams carved them out.

From this region, the range veers sharply to the east along Tollgate, Barren, and other associated ridges, culminating in the steep rugged volcanic formations in the vicinity of Cache Peak. These eastern ridges are long, narrow, and almost easterly trending. Miocene volcanic and sedimentary rocks, composed of tuffs and stream-laid sediments, are exposed from the region of Cache Peak northwest to Caliente Creek. Rhyolite plugs and dikes are also common between Monolith and Jawbone Canyon.

FIGURE 41. The flora of the east side of the summit of Tejon Pass has changed little since 1853 when Lt. R. S. Williamson made his observations and Dr. A. L. Heermann collected in the area.

South of Double Mountain, in the region of Tejon Pass, the Tehachapi Mountains narrow and swing rather sharply to the southwest. Here the ridges are steep and the deep canyons are often severely eroded but the summits are usually round and gentle and have an excellent soil cover, often of loose loam; there is little suggestion of mountain-top topography. At the southwest end, the Tehachapi Mountains are separated from the Mt. Pinos complex and the San Emigdio Range by Castac Valley and Grapevine Canyon.

Although at least seventy per cent of the surface rock is granite, the Tehachapi Range is rich in a variety of metamorphic rocks. A small area of lava is exposed at the southeast end on the border of Antelope Valley near the Los Angeles County line. Oligocene volcanic basalt and andesite occurs in scattered small areas at the edge of the valley from Pastoria Canyon to Grapevine Canyon. Calcareous formations are common, particularly in the regions northeast of Tehachapi and in the Monolith region. Roof pendants of various Paleozoic metamorphic rocks occur; some of these are several miles long; some only a few feet. Other non-granitic rocks include a large exposure of Cretaceous dioritic rocks on the west slope south of El Paso Creek. The Garlock Fault Zone from just north of White Oak Lodge at Tejon Pass south through Bear Trap Canyon nearly to Castac Lake is underlain with a long narrow exposure of pre-Cambrian Pelona schist. The

FIGURE 42. A Jeffrey pine (*Pinus Jeffreyi*) on the arid southwest rim of the Kern Plateau.

southwest slope of the range, from Pastoria Creek southwest to Mt. Pinos, is largely occupied by Cretaceous and older metamorphic rocks. Finally, non-marine Miocene sedimentary rocks are extensive in the region east of Cache Peak, along Tejon Creek near the border of the San Joaquin Valley, and in smaller exposures southwest to Grapevine Canyon.

At the termination of the Tehachapi Mountains at Grapevine Canyon and Castac Valley near Lebec, the trend of the ranges veers sharply from southwest to nearly west. This is the region of the Mt. Pinos-San Emigdio complex, where the highest point in Kern County is reached on Mt. Pinos at 8,750 feet elevation on the northwest side of the summit outcrop (the summit is in Ventura County). While the granitoid core of this region is probably a part of the original tremendous Sierra Nevada batholith, its history since Cretaceous time is not directly connected with the Sierran movement. The comparatively recent uplift of the region is directly connected with activity of the San Andreas Fault and its tributaries. The block of which Mt. Pinos, Sawmill Mountain, and Mt. Abel are the summits is apparently the result of compression between that fault, and its off-shoot, the Pine Mountain Fault, which extends from near Frazier Park to Pine Mountain in Ventura County. The southwest scarp of the San Andreas Fault forms the north face of the mountain block. Here it separates the Cretaceous metamorphic rocks that extend northeast to the Tehachapi Mountains and the pre-Cambrian schists and gneisses of Mt. Pinos.

At the base of the steep north face of Mt. Pinos and Sawmill Mountain the trough of the San Andreas Fault has been filled with sediments to form the historically interesting Mil Potreros, a series of narrow rapidly eroding meadows that separate the San Emigdio Range from the Mt. Pinos block. The San Emigdio Range is the result of violent folding, faulting, and upheaval during Pleistocene and Recent time, a process that is still actively continuing. It has deep and severely eroded canyons, such as San Emigdio, Blue, and Pleito. These developed rapidly from the unequal erosion of the sharply uplifted, folded, and fractured soft marine sedimentary rocks formed in the inland sea that occupied the region until late Pliocene or early Pleistocene time. The compressional forces of this region are of great magnitude. Two

FIGURE 43. Lodgepole pines (*Pinus Murrayana*) at Pine Flat on the southwestern Kern Plateau.

of the complex structures contributing to these forces are the San Andreas and White Wolf faults. The White Wolf fault is buried by sediments across the San Joaquin Valley from Comanche Point near Arvin (at the base of Bear Mountain) to the east end of the San Emigdio Range. Here the fault zone is clearly visible where Wheeler Ridge is sharply uplifted along its southeast side. The San Emigdio Range has an unusual assortment of exposed folded, faulted, and tilted Tertiary beds; many that are not only steeply tilted but actually overturned toward the San Joaquin Valley. While the most dramatic folds and faults are in the inaccessible Devils Kitchen area of San Emigdio Canyon, exposures can be seen from the road to Mt. Abel from Reyes Station at the summit of Maricopa grade. A notable feature of the complex faulting is the Pleito Thrust, a thrust fault whose steep vertical uplift is a dramatic feature of the twisted and contorted landscape extending from Pleito Canyon in a rough semi-circle around the east end of the range.

The soils of the Mt. Pinos region are mostly the erosion products of the underlying metamorphic rocks. These are coarse grained, in some areas relatively deep, and have excellent rapid drainage that results in a quick drying of the surface soil. The soils of the San Emigdio Range are often fine loams, many of uplifted old alluvial surfaces. Although considerable areas of heavy clay occur on the lower slopes, most of the soils are of high quality.

FIGURE 44. Limber pines (*Pinus flexilis*) on the north slope of Mt. Pinos.

The pre-Cambrian rocks of the northwest flank of Mt. Abel terminate at a thrust fault that extends around the northwest base of the mountain from Apache Saddle. Northwest of this thrust fault are a complex of Miocene marine formations; these merge with those of the Temblor Range, the southernmost of the arid inner south Coast Ranges. There is no clear break between the ranges, but for convenience, Grocer Grade from Maricopa to Cuyama Valley serves as a useful boundary.

FIGURE 45. This Piute cypress (*Cupressus nevadensis*) on the east slope of the Greenhorn Range is the only mature survivor of a once extensive colony.

The Temblor Range (roughly the eastern slope lies in Kern County, the west in San Luis Obispo County) is aligned along the east side of the San Andreas Fault. This range, and in the north, the neighboring southernmost extension of the Diablo Ranges at Antelope Mountain (Orchard Peak), are the only significant ranges in the county not directly associated with any of the uplifts of the Sierra Nevada. Both the Diablo and Temblor ranges are results of the major disturbances caused by compressive movements along the San Andreas Fault during Pliocene time. The orogenic forces that became active then have continued intermittently to the present.

Throughout much of the geological history of the region, the area now occupied by the Temblor Range was under various ancient seas and embayments. During those periods, great deposits of sediment were laid down. The uplifted tilted blocks that form the range are largely composed of beds of marine and fresh water sediments. A typical cross section of the Temblor Range shows a sequence of an unusually large number of sedimentary formations varying in age from Cretaceous time to the present. In parts of the Temblor Range and the western San Joaquin Valley, there are fully 30,000 feet of shale, sandstone, conglomerate, silt, and claystone. A characteristic feature of the relatively recent orogenic activity in the central Temblor Mountains is the massive outcrops of uplifted sandstone, such as the rocks in Carneros, Black, Garcia, and Agua Media canyons.

The soils, often of considerable depth, reflect the formations from which they were derived. Heavy adobes, rich in organic material, are most common in the hills in the northern part of the range. Those of the central portion are more complex. At lower elevations, there is much exposed shale, often severely

FIGURE 46. Mesquite Spring, a desert oasis in the El Paso Range. In the middle distance is a mesquite (*Prosopis juliflora* var. *Torreyana*); in the background are robust shrubs of common salt bush (*Atriplex polycarpa*); on the skyline is the only known surviving Kern County colony of carizzo grass (*Phragmites communis* var. *Berlandieri*).

FIGURE 47. San Emigdio Canyon from the summit of Mt. Pinos. The slopes are well covered with pinyon pines (*Pinus monophylla*). In the right foreground is a limber pine (*Pinus flexilis*).

fractured and weathered. Sandstones are most common along the summit; often they occur to the edge of the valley on the east slope of the range. In general, the soils are predominantly of redeposited sediments with a substantial content of organic material that forms an almost ideal growing medium for the winter and spring vegetation characteristic of the region.

Of great importance to the botanical history of the San Joaquin Valley, most of the San Emigdio Range, and the Temblor Range is that the entire region was submerged in the Pliocene and its floristic history dates from that period. Thus, this flora is distinctly more recent, at least in the region it occupies, than that of the remainder of the county. According to Robert F. Hoover (Endemism in the flora of the Great Valley of California. Ms., p. 25), the plants of local distribution in the upper San Joaquin Valley probably had their origins in the Temblor Range and the nearby Carrisa Plains and migrated to the east.

East of the San Joaquin Valley and its neighboring complex of mountains lies the Mojave Desert geological province; this

FIGURE 48. Joshua tree (*Yucca brevifolia*) on the desert plains northwest of Mojave.

great desert has its northwestern limits in Kern County along the Tehachapi Mountains and the desert El Paso Range, both uplifted along the Garlock Fault, which proceeds east from the El

Paso Range into the deep desert for at least one hundred miles.

The western part of the desert is a broad peneplain, the base of the almost completely eroded ancient land mass of Mohavea. Visible remnants consist only of low broad domes; from vantage points along the aqueduct road northwest of Mojave these can be seen to the east on the plains north of Boron and east of Mojave. Several rhyolite plugs, the remains of Tertiary volcanoes, are conspicuous desert landmarks. The largest and most impressive is Soledad Mountain, south of Mojave, which rises 1,550 feet above the desert plain. This mountain was the center of volcanic activity with a radius of about ten miles. Others are Desert and Castle buttes and Galileo Hill, which are landmarks on the broad plains north and west of Boron. These plains for the most part have soils of coarse sand, often much modified by wind; local beds of fine clay are common.

FIGURE 49. A grove of dormant valley white oaks (*Quercus lobata*) at Tejon Pass.

The region west and southwest of this old peneplain is occupied by two ancient drainage basins: Fremont Valley to the northwest and the Muroc Sink on the south.

FIGURE 50. The Kennedy buckwheat (*Eriogonum Kennedyi* var. *austromontanum*) covers large areas of the wind-swept western part of the summit plateau of Mt. Pinos; in the background are colonies of squaw currant (*Ribes cereum*) and Jeffrey and limber pines.

The Muroc Sink is the more impressive. Here Rogers, Buckhorn, and Rosamond dry lakes are separated by extensive sand dunes of considerable size that in many places have been stabilized as low gentle hills. However, this region was once a single broad lake that received great quantities of sediment from the east slope of the Tehachapi Mountains and the northern end of the Liebre Mountains of Los Angeles County during the Pleistocene erosion cycles. An oil well drilled on the bed of Rosamond Lake reached a depth of more than 5,000 feet without reaching the bottom of these sediments. The lakes now have hard surfaces of very fine playa clay, much impregnated with saline minerals that are tightly packed and only rarely and briefly have water (the surface of one, Rogers Lake, being the landing field for the U. S. Air Forces supersonic test aircraft).

Much smaller Koehn Lake, in Fremont Valley, where the sediments are not nearly so deep, receives the southeastern drainage of the Piute Mountains, the northeastern Tehachapi Mountains,

FIGURE 51. The subalpine Kennedy buckwheat on the summit of Mt. Pinos.

and the extreme southern Sierra Nevada by way of the Jawbone and Red Rock canyons drainage systems (as well as that of the nearby desert Rand and El Paso Mountains). Its surface is never completely dry and the lake often has a considerable depth of turgid water with an extremely high mineral (mostly sodium chloride) content.

The northeastern extension of Fremont Valley, which separates the desert Rand and El Paso mountains, is a depressed fault block; originally the two ranges and the intervening depressed area was the impressive southern limit of the Basin Ranges geologic province. Indeed, it does not require a keen imagination to match the steep southeast face of the El Paso Range with the equally abrupt northwest face of the Rand Mountains, several miles away.

The Rand Mountains are composed of a complex of primarily metamorphic rocks, some of great age. The older portion, of pre-Cambrian schist, is exposed on the northeast flank. The southwestern two-thirds has been intruded extensively by Cretaceous granite; at the west end this is largely overlain with Quaternary terrace deposits. Typical of desert ranges, the range is severely eroded and only locally has good soil cover.

Although the colorful El Paso Range marks the southern boundary of the Basin Ranges province, there is no accompanying floristic demarcation; the Mojavean creosote bush association extends unaltered north to the lower Owens Valley in Inyo County. The range is aligned against the Garlock Fault, which has a scarp of 280 feet at the mouth of Mesquite Canyon in the central part of the range. Relatively recent earthquake activity is shown by the clearly offset fans at the canyon mouths on the southeastern side of the range.

The El Paso Mountains are composed of an almost bewildering variety of rocks and exposed beds. Most are metamorphic, some of great age. This is particularly true of the eastern end of the range, where the pre-Cambrian Mesquite schist of the central part of the range is overlain by the Paleozoic Garlock Series of slate, chert, quartzite, and limestone. The northwest end has a great thickness of Pliocene sedimentary beds and lava flows, severely eroded to form colorful canyons. Particularly noteworthy is Red Rock Canyon, with its brilliantly colored Pliocene beds covered with late Quaternary sediments. The differential erosion of many different beds of uplifted Pliocene and Pleistocene sediments, largely by wind but occasionally by wind-driven rain, has resulted in colorful cliffs, buttes, and severely eroded badlands of many hues and textures. Two basalt flows occur in the central part of the canyon near Ricardo; these add a contrasting note to the brilliant hues of the canyon. Historic Last Chance Canyon in the central part of the range is less well known and more difficult of access but rivals Red Rock Canyon for diversity and beauty.

In the central northeast portion the El Paso Range reaches its 5,259 foot summit at somber Black Mountain, the cap of a large Pleistocene basalt flow that covers the entire north flank of the range.

Fossils are locally common. A Pliocene fossil flora has been described from Last Chance Canyon by J. E. Webber (Carnegie Inst. of Wash. Pub. No. 412:113–134) that establishes the composition of the Mojavean woodland of the time.

FIGURE 52. A good spring in the western San Joaquin Valley; during the preceding protracted drouth wind-deposited sand collected about a salt bush (*Atriplex polycarpa*); this is now the home of a thriving colony of Antelope ground squirrels (valley chipmunks).

North of the El Paso Range, Indian Wells Valley is the southernmost basin of the Basin Ranges geologic province. This valley, with dry China Lake at its lowest elevation on the east side, is part of the drainage system of the Pleistocene Sierra Nevada drainage that began at Mono Lake and eventually ended at prehistoric Lake Manly in what is now Death Valley. The ancient bed of China Lake is largely covered with stabilized wind-deposited sand, with drifting or loose sand, and with dunes, and only locally has the alkali sink vegetation typical of the dry lakes to the south. The broad plain on the southwest end of the valley has loose sandy soil that is ideally adapted to an especially rich development of the Mojavean annual desert flora in good years. Southeast of Walker Pass, on the edge of the plain, is Robbers Roost, a cluster of rhyolite plugs that are a well known local landmark.

Just as the basin ranges of the deep desert are approached, the San Bernardino County line is reached at the west base of

FIGURE 53. The varied soils and exposures of the east fork of Red Rock Canyon furnish habitats for an unusually large number of plant species.

the Argus Mountains. The lava flows of Little Lake at the south end of the Inyo Mountains stop little short of the Kern County line in Inyo County at the north end of the valley.

In summary, about half of the county's surface is covered with Quaternary alluvial sediments; these are primarily in the San Joaquin Valley and on the plains of the Mojave Desert. Of the mountain complex southwest to Mt. Pinos, at least seventy per cent of the rock is Sierra Nevadan granite of Cretaceous age; the other thirty per cent is largely Cretaceous or older metamorphic or basic rocks. Volcanic rocks occur in numerous places on the desert and locally in the mountains. Marine sedimentary rocks compose most of the Temblor and San Emigdio ranges. Both marine and non-marine sedimentary rocks underlie the San Joaquin Valley; non-marine sediments are common along the west base of the Sierra Nevada and Tehachapi Mountains. The desert ranges are composed of a bewildering variety of both metamorphic and granitic rocks and have considerable exposure of

Tertiary volcanic rocks. These desert formations range from the ancient pre-Cambrian beds of the Rand and El Paso mountains to the Quaternary gravels and sands.

In regions of little rain, the flora does not tend to have highly specialized edaphic demands; in general, the texture, depth, and drainage of the soil is much more important than its chemical composition (with the notable and obvious exception of alkali soils). However, there are local but widespread occurrences of soils that do have specialized floras adapted to their peculiarities. Gypseous soils are local but common in the low ranges about the San Joaquin Valley, in the upper Cuyama Valley, and in the region about Buena Vista Lake and Lost Hills. Soils with even a very low boron content are proving lethal to many cultivated crops; apparently nothing appears to be known about its effect on the native plants but it seems reasonable to assume that this element must be deleterious to some and perhaps many of the indigenous species. One of the odd soils of the county is the ultra-fine friable "dry bog" clay that occurs widely, sometimes in beds of as much as a mile or more, sometimes only a few feet across. This clay absorbs large amounts of water and becomes very slick when wet; when dry it is soft and crumbly. Believed to be wind deposited in ancient pools, it is common in the Temblor and San Emigdio ranges, and occasional east to Tehachapi and north-east to Bodfish Gap and to Dry Bog Knoll and White River in the Greenhorn foothills. This clay always has a distinctive flora. Finally, the deep talus slides in the desert canyons support highly specialized annual plants.

Climate and Weather

Except in the higher mountains, the Kern County climate can be summarized as hot and dry. However, this is an over-simplification; while the climate is arid it is also exceedingly complex and diverse. Although at least one-third of the county has a normal annual rainfall of less than 6 inches, one-half less than 9 inches, and two-thirds less than 12 inches, startling contrasts occur. The median annual rainfall at Inyokern on the desert is 2.6 inches; on the east slope of Sunday Peak in the Greenhorn Range it exceeds 40 inches. Armitage on the east side of the desert's Indian Wells Valley has recorded a seasonal low of less than 0.5 inch of precipitation; Portuguese Meadow just north of Sunday Peak has had more than 62 inches. Recorded temperatures range from a high of 117° F. at Cantil and Inyokern to lows of —4° F. at Tehachapi and —3° F. at Glennville and the desert station of Cantil. (Highs of 118° F. have been recorded at Angiola and at Dudley in the San Joaquin Valley, both just north of the Kern County line.) There are no adequate records for the high mountains; these would surely have lower minimum

FIGURE 54. Bohna Peak, the western spur of the west slope of Sunday Peak in the Greenhorn Range. The Sierran red fir forest reaches its southern limits along the summit; the upper slopes are covered with shin oak brush, the lower slopes with Kellogg and Douglas oaks. The dormant shrub in the foreground is a California buckeye (*Aesculus californica*).

records than those given above.) Spring flowering begins as early as February 15 in the valley; a few species are even in bloom in January. In late years, the peak of blooming on the Kern Plateau or at the higher elevations of the Greenhorn Range may not occur until early July. So, in discussions of climate (as in most other fields), generalizations should be applied with caution. But in general, the county has winters that are relatively warm and brief and summers that are hot and long, humidity is low, and the climate is dry.

The county has four climatic regions: the Mojave Desert, the San Joaquin Valley, the Temblor Range, and the mountains. For each of these, a brief discussion follows.

THE MOUNTAINS. There is a great variation in the climate in the mountains. In general, it becomes drier and somewhat warmer from north to south and drier and somewhat colder from west to east. Thus, the Greenhorn Range and Breckenridge Mountain have more precipitation than the Tehachapi Mountains; the Mt. Pinos region has least of the group. Eastward, the Piute Mountains are arid and the southern Kern Plateau very arid, supporting its open Jeffrey pine forest only because of its high altitude (mostly above 6,500 feet elevation). The southern main crest of the Sierra Nevada from the Tulare County line to its termination at Jawbone Canyon rapidly becomes drier to the south. South of Kiahvah Mountain (just south of Walker Pass) it is in all senses a desert range.

In the forest zones, freezing temperatures usually occur in early October and persist until as late as mid-May. Total precipitation varies from an average of at least 40 inches on the east slope of the summit of Sunday Peak to about 17 inches at Mt. Pinos (occurring primarily as snow at both places). Most of the precipitation occurs in winter but summer thunderstorms are not uncommon, especially on the Kern Plateau and in the Piute Mountains. The precipitation drops rapidly on the east slope; the aridity of these slopes is further increased by the prevalence of hot desert winds. Summer temperatures are mild; extremes of more than 90° F. are unusual.

With the exception of the higher elevations of the Greenhorn Range and the upper north slope of Breckenridge Mountain, all of the mountains have occasional years of severe drouth. While the forest trees and many of the shrubs and even herbaceous perennials survive these easily enough, it is the occasional severe drouths rather than the lower normal precipitation that has eliminated many of the mesic Sierran species from the Kern County mountains.

The Mojave Desert. The climate of the Mojave Desert is basically continental. There are wide variations in temperatures, both between summer and winter and between day and night. One of the most surprising phenomena to the desert visitor is the rapid chill that usually develops shortly after sunset, often even after uncomfortably hot afternoons. The broad treeless desert superficially would seem to have a uniform arid climate. Actually, a relatively wide range of rainfall occurs within the extensive creosote bush association. This ranges from a median of 2.6 inches at Inyokern to 5.5 inches at Backus near the western edge of the desert. The record annual rainfall for the desert is 14.2 inches at Backus (near Willow Springs). The deep desert at Randsburg has had as much as 12.4 inches and as little as 1.5 inches. Both Inyokern and Armitage in Indian Wells Valley have a median annual rainfall of less than 3 inches. These are extremes, in a typical year most of the creosote bush region has from 3 to 6 inches of precipitation, mostly as rain. At least 80 per cent falls between November 1 and April 30. However, occasionally there are heavy summer thunderstorms. These sometimes contribute materially to both the annual rainfall and the effective available soil moisture. The widespread occurrence of autumnally flowering plants, mostly in the *Compositae,* including at least two annuals, suggests (together with other evidence) that summer rainfall may have been more common in the geologically recent past.

Fog on the desert is extremely rare and occurs only in winter. Snow is not uncommon although it usually adds relatively little to the total precipitation. Winters are frequently cold and days with freezing temperatures are common. The average annual minimum is probably about 15° F. The earliest freeze rarely occurs before November 1; the latest may be in late April.

Summer temperatures are high, but not ordinarily strikingly so. While summer days of 100° F. or more occur frequently from June through September, temperatures of more than 110° F. are uncommon. The very low humidity makes even relatively high temperatures tolerable for man.

Strong winds, sometimes of extremely high velocity, are common, especially in the spring and fall. These are mostly westerly and usually occur in the afternoon and evening. By midnight,

FIGURE 55. In Temblor Canyon the north slopes are well covered with Douglas oaks and California junipers, the south slopes with Upper Sonoran grassland and scattered subshrubs.

the desert is normally calm, even after days with severe wind. Rain is infrequent and unpredictable and even in winter is most likely to occur as brief although sometimes hard showers.

THE SAN JOAQUIN VALLEY. The upper San Joaquin Valley, a region of broad arid treeless and often shrubless plains, is separated from the influence of the ocean by the Coast Ranges. This series of parallel ranges, together with the intervening valleys, are some 50 miles wide with crests to about 4,000 feet elevation. These effectively block many coastal storms and the upper San Joaquin Valley is in a broad rain shadow. Normally, about 90 per cent of the rain falls between December 1 and April 1.

This climatic isolation by the surrounding horseshoe of mountains results in enormous pressure systems blocked by these mountains, especially on the east, which frequently condense the atmospheric moisture at their base. This results in the formation of very dense usually very wet ground fogs in late November,

December, and January. These fogs often persist for days and sometimes for weeks. In a region of modest rainfall they greatly supplement the meager precipitation and are of enough significance to the vegetation to keep the region from becoming a desert of wind-drifted dunes.

Rainfall is variable and protracted drouth cycles are not uncommon. During the seven years between 1944 and 1950, Belridge received a total of only 16.49 inches. During this period, in no year was there more than 4.56 inches; in 1947 only 1.97 inches fell. In a typical year, the arid plains receive from 4 to 7 inches; in general, the east side of the valley receives about 2 inches more than the west side. At Bakersfield, the median rainfall is 6.36 inches; the record high was 11.17 inches in 1938; the low 1.87 inches in 1959 (there was only 1.98 inches in 1961).

Summers are long and hot. Bakersfield has an average of 110 days each year with readings of 90° F. or higher and temperatures well over 100° F. can be expected from May into early October.

FIGURE 56. A modified form of the Lower Sonoran grassland occurs at high elevations in the arid southern Temblor Range.

FIGURE 57. Dormant mesquite (*Prosopis juliflora* var. *Torreyana*); this shrub occurs over hundreds of acres northeast of Buena Vista Lake.

Significant amounts of rainfall rarely fall during the summer months. The Coast Ranges block summer maritime fogs, so summer in the San Joaquin Valley is a time of abundant sunshine with temperatures that usually are high both day and night.

The winters are mild. Although extreme lows of 20° F. have been recorded, winters with no readings below 28° F. are typical. Bakersfield averages only 16 days each year with frost. Freezes can be expected from late November until February; perhaps 50 per cent of them occur in January. The typical growing season is approximately 300 days.

THE TEMBLOR RANGE. The climate of the Temblor Range is somewhat different from that of the rest of the county. Although the most arid and far removed from the ocean of all the arid inner Coast Ranges, the climate is still modified by coastal influences.

Although summer temperature extremes are nearly as high as those of the San Joaquin Valley, nights are cooler. Further, the coastal morning fogs that moderate the valley hot spells although rarely actually reaching there, are relatively common in the Temblor Range, especially in the northern portion. Although few temperature records are available for the range, at the Dorothy Twisselmann Ranch in the northern portion, the record high from 1949 through 1966 is 112° F., reached on 2 different days in 1966; the record low is 20.5° F. in 1950. Both the high and the low temperatures would be more extreme in other parts of the range.

Winters are variable, with many warm spells broken by east or valley fogs and with frequent cold spells when the temperature falls to as low as 10° F. or less at the central crest of the range. Freezing temperatures occur from about November 1 through April 15 in most of the region. In general, the climate becomes drier and hotter in summer and drier and colder in winter progressively from north to south. Annual rainfall in the northern portion is about 11 inches, at the Temblor Summit of the

FIGURE 58. A creosote bush (*Larrea divaricata*) and low desert shrubs on the plains on the west side of Indian Wells Valley; the Walker Pass region is in the background.

FIGURE 59. Silver cholla (*Opuntia echinocarpa*) is a distinctive plant of the arid shrub association in Kelso Canyon.

central portion the average is 9 inches, and at the summit of the southern portion northwest of Taft no more than 6 inches.

SUMMARY. In general, the climate of the county is arid except for portions of the mountains. The margin between normal and drouth conditions is narrow and the amount and effectiveness of rain or snow is influenced by many unpredictable factors. Rainfall records are usually over-emphasized in studies of arid regions. On the arid San Joaquin plains, 4 inches of rainfall may result in very scant vegetation or it may be enough for abundant spring growth. In 1964, the desert town of Randsburg had an impressive rainfall of more than 8 inches. Most of this occurred as autumn thunderstorms; the hillsides the following spring were barren.

Slopes, exposures, and wind also greatly modify the climate. The amount of moisture available from the dense winter valley fog varies greatly, even in a small area. It is most effective under shrubs, where annuals sometimes make a vigorous growth while the intervening open areas have only depauperate plants or are completely drouth-stricken.

The meager available evidence suggests that sharp contrasts in humidity occur even within a short distance. Precise records from many stations might well reveal startling differences in even a

FIGURE 60. Foothill milkweed (*Asclepias erosa*) in the Upper Sonoran grassland in the Greenhorn foothills.

few hundred yards and would explain the abrupt differences in vegetation which are often difficult to ascribe to any other factors.

The climate, then, of all the county except the small portion in the Greenhorn Range from the Douglas oak woodland to the red fir forest is subject to frequent years of drouth, some of which are devastating. There is great variation in rainfall from year to year, and the winter rains do not follow a predictable pattern. Summers are hot and winters range from temperate to moderately severe, with at least some freezing weather in all parts of the county. Thus, the climate and the weather, particularly the precipitation, is by far the most important factor determining the character of the flora and the vegetation. Indeed, Kern County's specialized floristic assemblages are required to cope successfully with the region's erratic meteorology.

FIGURE 61. The eastern part of the summit plateau of Mt. Pinos has varied habitats for an impressive number of subalpine plants. The large shrub in the foreground is squaw currant.

TABLE 1. *Precipitation for selected stations given as inches of annual rainfall.*[2]

	Abnormally Dry	Dry	Normal	Wet	Abnormally Wet
San Joaquin Valley, west-side					
Lost Hills	2.2	3.3	4.9	6.9	10.7
Middlewater	2.5	3.5	4.8	6.5	8.3
Taft	2.9	3.3	4.6	5.7	6.7
San Joaquin Valley, east-side					
Bakersfield	3.6	4.7	6.1	7.8	9.6
Buttonwillow	2.5	3.4	4.7	6.2	7.8
Delano	3.4	4.5	6.1	8.0	9.9
Mt. Pinos Region					
Apache Saddle	4.7	6.4	8.2	11.6	14.8
Chuchupate	5.3	8.0	9.3	12.3	15.8
Mt. Pinos summit (estimate)	11.0	14.0	17.0	25.0	35.0

Table 1. *Precipitation for selected stations given as inches of annual rainfall.*[2] *(Cont'd)*

	Abnormally Dry	Dry	Normal	Wet	Abnormally Wet
Tehachapi Mountains					
Tehachapi	5.6	7.3	9.5	12.4	15.2
Old Tejon Ranch	7.5	9.4	11.8	14.6	17.2
Bear Valley	6.6	9.7	13.9	19.0	25.1
Tehachapi Peak (estimate)	10.0	12.5	14.0	19.0	24.0
Piute Mountains					
Caliente	6.5	8.2	10.6	13.2	16.0
Loraine	7.1	9.2	11.9	15.2	18.7
Piute Mountain (estimate)	10.0	12.0	15.0	20.0	25.0
Kern & South Fork valleys					
Kernville	5.4	7.1	9.5	12.4	15.4
Isabella	4.9	7.1	10.4	14.5	19.0
Weldon	3.4	4.6	6.4	8.5	10.7
Greenhorn Range					
Glennville	9.7	12.4	16.1	20.2	24.6
Morrow Ranch	12.9	15.7	19.2	23.1	27.0
Greenhorn Pass (estimate)	13.5	18.0	25.0	30.0	36.0
Portuguese Meadow (estimate)	25.0	32.0	41.0	48.0	60.0
Mojave Desert					
Backus	2.9	3.5	5.5	8.3	11.6
Mojave	2.4	2.9	4.6	6.6	9.0
Randsburg	1.0	3.1	5.0	7.3	10.3
Armitage	.8	1.3	2.3	3.5	5.0
Inyokern	1.1	1.7	2.6	3.8	5.1
Temblor Range					
Annette	4.9	6.7	9.3	12.3	15.8
Temblor Summit (estimate)	5.0	7.0	11.0	14.0	20.0
Crocker Pass (estimate)	3.5	4.0	6.0	11.0	14.0

Floristics

Kern County is an area of merger and transition of several of the regional floras of California. Here the flora of the Sierra Nevada reaches its southern limits; that of the southern Califor-

2. Records are from the United States Weather Bureau, as published in their annual reports, and from Elford (1964). The estimates are based on informaton supplied by reliable local authorities.

nia mountains its northern limits, and that of the inner south Coast Ranges its southeastern limits. A large part of the endemic rich flora of the Mojave Desert reaches its western and northwestern limits here. Finally, the distinctive flora of the arid upper San Joaquin Valley reaches its southern limit in the central part of the county.

A region with so many floristic boundaries is obviously also a region of floristic instability. Isolated occurrences of species or even groups of species is commonplace; these are frequently in situations quite different from those in which they usually grow. Valley white oak (*Quercus lobata*), ordinarily a tree of deep well-watered soils at low elevations, is common with Douglas oaks on the high slopes of the Tehachapi Range and occurs with Jeffrey pines along the ridges near Catskin Creek; from Cameron Canyon to the southeastern Antelope Valley it grows in scattered stands or as isolated individuals along washes at the desert's edge,

FIGURE 62. Chaparral below Pescadero Point on the southeast slope of the Tehachapi Mountains. In the distance is the desert Antelope Valley; the dark areas are groves of Joshua trees. The valley white oaks in the lower right foreground grow in the trough of a fork of the Garlock Fault.

FIGURE 63. Antelope Valley (Berenda Mesa) borders the western San Joaquin Valley; in the right background is Antelope Mountain (Orchard Peak), the southernmost extension of the Diablo Ranges. The shrubs on the plains are common salt bush (*Atriplex polycarpa*).

sometimes with Joshua trees and beaver-tail cactus. Digger pine (*Pinus Sabiniana*) grows in hot dry canyons in most of central and inner northern California; in the Kern County mountains it occurs as a forest tree at elevations of more than 6,000 feet, well above the winter snow line. Big sagebrush (*Artemisia tridentata*), a common plant of high dry meadows and the Jeffrey pine forest along the eastern Sierra Nevada and in the transmontane Basin ranges, grows in the Douglas oak woodland at Poso Flat and in the chaparral on the slopes around Cummings Valley. One of the annual Chinese houses (*Collinsia bartsiaefolia* var. *Davidsonii*) grows on shaded north slopes in the foothills around the valley; colorful colonies grow in full sun on the dunes on the desert floor at Rosamond Dry Lake. Such examples can be repeated almost endlessly for the county. Just as it is difficult to draw sharp lines between the plant associations, it is equally difficult to delimit the floristic regions.

The question immediately arises: how many of these displaced plant occurrences are an inevitable characteristic of the transitional areas between floras and how many are evidence of floristic migration? This is a profound botanical question that has no easy answer; before conclusive statements can be made, many complex problems must be studied. However, at this time, we can explore the relationships and note the trends that seem to be suggested even if final answers concerning floristic migration within the county must wait until much more data are available.

THE SIERRA NEVADA FLORA. The Sierra Nevada flora is dominant in the Greenhorn Range, the Kern Plateau, the Piute Mountains, and on Breckenridge Mountain. In this region a large number of plants that in California are largely restricted to the Sierra Nevada reach their southern limits; a surprising total of 188 species occur in the Greenhorn Range that are not known from elsewhere in the county. Thirty-one reach their southern limits on the Kern County portion of the Kern Plateau and an additional fifteen on Breckenridge Mountain and in the Piute Mountains. Thus, a total of 234 species occur in Kern County only within the geological southern limits of the Sierra Nevada; these represent slightly more than 16 per cent of the county's indigenous flora. Sixty-six of the species are endemic to the Sierra Nevada. The others extend beyond its borders; some to the mountains of northern California, others through the Cascade Ranges, some north to Alaska; a few even range through all the boreal regions of the northern hemisphere.

However, the southern limits of the Sierra Nevada flora are not as abrupt as these figures would indicate; this flora is predominant in its Jeffrey pine phase at the higher elevations southwest to Mt. Pinos (a substantial number of essentially Sierran species also occur in the high elevations of the southern California mountains and even in Baja California). Many of the Sierran plants that occur south to Mt. Pinos represent the eastern Sierran elements; these extend from the eastern Sierra southwest through the Piute and Tehachapi mountains.

South of the Piute Mountains, the Sierran flora occurs in discontinuous units, separated at the lower elevations by chaparral composed of shrubs with mixed affinities, and by Douglas oak woodland and Upper Sonoran grassland, both largely made up of plants that occur over wide areas of the California botanical province[3].

Sierran plants are locally common on the forested slopes of Tehachapi Peak and Double and Bear mountains, and are widespread at the higher elevations of the Mt. Pinos region. There are also scattered occurrences at such places as Tollgate Ridge and Cache Peak in the northeastern Tehachapi Mountains. The circumstances of all the occurrences south of the Piute-Breckenridge complex strongly indicate that here it is a relict flora. The Sierra Nevada flora seems to have been displaced, perhaps rather rapidly and recently, by plants adapted to more xeric and less stable climates. These have invaded the region occupied by the Sierran flora in late Pleistocene, Recent, and even historic time.

Sierran plants of the foothills on the west slope follow a similar pattern. A large number reach their southern limits in the Greenhorn Range south to Kern Canyon; a few continue on south through the Tehachapi foothills (here, El Paso Canyon seems to mark a minor floristic boundary), and plants of the Sierran foothills grow in isolated colonies as far southwest as Black Bob Canyon in the San Emigdio Range. In the Greenhorn Range, Cedar and Lumreau creeks and Poso Creek southwest to Poso Mine are a significant floral boundary for the Sierra Nevada plants of the lower elevations.

Thus the Sierra Nevada flora reaches its southern limits in Kern County, largely at the southern geological limit of that range at the Piute Mountains and Breckenridge Mountain. The Sierran flora seems to be in local retreat, with the outposts south and southwest of the geological boundaries of that range survivors from a wetter and probably cooler climate. Many occurrences are restricted to a single known colony; evidence from botanical collections suggest some may have been eliminated from the Kern flora within the past 100 or even 50 years. This is especially true of plants of wet or moist soils.

THE FLORA OF THE SOUTHERN CALIFORNIA MOUNTAINS. Ancestrally, the species of this flora are closely related to and largely

3. The California botanical province is the biological region defined by J. T. Howell (1957: 133) that includes California west of the main mountain crest and neighboring portions of Oregon and Mexico.

FIGURE 64. A dormant Douglas oak (*Quercus Douglasii*) in the hills between Granite Station and Glennville. Fog from the San Joaquin Valley fills the canyon in the middle distance.

derived from the plants of the Sierra Nevada flora. Probably much speciation took place in late Pleistocene time when the flora of the southern California mountains became isolated from that of the Sierra Nevada as the climate of California became drier and warmer. This southern flora is most common in Kern County in the Mt. Pinos region. While the Mt. Pinos florula is composed of species of several floras, including even those of the Mojave and Great Basin deserts, those of southern California and the Sierra Nevada are of greatest and about equal importance. The Mt. Pinos florula includes 38 southern California species, of which 18 are endemic from here to the higher mountains of the south. Although not a major component of the flora, plants of the southern California mountains are occasional through the Tehachapi Mountains, become scarce in the Piute Mountains and on Breckenridge Mountain, and relatively uncommon in the Greenhorn Range and on the Kern Plateau, although several extend well to the north into Tulare County. The evidence is not sufficiently developed to suggest whether the

Kern County occurrence of the southern montane flora is expanding, stable, or retreating.

An interesting group of plants of this regional flora are those that demark what Willis Linn Jepson called the Tehachapan Endemism Area (A Manual of the Flowering Plants of California, p. 12). These are plants that occur from the San Bernardino Mountains to the southern Sierra Nevada of Tulare and Kern counties, usually also occurring in the San Gabriel and Liebre mountains. Species from Jepson's list that occur in Kern County include *Astragalus pachypus, Collinsia Childii, Penstemon Rothrockii, Wyethia ovata, Chaenactis santolinoides,* and *Senecio ionophyllus.* Modern collecting records add to this list *Agrostis lepida, Ranunculus canus* var. *ludovicianus, Eschscholzia minutiflora* var. *rutaefolia, Phacelia mohavensis, Penstemon caesius,* and *Galium angustifolium* var. *onycense.*

THE SAN JOAQUIN VALLEY FLORA. This highly xeric flora covers the broad plain at the head of the San Joaquin Valley in Kern County. In addition, it extends over the southern Temblor Range

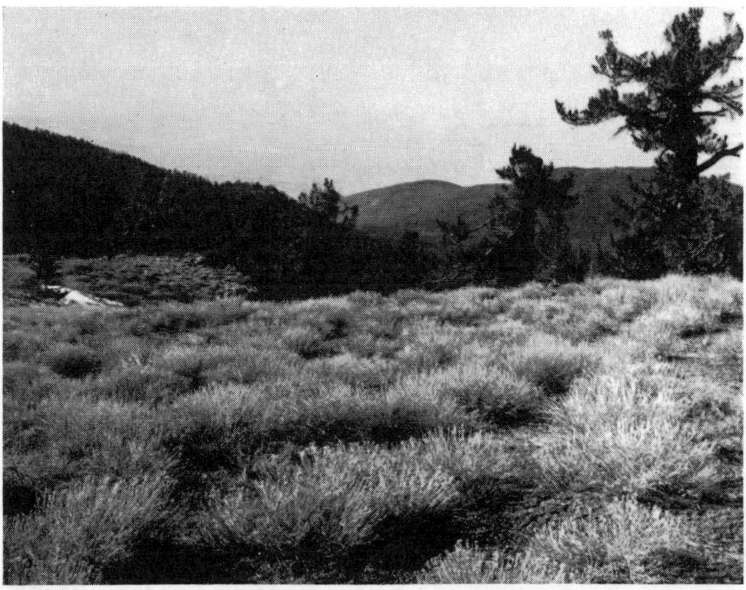

FIGURE 65. Rabbit brush (*Chrysothamnus nauseosus* ssp. *occidentalis*) near the summit of Mt. Pinos.

to the southern Carrisa Plains in San Luis Obispo County and to the upper Cuyama Valley in Santa Barbara County. (The flora also has local occurrences in other arid portions of the south inner Coast Ranges, such as the Panoche Plain of San Benito County.) Several taxa are endemic to the upper San Joaquin Valley; among these are *Eriogonum gossypinum, Eschscholzia caespitosa* ssp. *kernensis, Astragalus lentiginosus* var. *nigricalycis, Eremalche kernensis, Eriastrum Hooveri* (which also occurs in the central Temblor Range), *Trichostema ovatum, Hemizonia pallida,* and *Eatonella Congdonii.* Three alkali sink plants are also endemic to this southern part of the valley; these are *Atriplex tularensis, A. vallicola,* and *Delphinium recurvatum.*

Eighty-three species grow in Kern County only in the San Joaquin Valley; of these, thirty-three are San Joaquin Valley endemics. An additional one hundred nine have their Kern County occurrence restricted to the valley and the closely related Temblor Range.

This regional flora, composed largely of fast growing winter annuals adapted to success with scant and irregular rainfall, probably developed relatively recently in response to increasing aridity in late Pleistocene and Recent time. A large majority of the plants that are endemic grow on the arid plains; the flora of the wet lands of the valley's old central drainage system is largely made up of plants of widespread occurrence. Relatively few of these wet soil plants are restricted to the California botanical province; many occur widely in North America and some are even cosmopolitan in their ranges. There are, however, exceptions, such as *Cirsium crassicaule* that is known only from small colonies along canals northeast of Lost Hills and locally in the northern San Joaquin Valley.

THE COAST RANGE FLORA. The flora of the Coast Ranges is rather weakly represented in Kern County. Coast Range plants are locally common in the Temblor Range and on the lower slopes of Mt. Abel and in the Mt. Pinos region. There are scattered occurrences on the west slope of the Tehachapi Range and relict colonies on the southeast slope along the borders of the desert Antelope Valley; a few species grow north to the Greenhorn foothills. The isolated colonies of often senescent coastal shrubs in the Temblor Range, such as *Ribes californicum* and *Eriodictyon tomentosum,* reflect the post-Pleistocene aridity of the region that probably developed both recently and rapidly; these are undoubtedly remnants of a much more extensive flora in the range in the geologically recent past. Thus, similar to occurrences of the Sierra Nevada flora in the ranges south of the Greenhorn Range, the Coast Range flora in Kern County is probably a relict of wetter climates and is in the final stages of being replaced by more xeric aggregations. Seventy-three plants of primarily Coast Range distribution occur in Kern County; twenty-nine of these are endemic to those ranges.

FIGURE 66. Pinyon pines, chaparral, digger pines, and Douglas oaks grow on the lower east slope of the Tehachapi Mountains west of Willow Springs.

THE MOJAVE DESERT FLORA. The Mojave Desert flora occupies the greatest area of any in the county, covering approximately thirty per cent of the county's surface. It consists of a colorful and abundant group of annuals and an extensive list of shrubs. One hundred fifty-nine plants occur in the county only on the desert; twenty-five of these are endemic to the Mojave Desert;

FIGURE 67. The Upper Sonoran subshrub association merges with the Upper Sonoran grassland on McGovern Grade near Annette in the Temblor Range.

the remainder are mostly plants that occur widely in the North American deserts, largely to the southwest; a few represent southern extensions of the Great Basin flora. Ample evidence shows that the Mojave Desert was an arid woodland as recently as late Pliocene time. Thus, the present xeric flora is a fairly recent development. Probably most species migrated from the arid southwest. Others evolved from the old woodland flora; still others that have a close morphological relationship to the cismontane taxa suggest that many of the Mojavean species may have evolved from those now restricted to the neighboring western floras.

The Mojave Desert flora is made up of two associations. The first is that of the desert plains; the second of the borders of the desert and the desert ranges. Most of the species of the plains have their distribution restricted to the desert, although a number have infiltrated the San Joaquin Valley, mostly on its southwestern borders. Typical of these are *Malacothrix glabrata, Croton californicus* var. *mohavensis, Astragalus lentiginosus* var. *vari-*

abilis, and the isolated occurences of *Lycium Cooperi* and *Larrea divaricata.*

The second group of plants, those of the higher desert, are often common in the pinyon woodland; thirty-seven species occur in the yellow pine forest. A number extend to the mountains around the head of the San Joaquin Valley and some occur as far north as the mountains of San Benito County and the distant Mt. Hamilton Range. These desert plants with cismontane occurrences include such taxa as *Eriogonum trichopes, E. Heermannii, Chorizanthe Thurberi, Stanleya pinnata, Chrysothamnus nauseosus* ssp. *mohavensis,* and *Eriophyllum Wallacei.*

A Floristic Summary. While the five regional floras are well marked within the county, the largest single aggregation of plants are those of the California botanical province that occur widely in two or more of the county's regional floras; eighty-nine of these occur in all five floras and one hundred two in all except the Mojave Desert flora.

The Great Basin flora is not a distinct unit in the county but species of this flora are present, primarily in the pinyon woodland and the Jeffrey pine forest in the eastern part of the mountains south to Mt. Pinos. These include forty-one species, or 2.9 per cent of the county's indigenous flora.

While not as distinctive nor as widely recognized as the other California regional floras, that restricted to the lower elevations of cismontane southern California may well be noted. Many California phytogeographers include this southern California lowland flora in that of the Coast Ranges but these ranges reach their southern boundaries in San Luis Obispo County and their flora has its terminus in Santa Barbara County (*cf.* Clifton Smith, 1964:3). Twenty of these southern California species occur in Kern County, mostly in the San Emigdio Range, the upper Cuyama Valley, and on the borders of Antelope Valley along the southeastern Tehachapi Mountains. Some are of special interest, such as the highly local San Emigdio Range occurrence of *Haplopappus Palmeri* ssp. *pachylepis* in Salt Canyon and *Brickellia Nevinii* in San Emigdio Canyon and on the northwest flank of Mt. Abel. *Lathyrus laetiflorus* is known from a single plant in Black Bob Canyon in the San Emigdio Range and an-

FIGURE 68. Sunset over the southern Sierra Nevada from Indian Wells Valley.

other in Studhorse Canyon, a fork of Caliente Canyon at the northeastern limits of the Tehachapi Mountains.

In summary, the Kern County flora is made up of diverse elements, with five regional floras represented: those of the Sierra Nevada, the Coast Ranges, the southern California mountains, the Mojave Desert, and the San Joaquin Valley. In addition, there is a representation of plants of the Great Basin flora and of the southern California lowland regions. Two phytogeographical provinces are included: the California botanical province, which includes all the regional floras except that of the Mojave Desert, and the Sonoran Desert province, which is represented by the Mojave Desert flora.

THE PLANT ASSOCIATIONS

The generally accepted plant communities proposed by Munz and Keck (A California Flora) are quite adequate for the state as a whole; but for Kern County they are too broad, inclusive, and generalized for useful vegetational analysis. Several classification systems currently used in plant ecology have been tested in the field, but unrealistic rigidity was found to present a common obstacle in their application. At their present level of development they are of limited value in a region such as ours, however well they may work in regions with more stable climates and floras.

What is being proposed, then, is a series of less formal plant associations that occur as natural units in the county, even though they sometimes may be poorly defined in the field. Areas of transition between associations may be more extensive than the associations themselves and even the most carefully chosen plant indicators are likely to occur in quite atypical situations. On the other hand plant associations may change abruptly, not only from north slope to south slope but even on different parts of the same slope. Only slight differences in exposure, soil, or relative humidity may be reflected, often dramatically, in sharp vegetational changes, often in small areas.

This treatment, while lacking the precision that is the desired but elusive goal of plant ecologists, should prove to be adequate for the general student of the flora and vegetation yet detailed enough to suggest further ecologic studies for those interested. Although some of our associations could be combined to form certain of the "communities" defined by Munz and Keck, or on the other hand could be expanded and subdivided almost endlessly as finer and finer criteria are used for classifying vegetation, we believe that the ones recognized are a valid reflection of the actual major and conspicuous vegetational assemblages.

1. THE ALKALI SINK ASSOCIATION. Almost all of the county lacks oceanic drainage; only the small portion of upper Cuyama Valley that is in Kern County even theoretically has such drainage. This results in land-locked basins that receive the seasonal runoff. These are in three groups. The largest of these, the long drainage basin in the center of the San Joaquin Valley which lies between Buena Vista Lake on the south and Tulare Lake on the north, was originally a broad system of sloughs, marshes, playas, and shallow lakes. In the hills, the alkali sinks are often sag ponds, usually along the major faults. Some, such as the series along the San Andreas Fault in the Temblor Range and at the

western end of the San Emigdio Range, are usually quite small. Others, such as Castac Lake near Lebec and Proctor Lake near Tehachapi, cover several acres. Finally, the alkali sinks on the desert are the Pleistocene drainage basins of the mountains; these are the Muroc Dry Lakes, Koehn Lake, and China Lake.

All of these have soils permeated by soluble minerals (usually although not always alkaline). These vary from the crystallized minerals of Koehn Lake in the desert to pools and hollows in the foothills where the soils vary from alkaline to non-alkaline during drouth and wet years, and sometimes even in the same year. Some may have soils that are neutral or nearly so by late winter and in early spring; with evaporation they may become alkaline by late spring.

The traditional alkali sink occurs in areas in which the soil, even in the wettest of seasons, is fully mineralized and a sharp vegetational change separates the plants of such soils from those of the surrounding areas. Here the plants are often perennial, markedly halophytic, and highly specialized. Among these are *Nitrophila occidentalis, Kochia californica, Suaeda fruticosa, Salicornia subterminalis,* and, most common and most conspicuous, *Allenrolfea occidentalis.* In the areas with low alkalinity that is often much diluted in rainy seasons, the alkali sinks are colorful with spring annuals; most abundant are the goldfields group, *Lasthenia minor, L. chrysantha,* and *L. Ferrisiae.* Other conspicuous plants are alkali larkspur (*Delphinium recurvatum*) in the San Joaquin Valley and alkali wall flower (*Erysimum capitatum* var. *Bealianum*) on the Mojave Desert.

The alkali sinks have considerable variation both in vegetation and general aspect, from the stark barren dry lakes with their hard fine soils in the Edwards Air Base region (Rogers, Buckhorn, and Rosamond dry lakes) to the dense mesquite scrub that covers hundreds of acres east and northeast of Buena Vista Lake in the San Joaquin Valley. On the broad plains north of Semitropic, the alkali plains are covered with *Suaeda fruticosa;* the low-lying areas that are wet in winter have large colonies of *Allenrolfea occidentalis.*

Although all the vegetation of the alkali sinks is usually called "halophytic," the term may be misleading in suggesting an obligate need for mineralized soil. The plants seem to consist of two groups. The first will tolerate alkali soil but will grow, sometimes even more successfully, in neutral or even in moderately acid soils. These plants are apparently poor competitors, especially as seedlings, and in the wild are confined to mineralized soils where the more competitive plants of neutral soil will not grow. This group includes *Atriplex polycarpa, A. spinifera, A. lentiformis, Prosopis juliflora* var. *Torreyana, Sida hederacea, Heliotropium curassavicum* var. *oculatum,* and probably many of the spring annuals of the vernal alkaline poolbeds of the San Joaquin Valley and the Temblor Range.

The second group seems to require minerals found only in the alkaline soils. These, mostly perennials and shrubs with succulent leaves, include *Allenrolfea occidentalis, Salicornia subterminalis, Suaeda Torreyana, S. fruticosa, Nitrophila occidentalis,* and *Kochia californica.*

With the damming of Kern River to form Lake Isabella and the various other developments that impede run-off from the other mountain streams, combined with deep well pumping in the valley, many of the lands in the central valley that were once periodically flooded are now perennially dry. The result is that in some areas, such as the Buttonwillow region and the plains north of Semitropic, many lands formerly farmed have been abandoned because of increased salinity. On others, the productivity has been seriously lessened; some are now limited to crops that have a degree of alkali tolerance, such as sugar beets, cotton, and rice.

On the other hand, the greatly increased sophistication of agricultural techniques has made possible the use of land formerly too mineralized for farming. Thus, in the valley alkali sinks conflicting trends are at work. It will be interesting to observe the contest between man and alkali over the next few decades.

Typical plants of the alkali sink association are:

Distichlis spicata
Hordeum depressum
Sporobolus airoides
Scirpus paludosus
Allenrolfea occidentalis
Atriplex argentea var. *expansa*
A. coronata
A. Serenana

Bassia hyssopifolia
Kochia californica
Nitrophila occidentalis
Salicornia subterminalis
Suaeda fruticosa
S. Torreyana
Sesuvium sessile
Delphinium recurvatum

Lepidium dictyotum
Wislizenia californica
Astragalus Hornii
Prosopis juliflora var. *Torreyana*
Sida hederacea
Frankenia grandifolia var. *campestris*
Tamarix pentandra
T. tetrandra
Cressa truxillensis var. *vallicola*
Allocarya acanthocarpa
Heliotropium curassavicum var. *oculatum*
Aster intricatus
Lasthenia Ferrisiae
L. minor

2. The Lower Sonoran Grassland. The broad treeless plains at the head of the San Joaquin Valley that encircle the Buena Vista-Tulare drainage system are arid and often shrubless. Although only recently recognized as such, this region is a true desert; the annual rainfall averages less than six inches. The vegetation largely consists of winter annuals of rapid growth, many of them introduced. Only in years of exceptional rainfall do these plants grow with vigor.

Perennials are uncommon and only one shrub, common salt bush (*Atriplex polycarpa*), is at all widespread. Salt bush frequently occurs as a relatively dense shrubby scrub. Its place in the vegetational succession is not surely known; it tends to be ephemeral and its often weedy behavior would make it seem doubtful that it is a major component of a climax vegetation.

The annual flora is quite distinctive in normal years. In wet years, plants of the nearby Upper Sonoran associations are relatively common, occurring even in such extremely arid regions as the Elk Hills. In very dry years, few plants reach maturity. (Unlike the Mojave Desert, there is almost always enough rain for good germination; this is one of the chief differences in the annual floras of the two regions.) The most successful plants in years of scant rainfall are the native *Festuca microstachys* var. *simulans, Lepidium dictyotum* (which is often in bloom in January), *Lasthenia chrysostoma,* and the introduced Arabian grass (*Schismus arabicus*), red-stemmed filaree (*Erodium cicutarium*), red brome (*Bromus rubens*) and common foxtail (*Hordeum glaucum*). In years with dry winters followed by late spring rains there is a dense growth of Russian-thistle (*Salsola Kali* var. *tenuifola*) that sometimes covers thousands of acres.

Various authorities, including Munz and Keck, conclude that this region was once covered by a perennial grassland that has been destroyed by grazing. Impressive evidence can be marshalled to reject this assumption; doubt that the scant rainfall could ever have supported a perennial grassland is sufficient argument for the present purpose. It is probably safe to assume that the primitive flora was largely one of native annuals that still occur but whose number has been greatly reduced by the dominance of the immigrant annuals.

The normal rainfall is from 4.5 to 6 inches. Years of drouth with less than 2 inches are not unknown. The rainy season is usually between late November and early April. Some areas are nearly frost-free and temperatures below 28° F. are unusual anywhere. The meager rainfall is greatly supplemented by the dense tule fogs that fill the valley from December until February, sometimes lasting for weeks. The nearness to the coast results in a higher relative humidity in winter and spring than that of the more distant Mojave Desert, another factor which mitigates against the development of a typical desert flora.

The soils of the Lower Sonoran grassland on the west side of the valley are light alluvial loams of the finest quality and often of great depth. The east side has greater variety with coarser and heavier granite-derived alluvium being the most common.

The economic use of this association has had a profound influence on the plant life. The absence of water and the high summer temperatures make cattle grazing impractical but the feed can be used in winter and spring for sheep, which do not require water when their forage is green. In those seasons large numbers of sheep are moved into the region, and, in order to make such a short grazing season profitable, pasture is used to the fullest extent possible. This intensive grazing encourages fast-growing plants that set abundant seed under adverse conditions. The result is the widespread and often abundant occurrence of relatively few species.

Much of what was once Lower Sonoran grassland has been converted to irrigated farmlands; current projects will add many more thousands of acres. When the construction of the Feather River water project is completed, most of the broad plains from Coalinga to Taft will be irrigated, and only remnants of the present arid vegetation will remain.

Typical plants of the Lower Sonoran grassland are:

Bromus rubens
Festuca microstachys var. *simulans*
F. reflexa
Hordeum glaucum
Avena barbata
Schismus arabicus
Allium Howellii
Eriogonum gossypinum
Atriplex polycarpa
Lepidium dictyotum
Sisymbrium Irio
Astragalus lentiginosus var. *nigricalycis*
A. oxyphysus
Lupinus nanus var. *Menkerae*
Erodium cicutarium
Eremalche Parryi
E. kernensis
Opuntia Treleasei
Gilia tricolor ssp. *diffusa*
Linanthus liniflorus ssp. *pharnaceoides*
Trichostema ovatum
Hemizonia pallida
Eatonella Congdonii
Lasthenia chrysostoma
Malacothrix californica

3. THE UPPER SONORAN GRASSLAND. Most of the foothills above the Lower Sonoran grassland and below the Upper Sonoran woody associations are treeless and often also free of shrubs. This is a region of grassland that is distinctly different from that of the valley floor. Although the boundaries between the two are not sharp, especially because both associations are dominated by annuals and both have vegetation that is dry much of the year, there are great differences in the species and their habits. Unlike the Lower Sonoran grassland, perennial grasses are relatively common (although of secondary forage value) and scattered low shrubs grow in favorable sites. The soils of the Upper Sonoran grassland are usually good and relatively deep; heavy loam is the most common. North slopes, especially at the higher elevations, are often covered with woody vegetation. In these places, or in areas with more rainfall, Douglas and valley white oaks occur in scattered stands, making the limits of the Douglas oak woodland and the Upper Sonoran grassland sometimes difficult to define.

While the plants of the Upper Sonoran grassland grow well in winter, this growth is not as rapid nor is maturity in time of drouth as precocious as it is with the vegetation of the Lower Sonoran grassland. The plants stay green much later, usually into early May and in occasional years even lingering well into June.

The Upper Sonoran grassland occurs rather intermittently over a wide area, from as low as 900 feet elevation in the hills in the Granite Station region to nearly 6,000 feet elevation on the high rounded summits and slopes of the southwestern Tehachapi Mountains (here the hot desert winds of late spring and summer probably prevent the development of woodland or even forest). Although treated in this paper as a cismontane association, islands which are sometimes extensive occur on the high eastern slopes of the Tehachapi Mountains, on the west side of the desert Antelope Valley, and along the west side of Kelso Valley at the southeast base of the Piute Mountains.

Normal rainfall for the Upper Sonoran grassland varies from 6.5 inches at the lower levels to 10 inches where it blends with the Douglas oak woodland. Mean minimum temperatures are probably about 23° F. at the lower levels and probably approach 0° F. on the high Tehachapi summits. On typical summer days, temperatures of more than 100° F. are common. Dates of killing frosts vary widely, both from place to place in the association and from year to year. Generally, they can be expected from November 1 to April 15.

Historical changes in this grassland are difficult to assess. While one would expect that it is successionally transitional and perhaps much influenced by the activities of man, such scant early records as exist suggest there have been no marked changes in the vegetation in the past one hundred years, although the species and their abundance may have changed greatly. However, the advent of the rapidly growing naturalized annuals together with intensive grazing may have severely reduced the number of successful tree, shrub, or even perennial seedlings. The dense growth of the introduced plants and the highly flammable qualities of annual bromes and common foxtail undoubtedly contributed to fiercer hotter wild fires; this, too, would have a markedly deleterious effect on shrubs and trees.

The Upper Sonoran grassland is the basic range for Kern County's extensive cattle operations; the lower levels are often grazed in spring by sheep. Local areas, mostly in the northern and central Temblor Range, have been converted to dry-farm wheat and barley culture, often with only marginal success.

Typical plants of the Upper Sonoran grassland are:

Bromus arenarius
B. diandrus
B. mollis
B. rubens
Festuca confusa
F. dertonensis
F. megalura
F. pacifica
Poa scabrella
Hordeum leporinum
Avena fatua
Stipa cernua
Allium peninsulare
Bloomeria crocea var. *aurea*
Chenopodium californicum
Mirabilis Froebellii
Stellaria media
Delphinium gypsophilum
Astragalus asymmetricus
Medicago hispida
Trifolium tridentatum
Erodium cicutarium
E. moschatum
Clarkia cylindrica
C. purpurea ssp. *quadrivulnera*
Lomatium utriculatum
Nemophila Menziesii
Phacelia ciliata
Amsinckia Douglasiana
A. intermedia
Trichostema lanceolatum
Orthocarpus purpurascens
Plantago Hookeriana var. *californica*
Achyrachaena mollis
Monolopia lanceolata
Microseris Douglasii
Lagophylla ramosissima

4. The Upper Sonoran Subshrub Association. The arid hills around the head of the San Joaquin Valley from Adobe Canyon northeast of Bakersfield southwest through the Tehachapi Mountains and the San Emigdio Range, thence northwest along the east side of the Temblor Range, support a not commonly recognized vegetative zone that is transitional between the valley's grassland and the more typical Upper Sonoran associations. It is difficult to define floristically for the complex does not consist primarily of species that are limited to it, or that are even at their best in the association. The Upper Sonoran subshrub association is an assembly of soft-wood low shrubs of the neighboring plant associations that are most suited to limited moisture, usually poor or rocky soil, and that can endure long seasons of hot dry weather. These shrubs are not only summer dormant but most can endure long periods or even years of winter drouth. In the Temblor and western San Emigdio ranges, foothill golden bush (*Haplopappus linearifolius*) and gray California buckwheat (*Eriogonum fasciculatum* ssp. *polifolium*) are the most common shrubs. From Grapevine Canyon north through the lower slopes and canyons of the Tehachapi Mountains to the southwest flank of Breckenridge Mountain, bladder pod (*Isomeris arboreus* var. *globosa*) is dominant.

Four taxa are almost entirely limited to this association (which extends north along the slopes bordering the west side of the San Joaquin Valley as far as Corral Hollow in eastern Alameda County). These are: Eastwoodia (*Eastwoodia elegans*), Temblor buckwheat (*Eriogonum temblorense*), Temblor clarkia (*Clarkia tembloriensis*), and, in Kern County, *Stylomecon heterophylla* var. *micropetala* (which also ranges south to Baja California).

In this association, winters are warm and summers are hot. The minimum winter temperatures are probably rarely less than 26° F.; summer days are typically over 90° F., with those over 100° F. not at all uncommon. The rainfall averages from 5 to 7 inches; moisture is augmented by periods of dense winter fog. Elevations for the association range from 900 to 1,500 feet, ascending to as high as 2,200 feet in the extremely arid southern Temblor Range west of Taft and Fellows. Soils are largely sandstone and shale; in the Temblor Range often of pure white diatomaceous shale.

Plants typical of the Upper Sonoran subshrub association are:

Eriogonum fasciculatum ssp. *polifolium*
E. temblorense
Stylomecon heterophylla var. *micropetala*
Isomeris arborea var. *globosa*
Mentzelia pectinata
Camissonia californica
Clarkia tembloriensis
Amsinckia vernicosa
Eastwoodia elegans
Gutierrezia bracteata
Haplopappus acradenius ssp. *bracteosus*
H. linearifolius

5. DOUGLAS OAK WOODLAND. The slopes of the middle elevations of the mountains and those in favorable places in the Temblor Range are occupied by an open woodland which is characterized by the Douglas oak (*Quercus Douglasii*). This woodland is rich, diverse, and complex and occurs in three poorly defined phases. The first of these is dominated by the Douglas oak itself, which occurs as well-developed trees in broad parklands, usually in good soils on broad slopes and flats. In these parklands the trees are widely spaced with the open areas occupied by many of the annuals and perennials characteristic of the Upper Sonoran grassland. This sort of woodland is best developed between Granite Station, Woody, and Glennville.

In deep soils with good moisture, often in protected mountain valleys, the valley white oak (*Quercus lobata*) forms an open park-like woodland; the imposing trees are of great beauty. Vernal pools with their distinctive flora are often associated with the white and Douglas oaks in these valleys. Impressive groves grow in Castac Valley at Lebec, on the flats at Tejon Pass, in the valleys around Tehachapi, and at Lynns Valley in the Greenhorn Range.

In rocky and exposed places along ridges and in canyons, usually with poor or shallow soil, digger pine (*Pinus Sabiniana*) is dominant and the Douglas oak, although common, often grows in a stunted, dwarfed, or even shrubby form. In this digger pine phase of the Douglas oak woodland, California buckeye (*Aesculus californica*) is also common, and in such places as lower Kern Canyon, is often even dominant. The associated shrubs and herbs also tend to be characteristic groups in the three phases of the Douglas oak association. However, all three thoroughly overlap and the distinctions are not marked well enough to make discrete delineations of the vegetation that are meaningful for our present purposes.

With the exception of the region in the Greenhorn foothills between Granite Station and Glennville, the Douglas oak woodland is rarely extensive. It usually occurs in alternation with other associations, reflecting climatic and edaphic factors. At its lower levels, the woodland occurs on north slopes and in canyons with the Upper Sonoran grassland on the south slopes. At the middle and higher elevations it alternates with the chaparral, shin oak brush, and even the yellow pine forest. Here the Douglas oak woodland tends to occupy the sunny south slopes; the other associations the north slopes and the canyons. In the San Emigdio and Temblor ranges it occurs in a distinctive association with California junipers, and from the Piute Mountain region south through the Tehachapi Mountains with junipers and pinyon pines.

The Douglas oak woodland is not common on the desert slopes of the mountains but it does occur locally, particularly in the region from Tehachapi south to the west end of Antelope Valley. It is also well developed on the south end of the Piute Mountains at Kelso Valley, only a few hundred yards away from the broad sandy flats where Joshua trees and their typical desert associates grow.

The association occurs from about 1,000 feet elevation to 3,500 feet elevation in the Greenhorn Range; in the Tehachapi Mountains and the Mt. Pinos region it occurs from 2,000 to as high as 6,000 feet; in the Temblor Range from 1,500 feet to 4,300 feet.

This woodland has the greatest range of rainfall of any of the plant associations. In the Temblor and San Emigdio ranges, the median annual average is about 9.5 inches; it is probably as much as 19.5 inches in Bear Valley in the Tehachapi Mountains and in the Glennville region. Between these extremes, the typical amount would be 11 to 14 inches. Winter temperatures range from relatively mild in the Temblor and San Emigdio ranges, where winter days below 20° F. are uncommon and extremes are probably not less than 12°., to cold in places such as Lynns Valley where temperatures of 10° F. occur nearly every year. Some of the precipitation occurs as snow, which, however, is infrequent and usually melts within a few hours.

Typical plants of the Douglas oak woodland are:

Pinus Sabiniana
Stipa coronata
Brodiaea lutea var. *scabra*
Yucca Whipplei ssp. *caespitosa*
Quercus Douglasii
Q. lobata
Ranunculus occidentalis var. *Eisenii*
Eschscholzia Lobbii
Meconella linearis
Ribes quercetorum
Arabis sparsiflora var. *arcuata*
Astragalus lentiginosus var. *idriensis*
Trifolium ciliolatum
Aesculus californica
Sida malvaeflora ssp. *sparsifolia*
Viola quercetorum
Datisca glomerata
Clarkia Dudleyana
C. unguiculata
C. Xantiana
Daucus pusillus
Perideridia Pringlei
Linanthus montanus
Plagiobothrys nothofulvus
P. tenellus
Verbena lasiostachys
Penstemon heterophyllus
Mimulus pictus
Scrophularia californica var. *floribunda*
Hemizonia Heermannii
Microseris sylvatica

6. CHAPARRAL. Chaparral, perhaps the plant association most typically Californian, is not common in Kern County. There are extensive stands on the northwest spur of Mt. Abel and at other scattered localities in the Mt. Pinos region, in the central Temblor Range, in the southwestern and northeastern Tehachapi Mountains, on the southwest flank of Piute Mountain, and occasionally on other mountain slopes. But, with the exception of the

Blue Mountain region near Glennville and the ridges northeast of Tehachapi, it is nowhere common as the dense shrubby cover composed of many species typical of so much of the Coast Ranges and the Sierra Nevada foothills to the north.

Although typical chaparral species have many drouth-adaptive mechanisms to enable them to endure the long hot dry California summer, the association usually occurs only in areas of relatively high winter rainfall. As its growing season is late winter and early spring, chaparral is not well adapted to prolonged cold weather or places where spring is late. Areas, common in other parts of California, with warm winters and high rainfall are limited in the Kern County mountains.

As a result, the Kern County chaparral does not have the rich variety of species common elsewhere. The county has only four chaparral species of manzanita (*Arctostaphylos*), Monterey County has at least 14; Kern has four chaparral species of *Ceanothus,* Monterey at least 13. Greasewood (*Adenostoma fasciculatum*) forms extensive almost pure stands of hundreds of acres in many chaparral regions; it is a rarity in Kern County.

In addition to the Blue Mountain region, there are also extensive growths of chaparral from Tollgate Ridge east of Keene to the west slope of Cache Peak at the head of Jawbone Canyon high above the Mojave Desert. Here the chaparral is highly specialized and perhaps unique; by far the commonest components are only three species, dwarf oak (*Quercus turbinella* ssp. *californica*) and the two closely related species of buck brush, *Ceanothus vestitus* and *C. cuneatus.* These are among the most drouth-resistant species of their respective genera.

Interesting evidence of the increasing aridity of the county is the patches, sometimes of only an acre or two, of aged chaparral shrubs on the lower southeastern slopes of the Tehachapi Mountains northwest of Antelope Valley at such places as the mouths of Bronco and Lecheria canyons. These weather-beaten old shrubs are probably the remnants of former expanses of chaparral that have been replaced by Upper Sonoran grassland.

Rainfall in the Kern County chaparral belt is not precisely known; the best estimates from stations located near it suggest a median range from 12 inches to 18 inches. This is much less than the coastal chaparral receives; in Santa Barbara County the chaparral rainfall is as much as 30 inches. The association occurs at elevations considerably higher than typical; it grows from 3,200 feet to 4,200 feet in the Blue Mountain region, between 4,000 and 5,000 feet on Piute Mountain, and well over 5,000 feet on Mt. Abel. On Cache Peak its lower level is at about 4,000 feet; it ascends to over 6,000 feet. Winters are cold, with minimum temperatures from 0° F. to 10° F. Summers are hot; typical daytime temperatures are above 90° F. and readings of more than 100° F. are not uncommon.

Typical plants of the chaparral are:

Chlorogalum pomeridianum
Zigadenus exaltatus
Quercus dumosa
Q. Wislizeni var. *frutescens*
Q. turbinella ssp. *californica*
Silene californica
Adenostoma fasciculatum
Lotus scoparius
Cercocarpus betuloides
Ceanothus cuneatus
C. leucodermis
C. vestitus
Rhamnus californica ssp. *tomentella*
Fremontodendron californicum
Garrya flavescens var. *pallida*
Arctostaphylos glauca
A. glauca var. *puberula*
A. Parryana
Eriodictyon californicum
Marah horridus
Lonicera Johnstonii
Haplopappus arborescens

7. SHIN OAK BRUSH. Shin oak brush is a plant association that reaches its best development in the mountains of Kern and northeastern Los Angeles counties. Shin oak[4] (*Quercus Garryana* var. *Breweri*), a shrubby form of the large forest tree of the north, grows in dense almost impenetrable thickets on the west slope of the Greenhorn Range, on the east slope of Breckenridge Mountain, and, less extensively, on the north end of Piute Mountain. In the Tehachapi Mountains, particularly on the high steep slopes south of Tejon Pass and around the summit of Cummings Mountain, it densely covers the mountains, some colonies consisting of more than a thousand acres. The shrub does not occur in the San Emigdio or Temblor ranges.

Shin oak brush is often confused with chaparral. However, the characteristics of the two are very different. The shin oak brush is deciduous, winter dormant, and highly resistant to fire (a standard technique of the Kern County fire department is to head chaparral and grassland fires into the shin oak brush; here

4. Although the name "shin oak" is universally used in Kern County, elsewhere the shrub is known as Brewer oak.

they quickly burn themselves out). It requires substantial late spring and summer moisture and will endure and perhaps needs cold winters. It tolerates and even thrives in deep, rich, heavy soils. Not one of these things is true of chaparral. Further, although the Kern County chaparral is not richly developed, it is composed of a mixture of several species. Shin oak brush usually has other shrubs only in draws, ravines, wet places, disturbed areas, or in other special situations. The dense stands are usually pure except for scattered canyon live oaks (*Quercus chrysolepis*); these usually grow in a shrubby form.

It is not surpising that these two shrubby complexes apparently have a very different prehistory. Chaparral probably developed relatively recently. It migrated to California from the southeast in response to the increasing aridity and warmer winters (and especially drier summers) that developed late in the Pliocene. Shin oak is probably a remnant of a much older boreal flora that flourished in a colder, wetter climate. The plant probably evolved as a shrub after it was isolated from the parent arboreal species by the increasing aridity of California. The shrub may have originated in the Klamath highland.

Shin oak brush grows at elevations of 3,000 to 4,500 feet in the Greenhorn Range, at 4,000 to 6,000 feet on Breckenridge Mountain, and from 5,000 to 7,500 feet in the Tehachapi Mountains. Winter temperatures of less than 15° are common; extremes approaching 0° are probably not rare. Typical summer temperatures are about 85° F.; extremes are probably no more than 105° F. The median annual precipitation is approximately 17 inches; about 25 per cent occurs as snow.

Various spring annuals grow in the shin oak brush before its leaves appear, rather late in spring, but these are equally common in other nearby associations. While a plant association resting on a single plant approaches a contradiction in terms, still such a distinctive feature of the Kern County vegetation must be given recognition.

8. THE YELLOW PINE FOREST. The yellow pine forest is one of the Kern County plant associations that is well marked and easily delimited. This conifer forest is the typical vegetation of all the higher elevations of the mountains except a small area at Sunday Peak in the extreme northern part of the Greenhorn

Range where the Sierran red fir forest reaches its southern limits. But even such an obvious vegetational assemblage is not without its complexities. The common conifer in the Greenhorn Range and on Breckenridge Mountain is the ponderosa pine. In the other mountains ponderosa pine occurs only in relict colonies and is generally replaced by the Jeffrey pine (*Pinus Jeffreyi*). This is primarily because of the greater aridity of the other mountains, although the colder, harsher climate is also a factor.

The ponderosa pine is the dominant tree of the well developed, often dense forest in the Greenhorn Range and on Breckenridge Mountains. In the Piute Mountains, in local areas of the Tehachapi Mountain, and in the Mt. Pinos region, the Jeffrey pine forms a more open and more stately forest (which is particularly impressive along the crest of Piute Mountain). Sugar pine (*Pinus Lambertiana*) is common in the Greenhorn Range; it is only occasional on Breckenridge Mountains and is scarce at the north end of Piute Mountain. The tree is rare in the Tehachapi Mountains and is known in the Mt. Pinos region only from a colony on the east slope of Brush Mountain. Incense cedar (*Calocedrus decurrens*) is common in the Greenhorn Range but is inexplicably absent from Breckenridge Mountain. In the Jeffrey pine forest of the other mountains, it is rare and only occurs in deep canyons or in other very favorable places (making its only appearance in the San Emigdio-Mt. Pinos complex as a scattered grove of young trees along the ephemeral creek at the head of Black Bob Canyon). White fir (*Abies concolor*) is common in both forests; big cone spruce (*Pseudotsuga macrocarpa*) occurs in the Jeffrey pine forest in the Mt. Pinos region.

The Kellogg oak (*Quercus Kelloggii*) is a characteristic and common tree of both forests. At the lower borders, usually in areas with relatively good soil, it is especially common, often extending as a narrow woodland below the lowest yellow pines. Here many of the plants of lower elevations are common. In a more critical and elaborate classification of the vegetation, this Kellogg oak woodland would merit separate classification.

The ponderosa pine forest in Kern County is notable for the number of species that reach their southern limits within it; it is the southern outpost for no less than 48 plants. With the exception of the meadow species, the limits of these plants are usually abrupt; once the plant occurs at all, it is usually fairly common on to the north. On the other hand, the plants that reach their southern limits in the Jeffrey pine forest are often scattered or even rare. In some cases, such a plant is limited to a single isolated colony far from its nearest northern or northeastern continuous range.

The ponderosa and Jeffrey pine forests are colloquially yet appropriately called "east-side" and "west-side" vegetation (in reference to Kern Canyon) by the United States Forest Service personnel at Kernville. This is fitting as the ponderosa pine forest has its affinities with the great forests of the west slope of the Sierra Nevada to the north, extending on through the Cascade Mountain complex.

The Jeffrey pine forest, however, is more closely related to the vegetation of the Sierra Nevada's abrupt arid east slope, the northern Basin Ranges, and the other interior mountains of the western United States.

The chief economic product of the yellow pine forest is lumber. Timbering probably reached its peak in the early history of the county, primarily to serve the building needs of the gold mining towns of Havilah, Whiskey Flat, Keysville, and others in the 1860's. Later, much timber was cut on Breckenridge Mountain for the building trade in Bakersfield. In recent years, under the supervision of the U. S. Forest Service, intensive logging has resumed in the Greenhorn Range and on the Kern Plateau, and, to a much lesser extent, in the other forests. There has been no timber cutting in recent years in the privately owned forests of the Tehachapi Mountains. Grazing of cattle is important, particularly at the lower elevations and in the meadows. Recently there has been a great increase in recreational use, often with catastrophic results to both the flora and the fauna.

The annual precipitation, mostly as snow, is from 20 to 35 inches in the ponderosa pine forest; in the Jeffrey pine forest from 14 to 20 inches. At higher elevations the soils are usually porous decomposed granite or shallow fractured basement granite and an arid form of the Jeffrey pine forest survives with no more than 10 inches of precipitation; the porous soils that keep run-off to a minimum are perhaps the factor that makes this possible. Winter temperatures often approach 0° F.; a typical summer day may have a high between 80° F. and 90° F.; extremes are rarely more than 100° F. The frost free season in the ponderosa pine phase is usually from about May 1 to October 15; in the Jeffrey pine phase, from May 15 to October 1 (or often even earlier).

There are a surprising variety of soils, varying from extensive outcrops of basement rock to rich but shallow loams. Granitoid rocks and their erosion products are by far the most common but areas with metamorphic rocks for their base are common. These last are floristically much richer than the granite-derived soils.

The yellow pine forest occurs at elevations above 5,500 feet in the Mt. Pinos region, the Tehachapi Mountains (here it grows only at favorable sites, mostly on north slopes), and in the Piute Mountains. On Breckenridge Mountain and in the Greenhorn Range it makes its appearance between 4,000 and 5,000 feet; on the Kern Plateau at approximately 6,000 feet.

Although the two forests are distinctly different, the number of species that are common to both is far greater than the number that are not. So it is more realistic and convenient to treat them as two phases of the same plant association. This is done in both the list of indicator species and in the catalogue of plants.

Plants typical of both the ponderosa and Jeffrey pine forests are:

Pteridium aquilinum var. *pubescens*
Abies concolor
Bromus Orcuttianus
Stipa Lemmonii
Quercus Kelloggii
Calyptridium umbellatum
Ribes cereum
R. nevadense
Lotus crassifolius
Ceanothus cordulatus
Sarcodes sanguinea
Arctostaphylos Mariposa
A. patula
Gilia leptantha ssp. *Purpusii*
Phacelia mohavensis
Nicotiana attenuata
Castilleja Applegatei
C. disticha
Collinsia Wrightii
Penstemon speciosus
Galium Bolanderi
Madia Wheeleri
Cirsium Andersonii
Hieracium albiflorum

Plants typical of the ponderosa pine forest:

Cystopteris fragilis
Pinus Lambertiana
P. ponderosa
Calocedrus decurrens
Trisetum canescens
Carex alma
C. athrostachya
C. multicaulis
Stellaria Jamesiana
Arabis repanda
Rorippa curvisiliqua
Chamaebatia foliolosa

Rubus parviflorus
Sidalcea ranunculacea
Pterospora andromedea
Pyrola picta
Apocynum medium var. *floribundum*
Navarretia divaricata
Phacelia mutabilis
P. Purpusii
Cynoglossum occidentale
Cordylanthus Ferrisianus
Mimulus montioides
Symphoricarpos acutus
Haplopappus Whitneyi
Eriophyllum lanatum var. *croceum*

Plants typical of the Jeffrey pine forest:

Pinus Jeffreyi
Pseudotsuga macrocarpa
Muhlenbergia asperifolia
Carex Douglasii
Allium fimbriatum var. *denticulatum*
Calochortus invenustus
Fritillaria pinetorum
Eriogonum molestum var. *Davidsonii*
E. Wrightii var. *curvatum*
E. Kennedyi var. *austromontanum*
Delphinium patens var. *montanum*
Potentilla santolinoides
Astragalus Whitneyi
Lathyrus pauciflorus var. *Brownii*
Lupinus confertus
L. excubitus var. *austromontanum*
Viola purpurea ssp. *xerophyta*
Frasera neglecta
Leptodactylon pungens var. *Hallii*
Phlox diffusa ssp. *subcarinata*
Phacelia Davidsonii
Trichostema austromontanum
Symphoricarpos Parishii
Chrysothamnus nauseosus ssp. *occidentalis*
C. Parryi ssp. *vulcanicus*
C. viscidiflorus ssp. *pumilus*
Antennaria dimorpha
Chaenactis santolinoides
Crepis occidentalis ssp. *pumila*

9. The Red Fir Forest. Although occupying by far the smallest area of any association recognized, the red fir forest on the north and east slope of Sunday Peak near the summit is well worth recognition as it is the southern limits of a widespread and important forest zone of the Sierra Nevada. Growing for the most part on open slopes in thoroughly decomposed granite rich in organic matter, interspersed with open areas with extensive colonies of choke cherry (*Prunus emarginata*) and chinquapin (*Castanopsis sempervirens*), this small red fir forest is one of the most attractive places in Kern County. Particularly interesting is the summit ridge with its panoramic view of the southern Sierra Nevada north to Mt. Whitney and the Kaweah Peaks. Here the granite outcrops have colorful colonies of pride-of-the-mountains (*Penstemon Newberryi*), and Sierra manzanita (*Arctostaphylos nevadensis*), its prostrate stems picturesque against the gray rocks.

The red fir forest has the county's shortest growing season, probably no more than 120 days, and the highest precipitation (at least 40 inches, mostly as snow). Summer temperatures probably rarely equal 100° F.; winter extremes are unknown, but are undoubtedly below 0° F. The forest occurs from 7,600 feet elevation to the 8,400 foot elevation of the summit of the mountain.

Apparently spot fires caused by lightning near the summit are not uncommon; small burns are quickly taken over by chinquapin. The chinquapin offers protection for seedling firs; these probably could not survive the first few years of their life without such protection.

Plants typical of the area are (those that reach their southern limits here are marked with an asterisk):

Athyrium Filix-femina
**Abies magnifica* var. *shastensis*
Deschampsia elongata
Glyceria erecta
Carex feta
C. subfusca
Luzula comosa
Castanopsis sempervirens
**Aconitum columbianum*
Stellaria crispa
Arabis platysperma
**Ribes viscosissimum*
**Lupinus Breweri* var. *Breweri*
Prunus emarginata
Rosa pinetorum
Gayophytum heterozygum
Cornus stolonifera
**Arctostaphylos nevadensis*
**Hackelia velutina*
Penstemon Bridgesii
P. caesius
P. Newberryi
**Senecio Clarkianus*
**S. pseudaureus*
Chrysopsis Breweri

10. MOUNTAIN MEADOW ASSOCIATION. Meadows are a charactertic feature of the Sierra Nevada forests. Sometimes these are quite small, sometimes very large. These meadows have a rich and complex flora and each yields many species. Although seemingly uniform to the casual visitor, closer examination shows that their flora is varied and that the meadows have distinct zones of vegetation occupying diverse habitats. Some areas are quite wet all year, these are the home of the semiaquatic plants; completely aquatic plants grow in the occasional small permanent pools. Around these wet areas there is usually a belt of soil that is perennially moist but not wet. Finally, dry but disturbed soils are characteristic of the meadow borders. In Kern County, those that have areas with soil that is quite wet early in the summer but dry by fall are common. These soils favor a distinctive group of plants that mature by mid-summer. This is especially true of the meadows of the Jeffrey pine forest, such as Little Cannell and Cane meadows on the Kern Plateau and the broad and beau-

tiful Landers and Woolstaff meadows in the Piute Mountains.

In the Greenhorn Range, all of the meadows are quite small and are generally wet. They occur mostly on the east slope of the range. In the Piute Mountains they are broad, sunny, and in normal years are usually dry by mid-summer. Another sort of meadow has dry soil, although probably more than average subsurface moisture. These have been cut through by creeks, a normal process greatly hastened by improper grazing. On these, big sage (*Artemisia tridentata*) is the dominant plant. Typical examples are Pine Flat on the Kern Plateau and Woolstaff and Weldon meadows in the Piute Mountains. True mountain meadows do not occur in the Tehachapi Mountains although such cienegas as those at the head of Catskin and Tejon creeks simulate them floristically. Except for the long series of connecting meadows in the Mil Potreros that separate the San Emigdio Range from Mt. Pinos, they are unknown in that region and the meadow flora is limited to seeps and cienegas.

The meadows are of great economic importance for grazing. The rapid lush summer growth of the meadow vegetation at a time when the slopes below the forests are dry and often bare make them an important source of summer forage. Such grazing has a pronounced effect on the vegetation; it promotes sod-forming sedges and grasses at the expense of the tall herbaceous plants. Tall plants that are also palatable find survival especially difficult. These include such striking plants as *Lilium Kelleyanum* and the beautiful but deadly monkshood (*Aconitum columbianum*).

Extremely heavy grazing with trampling and pawing, especially by horses, of the sort once common sometimes cut and broke the sod. In many cases, in those whose lower end was not closed by a solid rock structure, the creeks that fed the meadows cut through the lower end and gradually drained them. The result was the dry or "sagebrush" type meadows, such as Pine Flat on the Kern Plateau and Landers, Woolstaff, Weldon, and Claraville meadows in the Piute Mountains. Changing marketing practices and other techniques of modern beef production require that livestock be maintained in a thrifty condition; this cannot be achieved with devastating grazing. Happily, this is now rare. The greatly decreased commercial use of horses and the regulation of their recreational use by the U. S. Forest Service has almost eliminated the highly deleterious pawing and "bare earth" grazing typical of the animal. However, the greatly increased use of four-wheel drive vehicles and especially of various two wheel vehicles of the "tote-gote" type has been more devastating in just a few years than livestock was in well over a century. The meadows are favorite play areas for the riders of these vehicles; their spinning and gouging can quickly leave results little short of appalling.

Typical plants of the mountain meadows are:

Glyceria elata
Poa pratensis
Muhlenbergia filiformis
Carex Bolanderi
C. feta
C. nebrascensis
C. subfusca
C. teneraeformis
Eleocharis acicularis
Juncus macrandrus
J. orthophyllus
J. tenuis
Veratrum californicum
Salix Scouleri
Polygonum bistortoides
Sagina saginoides var. *hesperia*
Stellaria sitchana var. *Bongardiana*
Potentilla glandulosa
P. gracilis ssp. *Nuttallii*
P. purpurascens
Lupinus latifolius
L. polyphyllus var. *superbus*
Trifolium monanthum
T. variegatum
Geranium californicum
Hypericum anagalloides
H. Scouleri
Perideridia Parishii
Allocarya hispidula
Mimulus moschatus
M. primuloides var. *pilosellus*
Galium trifidum var. *subbiflorum*
Helenium Bigelovii
Achillea lanulosa
Taraxacum officinale

11. THE PINYON WOODLAND. On the desert-facing slopes of the Sierra Nevada, the arid easterly slopes of the Piute Mountains and the northwestern Tehachapi Range, and about much of the Mt. Pinos region, the Douglas oak woodland of the western slopes is replaced by a sometimes well developed and sometimes sparse woodland of pinyon pines (*Pinus monophylla*), usually with large shrubs of California juniper (*Juniperus californica*) at the lower borders. This pinyon woodland is especially well developed along the Kern-Tulare county line at the southeast border of the Kern Plateau in the Lamont Peak region; from here it extends to Kiahvah (Scodie) Mountain south of Walker Pass. South of here, on the desert-like summits of the extreme southern Sierra Nevada, such as Gold, Dove, and Butterbread

peaks, it is poorly developed. Pinyons are scattered but hardly form a true woodland along the east slope of the Tehachapi Mountains, especially south of Tehachapi Pass. Finally, the woodland grows in a continuous belt, often of forest proportions, around Mt. Pinos and in the San Emigdio Range west to the canyons bordering the upper Cuyama Valley in Ventura and Santa Barbara counties.

Although the pinyon woodland usually has scant herbaceous cover, shrubs that are of prime importance as browse plants for livestock are common associates. The tree is a prolific bearer of nutritious and delectable nuts that were an important food for Indians, and are a basic food for rodents and numerous species of birds. Wild pigeons congregate, sometimes in great numbers, in nut season in the Mt. Pinos region.

The pinyon woodland has much of its floral affinities with the arid northeastern regions. It has a prominent intrusion of Mojavean species as well as those of the neighboring associations. Plants that are common in the Jeffrey pine forest of the eastern Sierra Nevada have scattered southwestern range limits on the high slopes of the pinyon woodland; some of these range northeast to the distant Rocky Mountains.

Rainfall records are scant for the pinyon woodland but the normal range seems to be from 7 to 12 inches. Winters are cold, with minimum temperatures of 10° F. not uncommon. Typical summer days have temperatures from 85° F. to 95° F.; temperatures over 105° F. are rare. The frost free season is usually from about April 20 to October 20, although these dates are highly variable (as they are for most of the county). Soils are usually shallow, ranging from sandy to rocky, and are almost uniformly well drained.

The economic use is mostly for browsing of the shrubs although in good springs the annual vegetation affords brief but excellent grazing for cattle and sheep.

Typical plants of the pinyon woodland are:

Ephedra viridis
Pinus monophylla
Juniperus californica
Allium fimbriatum var. *denticulatum*
Eriogonum baratum
E. Baileyi
E. Plumatella
Purshia glandulosa
Rosa Woodsii var. *ultramontana*
Lotus strigosus

Lupinus adsurgens var. *undulatus*
Rhamnus californica ssp. *cuspidata*
Tauschia Parishii
Eriastrum densifolium ssp. *elongatum*
E. sapphirinum
Gilia brecciarum ssp. *brecciarum*
Linanthus aureus
Cryptantha pterocarya var. *cycloptera*
Salvia Dorrii
Penstemon Grinnellii ssp. *scrophularioides*
Chrysothamnus nauseosus ssp. *consimilis*
Tetradymia canescens
Artemisia tridentata

12. THE ARID SHRUB ASSOCIATION. While fairly obvious in the field to even a casual visitor, one of the plant associations most difficult to define is the arid shrub formation that lies between the pinyon woodland and the desert creosote bush association. This occurs in a region of deep canyons with steep slopes that are hot and dry in the summer and cold and often windswept in the winter. Soils are coarse, often of granite sand and gravel on broad steep slopes, and are almost always well drained.

These slopes are most extensive on the west side of Indian Wells Valley (reflecting that this is the most arid region of the county) and form a more or less well defined zone south to Tehachapi Pass. Southwest of that pass, their occurrence is irregular and the vegetation less clearly marked to the west end of Antelope Valley. Here, on the southeast base of the Tehachapi Range, the canyons become more mesic, and the arid shrub association is gradually replaced by chaparral.

The vegetative cover is often open but some of the slopes have a dense cover of xerophytic shrubs and subshrubs that also grow in other associations. Perhaps the most common is gray California buckwheat (*Eriogonum fasciculatum* ssp. *polifolium*); this low gray subshrub often densely covers many acres of hot arid canyonsides. The steep slopes of usually deep sand have little annual growth in dry years; in wet years, they are covered with highly specialized colorful desert annuals.

With the exception of the ubiquitous gray California buckwheat, few of the species typical of this association are widespread within it; their occurrence is often sporadic. Further, many are more at home in other communities. Under the Munz and Keck classification, this vegetational complex would include many of the plants of the sagebrush scrub and of the Joshua tree woodland, but neither of these groupings is adequate for our purposes.

Indicator species are difficult to select, as some of the most common, such as gray California buckwheat, are also widespread in other regions. Perhaps the most typical is the clustered-stemmed Kern Joshua tree, *Yucca brevifolia* var. *Herbertii,* that grows in most of the places with deep sandy soil (although it is also common at the higher levels of the creosote bush association). Bladder pod (*Isomeris arborea* var. *arborea*) is widespread in the arid shrub association and not common in neighboring associations. (It is interesting to note that another variety of bladder pod, *Isomeris arborea* var. *globosa,* is a common plant of the Upper Sonoran subshrub association that borders the San Joaquin Valley and is a cismontane transitional association similar to the arid shrub association.)

Three plants are nearly endemic to this association: *Phacelia Nashiana, Gilia cana* ssp. *elongata,* and *Camissonia kernensis.*

Winters are cold in the arid canyons, snows are not infrequent. Summers are hot. Spring arrives in late April; the first frost usually occurs by November 1. Rainfall averages between 5.5 and 7.5 inches but is highly variable. The association typically occurs between 2,500 and 4,000 feet elevation.

Typical plants of the arid shrub association are:

Stipa speciosa
Allium fimbriatum var. *mohavensis*
Calochortus Kennedyi
Yucca brevifolia var. *Herbertii*
Chorizanthe Thurberi
Eriogonum fasciculatum ssp. *polifolium*
E. Heermannii
E. latifolium ssp. *Westonii*
E. microthecum
Eurotia lanata
Delphinium Parishii
Arabis pulchra
Isomeris arborea var. *arborea*
Prunus fasciculata
P. Andersonii
Coleogyne ramosissima
Lotus strigosus var. *hirtellus*
Lupinus excubitus
Camissonia kernensis
Eriastrum eremicum
Gilia brecciarum ssp. *neglecta*
G. cana ssp. *elongata*
Phacelia Nashiana
Salazaria mexicana
Salvia Dorrii
Castilleja chromosa
Penstemon incertus
Brickellia desertorum
Chrysothamnus nauseosus ssp. *hololeucus*
C. paniculatus
C. teretifolius
Haplopappus linearifolius var. *interior*
Tetradymia spinosa var. *longispina*

13. The Creosote Bush Association. The most extensive plant assemblage in Kern County is the creosote bush association. This

xeric vegetation covers all of the northwest corner of the broad Mojave Desert that lies within Kern County; it extends north to the Owens Valley in nearby Inyo County and for many scores of miles to the east and southeast across the great American deserts. Creosote bush (*Larrea divaricata*), which is everywhere the dominant shrub of this association, grows typically as rather evenly spaced shrubs that casually seem to be the indicators of an uncomplicated community. Closer study reveals it is much more complex. On higher slopes large groups of other shrubs occur, including many of the species of the arid shrub association. In low basins, common and spiny saltbush (*Atriplex polycarpa* and *A. spinifera*) are associates as the alkali sinks are approached. Groves and even desert woodlands of the picturesque Joshua tree grow in favorable areas with deep soils and more than average rainfall. In the rugged canyons of the desert ranges, particularly the canyon country of the El Paso Range, highly local out-of-range colonies of several plants grow in the creosote bush association. These include several western outposts of species common in the deep desert and the Death Valley region. In contrast, on the higher slopes near the summit of the range, the association contains such typically cismontane plants as *Platystemon californicus* and *Brodiaea pulchella.*

The Mojave Desert is capable of supporting a lush growth of highly specialized and mostly colorful annuals in seasons of unusual rainfall. The xeric character of these plants is often misunderstood. Unlike the flora of the upper San Joaquin Valley, many are not, paradoxically, particularly adapted to growth or survival in dry soil. Instead, the plants survive in a harsh environment by mechanisms that effectively inhibit germination in all but favorable times. Germination is followed by very rapid growth and prompt flowering of even very young plants. In years of little rainfall, only a few individuals of the native species may sprout but these usually survive to set at least some seed. In contrast, most of the annual plants of the upper San Joaquin Valley germinate even in the driest of years but often they do not survive the seedling or very young stage of their growth in times of extended drouth.

Although the creosote bush association is strictly limited to the desert, some of its annuals occur in the upper San Joaquin

Valley; others grow in the arid mountains south and west of the valley and as far to the northwest as the distant Mt. Hamilton Range.

The creosote bush association occurs at elevations from about 2,300 feet at Muroc, Boron, and Ridgecrest to more than 4,700 feet at Government Peak in the Rand Mountains. The climate is continental; winters are relatively cold (Cantil, at the mouth of Red Rock Canyon, has a recorded low of —3° F.) and summers are hot with very low humidity (Cantil and Inyokern share record highs for Kern County with readings of 117° F.). The association grows in a region with many clear days; the sun probably shines at least a portion of 350 days a year. There is little or no fog. Winter snow storms are common but rarely add substantially to the total precipitation.

The average annual precipitation for the creosote bush association is between 2.5 and 5.5 inches. The rainfall, however, is highly variable. A seasonal total of less than 0.5 inch has been recorded at Armitage on the east side of Indian Wells Valley; as much as 14.1 inches has fallen at Backus near Willow Springs and 12.4 inches at Randsburg. These latter figures reflect summer or early autumn thunderstorms and are not a true measure of the vegetationally significant rainfall. Significant rains do occur as summer thunderstorms, however; these are of particular importance to the shrubs that flower in the autumn, including *Brickellia desertorum* and subspecies of *Chrysothamnus nauseosus*. Two autumn annuals are also common after summers with good thunderstorms; these are *Dicoria canescens* var. *hispidula* and *Psathyrotes annua.*

In good years, very large numbers of sheep are grazed in the spring in the creosote bush regions. Historically, the desert has been economically most important for the mining of gold, silver, and tungsten. But the mines, largely in the Rand and El Paso mountains, have not been of consequence since 1946, with the exception of those extracting the very large borax beds and other deposited minerals in the Boron region. In recent years, desert subdivisions, military installations, and industry have become locally common. Recreation, often of a particularly destructive sort, has also greatly increased. Vandalism is everywhere a problem; it frequently has its impact on the natural scene as well as on the works of man. The poorly controlled recreational use of the creosote bush association and the yellow pine forest, both largely on public lands, is the county's most serious current conservation problem. When contemplating these regions, Eben McMillan often repeats his favorite quotation from Henry David Thoreau: "A man is rich in relation to the number of things he can afford to leave alone." If this is true, those seeking recreation in the wild lands of California and Kern County are singularly poverty stricken.

Typical plants of the creosote bush association are:

Allium fimbriatum var. *fimbriatum*
Chorizanthe Thurberi
Eriogonum inflatum
E. reniforme
E. trichopes
E. Watsonii
Oxytheca perfoliata
Grayia spinosa
Abronia pogonantha
Mirabilis Bigelovii var. *aspera*
Dithyrea californica
Lepidium flavum
L. Fremontii
Astragalus acutirostris
A. didymocarpus var. *dispermus*
A. Layneae
A. lentiginosus var. *variabilis*
Lupinus horizontalis var. *platypetalus*
Larrea divaricata
Croton californica var. *mohavensis*
Eremalche exilis
Mentzelia nitens
Camissonia claviformis
C. pallida
Gilia brecciarum ssp. *argusana*
G. latiflora ssp. *elongata*
G. latiflora ssp. *excellens*
G. ochroleuca ssp. *ochroleuca*
G. stellata
Linanthus Parryae
Nama demissum var. *deserti*
Tricardia Watsonii
Cryptantha dumetorum
C. nevadensis
C. oxygona
C. pterocarya
C. similis
Chrysothamnus nauseosus ssp. *hololeucus*
Filago depressa
Franseria dumosa
Coreopsis Bigelovii
Chaenactis carphoclinia
C. stevioides var. *stevioides*
C. stevioides var. *brachypappa*
Eriophyllum Wallacei
Syntrichopappus Fremontii
Dicoria canescens ssp. *hispidula*
Psathyrotes annua
Glyptopleura setulosa
Malacothrix glabrata

14. SHADSCALE SCRUB ASSOCIATION. The shadscale scrub association is a plant complex typical of much of the Mojave Desert but rather sparingly represented in Kern County. It occurs in relatively heavy often clayey soils, usually with a shallow hardpan which tends to restrict or eliminate the deep-rooted creosote bush and many of its associated shrubs. Thus, it consists of a small

group of highly specialized low shrubs; for the most part the annuals are the same as those of the creosote bush association. (The annuals that in Kern County are restricted to the shadscale scrub are *Eschscholzia glyptosperma, Cryptantha angustifolia, Abronia villosa,* and *Chaenactis macrantha.*)

The shadscale scrub is common on the lower northeastern slopes of the El Paso Range, is local in the Rademacher Hills southeast of Ridgecrest, in clay deposits at the heads of many canyons in the El Paso Range, on the plains north of Boron, and on some of the benches around the east and southern borders of Indian Wells Valley. Perhaps the most typical plant is the interesting desert-holly (*Atriplex hymenelytra*), which has holly-like leaves that are almost white from late summer until the first good rains.

The climate and weather is much the same as that of the creosote bush association; the shadscale scrub is based on edaphic factors.

The plants typical of this rather drab vegetation are:

Atriplex confertifolia
A. hymenelytra
Grayia spinosa
Abronia villosa
Eschscholzia glyptosperma
Cleomella obtusifolia var. *pubescens*
Cryptantha angustifolia
Gutierrezia microcephala
Chaenactis macrantha
Artemisia spinescens

15. The Freshwater Marsh Association. Marshlands once occupied vast areas in the Buena Vista-Tulare drainage system of the central San Joaquin Valley. The nearly level basin (Buena Vista Lake in the south is 252 feet above sea level; Tulare Lake, the northern terminus, 180 feet) was connected by numerous sloughs, marshes, and playas that received the waters from the Kern River and from less important streams that flowed out of the mountains and then meandered sluggishly northward.

The great cyclical variation in these marshlands is illustrated by the size of Tulare Lake. In 1873, it was estimated to cover 600 square miles. The deepest portion (on the northwest end near the present town of Kettleman City) had a depth of 25 feet. But after the great floods of 1868, it had an area of 1,200 square miles, with the deepest places reaching a depth of 50 feet. However, even in that historic wet year, the southern portion was no more than three to five feet deep, and it gradually passed into the extensive series of shallow sloughs, marshes, and playa lakes that extended south to the Kern Lake district southwest of Bakersfield (Anonymous, 1883:218).

The construction of major dams on the Tule and Kaweah rivers and particularly of Isabella Dam on Kern River, together with deep well pumping in the valley, has severely lowered the water table and, except for years of extraordinary flooding, almost eliminated the possibility of waters of any consequence reaching the valley. Fortunately, Mary Austin wrote (Land of Little Rain) a vivid if somewhat poetic contemporary description of the valley marshlands:

"Last and inevitable resort of overflow waters is the tulares, great wastes of reeds in sickly, slow streams. The reeds, called tules, are ghostly pale in winter, in summer deep poisonous-looking green, the waters thick and brown; the reed beds breaking into dingy pools, clumps of rotting willows, narrow winding water lanes and sinking paths. The tules grow inconceivably thick in places, standing man-high above the water; cattle, no, not any fish or fowl can penetrate them. Old stalks succumb slowly; the bed soil is quagmire, settling with the weight as it fills and fills. Too slowly for counting they raise little islands from the bog and reclaim the land. The waters pushed out cut deeper channels, gnaw off the edges of the solid earth."

Although the marshlands are gone, the present network of canals and low-lying places where irrigation water collects often simulate the old wet-place habitats. These have a good representation of the marshland plants. One of the most interesting developments of recent years is the Kern National Wild Life Refuge, northeast of Lost Hills, where approximately 16 square miles of the old slough lands will be restored as a winter stopping-place for birds, primarily water fowl, of the Pacific flyway. Incidental to this purpose, the Refuge is also becoming a splendid sanctuary for the marshland flora and vegetation.

The reclamation of this region for some of the world's richest farmlands has greatly disrupted the primitive flora; it is possible that some species have been extinguished. Unfortunately pioneer (and even recent) collectors concentrated on the mountain regions and almost entirely ignored the sloughs and marshes. (Mary Austin offered one possible explanation for this: "The tulares are full of mystery and malaria.") One plant was invariably mentioned in the early accounts that survive; this was the common tule (*Scirpus acutus*) which densely covered many thousands of acres. It is reasonable to assume that native species now common in wet soils were once widely prevalent and possibly abundant. Although the primitive vegetation may have been composed of relatively few species, there must have been scattered occurrences of many transitory species. Since the marshlands were on the Pacific flyway and were the winter quarters for great concentrations of waterfowl, one can assume they were an area where new species were frequently introduced. This process is still continuing; *Scirpus saximontanus* made its first appearance as a rare plant in the duck club region north of Semitropic in 1954; already it is a common plant of the wet lands of that region.

The climate of the marshlands was stable. Winters were relatively warm, with temperatures rarely below 28° F. Frosts before December 1 or after February 15 were rare. Rainfall was irrelevant to the vegetation, but was quite low, probably averaging no more than 4 inches. Dense persistent winter fogs were common; summers were long with predominantly clear weather. Summer days in excess of 100° F. were commonplace; record highs of 118° F. have been recorded.

Typical plants of the remnants of the freshwater marshes are:

Marsilea vestita
Typha domingensis
Echinodorus Berteroi
Sagittaria longiloba
Leptochloa fascicularis
Echinochloa crusgalli
Cyperus erythrorhizos
C. esculentus
Eleocharis palustris
Scirpus acutus
Salix Gooddingii var. *vallicola*
S. Hindsiana
Polygonum argyrocoleon
P. lapathifolium
P. coccineum
Rumex violascens
Amaranthus californicus
Rorippa islandica ssp. *occidentalis*
Ammannia coccinea
Ludwigia peploides
Bacopa Eisenii
Achillea gigantea
Cirsium crassicaule
Grindelia procera

16. VERNAL POOLBEDS. Depressions which collect rainfall in the winter but are dry by summer or late spring are common in the county. However, these are not the typical California vernal poolbeds that accumulate water nearly every winter. Many are

ephemeral; they may go many years without any water at all. Further, others, such as these in the Temblor and San Emigdio ranges, are at least somewhat subalkaline and their flora more closely approximates alkali sink vegetation than it does the California poolbed flora. This is especially true of the sag ponds along the San Andreas Fault.

Typical California poolbeds that do accumulate water every winter occur in the Glennville region, particularly in Lynns Valley. They were once common in the montane valleys in the Tehachapi region; with the exception of those in Bear Valley most of these have been destroyed by farming and other developments. These have an interesting and varied flora, often of great diversity in a small area. The plants are sporadic in their occurrence; many occur at only one poolbed or only in one region.

Typical plants of the Tehachapi and Glennville poolbeds are:

Deschampsia danthonioides
Phleum pratense
Zigadenus venenosus
Lilaea scilloides
Lepidium campestre
Allocarya bracteata
Pogogyne Douglasii
Limosella aquatica
Veronica arvensis
Downingia pulchella
D. cuspidata
Psilocarphus brevissimus

Typical plants of the poolbeds in the Temblor and San Emigdio ranges:

Deschampsia danthonioides
Juncus bufonius
Spergularia marina
Boisduvalia glabella
Allocarya leptoclada
Nicotiana Bigelovii
Veronica peregrina var. *xalapensis*
Gnaphalium palustre
Hemizonia pungens
Lasthenia Ferrisiae
L. minor
Matricaria matricarioides
Xanthium italicum
X. spinosum

17. Streambank Association. Streambank plants are conspicuous anywhere; in an arid land they are especially so. Kern River is the county's only stream of any consequence. A few creeks run most of the year; these include Poso Creek and its tributaries on the west slope of the Greenhorn Range, Bull Run Creek on the east slope, Erskine and Kelso creeks in the Piute Mountains, and El Paso Creek in the Tehachapi Mountains. Caliente, Tehachapi, and San Emigdio creeks normally have water until late in the season; in wet cycles these run all summer.

The streams support a characteristic flora on their banks. The common and conspicuous trees are Frémont cottonwood (*Populus Fremontii*), yellow willow (*Salix lasiandra*), and red willow (*Salix laevigata*). Oregon ash (*Fraxinus latifolia*) and buttonwillow (*Cephalanthus occidentalis* var. *californicus*) are common along Kern River and occasional along the streams in the Greenhorn Range. Along lower Kern River, naturalized trees of California fan palm (*Washingtonia filifera*) and common peppertree (*Schinus Molle*) are interesting additions to the native flora. Big-leaf maple (*Acer macrophyllum*) is local along El Paso Creek and near the head of Black Bob Creek. Perhaps the most conspicuous tree from Kern Canyon south is the California sycamore (*Platanus racemosa*). Black cottonwood (*Populus trichocarpa*) is local along Tejon and Bull Run creeks, and in a shrubby form along Kern River above Kernville. At higher elevations, Frémont cottonwood is replaced by the attractive white alder (*Alnus rhombifolia*).

Canadian waterweed (*Elodea densa*) is an aquatic that seems to grow only in running water; it is common in Kern River and in Poso Creek. The miniature greater duckweed (*Spirodela polyrhiza*) is common in quiet water and on damp sand.

The herbaceous perennials and annuals that grow along the streams are a mixture of those of the fresh water marsh association and those typical of the meadows of the yellow pine forest. Waifs that grow from seeds carried downstream, sometimes from quite high elevations, are an interesting feature of sand bars and winter-flooded flats of Kern River and Poso Creek and its tributaries.

Numerical Analysis of the Flora

The vascular flora of Kern County consists of 1,713 species and 162 recognized subspecies and varieties in 120 families and 573 genera for a total of 1,875 taxa. In addition, 58 taxa of less than specific rank are mentioned in the text. The native species number 1,463; 250 are introduced.

In the following table of the twelve largest families, the introduced species are given in parentheses.

TABLE 2. *The ten largest families.*

FAMILY	GENERA	SPECIES	SUBSPECIFIC TAXA
Compositae	94	257 (33)	35
Gramineae	52	158 (66)	6
Leguminosae	20	105 (22)	18
Scrophulariaceae	20	88 (4)	5
Polygonaceae	7	74 (8)	9
Polemoniaceae	11	71 (0)	12
Cruciferae	29	61 (17)	6
Cyperaceae	5	54 (3)	0
Boraginaceae	10	48 (1)	10
Onagraceae	11	46 (0)	5

It is interesting to note the comparison with the list of California's six largest families compiled by Gladys L. Smith and Anita M. Noldeke (1960: 117); the rank of the first four is the same for Kern County as it is for the state. In California, *Cruciferae* and *Cyperaceae* are fifth and sixth largest; in Kern, seventh and eighth. *Polygonaceae* and *Polemoniaceae* are impressive in the California flora, but not nearly of such high rank as they are in Kern County. This reflects the large number of species in the genera *Eriogonum* and *Eriastrum* in the county's xeric regions and the remarkable concentration of species of *Gilia* in the western Mojave Desert.

TABLE 3. *The fifteen largest genera.*

GENUS	TOTAL SPECIES	SUBSPECIFIC TAXA	INTRODUCED SPECIES
Eriogonum	37	9	0
Gilia	28	10	0
Carex	28	0	0
Mimulus	26	1	0
Lupinus	25	7	0
Phacelia	24	1	0
Cryptantha	23	8	0
Trifolium	18	2	5
Atriplex	17	0	2
Juncus	17	0	0
Lotus	16	4	4
Bromus	16	1	9
Astragalus	15	4	0
Festuca	15	1	4
Linanthus	14	0	0

Kern County and California have five of their six largest genera in common. However, the order is different; *Carex* is easily California's largest genus; *Phacelia* is third largest; *Lupinus,* fourth; *Eriogonum* fifth; and *Mimulus,* sixth. California's second largest, *Astragalus,* is thirteenth in Kern County.

It is interesting to note that the Compositae, easily the largest family in both California and Kern County, has no genus among the fifteen largest in Kern (or even in the first twenty-one genera). The very large grass family is represented by only two genera among the fifteen largest, both of which gain their rank only because of their introduced species.

Of California's total of 7,704 taxa of all levels, 1,933 or 25.1 per cent are represented in Kern County.

THE ENDEMIC PLANTS. For a region with a complex flora, the list of endemic plants in Kern County is surprisingly brief. The intensive field work needed for a local flora usually results in the discovery of new taxa. During the time of the collecting on which this study is based either the author or other collectors discovered the type material in Kern County for *Eriogonum Breedlovei, E. temblorense, Streptanthus cordatus* var. *piutensis, S. Farnsworthianus, Clarkia tembloriensis,* and *Nemacladus Twisselmannii.* However, searches in the regions bordering the county have resulted in finding new stations for plants that at the outset of this project were considered endemic to Kern County. The result, somewhat ironically, is a list of endemics shorter than the one in hand twelve years ago.

The plants endemic to Kern County are:

Brodiaea coronaria var. *kernensis*
Calochortus coerulea var. *Westoni*
Atriplex tularensis
Loeflingia pusilla
Eschscholzia caespitosa ssp. *kernensis*
E. procera
Streptanthus cordatus var. *piutensis*
Heuchera caespitosa
Camissonia kernensis
Opuntia Treleasei
Convolvulus tomentellus
Navarretia setiloba
Phacelia humilis var. *Dudleyi*
Nemacladus Twisselmannii
Eriophyllum lanatum var. *Hallii*
Hemizonia arida
Layia leucopappa

In addition, a number of species and subspecific taxa are primarily restricted to Kern County, although their ranges extend into neighboring counties. Most of the plants of this group are endemic to floristic units on the borders of the county, such as the floras of the Greenhorn Range and the Mt. Pinos region. These are listed with generalized range notes:

Cupressus nevadensis. In addition to the five Kern County groves, a recently reported extensive colony grows in the Stormy Canyon region just south of Baker Point in Tulare County.

Fritillaria Brandegei. Open forests in southern Tulare County and the Greenhorn Range.

Fritillaria striata. Foothill woodland of the Sierra Nevada of southern Tulare County through the Greenhorn Range.

Yucca brevifolia var. *Herbertii.* The range is mainly in Kern County; it extends north to Nine Mile Canyon in Inyo County and south to Antelope Valley in Los Angeles County.

Eriogonum Breedlovei. Local on the crest of Piute Mountain and on the north face of Baker Point in Tulare County.

Eriogonum gossypinum. Rare, head of the San Joaquin Valley in Kern County; known from a single collection in the triangle of San Luis Obispo County that approaches the valley on the east slope of the southern Temblor Range west of Fellows and another from Kettleman Hills in Kings County.

Eriogonum latifolium ssp. *Westoni.* Rand and El Paso mountains on the desert, west to the east slope of the Greenhorn Range, north to Rock House Basin in extreme southeastern Tulare County

Atriplex vallicola. Alkali playas north of Lost Hills; also reported from southern Fresno County.

Delphinium Parishii ssp. *purpureum.* Mt. Pinos region to Cuyama Valley, in Kern and Ventura counties.

Delphinium Purpusii. Colonies are rare in the Kern Canyon region, east to Erskine Creek, and along Poso Creek in the Greenhorn foothills. The range extends north to Roads End in Tulare County.

Lupinus horizontalis. An uncertain entity that as understood in this work is very rare in Kern County and extreme eastern San Luis Obispo County.

Eremalche kernensis. Known from the Belridge Plain in Kern County and from Soda Lake and Panorama Hills in southeastern San Luis Obispo County.

Sidalcea ranunculacea. Greenhorn Range north in the Sierra Nevada to southern Tulare County, probably reaching its northern limits in the Frog Meadow region.

Clarkia exilis. Endemic to the Sierra Nevada foothills from Kern Canyon to the Tule Indian Reservation in southern Tulare County.

Convolvulus fulcratus var. *deltoides.* Tehachapi Mountains to Mt. Pinos, where it is reported from extreme northern Ventura County.

Eriastrum Hooveri. Rare, Shafter (where it is possibly now extinct), at scattered stations in the central Temblor Range, reportedly ranges north to southern Fresno County.

Gilia cana ssp. *speciosa.* Red Rock Canyon to Nine Mile Canyon in extreme southern Inyo County.

Gilia latiflora ssp. *cuyamensis.* Upper Cuyama Valley in Kern, Ventura, Santa Barbara, and San Luis Obispo counties.

Gilia leptantha ssp. *pinetorum.* Jeffrey pine forest on Tehachapi Peak, Tecuya Ridge, and on Mt. Pinos in Kern and Ventura counties.

Phacelia Nashiana. Rare, Last Chance and Jawbone canyons north to Nine Mile Canyon in extreme southern Inyo County.

Galium angustifolium var. *onycense.* Walker Pass Station to the southern rim of the Kern Plateau at the head of Fay Canyon; also known from a few stations on the desert-facing slopes of the San Bernardino Mountains.

Grindelia camporum var. *australis.* Montane valleys of Kern County; Antelope Valley in Los Angeles County.

Hemizonia pallida. Abundant about the head of the San Joaquin Valley; extending north to Kings County and west to southeastern San Luis Obispo County.

Lessingia germanorum var. *Peirsonii.* Local on the west side and borders of the Mojave Desert in northern Los Angeles and southeastern Kern counties.

THE ADVENTIVE PLANTS. The adventive plants form a substantial part of the flora of the county, their importance varying greatly in different plant associations. Of the total of 1,713 species, 250 are not native. The percentage of introduced species, compared with that of other California regions, is as follows[5]:

5. The figures are taken from the following works: Howell (1949), Bowerman (1944), Smith (1952), Thomas (1961), Twisselmann (1956).

California[6]	11.0%
Kern County	15.6%
Temblor Range and the western San Joaquin Valley	24.9%
Santa Barbara region	38.5%
Mt. Diablo	17.3%
Santa Cruz Mountains	31.0%
Marin County	23.5%

The table shows that the percentage of introduced species in Kern County approaches that of the state as a whole much more closely than does that of the coastal regions. There are several reasons for this. The moister climates to the west and north are much more favorable for garden escapes; in Kern County these account for only 18 species, of these, just five are even locally established. In large areas of Kern County non-native species contribute little to the flora. In the yellow pine and red fir forests, few introduced plants are well established, and these only at such places as moist meadows or in disturbed areas around habitations. Non-native species are relatively few on the desert, although these few form an important part of the vegetation in some areas. Except for the annual grasses, their significance is not great in most of the pinyon, arid shrub, and chaparral associations. In the Douglas oak woodland the importance of the introduced plants is much greater, both in the number of species and in the dominance of the well-established ones. Finally, in the Upper and Lower Sonoran grasslands the influences of the immigrants is so great that it forms what could well be called an adventive flora. Such species as red-stem, white-stem, and in some places, broad-leaf filaree, Arabian grass, the wild oats, the annual fescues, red brome, and Russian-thistle are among the most abundant and widespread plants. Further, the irrigated farmlands and artificially or naturally wet places of the valley support a flora rich in weeds, often far outnumbering both in bulk of the vegetation and number of species the native plants of these places.

Six adventive plants of the county have not been reported as occurring spontaneously elsewhere in the state. These are: *Lep-*

6. Smith and Noldeke (1960).

tochloa viscida, Morus alba, Sphaerophysa salsula, Maclura pomifera, Vitex Agnus-castus, and except for a very narrow area of San Luis Obispo County along the Kern County line, *Papaver hybridum.*

TABLE 4. *Principal places of occurrence of adventive plants.*

HABITAT	NUMBER OF SPECIES
Naturalized, general flora	33
Naturalized, marshes, other places with moist soil	39
Naturalized, total:	72
Introduced (deliberately) and apparently established	9
Widespread but doubtfully naturalized	18
Irrigated farmlands	28
Dry farmlands	6
Ranchgrounds, waste places, vacant lots, etc.	45
Gardens	18
Waifs	29
Escapes from cultivation, not established	25

The table is generalized from sometimes arbitrary decisions; often a plant will be common in more than one of the groupings. The list of weeds includes four species native to California known to the author to have made their Kern County appearance as weeds. Doubtless this is true of other California plants, but to attempt to identify them would lead inevitably into a morass of conjecture.

CENTERS OF DISTRIBUTION AND ORIGINS OF THE NATIVE FLORA

An attempt has been made to analyze the origins and centers of distribution of the native plants. While such a survey involves many arbitrary decisions, the results expressed in percentages should have a high degree of reliability. In table 5 the California botanical province includes plants that have their principal range in that province, as defined by Howell (1957), although the range may extend to regions outside the state. The boreal plants are those whose principal distribution is to the north, northwest, and northeast of California. Most of these have their origins in the Arcto-Tertiary Geoflora (Axelrod 1958:433), a flora which was much more widespread and dominant in times of cooler and moister climates.

The plants of the Great Basin flora that occur here are of special interest. The influences of this flora extend along the east slope of the Sierra Nevada and at the higher elevations of the desert as far south as the San Bernardino Mountains. Several species reach their southwestern limits at Mt. Pinos. These plants are of mixed origin; most are of Arcto-Tertiary derivation and owe their ancestry to northern influences but some are probably of Madro-Tertiary origin (Axelrod, *loc. cit.*).

The southwestern desert species are to a very large extent recent derivatives of the Madro-Tertiary Geoflora, which was also probably ancestral to a large majority of the California plants (Axelrod, 1958). This assemblage originated in the highlands of northwestern Mexico and migrated to California in response to the increasing aridity since Pliocene time.

The small group listed under "other Austral" cannot be classified easily, at least as to immediate ancestry, as these are mostly species that occur over wide areas, some extending even to the tropics. This group probably has plants of both Madro-Tertiary and Neo-Tropical origin; the latter are probably the majority (Axelrod, 1958.).

The "miscellaneous" group includes those of widespread but erratic distribution, those for which definite distributional data are lacking, and those of wide range without any particular center of distribution.

TABLE 5. *Centers of distribution of the native species of the Kern County flora.*

FLORA	SPECIES	PER CENT OF THE FLORA
California Province	582	39.8
Boreal	260	17.8
Austral, southwest deserts	204	13.9
Other Austral	43	2.9
Great Basin	41	2.9
Continental (general North American)	128	8.7
Cosmopolitan (more one continent)	96	6.5
Miscellaneous	109	7.2

RANGE LIMITS IN KERN COUNTY

If it is true that Kern County is a region of significant floristic boundaries, it should have a large number of plants that have their geographical range limits here. That this is the case is established by the following lists that show that 327 taxa reach their southern limits within the county, 60 their northern limits, and 34 their western limits. Thus, a total of 421 plants, or 26.2 per cent of the indigenous taxa, have range limits of phytogeographic importance in the county.

Because the inner Coast Ranges and the Sierra Nevada are orientated in a northwesterly-southeasterly direction and because Kern County includes only the western part of the Mojave Desert, eastern range limits in the county are difficult to ascertain and are usually of little floristic significance. For this reason, they have not been listed.

SOUTHERN RANGE LIMITS IN THE KERN COUNTY MOUNTAINS. Two hundred fifty-nine plants of the Sierra Nevada and other regions to the north and northeast reach their southern limits in the mountains of Kern County. Of these, 184 occur in the geological terminus of the Sierra Nevada that is composed of the Greenhorn Range, the southern Kern Plateau, Breckenridge Mountain, and the Piute Mountains. This tends to verify that the southern limits of the Sierran flora coincide with its geological boundaries. Probably as many of the Sierran plants that extend south of that range reach their southern limits in the mountains of southern California as in Kern County's Tehachapi Mountains and the Mt. Pinos region. A number of these have their southern Sierran limits in north central Kern County but appear again south of the county in the southern California mountains.

Greenhorn Range. The yellow pine forest on the west slope of the Sierra Nevada reaches its southern boundary in the Greenhorn Range; in addition, the Sierran red fir forest has its southern limit on the north slope of Sunday Peak. This range is rich in southern range limits, not only in the coniferous forests but in the foothills and even to the borders of the San Joaquin Valley. The following plants have their southern limits in the Greenhorn Range:

Botrychium multifidum var. *silaifolium*
Cheilanthes intertexta
Pityrogramma triangularis var. *pallida*
Abies magnifica var. *shastensis*
Melica californica var. *nevadensis*
Trisetum cernuum var. *canescens*
Carex amplifolia
C. angustior
C. Brainerdii
C. Kelloggii
C. paucicostata
C. vicaria
Juncus ensifolius
Brodiaea volubilis
Calochortus amoenus
C. Vestae

Fritillaria Brandegei
F. striata
Lilium Kelleyanum
Iris Hartwegii
Eriogonum latifolium ssp. *deductum*
Polygonum Douglasii var. *latifolium*
Lewisia rediviva var. *rediviva*
Arenaria pusilla var. *pusilla*
Stellaria sitchana var. *Bongardiana*
Aconitum columbianum
Delphinium gracilentum
D. hesperium ssp. *pallescens*
Eschscholzia Lobbii
Meconella californica
Dicentra pauciflora
Idahoa scapigera
Lepidium densiflorum
Streptanthus Farnsworthianus
Thysanocarpus curvipes var. *longistylis*
Parvisedum Congdonii
Ribes viscosissimum
Chamaebatia foliolosa var. *foliolosa*
Prunus subcordata
Rosa pinetorum
Rubus leucodermis
Lupinus arbustus
L. Breweri var. *Breweri*
L. albicaulis var. *albicaulis*
L. albicaulis var. *shastensis*
Limnanthes montana
Ceanothus diversifolius
Sidalcea ranunculacea
Viola glabella
V. Sheltonii
Boisduvalia stricta
Clarkia exilis
Arctostaphylos nevadensis
Rhododendron occidentale
Dodecatheon Clevelandii ssp. *patulum*
D. Hanseni
Centaurium venustum var. *Abramsii*
Gilia inyoensis
Linanthus filipes
Nemophila maculata
N. pulchella var. *gracilis*
Phacelia Purpusii
Hackelia velutina
Cryptantha Torreyana
Cynoglossum occidentale
Monardella candicans
Scutellaria Bolanderi
S. tuberosa
Trichostema oblongum
Cordylanthus Ferrisianus
Gratiola ebracteata
Mimulus Douglasii
M. inconspicuus
M. Kelloggii
M. Whitneyi
Orthocarpus erianthus var. *micranthus*
O. linearilobus
Symphoricarpos rivularis
Nemacladus capillaris
N. interior
Downingia cuspidata
Calycadenia spicata
Chrysopsis oregona var. *scaberrima*
Eriophyllum confertiflorum var. *croceum*
Evax acaulis
E. caulescens var. *humilis*
Lessingia leptoclada
Pseudobahia Heermannii
P. Peirsonii
Senecio Clarkianus
S. Douglasii var. *tularensis*
S. pseudaureus

Piute Mountains. The following plants reach their southern limits in the Piute Mountains:

Poa nevadensis
Carex simulata
Brodiaea lutea var. *analina*
Zigadenus exaltatus
Sisyrinchium halophilum
Eriogonum Breedlovei

Huechera rubescens ssp. *alpicola*
Ribes Menziesii var. *ixoderme*
Potentilla glandulosa ssp. *Hansenii*
Rubus glaucifolius
Astragalus subvestitus
Lathyrus sulphureus
Lupinus Grayi
L. latifolius var. *columbianus*
L. polyphyllus var. *superbus*
Angelica lineariloba var. *Culbertsonii*
Arctostaphylos Mariposa
A. patula
Fraxinus latifolia
Frasera tubulosa
Asclepias cordifolia
Gilia leptantha ssp. *Purpusii*
Phacelia ramosissima var. *valida*
Antirrhinum cornutum var. *leptaleum*
Castilleja Breweri
C. disticha
Mimulus montioides
Penstemon Newberryi
Galium Matthewsii
Orobanche corymbosa
Agoseris glauca ssp. *dasycephala*
Aster Eatonii
Balsamorhiza sagittata
Chaenactis rubricaulis
Cirsium Andersonii
Haplopappus Whitneyi
Orochaenactis thysanocarpha
Senecio serra

Kern and South Fork Valleys. Plants that reach their southern limits in Kern and South Fork Valleys:

Myosotis laxa
Solidago spectabilis

Tehachapi Mountains. Plants that reach their southern limits in the Tehachapi Mountains:

Damasonium californicum
Puccinellia grandis
Stipa Lemmonii
Allium hyalinum
Eriogonum inerme var. *inerme*
E. inerme var. *hispidulum*
Salix melanopsis
Silene montana ssp. *sierrae*
Delphinium Hanseni
D. patens ssp. *Greenei*
Ranunculus hebecarpus
R. occidentalis var. *Eisenii*
Prunus Andersonii
Lupinus adsurgens var. *undulatus*
L. vallicola ssp. *vallicola*
L. vallicola ssp. *apricus*
Clarkia speciosa ssp. *polyantha*
Vitis californica
Ceanothus integerrimus var. *californicus*
Lomatium tomentosum
Arctostaphylos viscida
Asclepias vestita
Linanthus nudatus
L. serrulatus
Hydrophyllum occidentale
Nemophila parviflora var. *quercifolia*
N. pulchella var. *Fremontii*
Phacelia Congdonii
Cryptantha nemaclada
Limosella aquatica
Mimulus androsaceus
M. Dudleyi
M. pictus
Galium Bolanderi
Downingia pulchella
Calycadenia hispida forma *albiflora*
Chaetopappa fragilis
Cirsium cymosum
Erigeron foliosus var. *Hartwegii*
Lessingia nana
Microseris sylvatica

Breckenridge Mountain. The following plants reach their southern limits on Breckenridge Mountain:

Comandra pallida
Linanthus montanus
Nemophila pulchella var. *pulchella*
Eriodictyon californicum
Monardella odoratissima ssp. *parvifolia*
Collinsia tinctoria
Symphoricarpos acutus
Helianthella californica var. *nevadensis*
Wyethia invenusta

Kern Plateau. Plants that reach their southern limits on the Kern County portion of the Kern Plateau:

Eriogonum elatum
E. spergulinum var. *spergulinum*
Salix Geyeriana var. *argentea*
S. subcoerulea
Erysimum argillosum
Potentilla saxosa ssp. *sierrae*
Purshia tridentata
Oreonana Clementis
Ipomopsis aggregata ssp. *Bridgesii*
Collinsia Torreyi
Mimulus barbatus
M. primuloides var. *primuloides*
Penstemon Rothrockii
Haplopappus Bloomeri

Mt. Pinos Region. Plants that reach their southern limits in the Mt. Pinos region and the San Emigdio Range:

Stipa nevadensis
Allium Howellii
Brodiaea lutea var. *scabra*
Abronia turbinata
Lewisia rediviva var. *minor*
Silene montana ssp. *bernardina*
Dephinium gypsophilum
D. inopinum
Isopyrum occidentale
Caulanthus amplexicaulis
Erysimum moniliforme
E. perenne
Dudleya cymosa ssp. *cymosa*
Potentilla purpurascens
Rosa Woodsii var. *ultramontana*
Thelypodium Lemmonii
Astragalus Whitneyi
Lupinus Andersonii
L. Breweri var. *bryoides*
Trifolium cyathiferum
Nemophila pulchella var. *gracilis*
Phacelia cicutaria
P. curvipes
P. egena
P. vallis-mortae var. *heliophila*
Castilleja Applegatei
Cordylanthus rigidus
Mimulus Bolanderi
M. constrictus
Downingia bella
Chrysothamnus Parryi ssp. *vulcanicus*
Crepis occidentalis ssp. *occidentalis*
C. occidentalis ssp. *pumila*
Eastwoodia elegans
Layia heterotricha

SOUTHERN RANGE LIMITS IN THE SAN JOAQUIN VALLEY. The flora of the Great Valley of California, which becomes progressively more xeric to the south, reaches its southern limits on the plains at the head of the San Joaquin Valley in Kern County. Few species of this flora have a precise southern limit; their range usually ends irregularly where the valley's Lower Sonoran grassland merges with the less xeric vegetation of the San Emigdio Range or the southwestern Tehachapi Mountains. The plants that reach their southern limits in the San Joaquin Valley are:

Festuca Eastwoodae
F. microstachya var. *simulans*
Puccinellia simplex
Atriplex fruticulosa
A. vallicola
Eriogonum gossypinum
E. indictum
Hollisteria lanata
Rumex transitorius
Delphinium recurvatum
Astragalus lentiginosus var. *nigricalycis*
A. oxyphysus
Lupinus horizontalis
L. nanus var. *Menkerae*
Eryngium spinosepalum
E. Vaseyi
Eriastrum Hooveri
E. pluriflorum
Gilia tricolor ssp. *diffusa*
Cryptantha sparsiflora
Trichostema ovatum
Bacopa Eisenii
Cordylanthus hispidus
Mimulus tricolor
Cephalanthus occidentalis var. *californicus*
Nemacladus gracilis
Achillea gigantea
Chaenactis glabriuscula
Cirsium crassicaule
Eatonella Congdonii
Grindelia procera
Haplopappus racemosus ssp. *glomeratus*
Hemizonia obconica
H. pallida
Layia Munzii
L. pentachaeta
L. pentachaeta ssp. *albida*
Lasthenia chrysantha
L. debilis
L. Ferrisiae
Lessingia nana
L. nemaclada var. *albiflora*
Microseris campestris
Monolopia stricta

SOUTHERN LIMITS OF COAST RANGE PLANTS. The flora of the Coast Ranges is not well developed in Kern County. The Temblor Range bounds the county on the west, where several coastal species reach their central California eastern limits and a few their southern limits. Coastal species are occasional on the northwest flank of Mt. Abel, in the San Emigdio Range, and on the east side of upper Cuyama Valley, where Kern, Santa Barbara, and Ventura counties meet. A few coastal specis occur through the Tehachapi Mountains and even as far north as the Sierran foothills of Tulare County. As their number diminishes rapidly from the coast toward the interior, and as they generally have a north-south distribution, few have important range limits in Kern County. However, some taxa do have southern limits of significance in the Temblor Range. These are:

Eriogonum Covilleanum
E. latifolium ssp. *auriculatum*
E. temblorense
Montia gypsophiloides
Ribes californicum
Astragalus asymmetricus
A. macrodon
Clarkia tembloriensis
Gilia tenuiflora ssp. *amplifaucalis*
Eriodictyon tomentosum
Blepharizonia plumosa ssp. *viscida*
Madia radiata

NORTHERN RANGE LIMITS IN KERN COUNTY. The list of northern range limits in Kern County, largely of southern California plants, is less impressive than those of other regions. But it is of interest, especially when one considers the formidable barrier the Mt. Pinos region and the Tehachapi Mountains furnish to plant migration, the different climate, and the different geological history. Some of these plants are probably relicts whose distribution has been reduced by the changing climates of Pleistocene and Recent time and that have been restricted further by the uplift of the Tehachapi Mountains and the Mt. Pinos region. However, some of the plants are of probably relatively recent introduction, although the southern California natives of obviously weedy Kern County occurrence have been omitted.

Mt. Pinos Region. Northern range limits in the Mt. Pinos region and the San Emigdio Range:

Eleocharis pauciflora ssp. *bernardina*
Eriogonum Kennedyi var. *austromontanum*
Lotus Davidsonii
Lupinus elatus
Ceanothus integerrimus var. *puberulus*
Frasera neglecta
Eriodictyon crassifolium var. *denudatum*
E. crassifolium var. *nigrescens*
Mimulus Johnstonii
Penstemon labrosus
Brickellia Nevinii
Corethrogyne filaginifolia var. *pinetorum*
Haplopappus Palmeri ssp. *pachylepis*
H. squarrosus ssp. *obtusus*

Tehachapi Mountains. Northern range limits in the Tehachapi Mountains:

Lathyrus laetiflorus
Lupinus excubitus var. *austromontanus*
Arctostaphylos Parryana
Gilia leptantha ssp. *vivida*
Penstemon ternatus ssp. *septentrionalis*
Orobanche californica var. *Parishii*
Corethrogyne filaginifolia var. *Peirsonii*
Lasthenia glabrata ssp. *Coulteri*

Piute Mountains. Northern range limits in the Piute Mountains:

Cordylanthus Nevinii — *Chaenactis santolinoides*
Aster bernardinus — *Senecio ionophyllus*

Breckenridge Mountain. Northern range limits on Breckenridge Mountain:

Calochortus Palmeri — *Navarettia peninsularis*

South Fork and Kern Valleys. Northern range limits in South Fork and Kern valleys.

Selaginella asprella — *Grindelia camporum* var. *australis*

Greenhorn Range. Northern range limits in the Greenhorn Range:

Juncus macrophyllus — *Ranunculus canus* var. *ludovicianus*
Montia spathulata var. *viridis* — *Perideridia Pringlei*
Arenaria macradenia var. *arcuifolia* — *Gnaphalium beneolens*

San Joaquin Valley. Northern range limits in the San Joaquin Valley:

Aristida divaricata — *Baccharis Emoryi*
Eriochloa gracilis — *Pluchea sericea*

Temblor Range. Northern range limits in Kern County in the Temblor Range:

Montia spathulata — *Gilia jacens*
Dudleya lanceolata

RANGE LIMITS IN KERN COUNTY OF THE PLANTS OF THE MOJAVE DESERT[7]. Because the geological northwest boundary of the Mojave Desert geological province is the Garlock Fault at the south base of the El Paso Mountains, one would also expect a floral demarcation in the desert mountains aligned along that impressive fracture. Such is not the case; the northwestern limits of the Mojavean flora are in Inyo County's Owens Valley. However, a number of desert plants do reach their northern and northwestern limits in Kern County.

7. As most of the South Fork Valley west of Walker Pass nearly to Lake Isabella has a Mojavean rather than a cismontane flora, plants that reach their limits in that valley are included here.

A few reach these limits in the El Paso Mountains; these are:

Lepidium nitidum var. *Howellii*
Erodium texanum
Gilia aliquanta ssp. *aliquanta*
G. latiflora ssp. *elongata*
Phacelia pachyphylla

Places Other Than El Paso Mountains. Plants that reach their northern or northwestern limits at other places in the Mojave Desert in Kern County include:

Allium fimbriatum var. *denticulatum*
Canbya candida
Eschscholzia Parishii
Streptanthella longirostris var. *derelicta*
Nicolletia occidentalis
Perityle Emoryi

Kern County Part of Mojave Desert—Southern Limits. Similarly, there is no important southern floristic boundary on the Kern County part of the Mojave Desert. However, a few plants do reach their southern range limits in this region. These are:

Triglochin debilis
Euphorbia vallis-mortae
Eucnide urens
Cymopterus panamintensis
Gilia brecciarum ssp. *argusana*
G. cana ssp. *speciosa*
Phacelia Nashiana
Cryptantha decipiens
C. mohavensis
Mohavea breviflora

Kern County Part of Mojave Desert—Western Limits. Kern County shares the geographical western borders of the Mojave Desert with Los Angeles and Inyo counties. It would serve no floristic purpose to attempt to find the western limits of the plants that grow on the western part of the desert in these counties. However, there are a number that do have significant western limits in Kern County; several of these are plants of the deep desert that extend east through San Bernardino County.

The desert plants that have important western limits in Kern County are:

Cheilanthes viscida
Chorizanthe spinosa
Eriogonum Plumatella
Mirabilis Bigelovii var. *Bigelovii*
Eschscholzia glyptosperma
Astragalus didymocarpus var. *dispermus*
Lotus salsuginosus var. *brevivexillus*
Mentzelia tricuspis var. *brevicornuta*
Echinocactus polyancistrus
Gilia brecciarum ssp. *neglecta*
G. hutchinsifolia
G. latiflora ssp. *latiflora*
G. latiflora ssp. *excellens*
Coldenia plicata
Cryptantha angustifolia
Monardella exilis
Chrysothamnus paniculatus

CISMONTANE KERN COUNTY LIMITS OF DESERT PLANTS. A surprisingly small group of typical desert plants have their western limits west of the mountains in Kern County. These usually occur in isolated colonies, although a few, such as *Malacothrix glabrata* on the Antelope Plains, are fairly widespread.

A much larger group of desert species, mostly of those that occur from the high desert to the pinyon woodland, grow in the mountains around the head of the San Joaquin Valley, west to Santa Barbara and Ventura counties. Some extend to the La Panza Range and Carrisa Plains in San Luis Obispo County; a few to the north through the inner Coast Ranges as far as the Mt. Hamilton Range on the southeastern border of the San Francisco Bay region. Almost all of this group, which includes such plants as *Eurotia lanata, Stanleya pinnata, Camissonia Palmeri, Chaenactis Fremontii, Chrysothamnus nauseosus* ssp. *hololeucus, Eriophyllum Pringlei,* and *Hymenoclea salsola,* have their westernmost occurrence beyond the Kern County line.

The western limits for desert plants of cismontane occurrence in Kern County, with the locality of occurrence, follows:

Ephedra viridis. Central Temblor Range north to Cedar Canyon.

Allium Davisiae. East slope of the Greenhorn Range.

Calochortus striatus. Kern Hot Springs near Kernville; mouth of Chimney Creek Canyon.

Eriogonum maculatum. Cedar Canyon, Temblor Range.

Astragalus lentiginosus var. *variabilis.* From south of Bakersfield north to Seventh Standard Road near Shafter, scattered and scarce, possibly a recent adventive.

Lupinus horizontalis var. *platypetalus.* East slope of the Greenhorn Range.

L. pallidus. Kern River bed near Buena Vista Lake.

Croton californicus var. *mohavensis.* Sand Ridge, Caliente Canyon, Comanche Point.

Larrea divaricata. Point of Rocks, Sand Canyon near Poso Creek in the Poso Hills.

Opuntia basilaris. Kern Canyon, Caliente Canyon, near Woody, possibly north to Three Rivers in Tulare County.

Lomatium mohavense. Summit of Mt. Abel.

Cryptantha pterocarya. Messa Ridge, Temblor Range.

C. utahensis. Oak Flat, Greenhorn Range.

Lycium Andersonii var. *deserticola.* Poso Creek Gorge, Greenhorn Range.

Lycium Cooperi. Elk Hills, Belridge Plain near Gould Hill.

Aster pauciflorus. Scovern Hot Springs near Isabella.

Malacothrix glabrata. Belridge and Antelope Plains; Adobe Canyon in the Greenhorn foothills.

The Botanical Exploration of Kern County

Botanical collecting began in Kern County very late. It was not only long after the collections of such noted explorers as Chamisso, Douglas, and Nuttall had been made in the coastal region but even after San Francisco had its first resident botanists. Indeed, the coastal region and the Sacramento Valley were well settled when the first Kern County botanical collections were taken in 1853 in the beautiful valley where Ft. Tejon was soon to be located.

Even records of the vegetation before this time are scant. Father José Maria Zalvidea mentioned it in the most general sort of way in his journal while on an expedition to the interior to convert Indians in 1806. John Frémont, in his hurried trip out of California in April, 1844, wrote an excellent but brief description of the vegetation along his route through the arid hills from what is now Porterville to a point a short distance northeast of the present Bakersfield, then on to the desert by way of Caliente Canyon, Tehachapi, and Oak Creek Pass. His journal has the first recorded use of botanical names for Kern County plants. His comments on the widespread occurrence of the Mediterranean immigrant, *Erodium cicutarium,* are especially significant. Charles Preuss, Frémont's mapmaker, wrote in his journal on April 13, 1844 the first specific mention of a native Kern County plant when he described the Bakersfield cactus (which was to wait 52 years for a scientific description and name):

> "The hilly country is bleak, without any vegetation except a beautiful species of cactus whose magnificent red blossoms grace this sad, sandy desert in a strange manner." Gudde and Gudde (1958:123).

The first scientific collections were made in 1853 at Ft. Tejon by Colonel Leander Ransom, a member of the California Geological Survey and an early officer of the California Academy of Sciences. How many collections he may have made is not known, but two oaks he collected were described as new by one of the founders of the Academy and a pioneer California botanist, Dr. Albert Kellogg. The first, *Quercus Ransomi,* is now considered synonymous with *Q. Douglasii;* the second, *Q. acroglandis,* a synonym of *Q. agrifolia,* the California live oak. But Colonel Ransom may have had a keen eye for variations; Lebec is well known for hybrids among the native oaks. These types, as well as any other collections he may have made, were lost in the San Francisco fire of 1906.

Later in the same year, the botanical exploration of Kern County was launched more auspiciously with the collecting done by the United States Army's Pacific Railroad Survey of 1853, under the command of the remarkable Lt. R. S. Williamson. Lt. Williamson's careful description of the vegetation of the east side of the county, particularly that of the mountain passes, is lucid and detailed. It has great present relevance, as it shows that in the past 114 years there has been no substantial change in the vegetation of the mountains or the uncultivated lands of the valley, although the species composing that vegetation may have changed greatly.

The botanical collecting was in charge of Dr. Adolphus L. Heermann, surgeon-naturalist of the expedition. The party was in the county in August and September and the late summer season greatly reduced the material available, but extensive collections were made along Kern River to Buena Vista Lake, in the arid hills at the base camp near the mouth of Poso Creek, and at Walker Pass and in Kelso Canyon south of Weldon. Most of these base camp collections were simply labelled "Posé Creek"; the probable actual locality of the type for each new species is given in the appropriate place in the catalogue of plants. Collections were also made, in lesser numbers, at Tejon Pass and in the Lebec region. The beautiful engravings of many species that are included in the report of the botany of the expedition are particularly useful and interesting. Types collected by Heermann that are the basis of currently accepted taxa are: *Atriplex Serenana, Eriogonum roseum,* and *Euphorbia ocellata* from Poso Creek; *Eriogonum Plumatella* and *E. Heermannii,* credited to Poso Creek but probably from distant Walker Pass, and *Chrysothamnus nauseosus* ssp. *ceruminosus* and *C. teretifolius* from the mountains around Tejon Pass.

From May 18, 1857 until January 1859, the eccentric Hungarian naturalist, Xantus de Vesey, was stationed at Ft. Tejon, ostensibly as a hospital steward but actually getting the assignment at the instigation of Spencer F. Baird of the Smithsonian Institution for the purpose of making biological collections. However, the fort's commandants were not sympathetic with

Xantus' preoccupation with collecting and his duties as a hospital steward turned out to be more rigorous than he had anticipated. Xantus was apparently acting surgeon much of the time; in his correspondence he complained that the Post Surgeon spent most of his time "visiting the fleshpots of Los Angeles." (Madden, 1949: 60). Nevertheless, he made an impressive series of zoological collections within one mile of the fort. While he apparently made only 119 botanical collections at Ft. Tejon, eight currently recognized species are based on these. They are: *Chaenactis Xantiana, Chorizanthe perfoliata, C. Xanti, Clarkia Xantiana, Cryptantha microstachys, Asclepias californica, Solanum Xanti,* and *Penstemon laetus.* His findings (Gray, 1859–61) made known the biological diversity of the region and stimulated interest in the scientifically largely unexplored arid interior of California.

In 1863, Dr. George H. Horn, an entomologist, collected plants at Ft. Tejon and Ft. Independence in Inyo County for the distinguished Harvard botanist, Dr. Asa Gray. These were apparently made with scant notes and the locality of the collections is not always clear. Two species are based on types from these collections; these are *Stephanomeria cichoriacea,* which undoubtedly came from the canyon west of the fort, and *Astragalus Hornii.* While it is possible that *A. Hornii* may have grown around the nearby subalkaline Castac Lake, it is more likely that it was collected at one of the moist subalkaline places at the head of the San Joaquin Valley.

In 1875, Dr. J. T. Rothrock, the surgeon and botanist of the Pacific Railroad Survey under the command of Lt. George M. Wheeler, made a trip from Santa Barbara to Mt. Whitney. His collections, treated systemically by a number of botanists, including such eminent men as George Vasey, Sereno Watson, D. C. Eaton, William Boott, and other leading specialists of the day, were published in 1878. (Report on U. S. Geographical Surveys, Vol. VI. U. S. Printing Office. Washington. 1878.) The regions he collected included Mt. Pinos and Cuddy Valley, the Ft. Tejon region, Walker Basin, and along the South Fork of Kern River. Some of the South Fork collections attributed to Kern County almost certainly came from southeastern Tulare County; this would seem to be true of his only type collection attributed to Kern County, *Potentilla glandulosa* var. *nevadensis.* His collections are particularly valuable for the wealth of material from the Kern Plateau north to Whitney Meadows, especially from Olancha Peak and Monache Meadows.

Apparently while in Kern County, Rothrock met William L. Kennedy. In the 1850's and '60s, Kennedy was a well-known figure in the Kern mining regions. At various times he owned stores at Keysville, Bear Trap Flat (in Greenhorn Gulch), and Lynns Valley (Glennville). Further, he owned stamp mills near Keysville and at Havilah, and was considered "one of the two most enterprising business men in Kern County" (Boyd, 1852: 9, 18, 29, 44). In 1866, he was secretary of the group that promoted the formation of the county and in 1867 was an unsuccessful candidate for County Supervisor. At this point, he appears to have dropped from the historical record; nothing is known of his later life except the record of the botanical collections in 1876.

Efforts to discover whether he had a significant interest in botany or was merely doing a favor for his distinguished visitor of the previous year have been unsuccessful. It is perhaps most likely to have been the latter. At any rate, in 1876 he sent an impressive set of collections to Rothrock at the Smithsonian Institution, where they were added to those that Rothrock made the previous year. Although the majority were merely assigned to "the mountains of Kern County," some were attributed to Mt. Pinos and others to "the plains at Castac Valley." It is possible all were collected between Ft. Tejon and the summit of Mt. Pinos. Among the Kennedy collections were the types of the beautiful scarlet mariposa lily of the arid mountains and desert ranges, *Calochortus Kennedyi,* and the matted buckwheat of exposed rocky ridges, *Eriogonum Kennedyi.* While it cannot be said for certain where Kennedy collected these, a variety of the Kennedy buckwheat is common on the summit plateau of Mt. Pinos and the mariposa lily grows on the sandy slopes at lower elevations in the same region.

For eight years, Kern County collecting was quiescent. Then, in June, 1884, Katharine Layne Curran (later Brandegee) travelled from Bakersfield to Mojave; the trip must have been one of the most successful brief trips in California botanical history. On this trip, in addition to much other material, Mrs. Curran collected the types for no fewer than 14 currently accepted taxa!

This feat was even more remarkable when one considers that the month was June, not a particularly favorable one for collecting in the region, and that much collecting had been done for years in similar regions in counties to the south and the north. Further, almost all were collected close to the railroad right-of-way or in the vicinity of stations. Among them are such characteristic plants of the region as *Astragalus pachypus, Trichostema ovatum,* and *Purshia glandulosa.* Another was what was to be long considered one of the rarest of California plants; for 81 years hers was to be the only recorded collection of *Loeflingia pusilla.* The others include *Amsinckia intermedia* var. *californica, Bloomeria crocea* var. *montana, Cryptantha dumetorum, C. mohavensis,* the remarkable *Eriogonum gossypinum, Ranunculus canus* var. *ludovicianus, Tetradymia stenolepis, Gilia latiflora* ssp. *excellens, Mimulus androsaceus,* and *M. pictus.*

Mrs. Curran became Mrs. Townshend Stith Brandegee in San Diego on May 29, 1889. One is hard put to decide whether their wedding trip was one of botanical history's most romantic scientific or scientific romantic treks; they leisurely hiked to San Francisco, collecting in the coastal mountains as they went. Collections by the Brandegees attributed to Tehachapi Valley in May, 1889, are extant; whether these are correctly labelled is very doubtful. If so, the bridal pair added a long detour to their already lengthy trip by traveling from Saugus to Tehachapi and return. In any event, May cannot have been the correct month. (Cf. Ewan, 1942: 776.)

Mrs. Brandegee, Curator of Botany at the California Academy of Sciences from 1883 to 1894, was an indefatigable collector, often making lengthy trips to collect a single species. It was trips of this sort that probably led her back to Kern County. She collected at Tehachapi, Caliente, and Kramer in 1905. In 1909, she collected at what were apparently railroad stops from Bealville to Tehachapi; a collection of *Loeflingia squarrosa* from Keene indicates she had not forgotten the elusive *L. pusilla.* In 1910, Mrs. Brandegee collected between Caliente and Walker Pass; in 1911 she returned to the same region, collecting at Caliente, Hackberry Canyon above Caliente, Walker Basin, and from Havilah to Kernville (one suspects along the stage route of the time). Collections exist from Bissell Station (20 miles southeast of Mojave) taken on July 16, 1912, and from the Randsburg region in 1914.

Her husband, Townshend Stith Brandegee, a leading California botanist of his time who was long associated with the University of California, apparently first visited the county in 1891, if one discounts the detour on the honeymoon hike. In that year he botanized at Caliente, Walker Basin, Havilah, and Kernville from May 11 to May 16. A year later, he returned to much the same region, where he collected at Kernville, Kern Hot Springs, and Walker Basin from May 14 to May 16. His material that served for types includes collections of *Allocarya acanthocarpa* from Caliente, *Chaetopappa fragilis* and *Eschscholzia minutiflora* var. *rutaefolia* from Havilah and *Eschscholzia procera* from Kernville.

Both of the Brandegees were careless collectors; any records unique with them must be viewed with suspicion as localities and dates are often obviously wrong.

In 1888, the Greenhorn Range and Kern River were twice visited by the noted collector of the American west, Edward Palmer. Kernville was to be the base for a collecting trip to Mt. Whitney under the auspices of the United States Department of Agriculture. The departmental botanist, George Vasey, had commissioned Palmer for the exploration. On the first trip, Palmer was accompanied by William Greenwood Wright, the owner of a planing mill in San Bernardino and an important entomologist, being perhaps the leading authority of the time on California Lepidoptera. They travelled by train to Caliente and took the stage coach to Kernville, arriving on June 5 or 6 (the records are not clear which). Here they found the trip to Mt. Whitney had to be postponed; the higher mountains were still deep with snow. Apparently Palmer and Wright took advantage of the postponement to collect in the Greenhorn Range, where they made 97 collections, including the types of *Convolvulus simulans* and *Collinsia Wrightii.* Palmer, a rigidly righteous man, was appalled by the Kernville of that time; he wrote "Kernville is an abandoned mining town . . . It is a whiskey-blighted place; saloons occupy the two principal corners in this little hamlet, and on Sundays the streets are carpeted with playing cards and filled with gamblers" (McVaugh, 1956:96).

On this expedition 57 collections were made along Kern River; apparently all were from Kern County. One was the type of *Erigeron multiceps,* the only collection of this plant known with certainty.

In July, Palmer returned to Kernville where he engaged a guide and a pack train. The trip to Mt. Whitney was a fiasco; three days from Kernville Palmer realized his incompetent guide was hopelessly lost and he abandoned the journey at apparently what is now Long Meadow, northwest of Johnsondale in southern Tulare County. Before returning to Kernville he made 80 collections at the meadow. It cannot be determined with certainty that any plants were collected in Kern County on this second trip.

The tempo of early Kern collecting continued to quicken. In 1889, Professor Edward L. Greene of the University of California botanized extensively in the mountains around Tehachapi, apparently remaining in the region from June 22 to June 26. His trip was almost as fruitful as that of his severest botanical critic, Mrs. Curran, had been in 1884. He collected the type material for seven currently accepted taxa ("currently" is the important word; Greene was a prolific creator of new species, many of which have been considered of merit only by him). These are *Apocynum medium* var. *floribundum,* the beautiful mariposa of the Jeffrey pine forests, *Calochortus invenustus, Ceanothus vestitus, Convolvulus fulcratus* var. *deltoides, Rhamnus californica* ssp. *cuspidata, Senecio ionophyllus,* and *Malacothamnus orbiculatus.* While the specimens collected do not have specific localities given, all grow on the north slope of Tehachapi Peak between Water Canyon and the summit; it may be safe to assume that the types were collected there. Greene had visited the region briefly on September 9, 1881 with Dr. Charles C. Parry, a noted collector of the southwestern United States. Near Keene Station they collected the type of the tarweed widespread in the Kern County mountains, *Hemizonia Heermannii.*

Joseph Anton Purpus, the peripatetic professional western collector perhaps most important for his later collections in Mexico, visited the Piute Mountains in 1897 while en route to the Kern Plateau and the Mt. Whitney region. He was the first botanist to visit Piute Peak, where he collected the type of *Mimulus discolor,* now considered a synonym of *M. montioides.*

He also made the type collection of the very rare *Astragalus subvestitus*. Although attributed to Erskine Canyon along Erskine Creek, this interesting loco weed was presumably taken at Lookout Point, a northeastern spur of the main crest of Piute Mountain. Other significant collections were made along Erskine Creek. However, most of his 1897 collections were made on the Kern Plateau in Tulare County.

Miss Alice Eastwood joined the staff of the California Academy of Sciences in 1892 and later succeeded Katharine Brandegee as Curator of Botany. Her first extended collecting trip in her new position was in 1893 to the south end of the San Joaquin Valley, then considered "a remote part of California." She made extensive collections on the plains below the mouth of San Emigdio Canyon, in the canyon, on the broad plains at the head of the valley east to Rose Station near the Tejon Ranch headquarters, and about Bakersfield. Rugged San Emigdio Canyon furnished her with a wealth of material; this included the type of the oak still subject to botanical controversy, *Quercus Alvordiana.* A colorful account of the annual flora of the San Emigdio plains is preserved in Zoe (4:145). Old Rose Station, once a stop on the Butterfield Stage Route, was the home of the type of *Lessingia nemaclada* var. *albiflora.* While at Bakersfield, she collected the type of black loco (*Astragalus lentiginosus* var. *nigricalycis*), a widespread indicator plant of the San Joaquin Valley's annual grassland.

In October, 1894, she returned to San Emigdio Canyon to collect the autumn plants of the region; among them was the type of the rare *Haplopappus squarrosus* ssp. *obtusus.* Apparently while botanizing in colorful upper San Emigdio Canyon she met a miner's widow, Mrs. William Jasper, who was carrying on her late husband's mining project while proving a claim of her own. Little is known of Mrs. Jasper except that in 1896 she forwarded to the Academy from the remote and inaccessible head of San Emigdio Canyon the type material for *Heuchera caespitosa* and *Ribes ascendens* var. *Jasperae.* It is not known if Mrs. Jasper made further collections; if so, they were destroyed in the San Francisco fire of 1906.

In May, 1913, Alice Eastwood collected in the desolate Maricopa Hills, where she secured the type of spiny saltbush (*Atri-*

plex spinifera). She returned to collect in Kern County in 1914, 1917, and 1926. In 1930 John Thomas Howell joined the staff of the Academy; he accompanied Miss Eastwood to Kern County in 1935, 1937, 1938, and 1940. Accompanied by Dr. Isabel McCracken, an entomologist whose specialty was the solitary bees, Eastwood and Howell made their memorable collecting trip to southern Utah in 1941. They collected in Kern County en route in April and again in May on their return; during these visits, Alice Eastwood made her last Kern County collections.

Her long rich life spanned the time from the days when much of California was remote and botanically unknown to the days when almost all of the state was readily reached by automobile. In her later years when she visited the county she must have often remembered the pioneer trip in '93 when she stayed at robust Bakersfield's second hotel because she couldn't afford the good one, and took her meals in her room because the hostelry's combination saloon and lunch counter was no place for an unescorted lady.

Perhaps the currently most useful of the early Kern County collections are those of Frederick Vernon Coville and Frederick Funston, the leaders of the U. S. Department of Agriculture's Death Valley expedition of 1891. (Although Coville and Funston are usually cited as co-collectors, Funston was not with Coville at any time in the Kern County explorations.) Unlike those of most early expeditions, the collections were reported coherently and completely, with a detailed itinerary and a serial list of all numbers collected, together with the place and date of collection accurately and precisely given. In addition the main body of the reports has detailed ecological and geographical data for all of the places visited. (Coville, 1893.)

The expedition arrived in Kernville in June from Indian Wells Valley and Keeler on the return from Death Valley. While in Kern County, the party made its way from Kernville to Havilah, Walker Basin, and Caliente. From there, it continued through Tehachapi Valley and Tehachapi Canyon to Mojave; thence it proceeded along the western borders of the desert to the Liebre Ranch (in Los Angeles County near Gorman) and then to Ft. Tejon. The party remained in the region of the long-abandoned fort from June 28 to July 10. Local collecting trips were made in the region, including a trip to Tejon Pass at the head of Tejon Canyon.

On July 13 the expedition arrived in Bakersfield, where several days were spent. While there, Coville engaged in at least some local collecting, going as far north as Poso Station on the Glennville stage line and as far south as the alkali sinks at Weedpatch. In addition, numerous collections were made around the town.

In all, 168 collections were made in Kern County. In this diverse and carefully recorded collection were the types for the rare and perhaps even extinct endemic saltweed of the subalkaline plains south of Bakersfield, *Atriplex tularensis,* of *Navarrettia setiloba,* also a Kern County endemic, *Isomeris arborea* var. *globosa,* and *Euphorbia vallis-mortae.*

John Gill Lemmon, an indefatigable pioneer collector of California and Arizona, must have collected in nearby Kern County in the northern Temblor Range when he lived at Cholame in northeastern San Luis Obispo County in the 1880's. However, none of his collections from there can be specifically placed in Kern County. (Professor Lemmon is only one of many collectors whose enthusiasm in the field is greater than their precision with records.) He did collect at Keene in June and August of 1888 in the company of Charles C. Parry; here they collected the type of the Parry manzanita (*Arctostaphylos Parryana*). Mrs. A. E. Bush, apparently an enthusiastic collector of the deserts who lived in San Jose, collected the type of *Eriogonum gracillimum* "on the Mojave Desert in Kern County" in 1879. That Mrs. Bush had a continuing interest in botany is shown by her initiation into membership in the California Botanical Club in 1892.

Among the other 19th century collectors, the following made collections of special interest. William Russel Dudley, the professor of systematic botany at Stanford University for whom the University's herbarium is named, botanized in the vicinity of Bisses Station near Keene in 1895 (probably on the nearby northeast flank of Bear Mountain). Among his collections was the type for the very rare Tehachapian endemic, *Phacelia humilis* var. *Dudleyi.* On the same trip he also made collections at Mojave, Tehachapi, and Glennville. In 1896, Dudley, accompanied by Frank Haines Lamb, a student at Stanford about whom little is known, collected from June 4 to July 4 from Antelope Valley through the Liebre Mountains and the mountains of Santa Barbara and Ventura counties. Extensive Kern County collections were taken from Castac Lake and Cuddy Valley to the summit of Mt. Pinos and in upper San Emigdio Canyon. In the early 1870's George Davidson collected in the county, probably in the Mt. Pinos region; among his collections was the type of *Phacelia Davidsonii.* Cyrus G. Pringle made collections at Tejon Pass on May 13, 1882. Samuel Brannan, Jr., collected in the Oak Creek region between Mojave and Tehachapi on April 1, 1871; here he secured the type of *Lupinus densiflorus* var. *lacteus.* J. Burtt Davy, then a botanist at the University of California, collected extensively during most of May, 1896, at Bakersfield, Ft. Tejon, Rosamond, and the desert Antelope Valley. William Trelease, the Director of the Missouri Botanical Garden, collected the type of the Bakersfield cactus, *Opuntia Treleasei,* at Caliente in 1892.

After the turn of the century, the pioneering work for the botanical exploration of the county was completed. The growth of transportation and the construction of roads throughout the state made Kern County no longer a "remote part" of California. So almost every California botanist and many from elsewhere in the world have collected in the region with increasing frequency (although the routes followed show a certain monotony in their pattern; the result was that much of the county was not thoroughly explored until recently and it will still be many years before the county is really "completely known"). So, for the period following 1900 this resume will list only those whose work is of special significance. Dr. Harvey Monroe Hall, an authority on the California *Compositae,* collected extensively at Ft. Tejon and in the Mt. Pinos region from June 16 to June 19, 1905. Among his collections is the type of the extremely local *Eriophyllum lanatum* var. *Hallii,* collected in a canyon west of Ft. Tejon. His careful field notes, preserved at the University of California, are especially useful.

In April and May of 1905 a notable Kern County collecting trip was made by the well-known botanist Amos Arthur Heller of Los Gatos, later of Chico. His trip extended from Bakersfield to Mojave, and apparently as far west as the McKittrick Hills. His most noteworthy finds were in the Sunset District (near Arvin). These Sunset collections are of special importance as the extensive flower fields for which that region was noted have long been converted to farm lands. It was here that he collected the type of the rare *Lupinus horizontalis.* Other types collected at Sunset were *Eriastrum pluriflorum, Eriogonum viridescens,* and the rare *Phacelia vallis-mortae* var. *heliophila.* At Bakersfield Heller collected the type of the abundant and colorful sky lupine, *Lupinus nanus* var. *Menkerae.* Tehachapi yielded *Lupinus ruber* and Mojave what was to serve in the distant future as the type for *Lepidium nitidum* var. *Howellii.* Apparently on the same trip he made extensive collections in the Randsburg region.

Dr. Anstruther Davidson, a leading dermatologist and active amateur botanist and entomologist of Los Angeles, and Dr. Herman Hasse, a surgeon at the Soldier's Home in Sawtelle and an authority on lichens, on June 12, 1907, made one of the few collecting trips that has ever been made to Cummings Mountain; one of their important collections was material for the type of *Fritillaria pinetorum.*

Professor LeRoy Abrams, distinguished botanist at Stanford University and the author of the monumental "Illustrated Flora of the Pacific States," visited the county several times. In 1915 he travelled to the Piute Mountains to study the Piute cypress (*Cupressus nevadensis*) which he later described. (He had previously received collections from Leo Herzinger, apparently a resident of the Bodfish region; these probably provoked his interest in the tree.)

From 1930 until 1943, Dr. Carl B. Wolf, then botanist at the Rancho Santa Ana Botanic Garden, studied the Piute Cypress extensively at both the Bodfish and Back Canyon groves; his was the first detailed description of the latter grove. Wolf also collected extensively elsewhere in the county. On April 15, 1935, he collected the type of *Eschscholzia caespitosa* ssp. *kernensis* at Tejon Hills, together with many other valuable collections from that region and from Tejon Canyon. On another collecting trip, in April, 1937, he secured the type of the rare *Eremalche kernensis* on the plains near Belridge.

Ralph Hoffmann, Director of the Santa Barbara Museum of Natural History from 1926 until his untimely death in 1932, made numerous collecting trips to various regions in the county between 1927 and 1930. The Bodfish region, Caliente, and the

Mojave Desert are especially well represented in his collections. Of greatest value are his extensive collections from the Mt. Pinos region in Kern and Ventura counties; those taken in the summit area from June 17 to 19 in 1927 are an invaluable record of that mountain's flora before it was depleted by drouth, erosion of the cienegas, and recreational activity.

While Dr. Lyman Benson, now of Pomona College, was teacher of botany at Bakersfield Junior College from 1931 to 1938 he collected extensively in the county. His interests since teaching at Bakersfield have been diverse. He has collected widely in the western states; among his writings are two comprehensive texts on plant taxonomy and classification and, with Robert Darrow, a manual of the shrubs of the southwestern deserts. In addition, he is a specialist on the cactus family. But his interest in Kern County has held a special place in his busy and productive life. He has made numerous collecting trips to the county. Perhaps his locally most important contribution is that of awakening both the interest of the Nature Conservancy and the local residents to the need for preserving typical samples of the rapidly vanishing valley flora, particularly the annual flower fields.

In the late 1930's, Robert Francis Hoover collected extensively in the county as part of his research for his doctoral dissertation "Endemism in the flora of the Great Valley of California." His diligent field work resulted in an excellent collection of the valley flora; it is especially important as much of it is from vegetation since destroyed by the development of agriculture. The types that rest on his collections include *Atriplex vallicola, Brodiaea coronaria* var. *kernensis, Eriastrum Hooveri,* and *Festuca microstachys* var. *simulans.*

The California collections of John Thomas Howell, curator of Botany at the California Academy of Sciences, now number well over 40,000. This great collection has been enriched by his many trips to or through Kern County. His first trip in 1930 was unusually successful (even though it was cut short by illness at Glennville). With the veteran western collector, Frank Peirson, he collected extensively at Red Rock Canyon, Freeman Canyon, and numerous other localities west to Kernville and Glennville. This was an unusually good spring. The type of *Camissonia kernensis* was collected by them, while among their collections from

Red Rock Canyon and the Glennville region the only county records for several species are to be found. In 1931, Howell collected the type for the distinctive and extremely xerophytic *Allium Howellii* at the summit of Maricopa Grade. In the ensuing years he visited the county a number of times— in five different years with Alice Eastwood and several times alone. But his most significant trips were those of 1962, when he made systematic collections in the Greenhorn Range. Here he particularly concentrated on Black Mountain, at Shirley Meadows, and about Kernville, Wofford, and Isabella. In 1966, accompanied by Gordon H. True, formerly of the California Division of Fish and Game and now a Research Associate at the California Academy of Sciences, and by Tony Gasbarro, a United States Forest Service Ranger, he gathered the first Kern County collections from the rugged rim of the southern Kern Plateau at the head of Fay Canyon.

Mr. Lewis S. Rose, who has been a volunteer Research Associate in Botany at the California Academy of Sciences for many years, is a tireless and discriminating collector of the California flora. He has made numerous collecting trips to Kern County, sometimes with John Thomas Howell and Alice Eastwood or others, often alone. His beautifully prepared and carefully labelled collections are taken in large sets and widely distributed. Rose is easily the leading collector of Kern County material deposited in European herbaria and he ranks with Amos A. Heller as the botanist whose Kern collections are most widely distributed in the United States.

Several local botanists are of importance. Mr. E. Roy Weston collected in the county in 1925 and 1926 while living in Bakersfield. Weston, a draftsman with the Shell Oil Company, was an amateur photographer and enthusiastic hiker. He primarily collected his carefully prepared specimens as vouchers for his photographs of plants; these specimens were sent to Alice Eastwood at the California Academy of Sciences. His collections are particularly important as they yielded the types of four plants of narrow distribution. These are: *Calochortus coeruleus* var. *Westoni, Eriogonum latifolium* ssp. *Westoni, Fritillaria striata,* and the exceedingly rare *Layia leucopappa.* On his collecting trips he was frequently accompanied by two young geologists, G Dallas Hanna, and Alan J. Galloway. Galloway, after a career that led to an executive vice presidency of the Shell Oil Company, is now Chairman of the Board of Trustees of the California Academy of Sciences, and Hanna, long a geologist with the Associated Oil Company, is the Curator of the Geology Department. In addition to the collections made by Weston, many were made by Galloway, who also wrote articles about California wild flowers for various English journals.

Mrs. Allison Krames, a teacher at Bakersfield, collected extensively in the county for Willis Linn Jepson from 1934 to 1937. Desert plants are especially well represented in her collections; among them is the type for *Gilia cana* ssp. *speciosa* from Red Rock Canyon.

Mrs. Charlotte Nash Smith and her husband, Hugh, are cattle ranchers near Granite Station in the Greenhorn foothills. Both are members of pioneer families of the region. Either in their cattle operations or for recreation (with ranchers, it is sometimes hard to separate these two activities) the Smiths have ridden in virtually every region from lower Kern Canyon and South Fork Valley to Mt. Whitney. Mrs. Smith collected extensively in the southern Sierra Nevada and the Mojave Desert in the 1930's and 1940's for the late Willis Linn Jepson. She ceased active collecting in 1946 (to begin again in the early 1960's). Among all who have collected in Kern County, Mrs. Smith's collections are noteworthy for the precise and complete ecological notes that accompany her specimens. Her work has been commemorated by Dr. Jepson's dedication of *Phacelia Nashiana,* which friends have affectionately given the vernacular name "Charlotte's Phacelia," and by *Linanthus Nashianus.* Her discrimination and careful collecting can be measured by the frequency of the citation of her collections in the Jepson Flora and in this and other papers on the Kern flora.

Mrs. Freeland Farnsworth, the former Evalyn L. Klein, and her husband are ranchers at Glennville who are well known among California livestock people for their outstanding Ace Hi Hereford herd. Mrs. Farnsworth has collected in the Greenhorn Range since September, 1962, primarily on the ranches she and her husband own at Woody, White River, and Glennville; many of her collections are made while doing range riding for the family cattle operations. Her careful botanizing along the drainage of Cedar and Lumreau creeks is an excellent example of the value of intensive collection of a relatively small but critical area; her collections have yielded numerous Kern County and several Sierra Nevada records. Among them is the type material for the remarkable jewel flower *Streptanthus Farnsworthianus.*

Mrs. Noraine Meng, a secretary and keen amateur ornithologist of Inyokern (formerly of Glennville) has made more than 375 collections, largely in Indian Wells Valley and the desert mountains, the Greenhorn Range, and the Lake Isabella region.

Of the many botanists who have collected in the county, the collections of Dennis E. Breedlove, who was then a graduate student at Stanford University, from Piute Mountain in 1962 were of special value for the present project. These included the type for *Eriogonum Breedlovei* and *Streptanthus cordatus* var. *piutensis.* Other botanists who have supplied important Kern County duplicate material include Javier Peñalosa, who, as a student botanist at the California Academy of Sciences, made an interesting collecting trip to the El Paso Range and Indian Wells Canyon in 1960; Dr. Robert Thorne, taxonomist at the Rancho Santa Ana Botanic Garden; Mr. Clifton F. Smith, botanist at the Santa Barbara Museum of Natural History; Mr. Donald Myrick, a Santa Barbara business man closely associated with the Santa Barbara Botanic Garden; Dr. Thomas H. Fuller, botanist with the California State Department of Agriculture; and Mr. Henry M. Pollard, a teacher of Latin and a distinguished botanical collector of the Santa Barbara region.

Mr. Edward Pittman, while stationed with the U. S. Bureau of Land Management in Bakersfield in 1964 and 1965 sent collections from many remote and rarely visited localities in Kern and other counties. His collections from Mt. Owen and the Chimney Creek Gorge north of Onyx are especially useful.

Finally, much incidental collecting has been done by people with no particular interest in botany but who have been moved by innate intellectual curiosity or a generous spirit to supply material to the author. Many of these were of a single plant that the author wanted from a region inaccessible to him; some are plants of economic importance; others those that especially caught the collector's eye; all are of importance in the study of the Kern County flora. These locally collected specimens have been deposited in the herbarium of the California Academy of Sciences. Lest time bring the collectors the fate of botanical anonymity that befell William L. Kennedy and Mrs. William Jasper, it is appropriate that they be identified here.

Mr. Roy V. Parker, Kern County Farm Advisor of the University of California Extension Service and an authority on the production of beef cattle, has a keen eye for plants of interest. Mr. Eben McMillan, an accomplished naturalist, photographer, and rancher of Cholame, has contributed collections from various regions in the county. While a scientist at the China Lake Naval Ordnance Test Station, Mr. Frank J. Kirby collected desert plants for identification in connection with his avocation of plant photography. Mrs. Clare B. Hardham, a botanist and rancher at Paso Robles, made extensive collections in the mountains about Kernville and at other places in the county in 1957 and has since visited the county frequently.

While not numerous, of special value are the collections of Mr. Donald J. Bedell, who from 1958 until 1966 was a ranch manager in the Temblor Range. He is unusually perceptive of the relationship of the flora, the native fauna, and livestock. His collections include the first collection of *Howelliella* from the San Joaquin Valley, of *Photinia arbutifolia* from Kern County, and the third California station for the fungus *Gymnosporangium exiguum.*

Others who have contributed few (or even only one) collections which are important in completing the county's floral record are: Mr. Waldo Abbott and the late E. Z. Rett, naturalists at the Santa Barbara Museum of Natural History; Mr. Dick Rea, a U. S. Forest Service ranger at Kernville; Mr. Robert Luthey, a talented amateur photographer and employee of the Kern County Road Department at Kernville; Messrs. Carl F. and T. Franklin and Miss Dorothy Twisselmann, ranchers in the Temblor Range; and Mrs. Hazel Record, a Granite Station rancher.

My own serious interest in botany began in 1952 when nitrate poisoning from the lush growth of that wet year caused serious cattle losses in the Temblor Range. An attempt to interpret the strange and bewildering botanical names of plants known to concentrate lethal nitrate compounds led indirectly to a further interest in botany and taxonomy. In the summer of 1953, Eben

McMillan and Clifton Smith encouraged me to seek out John Thomas Howell of the California Academy of Sciences for advice and guidance. This led to my close botanical association with Mr. Howell and the Academy.

What started out in 1952 as an attempt to compile a simple list of the plants on the Twisselmann family ranches in the Temblor Range culminated in the completion of a formal flora of that range (Twisselmann, 1956). During the study, I became interested in the floras of arid regions. As a third generation resident of the hills along the San Luis Obispo-Kern County line with a deep affection for the Kern County region and its people, it was only natural for me to see if the county might not be a challenging yet rewarding project for florisitic study in the field, herbarium and library. This it has been.

Beginning with collection No. 1, *Juniperus californica* from Cedar Canyon in May, 1952, and ending with No. 12,901, *Alnus rhombifolia* from Kernville in February, 1967, roughly 9,500 collections have been made in Kern County and 3,500 in neighboring counties. Almost all have been deposited in the California Academy of Sciences. In addition, 12,000 duplicates have been taken for distribution to other herbaria.

Although collegiate training in political science may not be an ideal background for an avocational interest in botany in later life, lack of scientific training has not been an insurmountable handicap. I hope my experience will inspire others in the region to seek a similar interest in some branch of local zoology, botany, or geology that will give them rich personal rewards and at the same time have results that may be worthwhile contributions to science.

Thus, the work begun when Colonel Ransom snipped branchlets from the oaks at Ft. Tejon in 1853 is continuing. Perhaps the time has arrived when California botany has reached such a level of maturity that future projects should consist of intensive local field work to complete and verify the records established by the pioneering botanists. Those who seek the novel or sensational may be frustrated, but the California flora and vegetation will yield rich rewards for those who are satisfied to give order, coherence, and detail to that which is already known. One of the primary goals of this paper is to point to local areas within

Kern County that will be rewarding to the person who studies them more completely and intensively than is possible even in a county flora.

A NOTE ON THE ILLUSTRATIONS

Eben and Gladys McMillan have followed the progress of this project closely since its inception and many of my concepts of the flora and vegetation have developed during discussions with them. Thus, their photographs are interpretive in far more than the usual way. They planned their pictures after carefully reading the manuscript. Aesthetic considerations have been subordinated to scientific objectivity, and each picture has been chosen to serve a specific purpose in achieving their goal of a comprehensive and integrated presentation of the Kern County landscape and its flora. In every sense, the illustrations should be considered an amplification of the text; the author and illustrators hope the reader correlates the pictures and the narrative in order to have the best understanding of the plants and the places where they grow.

NOTES ON THE CATALOGUE OF PLANTS

Field work for the project began in the spring of 1956 and was completed in the autumn of 1966; collecting in the Temblor Range on the western border of the county began in 1952. The collections on which the catalogue is based are deposited in the herbarium of the California Academy of Sciences. Representative collections have been contributed to the herbaria of the Santa Barbara and Rancho Santa Ana botanical gardens, and, in the families from *Scrophulariaceae* through *Compositae,* to the Jepson Herbarium of the University of California. A representative set of grass collections has been deposited in the Agronomy Herbarium of the University of California at Davis and of the weeds and introduced plants in the Botany Department Herbarium of the same institution. Selected specimens of special groups have been distributed to numerous other institutions.

The county line has been rigidly followed as a boundary except for the Jeffrey pine forest at the summit of Mt. Pinos. Here the political line was too difficult to follow with any precision and in that region is floristically unimportant.

While the botanical exploration of the county is neither complete nor exhaustive, almost all areas have been visited at least once and most in different seasons in several years. Still, further collecting will yield results rich in records; one of my goals is to suggest various regions for intensive collecting; the Tehachapi Mountains or the Greenhorn Range should be especially interesting if collected intensively for either a floristic or ecological study.

While an attempt has been made to collect in all regions, inevitably some have received more attention than others. I have never been to the higher elevations of rugged Mt. Owen, north of Walker Pass; the record for that isolated peak is based on the collections of Edward Pittman. Similarly, the extreme southern portion of the Kern Plateau's Dome Lands has been collected only scantily, mainly by Mr. Pittman. Because of obvious collecting problems, the garden weeds of cities and towns have not been thoroughly collected.

The place names used are those on the Kern County map of the Automobile Club of Southern California; this organization's maps for southern California counties are almost a necessity for the field botanist. For names not on that map, local usage is followed. The context should aid the reader in avoiding confusion for three important duplicated place names: the Black Mountain of the Greenhorn Range and the high basaltic summit of the desert El Paso Range; the desert Antelope Valley along the Kern and Los Angeles county line and the one on the west side of the San Joaquin Valley (recently and possibly not successfully re-named "Berenda Mesa," an old Mexican name for the region); and El Paso Creek, which is not in the desert range of the same name but is the western drainage for the south central Tehachapi Mountains. Finally, the term "Tejon Pass" is used for the historic crossing from the head of Tejon Creek to the Mojave Desert; modern usage erroneously but almost universally applies the name to the Holland Summit between Lebec and Gorman.

In the notes on the ranges of the plants, the word "mountains" includes all the mountains from the Sierra Nevada east and west of Kernville south through the Piute Mountains, Breckenridge

Mountain, the Tehachapi Mountains, and the Mt. Pinos region. It does *not* include the Temblor Range.

As I anticipate that this catalogue will be used in conjunction with the Munz and Keck flora of California, keys have been omitted. Putting check marks after the Kern County species in the Munz and Keck keys will facilitate the quick and easy use of this work.

The taxonomic treatment is neither conservative nor liberal. Using the Munz and Keck California flora as a primary reference, each problem has been solved as it arose with the treatment that seemed best to portray the status of the plant as it grows. Synonyms are given only when this work departs from the treatment in that flora, in the Jepson manual, or occasionally when the plant is widely known by a name other than those given in either of these works.

Unless otherwise noted by standard herbarium symbols, all collections cited are deposited in the herbarium of the California Academy of Sciences. For my collections, the abbreviation "T" is used to save space.

ACKNOWLEDGMENTS

An early Kern County author once wrote "I suppose of all the people who are concerned with making a true book, the one who puts it to the pen has the least to do with it." This is peculiarly true of the compiler of a local flora; whatever his pretensions may be, the author is in a large sense only an assembler, annotator, and editor of information gathered or developed by others.

This is especially so for a non-professional botanist attempting a work that would be well beyond his competence were it not for the generous aid and assistance of many people in many fields. To list them perhaps will be tedious; not to do so would be boorish. I choose to run the risk of being tedious.

In my flora of the Temblor Range, I acknowledged my great debt to John Thomas Howell. Now, eleven years later, I can only add that in all respects it has greatly increased. Whatever merit this work may have is largely a result of his interest, guidance, and counsel. Among his many impressive contributions is the critical review of the determinations of more than 14,000 collections from this and neighboring regions. His careful reading of the manuscript has been vital to its successful completion.

It is pleasant to write that two others who received special mention eleven years ago have continued to enrich my knowledge of the natural scene in that long interval. Mr. Eben McMillan, who perhaps understands the essence of natural history in all its aspects better than anyone I know, has been a constant source of inspiration, information, and advice; quite apart from his great field knoweldge of flora, fauna, and man in Kern County, his frequent company on collecting trips turned what might have been dreary days into adventures. Mr. Clifton Smith has contributed in many ways, both large and small, to the completion of this work. He has generously shared his commanding knowledge of the flora of Santa Barbara and Ventura counties, and has been especially helpful with information on the plants of the Mt. Pinos region, Cuyama Valley, and the northern Ventura County mountains. Much of the information on the plants of these regions, not always acknowledged hereafter in the text, has been contributed by him.

In addition to Mr. Howell's review and counsel on all the collections from the county, I am indebted to many other botanists. Among them are Dr. Rimo Bacigalupi and Dr. Lawrence Heckard of the Jepson Herbarium of the University of California. Not only have they critically reviewed large numbers of collections from the county in the families that will be represented in Volume IV of the Jepson Flora but they have supplied much other information available only from the collections and records at the University. Dr. Bacigalupi has been especially helpful in clearing some of the obscurity from the botanical history of the county. Among his many contributions are his critical opinions of troublesome collections in the *Scrophulariaceae,* especially in *Mimulus, Collinsia, Cordylanthus,* and *Castilleja.* Dr. Heckard has been helpful with *Orobanche,* the perennial species of *Phacelia,* and miscellaneous other genera.

Dr. Peter H. Raven critically read the manuscript for the genus *Camissonia,* annotated the collections in that genus, and reviewed most of the others in the family *Onagraceae;* he also generously supplied much other Kern County information from the Dudley Herbarium of Stanford University.

Dr. Verne Grant has furnished or verified determinations for the complex and difficult genus *Gilia* and has guided me to an understanding of the Kern County representatives of this remarkable group of plants.

Among others, Dr. Wallace R. Ernst reviewed the collections in the family *Papaveraceae* and has been especially helpful with the genus *Eschscholzia;* he has been a generous source for detailed information on the poppies of California as well as other floristic concerns. Dr. Robert Ornduff reviewed the collections in the genus *Lasthenia,* Mr. Rupert Barneby in the genera *Astragalus* and *Dalea,* and Dr. Herbert A. Wahl in the genus *Chenopodium.* Dr. John Tucker has been helpful with the genus *Quercus,* Mrs. Lauramay Dempster with the genera *Galium* and *Plectritis,* and Mr. Beecher Crampton with problems in the family *Gramineae.* The late G. Thomas Robbins furnished lucid explanations for both the specific and subspecific distinctions in the genus *Chrysothamnus* and was also helpful with *Nemacladus.* Dr. Philip A. Munz kindly supplied the sources for various Kern County references in his "A California Flora." Dr. Robert F. Hoover has aided in solving numerous problems involving plants that occur along the border of San Luis Obispo and Kern counties. Mr. Edwin R. Chandler supplied valuable information on a number of plants, particularly in the *Boraginaceae.* Mr. James L. Reveal has been helpful with the genus *Eriogonum.* Dr. Thomas Fuller supplied useful information on various weeds of the county.

Collections made especially for this work, sometimes by helpful people with only a cursory interest in botany, have added greatly to its geographical coverage.

Mrs. Evalyn L. Farnsworth has contributed nearly 800 collections, mostly from the ranches she and her husband, Freeland, own at Glennville, Woody, and White River; several Kern County records rest solely on her collections. Mr. and Mrs. Farnsworth have been helpful in many, many other ways.

Mrs. Charlotte Nash Smith, a member of a pioneer Woody family, collected extensively in the Greenhorn Range and on the Mojave Desert for Willis Linn Jepson. In recent years, happily for both California botany and for Mrs. Smith, she has resumed collecting. In co-operation with Dr. Rimo Bacigalupi, she has furnished me lists of all her more than 1,400 collections and duplicates of many of those of special significance. She and her husband Hugh (whose family first settled in the region in 1853) have an encyclopedic knowledge of the natural and human history of the Greenhorn Range; it has been both a privilege and a joy to share in it.

Mr. Roy V. Parker, Kern County Farm Advisor and an authority on the production of beef cattle, has furnished numerous important collections and much field information, particularly on plants of forage importance. The active interest of Mr. Parker has been indispensable, especially for facilitating the field work; his co-operation has grealty aided in the solution of many and varied problems.

Mr. Donald J. Bedell has been helpful in many ways. Of special value are his collections from places difficult of access in the Temblor Range and his careful observations of the grazing habits and preferences of livestock, especially of the less common plants, a subject for which first hand information is meager. Without his co-operation and that of my sister, Miss Dorothy Twisselmann, time for much of the field work would not have been available.

Mrs. Noraine Meng forwarded more than 375 collections from the Indian Wells Valley region, Lake Isabella, the Greenhorn Range, and other areas in Kern and nearby portions of Tulare and Inyo counties. Among others who made available important collections are Mrs. Clare B. Hardham, Mr. Frank J. Kirby, Mr. Javier Peñalosa, Mr. Donald Myrick, Dr. Thomas Fuller, Mr. Dick Rea, Mr. and Mrs. Robert Luthey, Mr. T. Franklin Twisselmann, Mrs. Hazel Record, Mr. Clifton Smith, Mr. Waldo Abbott, Mr. John Thomas Howell, Dr. Robert Thorne, Mr. Eben McMillan, Mr. Dennis E. Breedlove, Mr. Jack Zaninovich, and Mr. Roger Meng.

Mr. Edward Pittman, whose duties while he was stationed at the Bakersfield office of the U. S. Bureau of Land Management took him into some of the most remote areas of Kern and neighboring counties, furnished valuable collections, plant lists, field notes, and distribution maps from rarely visited places such as Mt. Owen and the adjacent western canyons, Chimney Creek

Gorge, and the Kern Plateau southwest of Lamont Meadow. He also compiled extensive information on the plants of the Piute cypress groves on Piute and Breckenridge mountains and in Back Canyon.

Mr. Harold Thurber, Livestock Manager of the Tejon Ranch, has been of great help in making it possible to explore that historic ranching empire; he and Mr. Parker have been most instructive and congenial companions in field work in that region.

In a time when malicious trespass and vandalism are almost an acceptable form of outdoor recreation, locked gates are essential and a suspicion of even well-intentioned strangers is a natural attitude of rural people. So I am doubly grateful for the openhanded trust and generosity of the many landowners who freely gave me access to their property, and whose friendships have been one of the quite unexpected dividends of the field work. The list is far too lengthy to give completely, but common courtesy demands that in addition to those already listed, I recognize the special courtesies of Henry and Helen Bowen; Mr. Justinian Caire, Mr. Walter Slater, and the Kern County Land Company; Mr. Louis Rochford and the Tejon Ranch Company; Mr. and Mrs. Clifford Record, Mrs. Homer A. Hansen, Sr., Mr. Homer A. Hansen, Jr., Mr. John S. Broome, Mr. and Mrs. Ernest Still, and the various members of my own family who own ranches in the western San Joaquin Valley and the Temblor Range. For all the other good Kern County ranchers who gave me access to their property, I have equal gratitude.

The services of government agencies, I suppose, should be taken for granted, but the aid received from the most was beyond a routine response to a citizen's legitimate request. I am especially grateful to Airman Douglas L. Guerette and the Public Relations Office of Edwards Air Force Base; to Mr. James R. S. Toland, former Greenhorn District Ranger for the U. S. Forest Service; to the rangers of the Kernville Ranger Station, especially Mr. George Coombes and Mr. Tony Gasbarro; to Mr. Darrell Fullwilder and Mr. Edward Pittman of the U. S. Bureau of Land Management; and to Mr. Leon Snyder, manager of the Kern National Wild Life Refuge.

Mr. Thomas W. Dibblee, Jr., who has a profound knowledge of the geology of California, critically read the manuscript for the geology of the county. I greatly appreciate his help in my attempt to present this complex subject briefly yet accurately.

Mr. Richard Bailey, Director of the Kern County Museum, has been very helpful in clarifying some of the early history of the county that has botanical ramifications. Dr. John H. Thomas, Curator of the Dudley Herbarium of Stanford University, supplied the information concerning the Kern County collections of William Russel Dudley and Frank Haines Lamb.

The still and motion picture photography of Eben and Gladys McMillan is an unusual demonstration of art in the service of science; the contribution their work has made to this flora is obvious, so is my debt to them.

In addition to its scientific and economic aspects, the flora has a special dimension to those who live with and by it. My friendships with many old-time and some modern cowboys and sheepherders have supplied knowledge not attainable from any other source. Whatever the merits of progress in the industrialization of agriculture, it is sad to realize that theirs is a breed that is rapidly disappearing and that "we shall not see their likes again."

The friendships of my brother, Carl F. Twisselmann, with the Kern County livestock people have made access easy to private property throughout the county. His keen interest in the success of this project or anything else concerned with rural Kern County has facilitated its completion in many other ways. My sister, Dorothy, has patiently endured a great deal in deference to botanical research and study. Finally, beyond usual filial gratitude, I owe my mother, Mrs. Eleanora M. Twisselmann, a great debt; her help, of the most practical sort, made this project possible.

LITERATURE CITED

Anonymous

1883. History of Kern County, California. Wallace W. Elliott & Co., San Francisco. 226 pp.

Axelrod, Daniel L.

1958. Evolution of the Madro-Tertiary geoflora. Botanical Review, 24:433–509.

Bowerman, Mary L.

1944. The flowering plants and ferns of Mt. Diablo, California. The Gillick Press, Berkeley. 290 pp.

Boyd, Harlan
1952. Land of Havilah. Kern County Historical Society, Bakersfield. 54 pp.

Coville, Frederick Vernon
1893. Botany of the Death Valley Expedition. Contributions from the United States National Herbarium, vol. 4, U. S. Government Printing Office, Washington. 363 pp.

Elford, C. Robert
1964. The climate of Kern County. Kern County Board of Trade, Bakersfield. 73 pp.

Ewan, Joseph
1942. A bibliographical guide to the Brandegee botanical collections. American Midland Naturalist, 27:772–789.

Gray, Asa
1859–61. List of a collection of dried plants made by L. J. Xantus, at Ft. Tejon and vicinity, California. Proceedings of the Boston Society of Natural History, 7:145–149.

Gudde, Erwin G., and Elisabeth K. Gudde, trs. and eds.
1958. Exploring with Frémont, by Charles Preuss. University of Oklahoma Press, Norman. 162 pp.

Howell, John Thomas
1949. Marin Flora. University of California Press, Berkeley. 322 pp.
1957. The California flora and its province. Leaflets of Western Botany, 8:133–138.

Madden, Henry Miller
1949. Xantus, Hungarian naturalist in the pioneer West. Books of the West, Palo Alto. 312 pp.

McVaugh, Rogers
1956. Edward Palmer, plant explorer of the American West. University of Oklahoma Press, Norman. 430 pp.

Smith, Clifton F.
1952. A flora of Santa Barbara. Santa Barbara Botanical Garden, Santa Barbara. 100 pp.
1964. The vegetation of the passes. Noticias of the Santa Barbara Historical Society, 10:3–4.

Smith, Gladys L., and Anita M. Noldeke
1960. Statistical report on A California Flora. Leaflets of Western Botany, 9:117–123.

Thomas, John Hunter
1961. Flora of the Santa Cruz Mountains of California. Stanford University Press, Stanford. 434 pp.

Troxel, Bennie W., and Paul K. Morton
1962. Mines and mineral resources of Kern County, California. California Division of Mines and Geology. San Francisco. 370 pp.

TWISSELMANN, ERNEST C.
1956. A flora of the Temblor Range and the neighboring part of the San Joaquin Valley. Wasmann Journal of Biology, 14: 161–300.

WILLIAMSON, R. S.
1856. Reports of explorations and surveys. Beverly Tucker, Printer, Washington. 310 pp.

APPENDIX—THE CATALOGUE OF PLANTS

OPHIOGLOSSACEAE. Adder's-tongue Family

Botrychium multifidum (Gmelin) Ruprecht ssp. **silaifolium** (Presl) Clausen. California grape-fern. Once collected but apparently now extinct at Tiger Flat in the northern Greenhorn Range (Benson 7627, POM). (*B. silaifolium.*)

Botrychium simplex E. Hitchcock. Grape-fern. Local in a dense mat of *Eleocharis bernardinus* in a cienega at 8500 feet elevation on Mt. Pinos (Hoffmann, 18 July 1927, SBM). Grape-fern has since been sought at this station by several botanists but apparently has never been re-collected. It is possibly a victim of the severe drouths of the past forty years.

POLYPODIACEAE. Fern Family

Athyrium Filix-femina (Linnaeus) Roth var. **californicum** Butters. Western lady fern. A colony grows under rotting logs in deep shade in the red fir forest on the east slope of Sunday Peak (T6394) and another at a small meadow on the northwest slope (C. N. Smith 1011, JEPS).

Cheilanthes californica (Hooker) Nuttall. California lip fern. Rare in cracks and fissures in the granite walls of Cedar Creek gorge (T11,674, Farnsworth 551, 592) and in granite along Cedar Creek at the mouth of Fulton Creek (Farnsworth 18) in the Douglas oak woodland in the Greenhorn Range.

Cheilanthes Covillei Maxon. Coville lip fern. A common fern of rocky places below the yellow pine forest in the mountains; uncommon in the Temblor Range. It is scarce around granite blocks on the rim of the Kern Plateau in the Bartolas Country at more than 7,000 feet elevation.

Cheilanthes intertexta (Maxon) Maxon. Coastal lip fern. Scarce along the Cedar Creek drainage on the west slope of the Greenhorn Range, from the yellow pine forest at Cedar Creek Public Camp (T9784) to the lower levels of the Douglas oak woodland at the gorge below the Rock Corral (Farnsworth 499). (*C. Covillei* var. *intertexta.*)

Cheilanthes siliquosa Maxon. Cliff-brake. Rare in the yellow pine forest at Tiger Flat in the northern Greenhorn Range (Benson 12,495, POM). (*Onychium densum.*)

Cheilanthes viscida Davenport. A fern of the deserts that is scarce in shaded cracks that are moist in winter in the large metamorphic outcrop on the east side of the mouth of Last Chance Canyon in the El Paso Range (T11,971).

Crystopteris fragilis (Linnaeus) Bernhardi. Bladder fern, brittle fern. Occasional in moist shaded places, mostly in the yellow pine forest, but sometimes in the chaparral, in the Greenhorn Range and on Breckenridge Mountain; rare, south to El Paso Canyon in the Tehachapi Mountains.

Dryopteris arguta (Kaulfuss) Watt. Wood fern. Occasional, in shaded usually moist places, from the shin oak brush to the yellow pine forest in the Greenhorn Range and on Breckenridge Mountain; uncommon in similar places in the Tehachapi Mountains.

Notholaena Jonesii Maxon. Very rare and local on the limestone cliffs in the arid Douglas oak woodland in Erskine Canyon in the Piute Mountains (T8445).

Pellaea andromedaefolia (Kaulfuss) Fée. Coffee fern. A common and widespread fern, occurring in all the Upper Sonoran associations and reaching the desert's edge at Alphie Spring in Jawbone Canyon. In the southwest, sheep often get perforated intestines from grazing this fern; there it has the colloquial name of "stick gut". It is not common enough to be a livestock hazard in Kern County.

Pellaea mucronata (D. C. Eaton) D. C. Eaton. Bird's-foot-fern. Common in the Greenhorn Range, scarce but widespread in all the other cismontane Upper Sonoran associations, usually growing in rocky places.

Pityrogramma triangularis (Kaulfuss) Maxon. Goldback fern. The county's most widespread fern, goldback occurs in all the woody Upper Sonoran associations, usually growing around rocks. It reaches the desert's edge at Alphie Spring in Jawbone Canyon. (*Gymnogramme triangularis.*)

Pityrogramma triangularis var. **pallida** Weatherby. Silverback fern. A rare fern that grows in a large colony in deep protected recesses in the granite outcrops southwest of Grizzly Gulch on the Rag Gulch-White River Road in the Woody region (Farnsworth 92–1, 17, T9490) and is rare under a ledge of brown slate along Cedar Creek just below the mouth of Fulton Creek (Farnsworth 468) and under the large granite outcrops above the Rock Corral (Farnsworth 657). Its ample fronds with dense dusty white cover as well as its preference for protected heavily shaded places readily set it apart from the typical variety. (*P. pallida.*)

Polypodium californicum Kaulfuss. California polypody. A dense colony grows on a bank of heavily shaded moist metamorphic rock in El Paso Canyon at 3,500 feet elevation (T6162) and another in Tejon Canyon at 2,750 feet (T10,550) in the Tehachapi Mouutains. It is not known in the Sierra Nevada south of Tule River. (*P. vulgare* var. *occidentale.*)

Pteridium aquilinum (Linnaeus) Kuhn var. **pubescens** Underwood. Western bracken. Occasional, usually in dense colonies, in moist sunny places in the mountains south to the head of Catskin Creek in the Tehachapi Mountains. In other regions, bracken is poisonous to livestock; it is not considered a hazard in Kern County. (*Pteris aquilina* var. *lanuginosa.*)

Woodwardia fimbriata Smith. Western chain fern. This fern that is often so abundant in the Coast Ranges is known in Kern County only from three robust plants that grow at a spring on the south slope of Poso Creek, about one half mile above Poso Mine (C. N. Smith 1092, T8266). (*W. radicans.*)

SALVINIACEAE. Water Fern Family

Azolla filiculoides Lamarck. Duckweed fern, mosquito fern. A tiny plant that is occasional in quiet ponds and streams in the valley and is less common to the yellow pine forest. Hundreds of individual plants usually form dense masses that look like greenish scum that becomes reddish in the fall.

MARSILEACEAE. Clover Fern Family

Marsilea vestita Hooker & Greville. Clover-fern. Common in ponds, quiet streams, canals, and other wet places in the valley; a very common rice field weed. It occurs along Kern River above Lake Isabella. Early in succession, clover-fern was one of the first aquatics to become common when the alkali flats were flooded to form artificial marshes at the Kern National Wildlife Refuge in 1964.

EQUISETACEAE. Horsetail Family

Equisetum arvense Linnaeus. Common horsetail. Occasional in wet or damp places in the mountains and along Kern River to Bakersfield.

Equisetum × Ferrissii Clute. Scouring rush. Common on shady banks along Kern River at Hart Memorial Park and scarce along Kern River in Kern Valley and in Kern Canyon. The plants in Kern Valley are heavily grazed by deer. The plant is a common hybrid in California between *E. hyemale* and *E. laevigatum.*

Equisetum hyemale Linnaeus var. **robustum** (A. Braun) A. A. Eaton. Giant scouring rush. Scarce along Kern River in Kern Valley.

Equisetum laevigatum A. Braun. Braun's scouring rush. Our most common scouring rush, occurring in wet and moist places throughout the mountains and occasionally around Bakersfield. (*A. Funstonii.*)

SELAGINELLACEAE. Moss Fern Family

Selaginella asprella Maxon. Moss fern. Locally common around metamorphic outcrops in the arid Douglas oak association in the vicinity of River Kern (Kern Canyon above Kernville), east to the arid Jeffrey pine forest at the southern rim of the Kern Plateau at the head of Fay Canyon.

Selaginella Bigelovii Underwood. Bigelow moss fern. Common, sometimes locally abundant, around rock outcrops through the Douglas oak woodland in the mountains. It grows densely around the granite outcrops in lower Kern Canyon, where it is an attractive spring-time feature of the canyonsides.

ISOETACEAE. Quillwort Family

Isoetes Nuttallii A. Braun. Quillwort. Rare in Wagy Creek at the Wagy Flat Road on the east slope of the Greenhorn Range (Hardham 3662).

PINACEAE. Pine Family

Abies concolor (Gordon & Glendinning) Lindley. White fir. Common in the yellow pine forests. In burned or cleared areas, it is often the first of the montane conifers to reappear. It is apparently more drouth resistant than the others; in the Tehachapi Mountains in canyon bottoms or on high ridges swept by the hot dry desert winds, white fir grows in pure stands in these places that are too xeric for any other forest conifer. These trees are often stunted and picturesque. In deep canyon bottoms or on protected north slopes in the same region, isolated colonies or even lone individuals grow well below their normal range. The tree grows rapidly; the young trees with their gray to white bark are of exceptional beauty. White fir is of decidedly secondary value for lumber.

Abies magnifica Murray var. **shastensis** Lemmon. Shasta red fir. A variety of the dominant tree of the Sierran red fir forest, it reaches its southern limits on the north and east slopes of Sunday Peak, just within Kern County. This small sample of the extensive red fir forests of the Sierra Nevada, with its numerous openings covered with pine mat manzanita, chinquapin, with thickets of bitter cherry (*Prunus emarginata*) in moist places, and with light gray granite outcrops for accent, is perhaps the pleasantest place in Kern County in the hot dry summer months.

Pinus flexilis James. Limber pine. A common and beautiful timberline tree that descends a short distance into the Jeffrey pine forest on Mt. Pinos and Sawmill Mountain. The isolated occurrence of the tree here (and at nearby Frazier Mountain in Ventura County) is of interest; the nearest colonies are near Sirretta Peak on the Kern Plateau of Tulare County to the northeast and at high elevations in the San Bernardino Mountains to the southeast.

Pinus Jeffreyi Greville & Balfour. Jeffrey pine. The dominant tree at the higher levels of the yellow pine forest, especially along ridgetops and other exposed places in the Greenhorn Range and on Breckenridge Mountain, occurring mostly over 6,000 feet elevation. Except for isolated groves, it replaces *P. ponderosa* at all altitudes on the Kern Plateau, in the Piute and Tehachapi Mountains, and in the Mt. Pinos region. This stately tree is the principal component of the park-like forest along the summit of Piute Mountain. Individual specimens, such as the one near the ski lodge on Mt. Abel, are notably impressive. Although generally occupying more arid situations than *P. ponderosa,* the Jeffrey pine's tolerance of cold may be of more significance in its distribution relative to that of ponderosa pine than its greater drouth resistance.

An isolated colony of healthy trees of all ages grows at the western end of Tollgate Ridge in the arid Tehachapi Mountains east of Keene. A few old trees survive on the steep north slopes at the head of desert-facing Pine Tree Canyon in the same region. On nearby Sweetwater Ridge, a few fallen trees, long since dead, are surrounded by chaparral. Such fallen old trees, far from the nearest living ones, are found occasionally in the Tehachapi Mountains; this seems to indicate a recent response to the county's increasing aridity.

Pinus Lambertiana Douglas. Sugar pine. A common and impressive conifer of the ponderosa pine forest in the Greenhorn Range and on Breckenridge Mountain. Sugar pine is scarce in small groves on the north end of Piute Mountain and the north slope of Tehachapi Peak. Other small colonies grow on Tollgate Ridge east of Keene and on the southeast slope of Brush Mountain in the Mt. Pinos region. An interesting isolated occurrence of sugar pine is that on the northwest ridge of Mt. Owen at the Kern, Inyo, and Tulare counties boundary where a small colony grows with scattered Jeffrey pines in a most arid situation (Pittman 245). The tree is rare on the rim of the Kern Plateau above the South Fork of Kern River in the Bartolas country.

The long pendant cones and the tall straight trunk make sugar pine a distinctive member of the yellow pine forest. The lumber is of high quality. Seedlings are frequently abundant in disturbed places in the Greenhorn Range. Young trees grow rapidly.

Pinus monophylla Torrey & Frémont. One-leaf pinyon. Common, often forming distinct woodlands, on the desert slopes of the mountains west to the Piute Mountains and the east slope of the Greenhorn Range and southwest to Ventura and Santa Barbara counties at the east end of Cuyama Valley. It is easily the most xerophytic of our conifers and straggly twisted and bent individuals persist in the most inhospitable places, such as the ridgetops along the mountains bordering the desert in the Jawbone Canyon region. In contrast, on the lower slopes of Mt. Abel and the heads of Ballinger and Quatal canyons in the upper Cuyama Valley, the trees are exceptionally well developed and form a modest but impressive pinyon forest. It grows with digger pine on the east slope of the Piute Mountains, the southern Kern Plateau, and at the head of Tejon Canyon.

The nuts, often borne prolifically, are an important source of food for birds, mammals, and formerly the Indians. The tree grows well in cultivation and its dense branching and foliage make a pyramidal evergreen of great beauty that little resembles the weather-beaten specimens found along the desert borders. (*Pinus cembroides* var. *monophylla.*)

Pinus Murrayana Greville & Balfour. Lodgepole pine. A conifer of the red fir and lodgepole forests that reaches its southern Sierran limits with a colony at Pine Flat and another at Little Cannell Meadow, both on the southern end of the Kern Plateau and barely within Kern County. Pine Flat shows evidence of once having been a moist meadow that has rapidly dried in the past few decades and is now a sagebrush flat. As all of the lodgepoles here are mature, it is quite possible that the tree will eventually become extinct at this station. However, its common and widespread occurrence on the arid Kern Plateau as a forest tree just above the Jeffrey pine forest suggests that here it is adapted to less moisture than lodgepoles require in the western Sierra Nevada, where the tree is largely confined to the borders of meadows. Lodgepole pines grow rapidly under favorable conditions and are often invasive in meadows that are not well grazed.

Pinus ponderosa Douglas. Ponderosa pine. The dominant tree of the yellow pine forest below 6,000 feet elevation in the Greenhorn Range and on Breckenridge Mountain where it forms an extensive and well developed forest. It does not occur in the Piute Mountains and survives only in scattered relict colonies in the Tehachapi Mountains and the Mt. Pinos region (Tejon Canyon, Water Canyon, near Mil Potreros). It is the prime timber tree of the county. The common difficulty encountered in distinguishing ponderosa pine from Jeffrey pine is apparently based on misconceptions; the size of the cone (in ponderosa pine, 3 inches or less) and the outwardly turning cone prickles are almost always adequate characters to distinguish ponderosa from Jeffrey pine.

Pinus Sabiniana Douglas. Digger pine. A typical foothill tree in the Douglas oak woodland south to Tejon Canyon. It does not occur in the San Emigdio Range, nor, less surprisingly, in the Temblor Range. It is difficult to explain the tree's absence from the San Emigdio Range where the canyons seem typical of its general habitat. In much of California, the range and climatic and edaphic preferences of digger pine are well marked and even precise. It grows in a belt primarily below 4,000 feet and always below the permanent winter snow line, preferring hot dry canyons and ridges with rocky or thin soil. In Kern County, however, digger pine occurs to at least 6,000 feet on the east slope of Piute Mountain and on the east slope of the Greenhorn Range near Greenhorn Pass; in these places the tree grows with Jeffrey and ponderosa pine and becomes a well developed forest tree that little resembles the typical straggly foothill form. It approaches the desert in Sand and Cameron canyons in the Tehachapi Mountains and occurs at the heads of the very arid desert canyons north of Walker Pass to the summit of Nine Mile Canyon in Inyo County. Here digger pine often grows as scattered and even isolated individuals in the well developed pinyon woodland of the region. At Lamont Meadow, just north of the Kern County line, digger pines and western junipers (*Juniperus occidentalis*) often grow together; these trees ordinarily occur several life zones apart.

In its more typical cismontane occurrence, it forms a mixed woodland with Douglas oak, the digger pines generally preferring rockier shallower soils, the oaks at their best, heavier deeper soils. However, they are so generally mixed that it is impossible to attempt to segregate them as indicators of distinct plant associations. The tree has no value for timber and Glennville ranchers consider it undesirable for it uses moisture that might otherwise be available for forage plants.

Pseudotsuga macrocarpa (Vasey) Mayr. Big cone spruce. An occasional to common conifer of the Jeffrey pine forest in the Mt. Pinos region that reaches its northern limits on Tecuya Ridge west of Ft. Tejon. A fine colony grows near the road just east of the Jim Whitner Tree in Mil Potreros.

CUPRESSACEAE. Cypress Family

Calocedrus decurrens (Torrey) Florin. Incense-cedar. A common tree of the yellow pine forest in the Greenhorn Range. It is rare in the Piute Mountains and on the southern Kern Plateau and occurs only sporadically south to El Paso Canyon in the Tehachapi Mountains. A small isolated colony of young trees grows near the head of Black Bob Canyon in the San Emigdio Range. Incense-cedar does not occur in the well developed yellow pine forest on Breckenridge Mountain, an extremely curious hiatus in its distribution for which no explanation is available (it is abundant and even weedy directly across Kern River Canyon in the Davis Guard Station area of the Greenhorn Range). The tree grows rapidly and prolifically in some areas of the Greenhorn Range, where foresters and livestock men consider it a nuisance. Here young trees grow in dense colonies of spindly individuals that choke out the more desirable seedlings of ponderosa pines and shade out much annual and perennial growth. Apparently excessive fire control is responsible for this weedy growth. Wood from the tree takes a high polish, and is much used for ornamentation but is of secondary value for structural purposes. A former important use was for pencil wood. (*Libocedrus decurrens*).

Cupressus nevadensis Abrams. Piute cypress. A notable Kern and extreme southern Tulare County endemic, known from only six groves. The largest and most vigorous of these is on the north end of Piute Mountain, where several thousand trees grow over a fairly wide area in chaparral and arid Douglas oak woodland. In 1966, a grove of fine trees of all ages that occupied about 40 acres on the steep north slope of Hobo Ridge on Breckenridge Mountain was completely burned in a wild fire. A single mature tree and about 40 fire-following seedlings form the grove on the east slope of the Greenhorn Range, 2.9 miles east of Wofford. In the summer of 1966, Robert and Carol Luthey discovered about 12 trees, mostly fully mature, growing along Reagen Creek about a mile north of this colony. It is probable that these two groups are the remnants of an extensive grove that once covered the intervening area on the southeast slope of Black Mountain. A colony of several hundred mostly senescent trees grows at the south end of Piute Mountain near the head of Back Canyon. Increment borings by Jerry Decker (oral communication) show that trees in this colony are, at late maturity, slightly over 100 years old. A colony of perhaps two dozen trees has recently been discovered on the east side of Kern Canyon on the arid south slope of Cannell Creek Canyon, about 2.5 miles east of River Kern (Dick Rea, 21 December 1964). An extensive colony, nearly as large as the Bodfish colony, has recently been reported from Tulare County. Robert and Carol Luthey write (personal communication) that a grove of at least 1,000 trees, mostly of the same size, grows in Stormy Canyon, just south of Baker Point. This region was swept by a devastating fire in 1924. Apparently the pyrophytic nature of the trees, first noted by J. T. Howell (Leafl. West. Bot., 10:253), resulted in an excellent crop of seedlings in 1925 and thus almost all of the trees now present are about 42 years old. These scattered colonies are apparently relicts of the extensive woodland that until the late Pliocene covered the region now occupied by the Mojave Desert.

The Piute cypress does not compete well with other shrubby species. In open areas on the north end of Piute Mountain it grows as beautiful symmetrical specimens; further up the mountain where it competes with chaparral, the trees are spindly and poorly developed. It grows in many kinds of soil, from the heavy black clay of Back Canyon, the ultra-fine red clay of the Bodfish Grove, to the decomposed and fractured granite in the Greenhorn and Breckenridge groves. (Cf. Leafl. West. Bot., 10:248, 1962).

The tree grows rapidly in cultivation where it seems to thrive on garden culture and is not subject to the blight that affects other cypresses in inland regions. Thus, this rare tree is a good example of a species that can be preserved through widespread landscape use.

The type was collected at Red Hill, near Bodfish (LeRoy Abrams 5368, 29 July 1915).

Juniperus californica Carrière. California juniper. A common and picturesque shrub in all the arid Upper Sonoran shrubby associations except on the desert. It often grows in a distinct belt just below the pinyon woodland on the desert-facing slopes of the mountains. In the Temblor Range, it forms a distinctive local plant association (cf. Wasm. Journ. Biol., 14:174, 1956). California juniper was much used by early settlers for fence posts (many of these fences are still in service) and crude but serviceable corrals. The wood makes a hot rapidly burning fire with an unusually pleasant aroma. The berries are a staple fall and winter food of birds, rodents, and even coyotes. Seedlings are most successful in years of scant rainfall; apparently they are unable to compete with the lush herbaceous growth of good years.

Although extremely slow growing in the wild, California juniper grows rapidly in cultivation and is very well adapted to even severe pruning for interesting growth forms. A distinctive columnar form has been discovered by Robert and Carol Luthey in the Kernville and Bodfish regions. The upright form of the Luthey juniper should be especially valuable for horticultural use in arid regions.

Juniperus occidentalis Hooker. Western juniper, Sierra juniper. Tony Gasbarro reports (personal communication) that western juniper is common on the summit and north slope of Mt. Owen, where it reaches its southern Sierra Nevada limit.

EPHEDRACEAE. Ephedra Family

Ephedra californica Watson. (All of the ephedras are known generically by a bewildering number of "tea" names, such as sheepherder, cowboy, Mormon, Spanish, Mexican, etc., tea; the dried stems make a potable beverage that is an excellent substitute for coffee or tea and is thought by many to have medicinal value.) Occasional in the California juniper belt from Recruit Canyon and the Panorama Hills in the southern Temblor Range to the east end of Cuyama Valley (where unusually large plants grow on the sand flats) and along the sandy foothills on the east side of the valley from the mouth of Kern Canyon south to Comanche Canyon in the Tejon Hills.

Ephedra nevadensis Watson. A low shrub that is occasional to scarce, usually on rocky or gravelly slopes, in the Rand and El Paso mountains. It also occurs in the sand dunes around the Muroc dry lakes.

Ephedra viridis Coville. Canyon tia. Occasional, desert-facing slopes of the mountains; west to the east slope of the Greenhorn Range; south through the mountains, occurring in the Jeffrey pine forest in the Piute Mountains. Canyon tia is especially common in the pinyon woodland around Frazier Park and is occasional in the central Temblor Range. It grows mostly in the chaparral and pinyon and arid Douglas oak woodlands. Spanish Californians drank an infusion of the dried stems for relief from arthritic pains; Mexican vaqueros claimed it was also effective in the treatment of venereal disease.

TYPHACEAE. Cat-tail Family

Typha domingensis Persoon. Cat-tail. Common in cismontane wet places, mostly below the Douglas oak woodland; occasional at wet places to the edge of the desert. It often grows in dense colonies that become troublesome in canals and reservoirs.

Typha latifolia Linnaeus. Broad leaf cat-tail. Local in pools and slow-moving streams in and adjacent to the Greenhorn Range (Borel Intake east of Wofford along Kern River, Bull Run Creek northwest of Kernville, Glennville, Kern River below Lake Ming).

SPARGANIACEAE. Bur-reed Family

Sparganium eurycarpum Engelmann. Bur-reed. A common and handsome plant of pools and quiet canals in Lynns Valley near Glennville.

POTAMOGETONACEAE. Pondweed Family

Potamogeton crispus Linnaeus. Curly-leaved pondweed. Common along Kern River from Kernville to Buena Vista Lake and in adjoining canals; along Poso Creek in Lynns Valley. Native of Europe.

Potamogeton foliosus Rafinesque. Leafy pondweed. Rare in Kern River above Lake Isabella (T6496) and in Cedar Creek just below the mouth of Lumreau Creek (T11,667, 12,735).

Potamogeton nodosus Poiret. Common pondweed. Occasional in pools along Kern River above Lake Isabella. A dense colony grows in an old canal near Weedpatch.

Potamogeton pectinatus Linnaeus. Sego-leaved pondweed. Abundant in great masses in Lake Ming (T6589, 6593).

Potamogeton pusillus Linnaeus. Small pondweed. Abundant in Lake Ming (T6588, 6590, 6644).

Ruppiaceae. Ditch-grass Family

Ruppia maritima Linnaeus. Ditch-grass. Abundant in great masses in the brackish water of Castac Lake near Lebec (T9740). It was present at this station far inland from its usual occurrence as early as 1891, when it was collected by Coville and Funston (no. 1147).

Najadaceae. Najas Family

Najas marina Linnaeus. Water nymph. Rare in the dense growth of aquatics in Lake Ming (T6642).

Zannichelliaceae. Horned Pondweed Family

Zannichellia palustris Linnaeus. Horned pondweed. Common in tanks, troughs, and quiet pools in the mountains, especially in the Temblor Range; uncommon in similar places in the valley. It prefers quiet semi-stagnant water. Although the plant is not common in canals, it is abundant in the slowly moving water in those at the Kern National Wildlife Refuge.

Juncaginaceae. Arrow-grass Family

Triglochin debilis (Jones) Löve & Löve. Arrow-grass. Common on the subalkaline flats at Isabella, especially around Scovern Hot Springs; scarce in Red Rock Canyon (Hardham 4148) and at Rosamond (Davy 2234, UC). (*T. concinna* var. *debilis, T. maritima* var. *debilis.*)

Lilaeaceae. Flowering Quillwort Family

Lilaea scilloides (Poiret) Hauman. Flowering quillwort. Scarce in vernal pool beds in Bear Valley (T9674, 9685) and common around a vernal pool at the northwest end of Lynns Valley (T10,871). (*L. subulata.*)

Alismataceae. Water Plantain Family

Alisma triviale Pursh. Water plantain. Common in creeks, ponds, and irrigation ditches in Lynns Valley. (*A. plantago-aquatica.*)

Damasonium californicum Torrey. Star or fringed water plantain. Rare around vernal pools in Bear Valley (T9687). These scattered plants represent a southern range extension from Merced County (Howell 7, 1923). (*Machaerocarpus californicus.*)

Echinodorus Berteroi (Sprengel) Fassett. Upright burhead. A troublesome rice field weed that is common in other muddy or wet places in the valley, often growing in shallow water. (*E. cordifolius.*)

Sagittaria calycina Engelmann. Arrowhead. A rare weed in a rice field near Old River in 1962 (T7984). (*S. montividensis* ssp. *calycina.*)

Sagittaria latifolia Willdenow. Wappato, tule potato. Rare along a canal east of Greenfield (T8011).

Sagittaria longiloba Engelmann. Locally common in the canals in the Millux region east of Buena Vista Lake and in marshes and ponds at the Kern National Wildlife Refuge. A rare weed, perhaps becoming more common, in the rice fields south of Buttonwillow. (*S. Greggii.*)

HYDROCHARITACEAE. Frogbit Family

Elodea canadensis Michaux. Canadian waterweed. An occasional to common aquatic, mostly on gravel and sand bars, in Kern River from Richbar in Kern Canyon to Hart Memorial Park near Bakersfield. It is locally common in deep pools behind the wiers on Poso Creek in Lynns Valley. Probably native to the Rocky Mountains, eastern United States, and Canada. (*Anacharis canadensis.*)

GRAMINEAE. Grass Family

FESTUCEAE. Fescue Tribe

Arundo donax Linnaeus. Giant reed, locally and erroneously, bamboo. Occasional in wet places in the valley, often forming dense colonies. This, by far the tallest of our grasses, was once much planted as an ornamental at camps, homesteads, etc., that have long been abandoned. These old plantings are often thought to be spontaneous but spontaneous plants probably occur only along the remnants of the sloughs and marshes that once occupied the central valley. Native of Europe.

Bromus arenarius Labillardière. Australian chess. A common and widespread grass of the Upper Sonoran grassland that is occasional to the yellow pine forests. A forage plant of considerable but secondary importance. Native of Australia.

Bromus arizonicus (Shear) Stebbins. Arizona brome. Widespread in the valley grasslands, especially in light or disturbed soil in the Lower Sonoran grassland. It is scarce on the desert in the creosote bush association and is a common spring weed in Bakersfield.

Bromus breviaristatus Buckley. Occasional in the chaparral, pinyon woodland, and yellow pine forest in the mountains; rare in the Upper Sonoran grassland in the extreme northern Temblor Range.

Bromus carinatus Hooker & Arnott. California brome. Occasional but widespread from the Douglas oak woodland to the yellow pine forest in the mountains.

Bromus commutatus Schrader. Known from a meadow at the west end of Cummings Valley in the Tehachapi Mountains (Breedlove 5485 in 1963). Native of Europe.

Bromus diandrus Roth. Rip gut. Common in the grasslands on the western slope of the Greenhorn Range and somewhat less common in all the other cismontane associations to the lower levels of the yellow pine forest. A common and trying weed of gardens and moist waste places. While a fair forage plant when green, the stiff awns sometimes cause severe damage to livestock when the plant is dry. Veterinarians report that awns lodged in dog's nostrils are a common problem in cities and towns. Introduced from Europe. (*B. rigidus.*)

Bromus madritensis Linnaeus. Spanish brome. Occasional in the dense grasslands in the Greenhorn foothills south to Kern Canyon; scarce in the grasslands elsewhere. An abundant waste place and troublesome garden weed at the Ellen Newsom residence at Annette in the Temblor Range and a scarce weed in the yellow pine forest 1.5 mile south of Greenhorn Summit (Howell 38,325). It is often overlooked because of its very close resemblance to the ubiquitous *Bromus rubens.* Native of Europe.

Bromus marginatus Nees. Mountain brome. Common in the yellow pine forest; occasional in the shin oak brush and Douglas oak woodland in the mountains; rare in the Douglas oak woodland in the central Temblor Range.

Bromus mollis Linnaeus. Soft chess. Common in the cismontane Upper Sonoran associations, especially in the Upper Sonoran grassland where it is an important forage plant. Occasional in the valley and the yellow pine forest. A horticultural form called Blando brome is sometimes planted in re-seeding projects; although perhaps more robust it has no apparent morphological distinctions from the naturalized form. Seeds of soft chess are important food for wildlife, particularly for birds and rodents. Naturalized from Europe. (*B. hordeaceus.*)

Bromus Orcuttianus Vasey. Occasional in good soil in the ponderosa pine and red fir forests in the Greenhorn Range and on Breckenridge Mountain; scarce in the Jeffrey pine forest on the southern Kern Plateau.

Bromus pseudolaevipes Wagnon. Wagnon (Britt. 7:458) includes Kern County in the range for this brome; he cites no collections. The author knows of it no closer to Kern County than Parkfield in the inner south Coast Ranges in Monterey County but it is to be looked for in the Mt. Abel region.

Bromus rubens Linnaeus. Red brome. Common and often abundant in all associations below the yellow pine forest, especially in the Upper Sonoran grassland. It is occasional in disturbed places and in loam on open slopes in the forests. While often considered of secondary forage value, local livestock people praise red brome for its rapid growth and resistance to drouth and cold weather. (It is especially common and often dominant over wide areas of the foothills and the valley in years with cold winters and prolonged dry spells.) In some regions, red brome becomes especially common after prolonged heavy grazing. However, it also tends to be abundant, often crowding out the native plants, in light or sandy soils that are not grazed at all. A native of Europe long established as one of the most common plants of the Kern County flora.

Bromus tectorum Linnaeus. Downy brome. A very common often abundant weedy grass in the yellow pine forests, growing most densely in disturbed places. It is common in the Upper Sonoran grassland except in the Temblor Range where the grass is rare. Downy brome is occasional and sometimes locally common in the creosote bush association. Whether green or dry, it is an important forage plant, particularly valuable for its drouth resistance. Native of Europe.

Bromus tectorum var. **glabratus** Spenner. Smooth downy brome. The common form on the west slope of the Greenhorn Range that is frequent elsewhere through the mountains (Horse Canyon on Kiahvah Mountain; Hart Flat, Bear Mountain, and Tejon Pass in the Tehachapi Mountains). Although it is said to occur with the species, in Kern County the varieties grow separately with the typical variety adapted to generally but not always more arid situations. Native of Europe.

Bromus Trinii E. Desvaux. Chilean brome. Widespread in the desert ranges and the Temblor Range; scarce in the valley and on the desert plains. Although widely distributed, it is never abundant. Chilean brome usually grows through low shrubs that protect it from grazing.

Bromus unioloides Humboldt, Bonpland & Kunth. An occasional weed around Bakersfield and at Kernville; more common about the farmlands in Antelope Valley and at Cantil on the desert; a street weed at Mojave. It appears to be a recent introduction that is becoming more common. Native of South America. (*B. Haenkeanus* of California references.)

Bromus Wildenovii Kunth. Rescue grass. Occasionally adventive from pastures in the Panama region, it probably also escapes from irrigated pastures at other places in the valley. It grew as a rare escape along wet ditches in the farmlands north of North Edwards in the desert in 1966 (T12,043). Native of South America. (*B. catharticus* of California references.)

Dactylis glomerata Linnaeus. Orchard grass. Occasional in moist waste places in South Fork Valley and along lower Erskine Creek in the Piute Mountains. Native of Europe.

Distichlis spicata (Linnaeus) Greene. Salt grass. The common and characteristic grass of alkali sinks and other alkaline soils; often common in non-alkaline low places with light soil to the pinyon and Douglas oak woodlands. Several varieties have been described; these seem to merge insensibly.

Eragrostis cilianensis (Allioni) Lutati. Candy grass. A widespread weed of moist places in the valley; less common on the desert and in the Temblor Range; uncommon in the mountain valleys. Native of Europe. (*E. megastachya.*)

Eragrostis diffusa Buckley. An abundant weed of moist places in the valley; common in the mountain valleys and occasional in the Temblor Range. It is a scarce weed in gardens and irrigated fields on the desert.

Eragrostis mexicana (Hornemann) Link. An occasional weed of irrigated fields in the valley and a common street, garden, and field weed at Tehachapi.

Eragrostis Orcuttiana Vasey. Common in the mountain valleys and an occasional weed of valley croplands.

Festuca arundinacea Schreber. Alta fescue. An occasional escape from pastures that is especially common and well established in the marshy areas of South Fork Valley. It is also well established along lower Erskine Creek in the Piute Mountains. In 1966 it was seeded with only slight success on roadbanks and in disturbed areas in the Rhymes Toolbox region of the yellow pine forest of the Greenhorn Range. Native of Europe.

Festuca confusa Piper. Hairy-leaved fescue. Common and often abundant in the cismontane Upper Sonoran grasslands; occasional to the yellow pine forest; rare on the desert. All of the annual fescues are important early spring and late winter forage plants that are prized for their quick growth after the earliest rains.

Festuca dertonensis (Allioni) Ascherson & Graebner. Six weeks fescue. Common in the Upper Sonoran grasslands; less common in the valley; scarce on the desert. It is especially common in good soil in the Douglas oak woodland. Native of Europe.

Festuca Eastwoodae Piper. Eastwood fescue. Scarce in the Upper Sonoran grassland and in open Douglas oak woodland in the foothills around the valley.

Festuca Grayi (Abrams) Piper. Common in the lower levels of the ponderosa pine forest in the Greenhorn Range; widespread but less common in the mountains and the Temblor Range; scarce on the high plains on the valley west-side.

Festuca megalura Nuttall. Foxtail fescue. Common and often abundant from the cismontane grassland to the lower levels of the yellow pine forest.

Festuca microstachys Nuttall. Scarce on grassy slopes in the Greenhorn Range, mostly in the Douglas oak woodland.

Festuca microstachys var. **simulans** (Hoover) Hoover. Common, often abundant, in the valley and the southern Temblor Range where it is the most common native grass of the Lower Sonoran grassland. The type was collected 2 miles north of Blackwells Corner (Robert F. Hoover 451, 30 March 1935). (*F. pacifica* var. *simulans.*)

Festuca myuros Linnaeus. Rat-tail fescue. Rare: Messa Ridge, Temblor Range; ridge above Saucelito Corrals, Tehachapi Mountains; Bakersfield (Benson 3208, POM); infrequent in the Douglas oak woodland in the Greenhorn Range.

Festuca occidentalis Hooker. Western fescue. Rare in moist places in the ponderosa pine forest in the Greenhorn Range.

Festuca octoflora Walter. Scarce but widespread, usually in bare often alkaline places, in all associations to the lowest level of the yellow pine forest. Var. *hirtella* Piper, a form with pubescent lemmas, occurs with the species and is the more common form in the creosote bush association.

Festuca pacifica Piper. Pacific fescue. Common, often abundant, in the Upper Sonoran grassland; occasional to the yellow pine forest.

Festuca pratensis Hudson. Meadow fescue. An important grass of irrigated pastures that is an occasional escape in moist places. It is well established in the broad marshy flats just east of Weldon and at Grouse Meadow in the Piute Mountains. Native of Europe. (*F. elatior.*)

Festuca reflexa Buckley. Few-flowered fescue, side oats. Very common in the cismontane grasslands, particularly the Lower Sonoran grassland; frequent through the Douglas oak woodland and rare although widespread in the creosote bush association.

Festuca rubra Linnaeus. Red fescue. A persistent ranchground weed at the Temblor Ranch (T1291, 1369, 2014, 2726). Although adventive here, the grass is native in California.

Festuca Tracyi Hitchcock. Rare in the Upper Sonoran subshrub association in arid Chico Martinez Canyon in the Temblor Range (T1861).

Glyceria elata (Nash) Hitchcock. Tall manna grass. Occasional in wet meadows in the Greenhorn Range.

Glyceria erecta Hitchcock. Strict manna grass. Occasional in wet meadows in the Greenhorn Range.

Glyceria pauciflora Presl. Few-flowered manna grass. Occasional in moist soil along a fork of Cannell Creek near Pine Flat on the southwestern Kern Plateau (T8695, 8711) and scarce along Bartolas Creek near its summit (T12,446).

Lamarckia aurea (Linnaeus) Moench. Golden-top. Occasional, often in small dense colonies, along the east-side foothills south to Tejon Canyon and Gighas Ridge. It is most common in the hills between Woody and White

River but is becoming more frequent elsewhere, especially around rock outcrops and slides. Native of the Mediterranean region.

Melica aristata Thurber. A colony grows at a seep in the ponderosa pine forest 1.5 mile south of Greenhorn Pass (Howell 38,329, T9937).

Melica californica Scribner. California melic. Occasional in rocky places from the Upper Sonoran grassland to the lower levels of the yellow pine forest.

Melica californica var. **nevadensis** Boyle. Common from the chaparral to the lower limits of the ponderosa pine forest on the west slope of the Greenhorn Range, where it replaces the typical variety. It is occasional on the east slope of the range.

Melica imperfecta Trinius. Small-flowered melic. Common in the cismontane Upper Sonoran associations, extending to the pinyon woodland and in the desert-facing canyons to the borders of the desert. Several varieties occur; var. *refracta* Thurber grows apparently at random with the species; var. *flexuosa* Bolander occurs at rocky places in the mountains, reaching the desert's edge in Short Canyon, and var. *minor* Scribner is rare in fractured granite in the arid Douglas oak woodland in Erskine Canyon in the Piute Mountains (T8423).

Melica stricta Bolander. Rock melic. Scarce in rocky places from the pinyon woodland and chaparral to the Jeffrey pine forest where it often occurs along the highest rocky ridgetops. Rodents are very fond of the seed; their foraging probably limits reproduction.

Phragmites communis Trinius var. **Berlandieri** (Fournier) Fernald. Carizzo grass. Now known only from Mesquite Springs at the southeast base of the El Paso Mountains (T10,300), carizzo grass was once much more common in the county. Lt. R. S. Williamson (Report of Explorations in California for Rail Road Routes, IV:18) reported it was abundant in 1853 along Canebrake Creek (which was named for the grass) west of Walker Pass and along Kelso Creek south of Weldon where Williamson observed Indians harvesting the grass for the sugar incrustations on the leaves. According to Donald J. Bedell (oral communication) cattle graze it avidly. Its extermination, however, is more likely to have been caused by horses pawing out its roots in times of scant forage.

Poa annua Linnaeus. Annual bluegrass. A widespread weed of gardens and wet places in all associations, occurring to the edge of the desert. Native of Europe.

Poa Bigelovii Vasey & Scribner. Bigelow bluegrass was once collected in wet soil in Red Rock Canyon (M. N. Ackley 49, 15 April 1928, SBM).

Poa Bolanderi Vasey. Bolander bluegrass. Scarce in moist places in the yellow pine and red fir forests in the Greenhorn Range and on Breckenridge Mountain. Exceptionally large plants grew near Rhymes Toolbox in 1966 (T12,406).

Poa bulbosa Linnaeus. Bulbous bluegrass. Occasional in colonies in the Douglas oak woodland in the southwestern Tehachapi Mountains (Catskin Creek, Dry Field Corrals) and in the Glennville region and at French Ranch on the southwest flank of the Greenhorn Range. This recent introduction from Europe is apparently becoming more common.

Poa compressa Linnaeus. Canadian bluegrass. Occasional in the vicinity of Greenhorn Pass (Greenhorn Mountains at 6,000 feet, Hammerly 568, 569; near Greenhorn Mountain Park Headquarters, Howell 38,386). Presumably native of Europe although possibly some races are native in California.

Poa Howellii Vasey & Scribner. Howell bluegrass. Scarce in shaded canyons in chaparral and Douglas oak woodland south to Water Canyon on Bear Mountain. (*P. Bolanderi* ssp. *Howellii.*)

Poa incurva Scribner & Williams. Common in gravel and fractured rock about the broad summit of Mt. Pinos, where it frequently forms its typical fairy rings.

Poa longiligula Scribner & Williams. Mutton grass. Scarce from the chaparral to the yellow pine forest in the Greenhorn Range and on Breckenridge Mountain, usually growing around rock outcrops. It is rare along a fork of Cannell Creek on Pine Flat and near the summit of Bartolas Creek on the southwestern Kern Plateau. (*Poa Fendleriana* in part.)

Poa nevadensis Vasey. Nevada bluegrass. Common in moist soil at Weldon, Landers, and French meadows in the Jeffrey pine forest in the Piute Mountains; occasional on the subalkaline flats at Isabella and Kernville and in mineralized soil at Kern Hot Springs.

Poa pratensis Linnaeus. Kentucky bluegrass. Common in moist to dry soil in the higher mountains, often the most common grass of wet meadows. It is occasional in moist soil elsewhere (except the desert) where it is probably an escaped pasture or lawn grass. This form was introduced from Europe; the mountain form is probably native.

Poa scabrella (Thurber) Bentham. Pine bluegrass. Common everywhere except the desert plains and the valley floor. The species is highly variable and in need of study. Selected forms of this widespread bunchgrass would seem desirable for range re-seeding experiments.

Puccinellia grandis Swallen. According to A. A. Beetle (Leafl. West. Bot. 4:288) this grass occurs at Ft. Tejon (Davy 2367, Beetle 4323), far from its normal coastal range.

Puccinellia simplex Scribner. Alkali grass. A rare little annual grass of moist alkaline soils known from the alkali flats at Isabella (Howell 38,335) and around the alkaline vernal pools one mile southwest of McKittrick (Yates 6506, DS).

HORDEAE. Barley Tribe

Agropyron desertorum (Fischer) Schultes. Crested wheat grass, desert wheat grass. Known from a single plant near the summit of Quatal Canyon in the pinyon woodland in the Mt. Abel region (T2747). Although desert wheat grass is often seeded in burns, it is not adapted to our climate. Native of Russia.

Agropyron elongatum (Host) Beauvois. Tall wheat grass. Tall wheat grass is frequently planted in irrigated pastures and is occasionally spontaneous in moist places in the valley. Roy V. Parker reports (pers. comm.) it is locally established on brush burns in the Woody-Glennville region (cf. Leafl. West. Bot. 9:143). It is doubtfully established on Black Mountain in the Greenhorn Range where it was seeded following the wild fire of 1961 (Howell 38,266). Native of the Old World.

Agropyron intermedium (Host) Beauvois. Intermediate wheat grass. This, too, was seeded on the Black Mountain burn in 1961 and seems to be sparingly established in draws and ravines (Howell 38,227; T8755). At the Dorothy Twisselmann Ranch in the northern Temblor Range it has had good results in carefully selected range seedings. Native of Europe.

Agropyron Parishii Scribner & Smith var. **laeve** Scribner & Smith. Parish wheat grass. Scarce, mostly in the Jeffrey pine forest and the pinyon woodland, Mt. Pinos region, Tehachapi Mountains, Breckenridge Mountain, and Black Mountain in the Greenhorn Range. (*A. laeve.*)

Agropyron subsecundum (Link) Hitchcock. Scarce in shin oak brush along the Microwave Station Road in the Tehachapi Mountains northeast of Lebec (T4613). (*A. caninum.*)

Agropyron trachycaulum (Link) Malte. Slender wheat grass. Scarce in dry sandy soil at the borders of a broad sagebrush meadow in the pinyon woodland at the head of Cuddy Valley in the Mt. Pinos region (T6296) and in moist soil in French Meadow in the Jeffrey pine forest in the Piute Mountains (T7486). (*A. tenerum.*)

Agropyron trichophorum (Link) Richter. Pubescent wheat grass. Seeded and sparingly established on brush burns in the Woody-Glennville region (T4473), on Black Mountain (Howell 38,225, T8756), and along forest roads near Rhymes Toolbox (T12,409) in the Greenhorn Range. Although grass seedings have had little success in Kern County, pubescent wheat grass is the most promising of the species that have been tried. Native of Europe.

Elymus cinereus Scribner & Merrill. A distinctive coarse bunch grass that is occasional in large tussocks on the subalkaline clay flats around Rosamond and Buckhorn dry lakes.

Elymus glaucus Buckley. Blue wild rye, western rye. Occasional from the cismontane Upper Sonoran grasslands to the yellow pine forest. Before the naturalization of European annual grasses, blue wild rye may have been much more common in the primitive grasslands. Ssp. *Jepsoni* (Davy) Gould, segregated on the somewhat trivial basis of its pubescent sheaths and blades, grows on the south side of Black Mountain in the Greenhorn Range (Howell 38,239).

Elymus triticoides Buckley. Alkali rye. Common in low often alkali places from the valley through the cismontane Upper Sonoran associations; occasional in dry meadows in the Jeffrey pine forest in the Piute Mountains.

Elymus triticoides ssp. **multiflorus** Gould. This distinctive large-headed form is occasional in the pinyon and Douglas oak woodlands from the San Emigdio Range to the southwestern Tehachapi Mountains.

Holcus lanatus Linnaeus. Velvet grass. Local but apparently well established along Poso Creek in Lynns Valley. It is scarce about Kernville and in South Fork Valley. Native of Europe.

Hordeum brachyantherum Nevski. Meadow barley. Scarce in moist sunny soil along Cannell Creek in the very arid Douglas oak woodland 1.5 mile east of River Kern (T8745).

Hordeum californicum Covas & Stebbins. California barley. An occasional foxtail of moist sometimes subalkaline places in the Douglas oak and pinyon woodlands in the mountains. (*H. nodosum* of California references in part.)

Hordeum depressum (Scribner & Smith) Rydberg. Common in low usually subalkaline places in the valley and the Temblor Range. It is occasional in the mountains through the Douglas oak woodland and is rare on the desert (moist subalkaline sand, Red Rock Canyon).

Hordeum geniculatum Allioni. Mediterranean barley. Occasional, usually in small dense colonies, in wet often alkali places in the Temblor Range; less common through the Douglas oak woodland in the mountains. Native of Europe. (*H. Hystrix, H. Gussoneanum.*)

Hordeum glaucum Steudel. Foxtail. This ubiquitous immigrant is one of the most common grasses of the Lower Sonoran grassland and is frequent in all the cismontane Upper Sonoran associations, especially in disturbed places or in heavily grazed light soils. It is often especially dense even if depauperate following cold dry winters. The grass is common on the desert, where it most frequently grows in semi-stabilized dunes, as a weed about towns, or in irrigated farmlands. Native of Europe. (*H. Stebbinsii, H. murinum* in part.)

Hordeum jubatum Linnaeus. Common in wet marshy areas in South Fork and Kern valleys.

Hordeum leporinum Link. Farmer's foxtail. Widespread in all the cismontane associations below the yellow pine forests; infrequent on the desert. It can be distinguished from *H. glaucum* only with difficulty and occupies much the same range but *H. leporinum* is usually most abundant in somewhat less arid places and the plants are generally more robust and vigorous. Native of Europe. (*H. murinum* in part.)

Hordeum vulgare Linnaeus. Barley. Common as a waif below the yellow pine forest; possibly barely established in some areas. The large palatable seeds that are eaten by both wild and domestic animals rarely survive the summer. Barley is very commonly spontaneous on the desert in years of ample rainfall, especially where sheep have been fed hay or near highways. Native of Europe.

Lolium multiflorum Lamarck. Italian rye grass. Common in moist soils from the Lower Sonoran grassland through the Douglas oak woodland. It is often planted in pastures. Var. *muticum* De Candolle, a form without awns, is scarce in moist places around Lake Isabella (Howell 37,194) and in seeded road banks in the Greenhorn Range. Italian rye has frequently been seeded in burns and range re-seeding projects with scant success. Native of Europe.

Lolium perenne Linnaeus. Perennial rye, English rye. Rare in South Fork Valley near Weldon (T5550) and on an islet in Kern River near Kernville (T6440). It is scarce around the Temblor Ranch grounds and occurred as a waif in 1962 along Panama Road 2.5 miles east of Old River (T7972). Native of Europe.

Scribneria Bolanderi (Thurber) Hackel. An inconspicuous little annual grass that is scarce in bare areas, usually around gooseberry bushes, in the Douglas oak woodland in the Greenhorn Range. It grows in moist disturbed soil in the meadow at Breckenridge Public Camp in the yellow pine forest on Breckenridge Mountain.

Secale cereale Linnaeus. Cultivated rye, cereal rye. An occasional crop of higher altitudes, where it is often spontaneous. Rye is well established in waste places near Glennville, in the Tehachapi region, and along the Mt. Abel Road south of Grocer Grade. Native of southwest Asia.

Sitanion × Hansenii (Scribner) J. G. Smith. A hybrid between *Elymus glaucus* and *Sitanion Hystrix* that is rare in the Bartolas Country on the Kern Plateau (Howell & True 41,843).

Sitanion Hystrix (Nuttall) J. G. Smith. Squirrel tail. Occasional, sometimes common, from the desert ranges to the highest mountains, west to the east slope of the Greenhorn Range and southwest to Mt. Pinos. At lower elevations in the cismontane associations it is replaced by the following species.

Sitanion jubatum J. G. Smith. Occasional to common in the foothills around the valley, mostly in the Upper Sonoran grassland and the Douglas oak woodland.

Triticum aestivum Linnaeus. Wheat. A common dry-farm crop of the Temblor Range and occasionally elsewhere in the hills and mountains, wheat is often spontaneous in many associations but nowhere established. Native of the Old World, probably the Near East.

Aveneae. Oat Tribe

Avena barbata Brotero. Slender wild oat. A common annual in the Lower Sonoran grassland; occasional in the Upper Sonoran grassland; scarce to the lower levels of the yellow pine forest. It is an excellent forage plant when green. Native of the Old World.

Avena fatua Linnaeus. Wild oat, fat oats. Common and often abundant in the Upper Sonoran grassland; occasional in most other cismontane associations. Native of Europe.

Avena sativa Linnaeus. Cultivated oat. Occasionally adventive in the valley, rarely so in the surrounding hills and mountains. An occasional roadside weed on the desert. Native of Europe.

Deschampsia caespitosa (Linnaeus) Beauvois. Tufted hair grass. Scarce at moist places near the summit of Mt. Pinos (T12,594, Hoffmann, 19 July 1927, SBM, 18 July 1927, SBM). It is common in meadows on the Kern Plateau in Tulare County near the Kern County line. (*Aira caespitosa.*)

Deschampsia danthonioides (Trinius) Munro. Slender hair grass. Widespread in cismontane moist places, especially around vernal pools and in moist swales. It is rare in moist subalkaline soil on the desert (Red Rock Canyon). (*Aira danthonioides.*)

Deschampsia elongata (Hooker) Munro. A common meadow grass in the Greenhorn Range that is occasional on Breckenridge Mountain and in the Piute Mountains. (*Aira elongata.*)

Koeleria cristata (Linnaeus) Persoon. June grass. Scarce in the pinyon woodland and the lower levels of the Jeffrey pine forest from Mt. Pinos northeast to Tejon Pass.

Koeleria phleoides (Villars) Persoon. Colonies are scarce, although apparently becoming more common, in the low hills west of Woody, the Poso Hills near the mouth of Poso Creek, and the Antelope Plains and the nearby ridges of the central Temblor Range. Native of Europe.

Schismus arabicus Nees. Arabian grass, sheep grass. An abundant winter annual of the valley where it is the most common plant of the Lower Sonoran grassland except in wet years. It is occasional in the hills around the valley and common in the creosote bush association on the desert. Arabian grass was first collected in California as recently as 1935 at Kettleman Hills by Robert F. Hoover. By 1951 its range included hundreds of thousands of acres of arid California, Arizona, and Baja California. Locally, its explosive extension of range was associated with the great drouths in the San Joaquin Valley from 1945 to 1951. It seems unable to compete well with other annuals in years with good early rains that continue through the winter and is surprisingly scarce in those years (such as 1962). It is much prized by stockmen for its early growth and its ability to revive and make good growth when seemingly dead from drouth. The successful growth of Arabian grass in years when very little else survives to maturity also makes it valuable to wild life and for stabilizing the soil against wind erosion.

Inability to compete rather than intolerance of cold is apparently the primary factor limiting it largely to the Lower Sonoran associations; vigorous plants occasionally grow through the Douglas oak woodland in disturbed soil or bare places with even such cold winters as those of Glennville and Tehachapi. Native of southwest Asia.

Schismus barbatus (Linnaeus) Thellung. Apparently rare although widespread, usually growing in loose sandy soil. Perhaps it is not well collected because of its close field resemblance to the much more common *S. arabicus.* It is common in loose light soil at Sand Ridge and on the Caliente Canyon flood plain and in Ballinger Canyon in upper Cuyama Valley; is occasional in the creosote bush association, especially in the dunes about the dry lakes, and is known from northeast of Lake Isabella (Howell 38,193). Native of Eurasia and Africa.

Trisetum cernuum Trinius var. **canescens** (Buckley) Beal. Known from moist ground in the ponderosa pine forest 1.5 mile south of Greenhorn Summit (Howell 38,425, 38,322) and at Tiger Flat (Howell 38,789). (*T. canescens.*)

Agrostideae. Timothy Tribe

Agrostis alba Linnaeus. Red top. An occasional escape from lawns and pastures. Native of Europe. (*A. palustris.*)

Agrostis diegoensis Vasey. Scarce in meadows in the ponderosa pine and red fir forests of the northern Greenhorn Range.

Agrostis exarata Trinius. Western bent. Common in moist places in the yellow pine and red fir forests.

Agrostis exarata var. **pacifica** Vasey. Pacific bent. Rare along Kern River 3.4 miles south of Kernville (T6372.)

Agrostis idahoensis Nash. Idaho bent. Scarce in wet places in the ponderosa pine forest in the Greenhorn Range and on Breckenridge Mountain.

Agrostis lepida Hitchcock. Known from a spring in the Jeffrey pine forest in the Bartolas Country on the southern Kern Plateau (Howell & True 41,842).

Agrostis scabra Willdenow. Scarce in moist soil from the northern Greenhorn Range southeast to Kern River; rare near Pine Flat on the Kern Plateau. (*A. hiemalis.*)

Agrostis semiverticillata (Forskål) C. Christensen. Water bent grass. Common in cismontane usually subalkaline moist soil below the yellow pine forest. Native of Europe.

Agrostis stolonifera Linnaeus. Creeping bent. Known from Greenhorn Mountain Park (Howell 38,755).

Alopecurus aequalis Sobolevski. Rare along a fork of Cannell Creek at Pine Flat on the Kern Plateau.

Alopecurus saccatus Vasey. Scarce in wet years in the alkali playas about 10 miles north of Lost Hills (T1006).

Aristida divaricata Humboldt & Bonpland. Three awn. Scarce on the hills and flats along the valley east-side from Rag Gulch northwest of Woody south to the hills at the mouth of Pastoria Canyon in the Tehachapi Mountains. This is a disjunct occurrence of a grass that ranges from southern California to Texas and Oaxaca.

Heleochloa schoenoides (Linnaeus) Host. Swamp timothy. A rare grass in California that grew in a small colony in the Kern River channel in the dry bed of Lake Isabella in 1961 (T6488). It is becoming common at the Pixley National Wildlife Refuge in Tulare County just north of the Kern County line (T11,591). Native of southern Europe.

Muhlenbergia asperifolia (Nees & Meyen) Parodi. Scratch grass. Common in moist usually alkaline places in the pinyon and arid Douglas oak woodlands. It is local on the alkali plains near Weedpatch in the valley. (*Sporobolus asperifolius.*)

Muhlenbergia filiformis (Thurber) Rydberg. A common sod-forming grass of meadows and other moist places in the yellow pine forest south to Breckenridge Mountain and the Piute Mountains and at Mt. Pinos.

Muhlenbergia microsperma (DeCandolle) Kunth. Rare and transient in 1960 at a bare area at the mouth of Kern Canyon (T5950).

Muhlenbergia Richardsonis (Trinius) Rydberg. Occasional in stoloniferous colonies in dryish meadows in the upper levels of the pinyon woodland and in the Jeffrey pine forest in the Mt. Pinos region and at Landers Meadow in the Piute Mountains. (*M. squarrosa*).

Muhlenbergia rigens (Bentham) Hitchcock. Deer grass. Uncommon but widespread in the Upper Sonoran woody associations from Glennville to the east slope of the Sierra Nevada at the head of Short Canyon (north of Walker Pass) and around granite outcrops at Landers Meadow in the Jeffrey pine forest in the Piute Mountains. (*Epicampes rigens.*)

Oryzopsis hymenoides (Roemer & Schultes) Ricker. Indian rice-grass. Common on the desert plains in light sandy soils, extending west through the pinyon woodland to the crest of Piute Mountain. It is common along high summit ridges above the Jeffrey pine forest on the Kern Plateau in

Tulare County. In contrast, it occurs locally in light sandy soil in the hot, arid San Joaquin Valley (Sand Ridge near Bakersfield, Belridge). The plant is rarely found within a mile of stock watering places as it cannot endure even moderate grazing. The nutritious seeds are important food for birds and small mammals, and in primitive times, the Indians.

Oryzopsis miliacea (Linnaeus) Bentham. Smilo. Roy V. Parker writes (personal communication) that smilo is sparingly established in range land at the Benton Martin Ranch near Glennville. It is common about the old ranchgrounds and orchards at the San Emigdio Ranch where it was seeded long, long ago. A few plants grew from aerial seedings on the Black Mountain burn in 1961 (Howell 38,223). Smilo offers little promise for range re-seeding in Kern County. Native of the Mediterranean region.

Phleum pratense Linnaeus. Timothy. Occasional in moist soil in Lynns Valley, Bear Valley, and Cummings Valley. Native of Eurasia.

Polypogon interruptus Humboldt, Bonpland & Kunth. A small colony grew on a sand bar in Cedar Creek 0.25 mile below the mouth of Lumreau Creek in 1966 (T12,733). (*P. lutosus.*)

Polypogon monspeliensis (Linnaeus) Desfontaines. Beard grass, rabbit foot grass. Common in wet often subalkaline soil in all the associations below the yellow pine forest. Native of Europe.

Sporobolus airoides (Torrey) Torrey. Alkali sacaton. Common, forming large tussocks in the light subalkaline soils about Buena Vista Lake, north along the sloughs and wastelands to Buttonwillow. It is also common in light soil in the dunes around the desert dry lakes.

Stipa californica Merrill & Davy. Common on the slopes around Shirley Meadows, especially in the disturbed soil about the ski run.

Stipa cernua Stebbins & Love. Nodding stipa, purple needle grass. The most common bunch grass of the Upper Sonoran grassland, extending through the Douglas oak woodland. Contrary to much published opinion, it is of decidedly secondary value for forage. Ecologists frequently assert it was once the dominant grass in the valley grassland and that it has been replaced by annuals because of grazing; there is little evidence to support this view at least in the upper San Joaquin Valley and surrounding foothills. While the historical record is not detailed, the writings of early explorers strongly imply it was a region of winter annuals (Jose Maria Zalvidea in 1806; John C. Frémont in 1844; Charles Preuss in 1844; William Ingraham Kip in 1855; and William Brewer in 1863).

The scant rainfall furnishes an even stronger argument; it is extremely doubtful that a perennial grassland ever existed in a region with a median rainfall of less than six inches. Despite a century of grazing, nodding stipa still grows in dense colonies in the foothills in light soils on open slopes in regions with more than eight inches of rainfall. Dense and extensive stands may have grown on flats and valleys in the hills; if so, these were destroyed by tillage, not by grazing. Some of the best surviving colonies in eastern San Luis Obispo County grow in the Pinole Hills; here an entire colony of German settlers abandoned their homesteads when their last horse starved to death in the dry year of 1898 (Mrs. Nellie King Cooper, oral communication).

Thus, it is reasonable to assume that the present annual grassland of the upper San Joaquin Valley did not replace a perennial grassland in which *Stipa cernua* was dominant but a vegetation of native annuals; in this process grazing played an important but far from fully understood role. (*S. pulchra* in part.)

Stipa coronata Thurber. Giant stipa. Scarce in rocky places in the central Temblor Range. The occurrence of this bunch grass may have been reduced by grazing; its late growth and fruiting make it attractive to domestic and wild animals after annuals have dried. This probably accounts for its present occurrence in inaccessible places in large sandstone outcrops. An isolated colony grows on a gravelly slope in the Douglas oak woodland at the head of Lumreau Creek in the Greenhorn Range; this is one of the few Sierra Nevada stations for giant stipa (Farnsworth 657).

Stipa Elmeri Piper & Brodie. Known from the Jeffrey pine forest in the Bartolas Country on the southern Kern Plateau (Howell & True 41,786).

Stipa latiglumis Swallen. Scarce near the Greenhorn Mountain Park headquarters (Howell 38,394).

Stipa Lemmonii (Vasey) Scribner. Scarce in the yellow pine forest south to Tehachapi Peak.

Stipa nevadensis B. L. Johnson. A recently described species closely related to S. californica that, according to B. L. Johnson (Amer. Journ. Bot. 49:257, 1962), is known from the north spur of Mt. Pinos (Hall 6655, UC).

Stipa occidentalis Thurber. Western needlegrass. Occasional in the yellow pine and red fir forests, often growing around rock outcrops at the highest elevations.

Stipa speciosa Trinius & Ruprecht. Occasional in the arid Upper Sonoran associations in the Temblor Range and the Mount Pinos region; more common on the desert-facing slopes of the mountains and in the desert El Paso and Rand Mountains; occasional on the desert plains. In some areas, its occurrence may have been reduced by severe grazing.

Chlorideae. Chloris Tribe

Chloris Gayana Kunth. Rhodes grass. Known from a small marshy waste place near Kernville (Howell 38,913, 25 September 1962). Native of Africa.

Chloris virgata Swartz. Finger grass. A widespread and often abundant weed of irrigated farmlands in the valley, at Cantil on the desert, and in South Fork Valley. It is occasional and often not persistent at moist places in the Douglas oak woodland.

Cynodon dactylon (Linnaeus) Persoon. Bermuda grass. A widespread and extremely troublesome weed in moist soil below the yellow pine forest, especially in gardens, lawns, and valley farmlands. Introduced from the Old World.

Eleusine indica (Linnaeus) Gaertner. Goose grass. Scarce in wet often subalkaline places in the Weedpatch and Greenfield regions. Although well established in these places, it shows no tendency to increase its range. Native of Europe.

Leptochloa fascicularis (Lamarck) Gray. Sprangle-top. Common, often abundant, in irrigated fields, wet waste places, etc., in the valley and in the desert farmlands at Cantil. It is scarce but becoming more common in Kern and South Fork valleys.

Leptochloa uninervia (Presl) Hitchcock & Chase. A widely distributed but uncommon weed in valley farmlands and along streambanks and other moist places through the Douglas oak woodland in the surrounding foothills.

Leptochloa viscida (Scribner) Beal. The only known California station for this native of Arizona, New Mexico, and northern Mexico is a colony on a canal bank on the Kern Lake Ranch Road, 4.6 miles west of Highway U. S. 99 (T4765, 3 September 1958).

PHALARIDEAE. Canary-Grass Tribe

Phalaris canariensis Linnaeus. Canary grass. An adventive colony grew with seeded barley in 1961 on a new road fill on the Bear Trap Road 0.5 mile east of Lebec (T9757). Native of the Mediterranean region.

Phalaris Lemmonii Vasey. According to Beecher Crampton (pers. comm.) this grass was collected in an irrigation ditch near Buena Vista Lake (A. A. Beetle 3992, 24 April 1945, AHUC).

Phalaris minor Retzius. Lesser canary grass. Occasional in moist soils in the valley, sometimes growing in dense colonies in winter-wet low places or in ditches along irrigated fields. Native of the Mediterranean region.

Phalaris paradoxa Linnaeus. A roadside waif on the Devils Den Road in the hills north of Point of Rocks (T10,641, 2 May 1965). Native of the Mediterranean region.

Phalaris stenoptera Hackel. Harding grass. Seeded with scant success on the Black Mountain burn in 1961, although the plants in ravines with good soil may persist. Roy V. Parker writes (personal communication) that Harding grass is locally and sparingly established on the Indian Hills Ranch southwest of Tehachapi. It is scarce along Kern River above Kernville where it was seeded following wild fires. While the plant is quite drouth resistant when mature, its low germination rate seems to limit its potential for use in range re-seeding. According to J. T. Howell (Leafl. West. Bot. 10:40, 1963) "the origin of this impressive grass is uncertain; it may have originated through hybridization between P. aquatica and P. arundinacea at the Toowoomba Botanical Gardens in Australia or elsewhere". (*P. tuberosa* var. *stenoptera.*)

ORYZEAE. Rice Tribe

Oryza sativa Linnaeus. Rice. Widely planted and sometimes spontaneous but doubtfully persistent in wet places from Buttonwillow to Old River. Spontaneous plants are occasional in shallow pools at the Kern National Wildlife Refuge northeast of Lost Hills. Native of the Old World.

PANICEAE. Millet Tribe

Cenchrus longispinus (Hackel) Fernald. Sand bur. An occasional noxious weed in light sandy soils, especially in alfalfa fields, from Mettlers Station to Delano. It is becoming common in irrigated fields in South Fork Valley. The burs are particularly troublesome to sheep and a problem with all livestock. Hay containing the weed is considered worthless by most livestock men. (*C. pauciflorus.*)

Digitaria sanguinalis (Linnaeus) Scopoli. Crab grass. A widespread but not particularly troublesome cismontane weed of gardens, waste places, and stream banks, especially common at lower elevations. Native of Europe.

Echinochloa colonum (Linnaeus) Link. Water grass. An occasional weed of wet waste places and irrigated fields, mostly in the valley and the Temblor Range but occasional in similar places in Kern and South Fork valleys. Native of the Old World.

Echinochloa crusgalli (Linnaeus) Beauvois. Barnyard grass, water grass, sometimes called wild millet. Common, often abundant, in wet places in the valley where it is often an extremely troublesome weed in alfalfa and especially in rice fields. It is occasional in the Temblor Range and in the mountain valleys. Extensive plantings are made at the Kern National Wildlife Refuge for food for waterfowl. Vars. *zelayensis* (Humboldt, Bonpland & Kunth) Hitchcock and *mitis* (Pursh) Petermann often occur with the species and seem to be no more than random variations. Native of the Old World.

Eriochloa gracilis (Fournier) Hitchcock. A locally common weed in the vineyards around Delano. Scarce, mostly along canal banks, from the Weedpatch region west to the Greeley region. Known as a garden weed at the county road yards at Inyokern (Meng 189). Native from Imperial Valley to Texas.

Imperata brevifolia Vasey. Satin-tail. This impressive grass was collected long ago at Bakersfield (Davy 2917, UC); it is extremely doubtful that it is still present. (*I. Hookeri.*)

Panicum capillare Linnaeus var. **occidentale** Rydberg. Old witch grass. Scarce at scattered localities: roadside spring 3.5 miles west of Richbar, Kern Canyon (T6504); Cedar Creek on the flats just below the mouth of Lumreau Creek (T11,660); sand bar near Hart Park (Benson 8022, POM). (*P. barbipulvinatum.*)

Panicum dichotomiflorum Michaux. A small colony grows on the sandy flat along Cedar Creek 0.5 mile below the mouth of Lumreau Creek (T11,654, 12,727). Native of the eastern United States.

Panicum miliaceum Linnaeus. Broom-corn millet. An occasional garden weed, probably adventive from bird seed. Once collected in the Tehachapi Mountains 2.5 miles southeast of Keene (Howell 32,764, 25 June 1957). Native of the Old World.

Panicum occidentale Scribner. Scarce along Kern River above Lake Isabella (T6447, Howell 38,890, Hoffmann, 7 September 1930, SBM) and rare at the broad flat along Cedar Creek at the Rock Corral (Farnsworth 595).

Paspalum dilatatum Poiret. Dallis grass. Infrequent in wet places in the valley and around Rosamond on the desert, sometimes forming dense colonies on canal and ditch banks where it is an excellent soil binder. It is much planted as a pasture grass in the southeastern United States. Native of South America.

Paspalum distichum Linnaeus. Knot grass. Occasional in sunny wet places in all associations below the yellow pine forest.

Setaria geniculata (Lamarck) Beauvois. Bristle grass. Common in moist soils in the valley, especially in the Bakersfield region. Although it behaves as a weed, apparently it is native. (*Chaetochloa geniculata.*)

Setaria lutescens (Weigel) C. E. Hubbard. Yellow bristle grass. A common weed of moist soils in the valley; occasional to the Douglas oak woodland. It grows as a garden and irrigated farmland weed at Rosamond on the desert. Native of the Old World. (*S. glauca, Chaetochloa lutescens.*)

Setaria sphacelata (Schumacher) Stapf & C. E. Hubbard. According to Philip A. Munz (A California Flora, p 1648), it is an escape in wet places in Kern, Stanislaus, and Butte counties; however, Beecher Crampton writes (personal communication) that these scattered occurrences in California are usually not persistent. Native of Africa.

Setaria verticillata (Linnaeus) Beauvois. A common weed of moist soil in the valley; scarce in the Temblor Range. Native of Europe. (*Chaetochloa verticillata.*)

Setaria viridis (Linnaeus) Beauvois. Green bristle grass. A rare weed in Kern County, known from Kernville (Howell 38,914) and from the Rosamond school grounds (T3965). Native of Europe. (*Chaetochloa viridis.*)

ANDROPOGONEAE. Sorghum Tribe

Sorghum bicolor (Linnaeus) Moench. Milo, gyp corn. Often spontaneous, nowhere established, in the valley, lower Kern Canyon, Caliente, etc. Native of the Old World. (*S. vulgare.*)

Sorghum halepense (Linnaeus) Persoon. Johnson grass. A common weed of waste places and irrigated fields in the valley, in the mountains at South Fork Valley, and in the Temblor Range. When frosted it is highly toxic to livestock. Native of the Old World. (*Holcus halepensis.*)

Sorghum sudanense (Piper) Stapf. Sudan grass. A rare escape in waste places in the valley. Native of the Sudan.

TRIPSACEAE. Corn Tribe

Zea Mays Linnaeus. Domestic corn. An occasional escape in the valley, never persisting. Native of Mexico.

CYPERACEAE. Sedge Family

Carex. The genus *Carex,* the largest in California with 147 species, is represented by at least one of these bewildering and diverse species in nearly all moist places in the mountains and commonly elsewhere. Many of them spread from root stocks, often forming a dense turf. In heavily grazed mountain meadows, such as Little Cannell Meadow on the Kern-Tulare County line on the Kern Plateau, this mat-like turf becomes dense enough to exclude almost all other plants. The rhizomatous sedges are valuable in preventing washing and cutting of these heavily grazed meadows. Most species are excellent forage; indeed, they are the principal livestock feed of many of the lush mountain meadows.

Although often thought to be of almost universal occurrence in California, no species grows on the Kern desert and only one grows at a single spring in the Temblor Range. *Carex* is feebly represented in the valley by one small colony on a canal bank near Weedpatch. Even so, the 28 species that occur here make it the second largest genus in the county.

Taxonomically the genus is complex and dealing with it requires a specialized knowledge possessed by few botanists. All of the determinations of our species have been made by John Thomas Howell, the leading authority on the California sedges.

Carex alma Bailey. Occasional, mostly on protected slopes, often in moist rock clefts in canyon walls, from the Douglas oak woodland to the lower limits of the yellow pine forests.

Carex amplifolia Boott. A few plants grow in moist heavily shaded soil along Lumreau Creek at the Borton Place in the Douglas oak woodland in the Greenhorn Range (T12,745).

Carex angustior Mackenzie. Scarce in wet soil around a broad seep near Mountain Meadow in the red fir forest on the east slope of Sunday Peak (T7761), a southern range extension from Mariposa County.

Carex athrostachya Olney. One of the most common sedges of the meadows in the yellow pine forest, growing in both moist and dryish soils.

Carex barbarae Dewey. Occasional in meadows in Lynns Valley in the Greenhorn Range.

Carex Bolanderi Olney. Occasional in moist places in the Greenhorn Range south to Shirley Meadows.

Carex Brainerdii Mackenzie. Known from 1.5 mile south of Greenhorn Pass (Howell 38,439), a southern range extension from Mariposa County.

Carex densa (Bailey) Bailey. Occasional in moist meadows in Lynns Valley.

Carex Douglasii Boott. Common in dry gravel on the broad summit of Mt. Pinos. It grows in dense colonies in dry loam on the sage brush flats at the borders of Woolstaff Meadow in the Piute Mountains.

Carex feta Bailey. Occasional in moist meadows in the ponderosa pine and red fir forests in the northern Greenhorn Range.

Carex fracta Mackenzie. Scarce in meadows in the yellow pine and red fir forests on Breckenridge Mountain and in the Greenhorn Range.

Carex gracilior Mackenzie. Common in mud around a pond on the west side of Lynns Valley in the Greenhorn Range (T9466).

Carex Jonesii Bailey. Occasional in moist places in the ponderosa pine forest in the Greenhorn Range south to Shirley Meadows.

Carex Kelloggii W. Boott. Scarce at Tiger Flat in the Greenhorn Range (T3764), the southern limit for the species.

Carex lanuginosa Michaux. Extensive at Shirley Meadows (T3921, Howell 38,607); Browns Meadow in the Piute Mountains (Breedlove 3983A); and Mt. Pinos at 8,200 feet elevation (Hoffmann, 18 July 1927, SBM). It occurs in a completely different sort of place along a canal near Weedpatch (T5002), the only occurrence of a Carex in the San Joaquin Valley in Kern County.

Carex Lemmonii W. Boott. Known from Shirley Meadows (Howell 38,603) and from a moist bank 1.5 mile south of Greenhorn Summit (Howell 38,443, 38,444).

Carex mariposana Bailey. Occasional in wet meadows in the Greenhorn Range south to Shirley Meadows and from springs, creeks, and moist meadows from Pine Flat east through the Bartolas Country on the Kern Plateau, the southern limits for the species.

Carex multicaulis Bailey. Common on dry slopes in the ponderosa pine forest in the Greenhorn Range; scarce on Breckenridge Mountain.

Carex nebrascensis Dewey. This distinctive sedge of many southern Sierran meadows, with black spikes and cord-like roots, is locally abundant along a fork of Cannell Creek at Pine Flat and at Little Cannell Meadow and grows in small colonies along Bartolas Creek near its summit on the Kern Plateau.

Carex paucicostata Mackenzie. Reaches its southern limit at Shirley Meadows (Howell 38,459, 38,605).

Carex praegracilis W. Boott. Occasional to common in moist places from the Douglas oak woodland to the yellow pine forest in the mountains, sometimes growing in subalkaline soil.

Carex Sartwelliana Olney. Scarce in moist places in the ponderosa pine and red fir forests in the Greenhorn Range south to Shirley Meadows (northwest slope of Black Mountain, Howell 38,561; Shirley Meadows, Howell 38,506, T3732). (*C. yosemitana.*)

Carex senta Boott. Scarce but widespread in the Douglas oak woodland in the Greenhorn Range; along Kern River in Kern Canyon; rare, southeast to Walker Basin. The very large tussocks of this sedge are an impressive feature of the broad sunny moist meadow at the southeast end of Walker Basin.

Carex serratodens W. Boott. Occasional along upper San Emigdio Creek and its tributaries in the pinyon woodland and the Jeffrey pine forest.

Carex simulata Mackenzie. Scarce and very local along Rancheria Creek in the chaparral on the west slope of Piute Mountain (T2851), an isolated occurrence that is the southern limit for the species.

Carex subfusca W. Boott. Occasional in the meadows in the mountains from the Douglas oak woodland to the Jeffrey pine forest.

Carex teneraeformis Mackenzie. Common in wet soil in the mountains from the Mt. Pinos region north to Kern Canyon.

Carex vicaria Bailey. Rare at Pettit Meadow in the ponderosa pine forest on the west slope of the Greenhorn Range.

Cyperus aristatus Rottboell. Occasional in wet soil or shallow water in the Upper Sonoran associations from Kern Valley west through the Greenhorn Range.

Cyperus difformis Linnaeus. A scarce rice field weed in the Buttonwillow region and around the shore of Lake Woolloomes near Delano. It is apparently a recent introduction in the county, perhaps arriving with the cultivation of rice. Native of Asia.

Cyperus Eragrostis Lamarck. Scarce in wet sometimes subalkaline places in the valley, most frequently in the Bakersfield region.

Cyperus erythrorhizos Muhlenberg. Common in wet places in the valley; scarce along Kern River in Kern Valley.

Cyperus esculentus Linnaeus. Chufa, yellow nut grass. Occasional in wet places in the valley. It is scarce along Cedar Creek in the Douglas oak woodland in the Greenhorn Range and local about Kernville and at a roadside spring two miles west of Wofford on the Greenhorn Pass Road.

Cyperus ferax Richard. Scarce along canals and in other very wet soils in the valley.

Cyperus laevigatus Linnaeus. Rare, although sometimes occurring in extensive colonies in wet subalkaline soils, from Canebrake Creek below Walker Pass west to Scovern Hot Springs near Isabella.

Cyperus niger Ruiz & Pavon var. **capitatus** (Britton) O'Neill. Occasional along Kern River from Kernville to Hart Park (near Bakersfield) and at wet places in the Douglas oak woodland on the west slope of the Greenhorn Range. (*C. melanostachys.*)

Eleocharis acicularis (Linnaeus) Roemer & Schultes. Needle spike rush. Common in moist places in the ponderosa pine forest in the Greenhorn Range and along Kern River west to Hart Park. It is rare at Lake Woolloomes near Delano. (*Heleocharis acicularis* var. *acicularis.*)

Eleocharis bella (Piper) Svenson. Known from a seep in the ponderosa pine forest 1.5 mile south of Greenhorn Pass (Howell 38,421A). (*Heleocharis acicularis* var. *bella.*)

Eleocharis macrostachya Britton. Common spike rush. Common in wet places, often growing in water to as much as three feet deep, in all cismontane associations below the yellow pine forest; occasional in the Jeffrey pine forest. It is especially common in the valley. (*Heleocharis palustris.*)

Eleocharis montevidensis Kunth. Scarce in wet and moist places in the mountains, mostly in the Douglas oak woodland. (*E. montana, Heleocharis montevidensis* var. *montevidensis.*)

Eleocharis Parishii Britton. Common in wet often subalkaline places in the mountains, especially in the meadows in the Jeffrey pine forest; occasional in the Temblor Range. While it is scarce in the desert, it is the only species of *Eleocharis* that grows there (Red Rock Canyon, Mesquite Springs, Cudahy Camp in Last Chance Canyon, Dove Spring). (*Heleocharis montevidensis* var. *Parishii.*)

Eleocharis pauciflora (Lightfoot) Link var. **bernardina** (Munz & Johnston) Svenson. Common in wet meadows in the Jeffrey pine forest on Mt. Pinos, often forming dense mats. (*Heleocharis pauciflora* var. *bernardina.*)

Eleocharis radicans (Poiret) Kunth. Scarce along Kern River east of

Wofford (T6373, 6373B). The plants are atypical with bifid rather than trifid styles. (*Heleocharis acicularis* var. *radicans.*)

Eleocharis rostellata (Torrey) Torrey. Walking sedge. Scarce in locally extensive colonies in wet often highly mineralized soils in the pinyon and Douglas oak woodlands in Black Bob and San Emigdio canyons. It is also known from near Kernville (Howell 38,918) and nearby Kern Hot Springs and Bath House Canyon (T12,213) where it grows in dense colonies.

Fimbristylis thermalis Watson. According to Philip A. Munz (A California Flora, p. 1422), it is recorded from marshes and hot springs in San Bernardino, Kern, and Inyo counties. The commercial development of the hot springs in Kern County about the turn of the century probably eliminated our colonies of the plant.

Fimbristylis Vahlii (Lamarck) Link. Scarce, usually on drying mud, along Kern River from Hart Park west to the Buena Vista Lake region. Rare in Kern Canyon (1,400 feet elevation, Howell 38,687).

Hemicarpha occidentalis Gray. Scarce along Wagy Creek in the chaparral at the Wagy Creek fire road (Hardham 12,987); known from a sand bar in Kern River near Hart Park (Benson 8021).

Scirpus acutus Muhlenberg. Common tule. Widespread and often abundant at wet places in the valley, less common in the surrounding foothills. The "tule" of Tulare Lake, it once densely covered thousands of acres of shallow water in Tulare Lake and formed dense almost impenetrable colonies in the connecting sloughs and marshes south to Buena Vista Lake. These vast colonies were havens for hordes of wild fowl, for Indians eluding missionaries, and, later, for outlaws fleeing lawmen. Hudson Bay trappers illicitly entered Mexican California to trap beaver in the tule marshes; Peter Lebec, of whom almost nothing is known, may have been one of them.

Scirpus americanus Persoon. Three square. Uncommon but widespread in wet often subalkaline soils below the yellow pine forest. It reaches the edge of the desert at a spring along Freeman Creek below Walker Pass (Onyx Ranches line camp) and in Red Rock Canyon.

Scirpus californicus (C. A. Meyer) Steudel. Known from Bakersfield (Benson 7636, POM).

Scirpus microcarpus Presl. Rare at Shirley Meadows in the Greenhorn Range (T3919); Greenhorn Mountain Park (T12,404); at a spring at 8,200 feet elevation on Mt. Pinos (Hoffmann, 19 July 1927, SBM); and at Bakersfield (Benson 3294, POM).

Scirpus Olneyi Gray. Occasional in often highly mineralized wet places in the San Emigdio Range east to Ft. Tejon. An extensive colony grows along a shallow oil field waste water creek at the old Temblor Oil Field at the mouth of Temblor Canyon.

Scirpus paludosus A. Nelson. Prairie bulrush. Common in sunny wet places, especially in subalkaline soil, in the valley and the Temblor Range. It occasionally grows in similar places on the desert. (*S. robustus, S. campestris.*)

Scirpus saximontanus Fernald. Rocky Mountain bulrush. A recent introduction in California, first collected in Kern County in 1954, that is already becoming common around the duck clubs north of Semitropic. Rocky Mountain bulrush appeared at the Kern National Wildlife Refuge less than two months after the arid subalkaline plains were flooded to form marshes in 1964; in the summer of 1965 it grew in extensive colonies in mud and shallow water. Native to the Pacific Northwest and the Rocky Mountains.

Scirpus tuberosus Desfontaines. Tuberous bulrush. Becoming common in wet subalkaline soil and shallow water at the duck clubs north of Semitropic. Native of the Old World.

PALMAE. Palm Family

Washingtonia filifera (Lindley) Wendland. California fan palm. Eight spontaneous trees grow along Kern River near the canyon mouth. Apparently this is the only spontaneous occurrence of California fan palm in the western United States. It is native in similar places about the Colorado Desert; the Kern colony indicates the tree has the ability to establish itself in favorable situations with a minimum of assistance from man. (Cf. E. C. Twisselmann. Leafl. West. Bot. 9:187.)

LEMNACEAE. Duckweed Family

Lemna gibba Linnaeus. According to E. Landolt (Ber. Schwiz. Bot. Ges. 67:271–410), gibbous duckweed is known from Bakersfield (Davy, UC).

Lemna minima Philippi. Least duckweed. One of the smallest of all flowering plants, least duckweed is common in pools, quiet streams, etc., in all the cismontane associations.

Lemna valdiviana Philippi. E. Landolt (Ber. Schwiz. Bot. Ges. 67:271–410) cites as *L. cyclostasa,* a synonym, collections from Kern River (Dudley, DS) and Bakersfield (Thompson, DS).

Spirodela polyrhiza (Linnaeus) Schleiden. Scarce in ponds and quiet streams in the valley, sometimes locally common in similar places through the Douglas oak woodland in the mountains.

Wolffiella lingulata (Hegelmaier) Hegelmaier. Known from 3 miles west of Bakersfield (J. Walter Thompson 204, DS). H. L. Mason (Madroño 4:245) reports another collection from near Bakersfield (C. H. Thompson, 7 October 1895, M). A minute aquatic plant that may be overlooked.

PONTEDERIACEAE. Pickerel Weed Family

Eichhornia crassipes (Martius) Solms-Laubach. Water hyacinth. Once common and a pest in canals, etc., about Bakersfield, water hyacinth was subjected to an eradication program in the 1920's and 1930's and is presumably no longer present in the county. Native of tropical America.

JUNCACEAE. Rush Family

Juncus balticus Willdenow. Baltic rush. Common in sunny wet places everywhere. It is very tolerant of subalkaline soils. Baltic rush often grows in soil that appears dry but has subsurface moisture.

Juncus bufonius Linnaeus. Toad rush. By far the county's most common rush, the annual toad rush is almost universal in wet places.

Juncus dubius Engelmann. Occasional in wet places from the Upper Sonoran associations in the mountains to the yellow pine forest.

Juncus effusus Linnaeus var. **pacificus** Fernald & Wiegand. Pacific rush. Occasional in wet places in the northern Greenhorn Range.

Juncus ensifolius Wikstrom. Known from moist places in the ponderosa pine forest 3 miles north of Greenhorn Pass (Howell 38,786), a southern range extension from Fresno County.

Juncus Kelloggii Engelmann. Rare in mud and wet granite sand in the Greenhorn Range and at the southwest rim of the Kern Plateau. (*J. triformis* var. *brachystylus.*)

Juncus macrandrus Coville. Occasional in meadows in the yellow pine forest south to the Piute Mountains and Breckenridge Mountain. (*J. phaeocephalus* var. *paniculatus* in part.)

Juncus macrophyllus Coville. Occasional on Black Mountain in the Greenhorn Range. (*J. canaliculatus.*)

Juncus mexicanus Willdenow. Mexican rush. Common in wet sunny places, often in low areas that are wet only in winter; occasional to the Jeffrey pine forest. It occurs in moist subalkaline soil on the desert and a few old colonies survive on the hard dry subalkaline clay bed of Rosamond Dry Lake, where there is no other plant life for scores of yards around.

Juncus nevadensis Watson. Scarce in moist or wet places in the yellow pine forest south to Grouse Meadow in the Piute Mountains.

Juncus orthophyllus Coville. Scarce in dryish soil on the borders of meadows in the Jeffrey pine forest (Kern Plateau, Piute Mountains, Mt. Pinos).

Juncus oxymeris Engelmann. Common in wet places in the Temblor and San Emigdio ranges; scarce in the Upper Sonoran associations in the other mountains.

Juncus rugulosus Engelmann. Scarce in wet places in the Douglas oak woodland from Kern Valley west through the Greenhorn Range.

Juncus sphaerocarpus Nees. Occasional in wet places in the mountains south to Bear Valley.

Juncus tenuis Willdenow. Slender rush. Scarce in meadows and along streams in the ponderosa pine forest in the Greenhorn Range and in the Douglas oak woodland in Lynns Valley. Var. *congestus* Engelmann grows on the south side of Black Mountain 2 miles east of Greenhorn Mountain Park (Howell 38,731).

Juncus Torreyi Coville. Scarce along Kern River from Kern and South Fork valleys to Bakersfield.

Juncus xiphioides E. Meyer. Iris-leaved rush. Common in wet soil and shallow water in the Temblor Range and in the mountains, mostly below the forests.

Luzula comosa E. Meyer. Wood rush. Scarce in wet places in the forests south to Shirley Meadows and Tillie Creek in the Greenhorn Range; rare near Pine Flat on the Kern Plateau. It grows around a permanent pool in Douglas oak woodland along Fugitt Road on the west side of Lynns Valley (Farnsworth 530). (*L. campestris.*)

Liliaceae. Lily Family

Allium Burlewii A. Davidson. Scattered colonies grow in gravel on the northeast slope of Mt. Pinos (and on nearby Frazier Mountain in Ventura County). It is also known from Shepherds Peak, vicinity of Bisses Station, Tehachapi Mountains (Dudley 417, 26 June 1895, DS) and from Pah Ute [Piute] Peak (Purpus 5333, June, 1897, UC).

Allium campanulatum Watson. Occasional in good soil from the chaparral to the Jeffrey pine forest in the mountains.

Allium crispum Greene. Common in good soil in the Temblor Range; less common in the Upper Sonoran associations southeast to the Tehachapi Mountains. (*A. peninsulare* var. *crispum.*)

Allium Davisae Jones. Occasional, sometimes locally common, in gravelly soil from the desert ranges west through the mountains to the east slope of Piute Mountain and the Greenhorn Range, the northwest limits for the species. Plants from the east slope of the Greenhorn Range seem to be close to *A. lacunosum* Watson var. *micranthum* Eastwood, an onion of the inner south Coast Ranges.

Allium fimbriatum Watson. Occasional from the desert to the west slope of the Piute Mountains and the Greenhorn Range; southwest to the Tollgate Ridge region east of Keene. It is most common in the pinyon woodland but reaches the edge of the desert in the arid shrub association and grows in the arid Douglas oak woodland on the east slope of the Greenhorn Range.

Allium fimbriatum var. **denticulatum** Ownbey & Aase. The common form in the Jeffrey pine forest from the Piute Mountains southwest to the high exposed summits of Mt. Pinos and Sawmill Mountain.

Allium fimbriatum var. **mohavense** Jepson. Occasional on the desert, mostly in canyons or on high slopes. It grows in friable clay in the pinyon woodland at the head of Quatal Canyon in the Mt. Abel region in a form that, according to Marion Ownbey (personal communication), is intermediate with var. *diabloense* Ownbey & Aase (T7200).

Allium Howellii Eastwood. Howell onion. Scarce on the plains on the west side of the valley south to the San Emigdio Range; in arid places in the southern Temblor Range; local in heavy soil in Poso Hills and reaches its northeastern limits at White River, Tulare County (Farnsworth 346). An isolated colony grows in clay on the divide between Tehachapi and Keene. The type was collected at the summit of Maricopa Grade, 7 miles from Maricopa (John Thomas Howell 5915, 21 March 1931).

Allium hyalinum Curran. Paper-flowered onion. Occasional, in dense colonies, usually in shallow moist soil on sheets of granite, in the Douglas oak woodland and chaparral south to Tehachapi. All of the wild onions are edible; this one is mild and particularly tasty. Hugh Smith says (oral communication) it often served as an emergency lunch for cowboys in the Greenhorn Range.

Allium peninsulare Lemmon. Occasional in chaparral and Douglas oak woodland at higher elevations in the Temblor and San Emigdio ranges; more common in the east side foothills, extending east to South Fork Valley.

Allium tribracteatum Torrey. Scarce in decomposed granite on Black Mountain in the Greenhorn Range (saddle between Black Mountain and Cane Peak, T9797; northeast slope of Black Mountain, Howell 38,556).

Asparagus officinalis Linnaeus. Garden asparagus. Asparagus is well established along Kern River from Bakersfield west to Buena Vista Lake. It is often spontaneous about ranchgrounds, gardens, towns, etc., and is also established along old canals in the Greenfield region. Native of Europe.

Bloomeria crocea (Torrey) Coville. John Ingram (Madroño 12:22) reports a collection from the mesas near Bakersfield (Osborn in 1930, LAM). Considering the geographical range of the typical subspecies and the ecological situation on the Kern mesas, this is a most anomolous station, and it has not been verified by other collections.

Bloomeria crocea var. **aurea** (Kellogg) Ingram. Golden stars. Common in grasslands in the northern Temblor Range, extending through the Douglas oak woodland in heavy or deep soils.

Bloomeria crocea var. **montana** (Greene) Ingram. Mountain golden stars. Occasional in the Greenhorn foothills, more frequent southwest to the Tehachapi Mountains, the San Emigdio Range and the Mt. Pinos region, northwest to the southern Temblor Range. It grows mostly in heavy soil in the grasslands but extends to the pinyon woodland and is rare in the arid shrub association on flats between upper Jawbone and Pinetree canyons. Both varieties of *Bloomeria* have colorful flowers that bloom after most spring growth has dried. The type of var. *montana* was collected near Tehachapi (Katharine Layne Curran in June, 1884).

Brodiaea coronaria (Salisbury) Engler. A colony grows in the dense vegetation in the highly mineralized soil at Kern Hot Springs (T12,205).

Brodiaea coronaria (Salisbury) Engler var. **kernensis** Hoover. Kern brodiaea. Endemic to the arid low hills along the east side of the San Joaquin Valley from southern Tulare County to the north end of the Tehachapi Mountains. It is occasional in the Douglas oak woodland in the Glennville and Tehachapi regions and in the ponderosa pine forest at Golf Meadow on Breckenridge Mountain. The type was collected at Poso Creek Oil Field (Robert Francis Hoover 4083).

Brodiaea laxa (Bentham) Watson. Ithuriel's spear. A distinctive flower of late spring that is common from the Douglas oak woodland through the lower levels of the yellow pine forest south to the Tehachapi Mountains. It is rare in the San Emigdio Range (Pleito Canyon).

Brodiaea lutea (Lindley) Morton var. **analina** (Greene) Munz. Scarce in small colonies in the Jeffrey pine forest in the Piute Mountains (T3646, 7347, 7353).

Brodiaea lutea var. **scabra** (Greene) Munz. Golden brodiaea, pretty face. A common and colorful spring flower from the foothills to the lower levels of the yellow pine forest south to the Tehachapi Mountains and on the higher plains of the valley east-side; rare at Neasons Flat in the San Emigdio Range.

Brodiaea pulchella (Salisbury) Greene. Common brodiaea, blue dicks, and locally, pig nut. Common in all the associations below the yellow pine forest, especially in the Upper Sonoran grassland. Occasional on the desert mountains and slopes.

Brodiaea volubilis (Morière) Baker. Snake lily. An interesting twining plant with attractive rose-pink flowers that is scarce in the higher Douglas oak woodland and the lower ponderosa pine forest on the west slope of the Greenhorn Range and rare on the east slope. (*Dichelostemma californicum.*)

Calochortus amoenus Greene. A very common nodding pink calochortus of the Greenhorn Range, extending from the foothills to open areas in the ponderosa pine forest. (*C. albus* var. *amoenus.*)

Calochortus clavatus Watson. Yellow mariposa. Scarce in colonies in light soil on arid slopes from the California juniper association in the Temblor Range southeast to the pinyon woodland in the Mt. Pinos region. It is one of the county's most striking flowers, the yellow of the petals having exceptional depth and clarity. (Our plants have been described as *Mariposa clavata* (Watson) Hoover var. *pallida* Hoover.)

Calochortus coeruleus (Kellogg) Watson var. **Westoni** (Eastwood) Ownbey. Weston mariposa. The type colony, which grew at the saddle above Shirley Meadows, has been destroyed by the construction of a ski run. (Type: E. Roy Weston 680, 20 May 1927; exact topotype: T5216.) Recently, a very small colony was discovered in the incense-cedar grove just above the toboggan slide at Shirley Meadows; this will probably not survive for long. Thus, the Weston mariposa seems likely to be one of the first casualties of the despoilation of the southern Sierra for recreational facilities.

Calochortus invenustus Greene. Occasional through the arid Douglas oak and pinyon woodlands to the Jeffrey pine forest from the Kern Plateau and Walker Pass regions south to Mt. Pinos and Sawmill Mountain, growing most often on high arid slopes and exposed summit ridges. The type was collected in the mountains west of the Mojave Desert, probably near the summit of Tehachapi Peak (Edward L. Greene, 25 June 1889).

Calochortus Kennedyi Porter. Desert mariposa. A colorful bright red mariposa that is occasional on the desert mountain and slopes (especially common in the Randsburg region) and frequent west to the Mt. Pinos region, reaching Santa Barbara County at the east end of Cuyama Valley (T7176). The type was collected in the Sierra Nevada Mountains in Kern County (William L. Kennedy, spring, 1876). The plant was named for the collector of the type, a pioneer merchant, miner, and financier of Havilah, Keysville, and Kernville who collected in the Kern County mountains for J. T. Rothrock in 1876.

Calochortus luteus Douglas. A beautiful mariposa lily common in the Greenhorn Range from the Douglas oak woodland to the lower levels of the ponderosa pine forest.

Calochortus Palmeri Watson. Abundant and colorful at Weldon Meadow and less common at Landers Meadow in the Piute Mountains. It is occasional in meadows and moist swales in the Jeffrey pine forest on the high north slope of Breckenridge Mountain.

Calochortus striatus Parish. Alkali mariposa. Rare in low winter-wet subalkaline places in the dunes south of Rogers Dry Lake (T10,814), at an alkali seep at the mouth of Chimney Creek Canyon near Weldon (T10,936), and at Kern Hot Springs (T12,200). The last station is the western limit for this mariposa of moist alkali places which ranges across the deserts to Las Vegas.

Calochortus superbus Purdy. Occasional from the west slope of Blue Mountain in the Greenhorn Range west to Delano and Richgrove, growing in the Douglas oak woodland and Upper Sonoran grassland.

Calochortus venustus Douglas. Butterfly mariposa. Common from the arid shrub association along the desert borders to the pinyon woodland, occasional in the woody Upper Sonoran associations south of Kern Canyon to the Mt. Pinos region and in the Temblor Range. The colors and shading of the flowers are diverse and of great beauty; those on the high slopes in the southwestern Tehachapi Mountains are exceptionally large and variably colored. An un-named purple-flowered race occurs in the Mt. Pinos region. Var. *purpurascens* Watson is the form that occurs in clay in the central Temblor Range and at Davis Guard Station in the Greenhorn Range.

Calochortus Vestae Purdy. Scarce in heavy soils in the Douglas oak woodland in the Greenhorn Range. (*C. luteus* var. *oculatus.*)

Chlorogalum pomeridianum (DeCandolle) Kunth. Soap plant, amole. Scarce in the Upper Sonoran grassland, occasional in the Douglas oak woodland, locally common in openings in the chaparral, in the east-side foothills south to Tejon Canyon and southeast to Cummings Valley. The Indians are said to have used the bulbs for soap.

Fritillaria agrestis Greene. Mission bells, chocolate lily. Scarce in fine compacted sand around sandstone outcrops in the central Temblor Range.

Fritillaria Brandegei Eastwood. Greenhorn fritillary. Scarce in colonies in heavy soil on north slopes in the ponderosa pine forest and rare in the shin oak brush in the Greenhorn Range (the plant is endemic to this region and extreme southern Tulare County). These colonies are usually browsed by deer and the flower stalks are completely eaten so that the plant is rarely found in bloom.

Fritillaria pinetorum A. Davidson. Pine fritillary. Occasional in humus in shade along the crest of Piute Mountain; rare in a moist opening in the chaparral at the head of Stevenson Creek east of Keene and in the yellow pine forest in the Tehachapi Mountains; known from the Jeffrey pine forest on Mt. Pinos at 8,500 feet elevation (Hoffmann 25 May 1928, SBM). The type was collected at Cummings Mountain in the Tehachapi Mountains (Herman Hasse & Anstruther Davidson 1739, 12 June 1907).

Fritillaria striata Eastwood. Adobe lily. Scarce in colonies in heavy adobe in the Douglas oak woodland on the west slope of the Greenhorn Range. It is endemic to this region and southern Tulare County. The type was collected on Rattlesnake Grade (E. Roy Weston 524, 13 March 1927).

Lilium Kelleyanum Lemmon. Sierra lily. Charlotte Nash Smith reports (personal communication) that the Sierra lily once grew near Cane Spring on Black Mountain and near Alta Sierra Lodge below Greenhorn Pass. J. T. Howell collected a very mature lily (38,791) just north of Greenhorn Pass; although it is impossible to determine it definitely, it is apparently this species. It is probable that both grazing and collecting for gardens have greatly reduced the occurrence of this species in the southern Sierra Nevada; the drouths of the last 40 years also may have been a factor. (Tiger Flat, in the Greenhorn Range, is said to have been named for the plant, which has not been present there in the last 30 years.) *(L. pardalinum.)*

Muilla coronata Greene. A colony grows on a small sandy flat in the creosote bush association between Horse and Sage canyons on the west side of Indian Wells Valley (Meng 23, 253). Mrs. Meng found that the sand, wind deposited over a very long period, was about six inches deep and deceptively covered a bed of compacted fine clay. The Muilla bulbs were about four inches below the surface of the buried clay bed. The plant is also known from Iron Canyon in the El Paso Mountains (Weston 488).

Muilla maritima (Torrey) Watson. Scattered colonies occur on the high plains on the west side of the valley. It is rare near the summit of Mt. Pinos and on the summit of Mt. Abel. The montane plants are much smaller and superficially little resemble the taller plants with open inflorescences of the hot arid plains far below.

Nolina Parryi Watson ssp. **Wolfii** Munz. Bear-grass. Occasional, in some places locally common, on the steep rocky slopes from Fay and Chimney Creek Canyons to the summit of the south rim of the Kern Plateau, east to Short Canyon on the desert border in the Mt. Owen region. When in bloom, the plants are dramatic accents for the steep dry granite slopes and walls. This is a disjunct occurrence of a plant of the mountains around the southern Mojave and Colorado deserts.

Smilacina stellata (Linnaeus) Desfontaines. Slim solomon. Rare at Iris Point on Mt. Pinos (T3782) and in wet soil under the willows at the north end of French Meadow in the Piute Mountains (C. N. Smith 870, JEPS, Breedlove 3964, T7748).

Veratrum californicum Durand. Skunk cabbage, corn lily, false hellebore. A common and striking plant that grows in dense colonies in sunny wet meadows in the Greenhorn Range and on Mt. Pinos.

Yucca brevifolia Engelmann. Joshua tree. This characteristic tree of the high desert forms an extensive woodland on the plains between Mojave and Soledad Mountain and the east base of the Tehachapi Mountains. It occurs sporadically on the rest of the desert plains and in the canyons and mesas of the El Paso Range. The trees with their angular limbs are said to have reminded the early Mormon explorers of Joshua with his arms outstretched in supplication, hence the common name. To those new to the desert, the Joshua tree often seems weird and grotesque, but to others it is strangely beautiful and breaks the monotonous landscape of the arid slopes and plains where it grows in a fashion most fitting for a desert where many other forms of life might seem to the casual wayfarer equally weird and grotesque.

According to Edmund Jaeger (Desert Wildflowers, pp. 15–19), at least 25 species of birds nest in the tree yuccas; it is the host for the larvae of a butterfly, the Navajo yucca borer (*Megathymus yuccae navaho*), and of the yucca-boring beetle (*Scyphophorus yuccae*). It also furnishes the obligate habitat of an insect-eating night lizard (*Xantusia vigilis*).

The pollination of the Joshua tree is a remarkable example of symbiosis. Yuccas (excepting only one species) are pollinated only by species of small unimpressive silver-gray night-flying moths of the genus *Tegeticula (-Pronuba)*. The female lays her eggs in the pistil of a flower and then deliberately deposits a ball of previously gathered pollen on the stigma; this she carefully rubs in place with her head. The larvae, which apparently can exist only on a diet of immature yucca seeds, hatch as the fertilized flower matures. These larvae eat some but never all of the immature seeds before boring out of the pod and pupating in the soil. Pollination structures and mechanics of plants are infinitely diverse and specialized. These are usually vital to the plant, but are only incidental to the insect in its search for nectar. However, the continued existence of the stately Joshua tree and the lowly moth are completely interdependent. Joseph Wood Krutch (The Desert Year, pp. 79–97) relates how he actually watched this pollination; this leads him to an unusually interesting philosophical discussion based on this symbiotic relationship.

Although well authenticated, the moth and yucca relationship seems an almost incredible example of parallel evolution. As probably every desert naturalist has done, Eben McMillan and the writer, as a lunch time diversion at the summit of Jawbone Canyon in July, 1958, cut open about 150 pods. Every pod had some seeds eaten, but in all some remained. In the average pod, about two thirds had been consumed.

Yucca brevifolia var. **Herbertii** (J. M. Webber) Munz. Kern Joshua. This distinctive Joshua tree that reproduces freely from underground rootstocks and forms dense colonies of usually unbranched trees is the common form from Antelope Valley in northeastern Los Angeles County to Walker Pass and upper South Fork Valley. It usually grows in deeper looser soil than *Y. brevifolia* and apparently requires more moisture. Perhaps the most impressive grove is in Freeman Canyon on the east side of Walker Pass; another extensive and especially well developed colony grows in Kelso Valley. Small colonies that can easily be observed by desert travellers grow on the sand flats in Red Rock Canyon.

In 1966, Eben McMillan and the writer examined a colony below Tejon Pass that had been burned by a wild fire the year before. About 35 mature stems had been killed; at least 70 vigorous new stems had grown from the underground root system. Some were more than 40 feet from the perimeter of the burned colony. While few Kern Joshuas now grow in places that would support a hot wild fire, this response to burning is another piece of evidence that the Mojave Desert's present arid climate is a relatively recent development.

Yucca Whipplei Torrey ssp. **caespitosa** (Jones) Haines. Lord's candle, yucca. The tall flowering stalks and their white flowers are a distinctive feature of rocky slopes in the Douglas oak and pinyon woodlands in the mountains. Plants on the dry slopes east and north of Weldon have short stocky flowering stalks; this form is abundant and makes a dramatic display when in bloom on the granite cliffs and slopes at the head of Spanish Needle Creek and in Chimney Creek gorge. The plant grows at elevations of more than 7,000 feet on the rim of the Kern Plateau in the Bartolas Country west of the South Fork of Kern River.

A form that is probably ssp. *percusa* Haines is occasional, sometimes in fairly sizeable colonies, in the Upper Sonoran subshrub association in the southern Temblor Mountains and is rare and local on Orchard Peak, the southernmost extension of the Diablo Ranges.

Zigadenus brevibracteatus (Jones) Hall. Desert zigadene. Common in sandy washes in the pinyon woodland from Mt. Pinos northwest to Cuyama Valley, scarce on the desert slopes of the Tehachapi Mountains northeast to Red Rock Canyon, and rare in the dunes around Buckhorn Dry Lake. There is no evidence that either this species or the next is poisonous to livestock.

Zigadenus exaltatus Eastwood. Chaparral zigadene. Common on dry slopes, mostly in chaparral but often in Douglas oak woodland, on Breckenridge Mountain, in the Piute Mountains, and in the Greenhorn Range.

Zigadenus venenosus Watson. Death camas. Common in winter-wet and marshy areas on the west side of Lynns Valley and scarce in Bear Valley in the Tehachapi Mountains. Although there is no record of Kern County losses, the plant is highly toxic to livestock. Pigs rooting up the bulbs for food are especially vulnerable.

IRIDACEAE. Iris Family

Iris Hartwegii Baker. Wild iris. Occasional in colonies on moist shaded slopes north of Greenhorn Pass, the southern limits for the species.

Iris missouriensis Nuttall. Missouri iris. Common in wet meadows and around seeps in the Jeffrey pine forest on Mt. Pinos.

Iris Pseudacorus Linnaeus. Yellow iris. Spontaneous and apparently established along New Rim Ditch (near Highway U. S. 99), an irrigation canal in the Mettlers Station region (T8028). Native of the Old World.

Sisyrinchium bellum Watson. Blue-eyed grass. Colonies grow in wet soil at the spring below the mouth of Alamo Canyon at the southeast end of the Tehachapi Mountains (T6950), at Basque Encina (T12,563), and in the cienegas at the head of Tejon Creek (T8875).

Sisyrinchium halophilum Greene. Occasional in wet soil at French and Weldon meadows in the Piute Mountains, at Kern Hot Springs near Kernville, and at Bear Valley in the Tehachapi Mountains. It is otherwise known in California from the east base of the Sierra Nevada in Inyo and Mono counties.

Sisyrinchium idahoense Bicknell. Common in moist meadows in the Jeffrey pine forest at Mt. Pinos and in the Piute Mountains (although it does not occur with *S. halophilum* at French and Weldon meadows).

ORCHIDACEAE. Orchid Family

Corallorhiza maculata Rafinesque. Coral root. A saprophyte that is occasional in the upper levels of the ponderosa pine forest and in the red fir forest in the northern Greenhorn Range.

Epipactis gigantea Douglas. Stream orchis. Rare at the base of the limestone cliffs along Erskine Creek at 2,900 feet elevation in the arid Douglas oak woodland (T8442) at the north end of the Piute Mountains. These plants have yellow flowers, the albino form for this orchid.

Habenaria leucostachys (Lindley) Watson. Rein orchis. Rare in wet soil in Mountain Meadow on the shaded east slope of Sunday Peak (T7748).

Habenaria sparsiflora Watson. Rare in wet soil under willows at the north end of French Meadow in the Jeffrey pine forest in the Piute Mountains (Breedlove 3955) and very rare (one plant along Cedar Creek at the Rock Corral) on the west slope of the Greenhorn Range (Farnsworth 278). Also known from a creek bank and spring at 8,200 feet elevation on Mt. Pinos (Hoffmann, 17 July 1927, SBM).

Habenaria unalascensis (Sprengel) Watson. Colonies are scarce in good loam on north slopes in the ponderosa pine forest in the Greenhorn Range.

SALICACEAE. Willow Family

Populus acuminata Rydberg. Common along Kern River above Kernville, mostly in a shrubby form possibly caused by repeated breakage in times of high water.

Populus Fremontii Watson. Fremont cottonwood. Common in moist sometimes somewhat alkaline soil from the valley floor to the Douglas oak woodland and in the desert-facing canyons of the mountains. Two groves grow on Rosamond Dry Lake on the desert. The one along the highway at the north end of the lake may have been planted. The second grows on a hillock that was once a small islet in the Pleistocene lake and is obviously indigenous. The often short-lived Fremont cottonwood establishes itself readily in moist soil and is sometimes weedy. It was widely planted for shade by the early settlers who started trees easily from cuttings. Since it transpires very large quantities of water it often significantly lowers the water table at springs and along creeks. Trees of unusual size grow at the Temblor Ranch in Temblor Canyon, the trunk of the largest measuring 23 feet ½ inch in circumference at four feet above ground; the much broken crown is about 60 feet tall.

Ranch people used cottonwood when they wanted a quick hot fire, often as starter for oak or other more slowly burning fuels.

Populus trichocarpa Torrey & Gray. Black cottonwood. An extensive colony of trees of all ages grows for nearly a mile in Tejon Canyon from 3,000 to 3,600 feet elevation and a single very large tree grows in upper El Paso Canyon. Charlotte Smith writes (personal communication) that a few black cottonwoods grow along Bull Run Creek in the northern Greenhorn Range.

Salix exigua Nuttall. Narrow-leaf willow. Occasional to common in sunny wet places (sometimes wet only in winter and spring) in all cismontane associations through the yellow pine forest and at springs in the mountains bordering the desert.

Salix Geyeriana Andersson var. **argentea** (Bebb) C. K. Schneider. Silver willow. Common along the fork of Cannell Creek southwest of Pine Flat on the Kern Plateau, the southern limit for the species (T8679, 8707).

Salix Gooddingii Ball. Goodding's black willow. Common, often abundant, at wet places in the valley. It is rare on the bed of Rosamond Lake on the desert. Var. *variabilis* Ball occurs in the Kern Island region (about 10 miles north of Mettlers Station, T9722) and a single tree grows at a seep at the southeast base of the El Paso Range about 1.5 mile east of the mouth of Last Chance Canyon (T11,988).

Salix Hindsiana Bentham. Sandbar willow. Occasional in the valley, often in dense thickets that are frequently troublesome along canals. Var. *leucodendroides* (Rowlee) Ball is rare with the typical variety (Poso Creek northwest of Wasco, T8971). (*S. sessilifolia* var. *Hindsiana.*)

Salix laevigata Bebb. Red willow. Occasional in the foothills around the valley, occurring along Poso Creek on the valley floor as far west as Famosa, and extending through the Upper Sonoran associations in the mountains to the desert ranges, where it is common at seeps and springs. It is the common willow of the Temblor Range. Var. *arequipa* (Jepson) Ball grows along Cannell Creek east of River Kern (T8750) and at Basque Encina in the Tehachapi Mountains (T12,555).

Salix lasiandra Bentham. A common foothill tree of wet places in the mountains south to El Paso Canyon and east to Spanish Needle Creek north of Walker Pass. Var. *Abramsii* Ball, a shrubby form, is occasional, usually in dense colonies, along Kern River from Lake Isabella to the Tulare County line.

Salix lasiolepis Bentham. Arroyo willow. Common in the mountains from the foothills to the yellow pine forests, east to Erskine Creek; scarce in the Temblor Range. Specimens growing with ample sun and water are beautiful trees and the leaves that are rich green above and silvery below are especially attractive when shimmering in a breeze.

Salix lasiolepis var. **Bracelinae** Ball. Interior arroyo willow. This shrubby variety (the typical variety is also often shrubby) replaces the species (which has somewhat broader oblanceolate leaves) from Indian Wells Canyon and the southeastern Kern Plateau through the mountains southwest to Mt. Pinos. It grows most frequently in the pinyon woodland but is also common in the Jeffrey pine forest where it often forms dense thickets in broad sunny meadows. The thicket along the creek at Claraville Meadow in the Piute Mountains is nearly a mile long.

Salix Lemmonii Bebb. Known from a spring at 7,200 feet elevation in the Jeffrey pine forest in the Bartolas Country on the southern Kern Plateau (Howell & True 41,485).

Salix lutea Nuttall var. **Watsonii** (Bebb) Jepson. A colony of several robust plants form a thicket on a sandy flat along upper Bartolas Creek in the southern Kern Plateau (T12,458).

Salix melanopsis Nuttall. Dusky willow. Common along Kern River above Lake Isabella; rare at Cache Creek northeast of Monolith in the Tehachapi Mountains.

Salix Scouleriana Barratt. Scouler willow. An attractive willow with gray to white bark that is scarce, often growing in small dense clusters, in shaded often moist places, mostly on north slopes, in the ponderosa pine forest in the Greenhorn Range and on Breckenridge Mountain.

JUGLANDACEAE. Walnut Family

Juglans nigra Linnaeus. Black walnut. Black walnut was often planted in the past for shade, ornament, and fruit. Spontaneous individuals are occasional and spontaneous colonies grow in the ravine on the Glennville Road 0.4 mile east of Woody and along the cienega below Ft. Tejon. Native of the eastern United States.

BETULACEAE. Birch Family

Alnus rhombifolia Nuttall. White alder. Common along streambanks from the upper limits of the Douglas oak woodland through the red fir forest south to El Paso Creek. An isolated colony, surprisingly out of range, grows along Kern River near Lake Ming in the hot dry low hills northeast of Bakersfield.

MORACEAE. Mulberry Family

Cannabis sativa Linnaeus. Marijuana. Apparently once a widespread weed (Mariannas Spring and Canyon in the Temblor Range were named for the plant), rigorous control under the California narcotics laws has greatly reduced its occurrence and perhaps eliminated it. It was once a common weed in moist soils in waste places around Bakersfield. Native of the Old World.

Ficus carica Linnaeus. Domestic fig. Scarce along creeks in canyons, about springs, etc., through the Douglas oak woodland. Figs start easily from cuttings stuck in wet soil and were once frequently planted by cowmen and early settlers for fruit and shade at wet places. Evalyn Farnsworth reports (personal communication) that seedlings occasionally grow until eaten by gophers at moist places in the Greenhorn Range. Some mature trees would appear to be almost certainly spontaneous (Oiler Canyon, a fork of Caliente Canyon; the canyon above Woody). Native of the Mediterranean region.

Maclura pomifera (Rafinesque) C. N. Schneider. Osage-orange. A scattered colony of trees of various ages, the younger ones obviously spontaneous, grows along a drainage ditch on the alkali plains east of Kern Lake (near old Highway U. S. 99, 10.3 miles north of Mettlers Station, T9720, 9721). There seems to be no other record of this widespread tree of the Mississippi Valley being spontaneous in California.

Morus alba Linnaeus. White mulberry. Established and common along Kern River near the Stockdale Country Club (T4752, 5000). (Cf. Leafl. West. Bot. 9:142). Native of China.

Fagaceae. Beech Family

Castanopsis sempervirens (Kellogg) Dudley. Chinquapin. Dense and often extensive colonies are occasional at the upper levels of the yellow pine forest and in the red fir forest in the Greenhorn Range, the Kern Plateau, the Piute Mountains and on Breckenridge Mountain. On Sunday Peak, chinquapin grows quickly after spot lightning fires; the mature shrubs give protection to Shasta red fir seedlings.

Quercus agrifolia Neé. Coast live oak. An isolated senescent colony of this beautiful coastal evergreen oak grows at Bill Little Spring in upper Chico Martinez Canyon and a few scattered trees grow in swales on Drake Ridge in the Temblor Range, the local eastern limits for the species.

Quercus Alvordiana Eastwood. A distinctive sometimes round-headed but often straggly tree with persistent leaves that is common from the Temblor Range southeast through the San Emigdio Range, often forming a specialized woodland and sometimes growing in dense thickets. The tree is a fertile hybrid between *Q. turbinella* var. *californica* and *Q. Douglasii.* The type was collected in San Emigdio Canyon (Alice Eastwood 122, 2 October 1894). (*Q. dumosa* var. *Alvordiana.*)

Quercus chrysolepis Liebmann. Canyon live oak, maul oak, golden cup oak, in much of Kern County erroneously called water oak. Common, mostly in canyons, in the mountains to the border of the yellow pine forest. It is occasional in a shrubby glaucous form in the desert-facing canyons on the east slope of the mountains. The very shrubby var. *nana* (Jepson) Jepson grows on exposed high ridges on Black Mountain in the Greenhorn Range and along the rimrock at the southwest end of the Kern Plateau.

In deep canyons in the Tehachapi Mountains and to a lesser extent elsewhere, it tends to replace Douglas oak and become the dominant tree of a plant community more mesic than that of the surrounding slopes but not markedly enough to warrant formal recognition as a plant association.

The Chandler Oak on the broad bottom of Rising Canyon northeast of Lebec is a notable specimen of canyon live oak; this huge densely branched round-headed tree is 70 feet tall, has a spread of 120 feet, and trunk diameter of six feet.

Quercus Douglasii Hooker & Arnott. Douglas oak, blue oak. The common tree of the cismontane foothills and the Temblor Range that forms a distinctive woodland at the middle elevations, where it usually grows with digger pine (*Pinus Sabiniana*). Douglas oak is also occasional east of the crest of the mountains, especially in the canyons of the Tehachapi Range and the other desert-facing canyons north to Jawbone Canyon and the west side of Kelso Valley.

The type of a form described as *Q. Ransomi* by Albert Kellogg (which was presumably lost in the San Francisco fire of 1906) was collected in 1853 at Ft. Tejon by Colonel Leander Ransom. This was probably the first botanical collection made in Kern County. Colonel Ransom was an officer of the U. S. Geological Survey. His Ft. Tejon collections were deposited in the newly founded California Academy of Sciences, where he was Chairman of the Library Committee; in 1855 he was elected Vice President of the institution. Thus, the current close association of the Academy and Kern County botany has an interesting historical precedent.

Douglas oak was the chief source of firewood for the early settlers. The acorns were sometimes used as an unpalatable substitute for coffee. The balls formed by gall wasps were the most common and often the only toys of pioneer children.

Seedlings and young trees are uncommon in many regions. This is probably because of a combination of the effects of the drouths of recent decades, the browsing and breakage by cattle and other livestock, and the inability of very young seedlings to compete with the introduced annuals. The abundance of young trees in the Glennville region would suggest that the prolonged drouths in other regions is the most significant of these factors.

Quercus dumosa Nuttall. California scrub oak. This abundant and widespread chaparral shrub of the Coast Ranges is occasional but not at all common in the chaparral in the Tehachapi Mountains and the San Emigdio Range.

Quercus Garryana Douglas var. **Breweri** (Engelmann) Jepson. Shin oak, Brewer oak. Common in dense and sometimes extensive thickets just below the yellow pine forest from Glennville to the slopes northeast of Lebec. In the Tehachapi Mountains it sometimes forms impenetrable thickets covering hundreds of acres. Deciduous shin oak is a mesic plant with relatively high moisture requirements. It burns only with difficulty and is not a member of the real chaparral (although often considered so). A large old tree on Tunis Ridge (elevation 3,500 feet) in the Tehachapi Mountains seems to be a rare hybrid between *Q. Garryana* var. *Breweri* and *Q. Douglasii* (T7408). (*Q. Garryana* var. *semota.*)

Quercus Kelloggii Newberry. Kellogg oak, California black oak. A large spreading tree of the lower levels of the yellow pine forest, often forming a woods just below it. Kellogg oak is especially attractive for its red buds and young leaves in the spring and the red or orange foliage in the fall. It grows well in cultivation and is an excellent large tree for parks, estates, and ranchgrounds.

Quercus lobata Neé. Valley oak, California white oak. Common in the foothills of the mountains, forming park lands in the valleys around Glennville, in the Tehachapi Mountains, and at Lebec. Valley oak is occasional on the edge of the desert in the northwest end of Antelope Valley and on the broad slopes of the southeastern Tehachapi Mountains north to Cameron Canyon. It occurs in a singular association with Jeffrey pines on the ridges southwest of the summit of Tejon Pass at an elevation of 5,600 feet, perhaps an altitude record for the species. In the central Temblor Range, three aged trees grow at White Oak Spring at the head of Agua Media Canyon. An erect sparsely branched tree along Lumreau Creek just below the Basket Pass road is a hybrid between *Q. lobata* and *Q. Douglasii* (T12,724).

Valley oak grows well in cultivation if it is planted in deep soil and given ample water. Horticultural plantings are especially desirable as the tree is being eliminated by agricultural and urban development in much of its native range.

Quercus moreha Kellogg. Oracle oak. A hybrid between *Q. Kelloggii* and *Q. Wislizeni* that is occasional in the mountains south to Mt. Pinos. Eben McMillan reports (oral communication) that germination experiments with acorns from a tree on Eugene Grade in the Greenhorn Range show that the nuts have very low viability.

Quercus turbinella Greene ssp. **californica** Tucker. A common shrubby oak, often becoming a picturesque small tree, of the arid Douglas oak woodland in the San Emigdio and Temblor ranges, southwest to the north flank of Mt. Abel, occurring occasionally northeast through the Piute Mountains to Erskine Canyon. On the arid slopes in the northeastern Tehachapi Mountains in the Cache Peak, Weldon Pond, and upper Pine Tree Canyon regions it forms an extensive chaparral with *Ceanothus cuneatus* and *C. vestitus* that often covers hundreds of acres of the steep slopes.

Quercus Wislizenii A. DeCandolle var. **frutescens** Engelmann. Interior scrub oak. A xerophytic shrub that is a common element of the chaparral from Mt. Abel northeast through the mountains to the Tulare County line. Small trees that grow at Hobo Hot Springs in Kern Canyon and near Keene vary toward the typical arboreal variety of the north and west. However they are probably only robust forms of the shrubby variety responding to very favorable soil, climatic, or moisture conditions. On Ednas Mountain, in the Greenhorn foothills south of Glennville, the shrubs also tend to be arboreal; these apparently are hybrids with *Q. Kelloggii.* These oaks have most of the characteristics of *Q. Wislizeni* (unlike *Q. moreha,* which more closely resembles *Q. Kelloggii*) (Farnsworth 710, 711).

URTICACEAE. Nettle Family

Hesperocnide tenella Torrey. Western nettle. Scarce in shady places in the cismontane woody Upper Sonoran associations. The plant's delicate appearance belies its sharply stinging hairs.

Parietaria floridana Nuttall. Florida pellitory. Known from 2 miles below Poso Mine, near Poso Creek (C. N. Smith 762, JEPS), Long Tom Mine (T10,635), and near the lower end of Kern Canyon (Isabel McCracken, 20 April 1935). (*P. debilis.*)

Urtica holosericea Nuttall. Nettle. Common at wet places in all the cismontane associations. (*U. gracilis* var. *holosericea.*)

Urtica urens Linnaeus. Potato nettle. Occasional in moist shady places in the Douglas oak woodland in the east-side foothills, in lawns and gardens in Bakersfield, and a common weed of potato fields around Edison. Native of Europe.

ULMACEAE. Elm Family

Celtis Douglasii Planchon. Arizona hackberry. Scattered colonies occur in Caliente Canyon about three miles above Caliente, a weather-beaten colony of old shrubby plants grows on a windswept ridge five miles southeast of Keene overlooking Tehachapi Valley, and a small colony grows at Democrat Hot Springs in Kern Canyon. This disjunct occurence of a tree common in Arizona may be a relict of the Pliocene Mohavean woodland. (*C. mississippiensis* var. *reticulata.*)

Ulmus parvifolia Jacquin. Chinese elm. Spontaneous trees are occasional along Kern River at Hart Park and seedlings are common in the park. It is a frequent ranchground, waste place, and vacant lot seedling. Native of eastern Asia.

Ulmus procera Salisbury. English elm. Root suckers are frequent in waste places and vacant lots in Wasco. Native of Europe.

Ulmus pumila Linnaeus. Siberian elm. Occasional as a weed or from root suckers at Kernville, Mojave, and Rosamond. (Cf. J. T. Howell. Leafl. West. Bot. 10:328).

LORANTHACEAE. Mistletoe Family

Arceuthobium campylopodum Engelmann. Pine mistletoe. An occasional parasite on *Pinus Jeffreyi;* forma *Blumeri* (A. Nelson) Gill grows on *Pinus Lambertiana;* and forma *abietinum* (Engelmann) Gill on *Abies concolor.*

Phoradendron Bolleanum (Seeman) Eichler var. **densum** (Torrey) Fosberg. Juniper mistletoe. Common on California juniper, sometimes densely infesting old or drouth stricken shrubs.

Phoradendron Bolleanum var. **pauciflorum** (Torrey) Fosberg. Scarce on white fir (*Abies concolor*).

Phoradendron flavescens (Pursh) Nuttall var. **macrophyllum** Engelmann. Occasional, sometimes locally common and dense, on *Populus Fre-*

montii, Platanus spp., *Robinia Pseudo-Acacia;* and on *Forestiera* near the alkali flats at Kernville (Howell 38,905).

Phoradendron flavescens var. **villosum** (Nuttall) Engelmann. Oak mistletoe. Common on oaks, especially *Quercus Douglasii.* (*P. villosum.*)

Phoradendron juniperinum Engelmann var. **Libocedri** Engelmann. Scarce on *Calocedrus decurrens* in the Greenhorn Range.

SANTALACEAE. Sandalwood Family

Comandra pallida DeCandolle. Occasional in the ponderosa pine forest and scarce in the Douglas oak woodland in the Greenhorn Range, rare on Breckenridge Mountain. A root parasite with apparently numerous hosts. (*C. umbellata.*)

SAURURACEAE. Lizard-tail Family

Anemopsis californica Hooker. Yerba mansa. Occasional in colonies in sunny moist subalkaline places in all the associations below the yellow pine forest. The large white bracts, often blotched with purplish pink, are colorful around seeps in the summer when the surrounding areas are usually dry and even barren. A paste made from the leaves is said to be very effective in treating cuts and wounds in livestock and man (an interesting combination of the old and new is that ranch wives find it most easily prepared in a Waring blender).

POLYGONACEAE. Buckwheat Family

Chorizanthe brevicornu Torrey. Rather common, mostly in rocky or gravelly places, from the desert west to the arid slopes around South Fork Valley.

Chorizanthe Clevelandii Parry. Known from colonies at Evans Flat in the ponderosa pine forest in the Greenhorn Range (T3912) and in the arid Douglas oak woodland in Rancheria Canyon north of Weldon (Hardham 2191).

Chorizanthe coriacea Goodman. Occasional in dry fine often packed sandy soil in the Upper Sonoran associations in the Temblor Range and in sandy places around rock outcrops in the east-side foothills north to White River; rare in the western San Joaquin Valley at Point of Rocks. (*Lastarriaea chilensis.*)

Chorizanthe membranacea Bentham. Pink chorizanthe. Occasional around rocky places in the chaparral and the Douglas oak woodland in the Greenhorn Range, on Breckenridge Mountain, in the Tehachapi Mountains, and in poor soil and decomposed serpentine in the northern Temblor Range.

Chorizanthe perfoliata Gray. Common in light or rocky soils in the Temblor Range and the low hills about the valley north to Adobe Canyon in the Greenhorn foothills, occasional in the Tehachapi Mountains. The type was collected at Fort Tejon (Xantus de Vesey in 1857–58).

Chorizanthe rigida (Torrey) Torrey & Gray. Rigid chorizanthe. Uncommon on the desert, often in the compacted gravel of pavement terraces. The rigid plants persist for as long as two years after they have dried.

Chorizanthe spinosa Watson. Spiny chorizanthe. Occasional in loose light sand in the dunes at the northeast end of Buckhorn Dry Lake (T10,841); rare in compacted basalt on a steep slope in the east fork of the Red Rock Canyon (T3503).

Chorizanthe Thurberi (Gray) Watson. Common in sandy soil in the creosote bush association, extending west to the Mt. Abel region and upper Cuyama Valley and to the east slope of the Piute Mountains and the Greenhorn Range.

Chorizanthe uniaristata Torrey & Gray. Occasional in dry or disturbed places in the Temblor Range and from the Mt. Abel region northeast to Breckenridge Mountain. It also occurs on the very arid mesas east of Bakersfield (Howell 37,097).

Chorizanthe Watsonii Torrey & Gray. Occasional in dry places in the mountains, mostly in the Douglas oak and pinyon woodlands.

Chorizanthe Xanti Watson. Common in the cismontane Upper Sonoran associations. The type was collected at Ft. Tejon (Xantus de Vesey, 1857). The brilliant red dried plants of this and C. perfoliata are great favorites for dried plant arrangements.

Eriogonum. The genus *Eriogonum,* the wild buckwheats, is the largest in Kern County, with 37 species and 9 lesser taxa. Superbly adapted to poor or light soils, drouth, and long hot summers, the genus is represented by at least one species in almost every region of the county. These distinctive late spring, summer, and fall blooming plants are among the most characteristic of California's lands of little rain, often growing in places too arid for the survival of any other plant. Some, such as the annual *E. gracillimum* and *E. angulosum,* are widespread, occurring in many of the plant associations. Others, such as *E. gossypinum* of the upper San Joaquin Valley and *E. Breedlovei* of Piute Mountain and Baker Point, are rare and restricted in their distribution.

All are plants with flowers of unusual although often subdued beauty. The delicate shadings and structural detail of many of the annual species must be viewed with a lens for full appreciation but some, such as the perennial *E. umbellatum* which is a mound of yellow in summer and early fall, are striking plants when in bloom.

The annual buckwheats are important food for deer and contribute substantially to the livestock forage in some areas. The flowers furnish nectar for honey of premium quality. Some of the perennials make colorful and easily maintained rock and natural garden plants.

However, it is in the field of taxonomy that they have their greatest current interest. Despite persistent study for more than a century, the delineation of the species is neither clear nor complete. Botanists rarely agree on the treatment of the genus and the findings of new research frequently revises or expands our knowledge of the plants. Hardly a year passes without the discovery of new species, usually of rare and local distribution but sometimes of fairly widespread distribution. Perhaps more than any other genus that occurs in the county, the wild buckwheats demonstrate that taxonomic knowledge of California plants is far from complete, and that taxonomy is still a fertile and challenging field for a scholar with a diligent, perceptive, and lively mind.

Eriogonum angulosum Bentham. A very common late spring and summer annual of the west side of the desert and the Upper Sonoran associations in the mountains and the Temblor Range; scarce in the valley. It is preferred food of deer in summer and fall.

Eriogonum Baileyi Watson. Common, often abundant, in sandy soils, dry washes, etc., in the pinyon woodland and arid shrub associations; occasional in the Douglas oak woodland on the eastern slopes of the mountains; extending southeast to Cuyama Valley and west to Kernville, Havilah, and upper Tejon Canyon.

Eriogonum baratum Elmer. Skeleton weed. Occasional in the pinyon woodland, mostly in the Walker Pass region, less frequent southwest to Cuddy Valley and west to the arid slopes east of Kern River above Kernville. (*E. deflexum* ssp. *Watsonii.*)

Eriogonum brachyanthum Coville. Occasional in sandy places, mostly in the arid shrub association, scarce in the creosote bush association, in the desert-facing slopes of the mountains northeast to Indian Wells Valley, and rare in the El Paso and Rand mountains. (*E. Baileyi* var. *brachyanthum.*)

Eriogonum brachypodum Torrey & Gray. Scarce in canyons and fans in the creosote bush association in and along the El Paso and Rand mountains. (*E. deflexum* var. *brachypodum.*)

Eriogonum Breedlovei (J. T. Howell) Reveal. Piute buckwheat. Very local in quartz outcrops in the Jeffrey pine forest along the crest of Piute Mountain (Dennis E. Breedlove 3399, 14 July 1962, the type collection; other collections are T8906 and 7338). The only other station known for this rare matted buckwheat is the rugged north face of Baker Point in southern Tulare County (T8996, 9618). The Baker Point plants have flowering stems that are from divergent to mostly prostrate and the inflorescence is more lax than that of the Piute plants. (*E. ochrocephalum* var. *Breedlovei.*)

Eriogonum Covilleanum Eastwood. Rare in an atypical form in brown shale slides in Cedar Canyon in the Temblor Range (2 miles south of Bitterwater Road, T1207; 5 miles south, T11,768; 6 miles south, T1413, 11769).

The typical form grows in the inner south Coast Ranges from San Benito County to Alameda County.

Eriogonum elatum Douglas. Known from the rim of the Kern Plateau east of Fay Creek Canyon, a southern range extension from Mono County (Howell & True 41,836).

Eriogonum elongatum Bentham. Occasional, sometimes locally common, in the Upper Sonoran associations from the Temblor Range southeast to the hills at the west end of Antelope Valley in the Gorman region.

Eriogonum fasciculatum Bentham ssp. **foliolosum** (Nuttall) Stokes. California buckwheat. Common in the northern Temblor Range where it occasionally reaches Kern County in the Annette region.

Eriogonum fasciculatum ssp. **polifolium** (Bentham) Stokes. Gray california buckwheat. Common, often densely covering arid slopes at the lower elevations of the Temblor Range and along the desert-facing slopes of the mountains. It is also common in the mountains below the chaparral, particularly east of the crest, and in the desert Rand and El Paso Mountains.

Eriogonum gossypinum Curran. Cotton-flowered buckwheat. A rare and remarkable plant endemic to sandy places in Kern and Kings counties and extreme southeastern San Luis Obispo County (Sand Ridge near Edison, Buena Vista Hills, bluffs across the river from Hart Park, Kettleman Hills in Kings County, east slope of Crocker Pass in San Luis Obispo County). The type was collected at Bakersfield (Katharine Layne Curran in July, 1884).

Eriogonum gracillimum Watson. A widespread late spring and summer annual that is common in the valley and the Temblor Range, occasional in the mountains through the chaparral, and common from the higher creosote bush association through the arid shrub association on the desert. The type was collected on the Mojave Desert in Kern County by Mrs. A. E. Bush in June, 1879. Mrs. Bush was apparently a resident of San Jose who collected for Asa Gray.

Eriogonum Heermannii Durand & Hilgard. Rare, usually in small colonies, on rocky slopes in the pinyon woodland from Walker Pass southwest to Cuddy Valley (and thence northwest to San Benito County). The type was collected in 1853 by Adolphus L. Heermann, surgeon-naturalist attached to the Williamson Expedition; the type locality, given as Poso Creek, is probably Walker Pass. (The name of the base camp near the mouth of Poso Creek was used as the locality for all of the plants collected in the mountains east to Walker Pass and the head of Kelso Canyon.)

Eriogonum hirtiflorum Gray. Scarce, mostly in disturbed places, in the yellow pine forest in the Greenhorn Range. It is rare in the Douglas oak woodland (French Ranch, C. N. Smith 1118, JEPS; a waif along Cedar Creek near the mouth of Lumreau Creek, Farnsworth 614).

Eriogonum indictum Jepson. Occasional in barren places or on cuts;

slides, etc., in the Temblor Range south to Maricopa. A large colony of exceptionally robust plants grows in barren asphaltic soil just west of McKittrick.

Eriogonum inerme (Watson) Jepson. Occasional in disturbed often moist places, particularly on the borders of meadows, from the chaparral to the yellow pine forest in the mountains south to Tehachapi Peak. It is our only buckwheat that is tolerant of wet soil.

Eriogonum inerme var. **hispidulum** Goodman. Scarce in disturbed places in the ponderosa pine forest south to Shirley Meadows and rare on a sagebrush slope on the south side of Bear Mountain (T9651).

Eriogonum inflatum Torrey & Frémont. Desert trumpet. Common in rocky places from the desert southwest to Mt. Pinos.

Eriogonum Kennedyi Porter ssp. **austromontanum** (Munz & Johnston) Stokes. Common in extensive mat-forming colonies on the summit of Mt. Pinos; scarce with sagebrush in sand on the high ridge south of Purdy Canyon in the Tehachapi Mountains; locally common about Lookout Point in the Piute Mountains, Noonday Rest near Cache Peak, the higher slopes of rugged Mt. Owen at the head of Indian Wells Canyon, and on the southern Kern Plateau.

Eriogonum latifolium Smith ssp. **auriculatum** (Bentham) Stokes. Rare in rocky places and in shale banks in the northern Temblor Range, reaching its southern limits in Cedar Canyon (T1401). (*E. nudum* var. *auriculatum.*)

Eriogonum latifolium ssp. **nudum** (Douglas) Stokes. Tibinagua. Common in the Temblor Range, especially in the Upper Sonoran grassland; occasional, south to the Mt. Pinos region and in the foothills of the Greenhorn Range. (*E. nudum.*)

Eriogonum latifolium ssp. **pauciflorum** (Watson) Stokes. Common around rocky places in the red fir forest on Sunday Peak (T3854). (*E. nudum* var. *pauciflorum.*)

Eriogonum latifolium ssp. **saxicola** (Heller) Stokes. Common in the mountains at the middle elevations south to Mt. Pinos and Brush Mountain; occasional to the Jeffrey pine forest, rare from the rocky desert-facing canyons to the creosote bush association.

Eriogonum latifolium ssp. **Westoni** Stokes. Weston buckwheat. Scarce in rocky places in the desert ranges and the desert-facing slopes of the mountains, west to the slope northeast of Lake Isabella and the east slope of the Greenhorn Range. Endemic to Kern County and extreme southeastern Tulare County (Rock House Basin, T10,370). The type was collected in the Greenhorn Mountains (E. Roy Weston, April, 1926).

Eriogonum maculatum Heller. Spotted buckwheat. A fairly common plant of the creosote bush association that is occasional through the pinyon woodland west to Kern Canyon and the Kernville region, southwest to the

Mt. Pinos region, and in the Temblor Range. (*E. angulosum* var. *maculatum.*)

Eriogonum microthecum Nuttall var. **laxiflorum** Nuttall. An interesting gnarled and twisted dwarf shrub that resembles a miniature timberline tree that is scarce on dry ridges and washes in the pinyon woodland south to Jawbone Canyon.

Eriogonum mohavense Watson. Occasional in the creosote bush association, mostly in sandy places at lower elevations.

Eriogonum molestum Watson var. **Davidsonii** (Greene) Jepson. Scarce in sand, gravel, and decomposed granite in open areas in the Jeffrey pine forest in the Mt. Pinos region (T2189, 2399); rare at the head of Lecheria Canyon in the southeastern Tehachapi Mountains (T11,065).

Eriogonum nidularium Coville. Locally, bird's nest. Rare although widespread on desert slopes, usually growing in gravel.

Eriogonum Ordii Watson. Occurs in widely separated but usually dense colonies in the Temblor Range, almost always in friable (dry bog) clay; in fine clay or sand in the upper Cuyama Valley, the San Emigdio Range, the west slope of the Tehachapi Mountains; and rare in the low foothills of the southwestern Greenhorn Range (Adobe Canyon, Poso Mine).

Eriogonum Plumatella Durand & Hilgard. An interesting subshrub that is occasional on the desert-facing slopes of the mountains south to the Tehachapi Mountains and east to Red Rock Canyon, mostly in the arid shrub association and the pinyon woodland. The type was collected by Adolphus L. Heermann in 1853 and the type locality, given as Poso Creek, is probably the Walker Pass region.

Eriogonum pusillum Torrey & Gray. Common and widespread on the desert, extending through the pinyon woodland southwest to Cuddy Valley and west to Kelso Canyon.

Eriogonum reniforme Torrey & Frémont. Occasional in Red Rock Canyon, usually in bare areas at the base of cliffs; less common in the other canyons of the El Paso Range and on the sandy flats on the east side of Indian Wells Valley; scarce on a recently bulldozed vacant lot at the southeast corner of Burrles Street and Drummond Drive in Ridgecrest in 1965 (Meng 133).

Eriogonum roseum Durand & Hilgard. Common in the cismontane Upper Sonoran associations and occasional on the desert-facing slopes of the mountains north to Morris Peak in the southeastern Sierra Nevada. The flowers range from white through buff to pink; those from the Tehachapi region northwest to Glennville are frequently bright red; in Kern Canyon they are commonly bright yellow. The type was collected at Poso Creek by Adolphus L. Heermann in 1853. (*E. virgatum.*)

Eriogonum roseum var. **leucocladon** (Bentham) Hoover. Often common south to McKittrick Summit in the Temblor Range, growing mostly in light arid soils in the Upper Sonoran grassland. (*E. gracile* var. *leucocladon*. Cf. Leafl. West. Bot., 10:343).

Eriogonum saxatile Watson. Rock buckwheat. Occasional to scarce in rocky places from the Douglas oak and pinyon woodlands to the yellow pine forest in the Sierra Nevada north of Walker Pass, in the Greenhorn Range, and on Breckenridge Mountain; rare on the southwest flank of Mt. Abel.

Eriogonum spergulinum Gray. Rare in dry disturbed soil at the south end of Little Cannell Meadow in the arid Jeffrey pine forest on the Kern Plateau (T9840).

Eriogonum spergulinum var. **Reddingianum** (Jones) J. T. Howell. Occasional in colonies in disturbed places, usually in dryish loam at the borders of meadows, from the Greenhorn Range and the Kern Plateau through the mountains southwest to Mt. Pinos. It grows mostly in the yellow pine forest, but occasionally descends to the chaparral.

Eriogonum temblorense Howell & Twisselmann. Temblor buckwheat. This extremely xerophytic plant grows in two colonies in barren white shale at the type locality in lower Chico Martinez Canyon (type: Ernest C. Twisselmann 1619, 14 October 1954; other collections from the type locality: T1508, 1848, 2085, and 11,111). Another colony grows in the desolate Shale Hills west of Blackwells Corner (T. Franklin Twisselmann, 1 August 1955). The plant is also known from Polonio and Cottonwood passes in extreme eastern San Luis Obispo County and from Indian Valley in Monterey County; all of the stations but the last are in fractured shale. The remarkably drouth tolerant Temblor buckwheat grows in these places when the spring annuals have dried. (Cf. Leafl. West. Bot., 10:42, 1963).

Eriogonum Thomasii Torrey. Rare in metamorphic talus in Last Chance Canyon in the El Paso Mountains (T11,856, 11,868).

Eriogonum trichopes Torrey. Abundant in good years on basalt in the Black Mountain region of the El Paso Mountains, coloring the high black rocky slopes of that somber mountain a bright greenish yellow. These years, it is also abundant in metamorphic gravel and fractured rock in the Rand Mountains and is scarce in terrace gravel in the Kramer Hills. Dense colonies grow in the scattered beds of ultra-fine friable (dry bog) clay in the Temblor Range. (*E. trichopodum.*)

Eriogonum umbellatum Torrey ssp. **bahiaeforme** (Watson) Munz. This subspecies, in its typical form, grows around rock outcrops in the Jeffrey pine forest from Lamont Meadow in southeastern Tulare County and Mt. Owen southwest through the Piute Mountains and on Breckenridge Mountain, from the pinyon woodland through the Jeffrey pine forest. However, over most of the county, ssp. *bahiaeforme* seems to be strongly introgressed with ssp. *polyanthum* (Bentham) Stokes and the two can only be separated arbitrarily. Plants in the Mt. Pinos region have the simple bractless flowering umbels of ssp. *polyanthum* but the densely tomentose gray leaves of ssp. *bahiaeforme.* Plants with leaves that are variously glabrous, glabrate, and tomentose and with both bractless simple and bracted compound umbels grow at Lookout Point in the Piute Mountains. Until the complex is studied with conclusions more lucid than those achieved to date, it would perhaps be best to treat all of our plants as ssp. *bahiaeforme* strongly introgessed with ssp. *polyanthum.* Plants of rocky ridges in the northern Greenhorn Range are nearest ssp. *polyanthum* but it would appear that this subspecies in its typical form with short simple bractless rays and rather dense flowering heads does not occur in the southern Sierra Nevada.

Eriogonum viridescens Heller. Common in the Lower Sonoran grassland, occasional to the pinyon woodland and higher creosote bush association, northwest to Red Rock Canyon, west to the Temblor Range and Cuyama Valley. The type was collected at Sunset [near Arvin] (Amos Arthur Heller 7733). (*E. angulosum* ssp. *viridescens.*)

Eriogonum Wrightii Torrey ssp. **subscaposum** (Watson) Stokes. Common, mostly along exposed ridgetops but also in the pinyon woodland, in the Mt. Pinos region, at the summit of Breckenridge Mountain, in the Piute Mountains, and on Black Mountain in the Greenhorn Range (Howell 38,573), and on the Kern Plateau. Var. *curvatum* (Small) Munz, a form with arching stems, is occasional in open areas in the Jeffrey pine forest in the Mt. Pinos region. The plant is much used in making the currently popular Ming trees.

Eriogonum Wrightii ssp. **trachygonum** (Torrey) Stokes. Common, mostly in the arid shrub association and pinyon woodland but also frequent in the Douglas oak woodland and chaparral in the southern Sierra Nevada and Piute Mountains south to the head of Caliente Canyon and Walker Basin. Both subspecies of *E. Wrightii* grow well in cultivation and are attractive additions to the rock garden.

Hollisteria lanata Watson. Common in dry soil and often on barren slopes in the Temblor Range and on the borders of the valley and in the surrounding low hills north to Poso Hills and Adobe Canyon in the Green horn Range.

Oxytheca luteola Parry. A delicately beautiful prostrate yellow-green plant that is scarce on the barren sand dunes and clay flats around the Muroc dry lakes. According to Philip A. Munz (A California Flora, p. 331), it also occurs in Kern County in the San Joaquin Valley.

Oxytheca perfoliata Torrey & Gray. Saucer plant. Common in sandy soils in the upper levels of the creosote bush association through the arid shrub association.

Polygonum arenastrum Boreau. Wire weed, pig weed. A common weed in beaten areas, waste places, cracks in pavement, and disturbed places, mostly at lower elevations. However, it grows in meadows in the Jeffrey pine forest at Catskin Creek in the Tehachapi Mountains and at Landers

Meadow in the Piute Mountains. Wire weed is common in sandy-adobe soil in grainfields in the northern Temblor Range. Livestock are very fond of it and ranchers consider it excellent forage. The form that has been known as *P. aviculare* Linnaeus var. *erectum* Roth grows at Greenhorn Mountain Park (Howell 38,756) and near the south fork of Lucas Creek on Breckenridge Mountain (T11,182). Native of Eurasia. (*Polygonum aviculare.*)

Polygonum argyrocoleon Steudel. Occasional in valley, usually in dense colonies in wet places and in shallow water. Native of Asia.

Polygonum bistortoides Pursh. Common in wet meadows in the yellow pine forest in the Greenhorn Range and on Breckenridge Mountain; scarce at Browns Meadow in the Piute Mountains (Breedlove 3991).

Polygonum coccineum Muhlenberg. Swamp knotweed. Occasional in very wet soil and in water in the valley and along Las Yeguas Creek in the central Temblor Range, often forming dense lush masses on the borders of canals.

Polygonum Douglasii Greene. Occasional in meadows and along streams in the yellow pine forest on Breckenridge Mountain and at Shirley Meadows in the Greenhorn Range; also known from Mt. Pinos (Hoffmann, 17 July 1927, SBM).

Polygonum Douglasii var. **latifolium** (Engelmann) Greene. Common in wet meadows in the yellow pine forest in the Greenhorn Range and on the Kern Plateau. This variety and the typical variety are not always distinct in their Kern County occurrence.

Polygonum Kelloggii Greene. Small dense colonies grow in wet soil at Shirley Meadows and Mountain Meadow in the Greenhorn Range and at wet places in the yellow pine forest on Breckenridge Mountain, the Kern Plateau, and, in wet years, on Mt. Pinos.

Polygonum lapathifolium Linnaeus. Willow-weed. Common, often abundant in great masses, in wet soil and floating on water on the borders of canals in the valley. The extensive growth of willow-weed is especially impressive in the drainage pool of about 40 acres that is all that is left of Buena Vista Lake. It is infrequent east to Glennville and in South Fork Valley and is local along Las Yeguas Creek in the Temblor Range.

Polygonum mexicanum Small. Mexican knot-weed. Known from a dense colony at the base of Elk Hills, 3 miles north of Tupman [Buena Vista Slough] (Nobs & Smith 451, 24 August 1948). These plants grew in a canal that has since been abandoned and the species apparently is no longer present in Kern County. Native of Mexico.

Polygonum Persicaria Linnaeus. Lady's thumb. Occasional in wet places in the Upper Sonoran associations from the Greenhorn Range east to Kern River above Lake Isabella. It is local along the South Fork of Kern River at Kennedy Meadows on the Kern Plateau in southeastern Tulare County. Native of Europe.

Polygonum punctatum Elliott. Water smart weed. Scarce in wet places in the Buena Vista basin. It was once collected along Kern River above old Kernville (C. N. Smith 909, JEPS). (*P. acre.*)

Polygonum sawatchense Small. Known from Greenhorn Mountain Park (Howell 38,756). (*P. Douglasii* var. *Johnstonii.*)

Rumex Acetosella Linnaeus. Sheep sorrel. Common at Shirley Meadows and around the Shirley Meadows ski run (where it is a recent adventive) and in wet meadows in the oak woodland in Lynns Valley. Native of central Europe and Asia.

Rumex californicus Rechinger fil. Known from the Jeffrey pine forest in the Bartolas Country of the southern Kern Plateau (Howell & True 41,846). According to J. T. Howell, the valves lack callus-grains.

Rumex conglomeratus Murray. Wild rhubarb. Occasional to common in soils that are wet (at least in the winter) in the cismontane Upper Sonoran associations. Native of Europe.

Rumex crispus Linnaeus. Curly dock. Common in wet places in the Lower Sonoran grassland; occasional at higher elevations, especially in the meadows in the Jeffrey pine forest. Native of Eurasia.

Rumex hymenosepalus Torrey. Canaigre. Occasional in dry washes and other sandy places in the Temblor Range, on the high plains about the valley, and common in dry washes along the desert ranges. The pioneers used the stems for pie.

Rumex salicifolius Weinmann. Willow dock. Scarce but widespread in cismontane wet places from the valley floor to meadows in the yellow pine forest.

Rumex transitorius Rechinger fil. A small colony grows around the pond at the headquarters of the Kern National Wildlife Refuge (T10,284, 10,486).

Rumex violascens Rechinger fil. Occasional in wet places in the valley, especially from Semitropic to the Tulare County line. It is local along Las Yeguas Creek in the Temblor Range.

CHENOPODIACEAE. Goosefoot Family

Allenrolfea occidentalis (Watson) Kuntze. Iodine bush. A common shrub of alkali sinks in the desert and the valley, occurring sporadically west to Soda Lake and northwest to the Cholame Valley in San Luis Obispo County. This is the most markedly halophytic of our plants, often extending well beyond any other vegetation on the salt flats at places such as Koehn Lake. The compactly branched shrub serves here as a collector of fine wind-blown sand and forms small sandy hillocks on the barren flats. These mounds often support annuals; thus the plant is an important soil builder in such highly alkaline places.

Atriplex argentea Nuttall ssp. **expansa** (Watson) Hall & Clement. Fog weed. Common to abundant, often forming dense colonies on sandy flood plains and in alkali sinks in the valley, especially on the west-side; in similar places in the Temblor Range; and in moist places and well-watered flood plains on the desert (where it is an especially common weed of irrigated farmlands). (*A. expansa.*)

Atriplex Breweri Watson. Quail brush. A few plants grow in wet alkali soil along Bitterwater Creek 3.5 miles southwest of Maricopa (T4125), the only known San Joaquin Valley occurrence of the plant. (*A. lentiformis* ssp. *Breweri.*)

Atriplex canescens (Pursh) Nuttall. Four-wing. Occasional in the dunes around the desert dry lakes and on sandy flats in the pinyon woodland west to Kernville and southwest to Cuyama Valley. The rather indistinct var. *laciniata* Parish apparently is the form that grows on the stabilized dunes west and south of Koehn Lake (T4938).

The fruits are an important fall and winter food for large birds, particularly doves. There are numerous plantings of four-wing around ranches in the northern Temblor Range for quail food and shelter and for ornament. In barren or weeded ranchground or garden soils, spontaneous seedlings are common and sometimes abundant. However, experiments with plantings for game management in fenced plots in pasture lands have established that the seedlings cannot compete with the annuals of those places.

Atriplex confertifolia (Torrey & Frémont) Watson. Shadscale. A common low shrub of the alkali sinks and shadscale scrub that is occasional in other sandy places in the desert.

Atriplex cordulata Jepson. Occasional in the northern Temblor Range, usually in bare often subalkaline soil around vernal pools.

Atriplex coronata Watson. Occasional around alkali sinks from the plains south of Tulare Lake west through the Temblor Range to Soda Lake in San Luis Obispo County.

Atriplex hastata Linnaeus. Arrow weed. Locally common along the flood plain of the South Fork of Kern River in South Fork Valley. (*Atriplex patula* ssp. *hastata.*)

Atriplex hymenelytra (Torrey) Watson. Desert-holly. Extensive colonies grow in the shadscale scrub in the low hills east and south of Ridgecrest through the El Paso Range. The plant is gathered in quantities at Christmas for the decorative use of its white holly-like leaves. (Such collections are illegal under Kern County Ordinance no. 250.)

Atriplex lentiformis (Torrey) Watson. Quail brush. Common in the central valley from Buena Vista Lake to Tulare Lake along sloughs, canals, and in waste places. Quail brush is occasional about the dry lakes on the desert and in waste places in South Fork Valley. A form with attractive lavender-pink bracts is common in the Tupman region. The fruits are very important food for rodents and birds. The plant grows well in cultivation, forming a large rounded shrub, as much as 12 feet tall. While native to subalkaline soil, in cultivation it succeeds admirably in neutral and even in slightly acid soil.

Atriplex phyllostegia (Torrey) Watson. Arrowscale. Occasional in alkali soils in the valley and around Kernville. It is common about the Muroc dry lakes. A form with succulent narrow scarcely or not at all hastate leaves is common in subalkaline places from Antelope Valley (Los Angeles County) to Rogers Dry Lake and is rare in Red Rock Canyon.

Atriplex polycarpa (Torrey) Watson. Common saltbush. A widespread and characteristic shrub of the Lower Sonoran grassland in the valley and the southern Temblor Range, where it often forms a scrub that covers hundreds of acres. It is also common on the lower desert plains from Mojave east to the San Bernardino County line and in Indian Wells Valley. Although it is often said to be the dominant plant in a climax association, it is actually rather weedy and seedlings are most abundant in heavily grazed lands, often after long periods of severe drouth when the wind has severely modified the soil. Immigrant annuals, especially grasses, that grow densely under common saltbush, have made the plant much more susceptible to fire damage than it was in primitive times. While the shrub is extremely xerophytic, even this sturdy inhabitant of the arid plains cannot stand prolonged drouth as is evidenced by the many dead plants on the low hills at Edwards Air Force Base and near Boron.

The shrub is important to the local livestock industry; both cattle and sheep browse it with relish. It is excellent forage; the average protein content is higher than that of alfalfa. The seeds are a prime source of food for birds and rodents, particularly chipmunks (the Antelope ground squirrel and related desert species).

Experimental plantings indicate that it can be an interesting and novel gray-green clipped or sheared shrub in areas where subalkaline soil makes conventional landscape plantings difficult.

Atriplex rosea Linnaeus. Red scale. Common in waste places and flood plains in the valley, the Temblor Range, and from Monolith to the Proctor Lake bed in the east end of Tehachapi Valley. Native of Eurasia.

Atriplex semibaccata R. Brown. Australian salt weed. Common in light soils on flood plains, in waste places, etc., in the valley and the Temblor Range and rare in the low foothills on the west slope of the Greenhorn Range. The plant was seeded in the northern Temblor Range in the 1920's; at least some of the Kern County plants are descended from those seedings. It is excellent food for livestock and the bright red bracts are relished by larger birds. Native of Australia.

Atriplex Serenana A. Nelson. A vile smelling plant that is common, often in dense colonies, in subalkaline soils in the valley and the Temblor

Range and is occasional in waste places in the Tehachapi Mountains. The type was collected at Poso Creek [probably near its debouche on the Bakersfield plains] by Adolphus L. Heermann in 1853. (*A. bracteosa.*)

Atriplex spinifera Macbride. Spiny salt bush. Occasional, usually in sizeable colonies, in the valley and on the desert plains and sinks east of Mojave and Rosamond. It is especially common about Boron and on the subalkaline plains northeast of Lost Hills.

While generally preferring somewhat more alkaline soils, it often grows with *A. polycarpa,* from which it can be distinguished by its very large fruiting bracts. Some plants are distinctly thorny, others not at all so; apparently the degree of thorn development is determined by the amount of browsing of young branchlets. The type was collected in the Maricopa Hills (Alice Eastwood 3269, 15 May 1913).

Atriplex tularensis Coville. Endemic around the borders of the alkali sinks south of Bakersfield in the Greenfield and Weedpatch regions. Repeated searches by the writer have failed to rediscover the plant; it may grow only in highly favorable years, the searches may have erred in time or place, or because of the great increase in reclaiming for farming of what was only recently considered subalkaline waste land, the plant may be extinct. The latter possibility is suggested by the notation on the label of a collection by Severin & Hall (no. 11,782) made on 15 October 1921 that the plant grew on the borders of the alkali sinks but not actually in the Salicornia belt. The type was collected on the Tulare Plains about 25 kilometers south of Bakersfield (Frederick Vernon Coville 1235, 14 June 1891).

Atriplex vallicola Hoover. A rare endemic at dried rain pools and flats in northern Kern County and Fresno County. The type was collected 5 miles north of the Lost Hills Oil Field (Robert Francis Hoover 2666, 10 August 1937).

Bassia hyssopifolia (Pallas) Kuntze. Fivehook. Common and often abundant in subalkaline soils in the valley and about the Cantil farmlands on the desert and occasional in the Temblor Range and the mountain valleys. It is a very common street and waste place weed about Bakersfield and Mojave and the other cities and towns. Native of Eurasia. (*Echinopsilon hyssopifolium.*)

Beta vulgaris Linnaeus. Sugar beet. Occasionally escaped in waste places about sugar beet farms in the valley, usually in subalkaline soil. Native of Europe.

Chenopodium album Linnaeus. Lambs quarters. A common cismontane weed of grain fields, waste places, etc., through the Upper Sonoran associations. Native of Europe.

Chenopodium ambrosioides Linnaeus. Mexican tea. A vile smelling weed that is rare along Kern River from just above the canyon mouth west to

Lake Ming and above Kernville to the Tulare County line. Native of tropical America.

Chenopodium Berlandieri Moquin-Tandon. Slender lamb's quarters. A widespread weed in the valley that is scarce in grain fields and roadsides in the Bitterwater Valley region of the northern Temblor Range; common in South Fork Valley. Var. *Zschackei* (Murr) Murr occurs with the species.

Chenopodium Botrys Linnaeus. Jerusalem-oak. Occasional along Kern River above Lake Isabella. A small but recurrent roadside colony grows along the Breckenridge Mountain road two miles west of the Havilah Road. Native of Eurasia.

Chenopodium californicum (Watson) Watson. Soap plant. Common in the cismontane Upper Sonoran associations, especially in the grasslands; scarce on transmontane slopes in the pinyon woodland and the higher arid shrub association, east to Randsburg. The root is reported to have been used by the Indians and early settlers as a substitute for soap.

Chenopodium dessicatum A. Nelson var. **leptophylloides** (Murr) Wahl. A common weed of grainfields, waste places, etc., in all associations below the yellow pine forests; scarce in sandy usually disturbed places in the pinyon woodland on the east slope of the mountains, southwest to Cuddy Valley.

Chenopodium Fremontii Watson. Fremont's goosefoot. Occasional in colonies in sandy or disturbed places in the yellow pine forest, especially on the borders of meadows.

Chenopodium incognitum Wahl. A common grain field weed in the Temblor Range.

Chenopodium murale Linnaeus. A very common weed below the yellow pine forest, especially in gardens and orchards; scarce in towns and in irrigated farmlands on the desert. It grows very early in winter and is often in seed by February. Native of Europe.

Chenopodium rubrum Linnaeus. Rare in the dense growth in the moist subalkaline soil on the south side of Castac Lake near Lebec (T9745, 9744, 12,793).

Chenopodium strictum Roth. An occasional weed of farm lands in the valley.

Chenopodium strictum var. **glaucophyllum** (Aëllen) Wahl. A common weed of irrigated croplands in the valley and in the mountain valleys; uncommon in the dry farmlands in the Temblor Range.

Chenopodium Vulvaria Linnaeus. Known from low ground in Cuddy Valley (Hoffmann, 13 June 1927, SBM). Native of Europe.

Eurotia lanata (Pursh) Moquin-Tandon. Winter fat. Occasional on the higher plains and in the desert ranges; scarce on the valley east-side (Sand

Ridge, Comanche Hills); occasional in the west-side shale hills, almost common on the arid shale slopes of the southern Temblor Range. In the southwestern United States it is an important browse plant and research to find ways to increase its occurrence here might be economically worthwhile.

Grayia spinosa (Hooker) Moquin-Tandon. Hop-sage. Occasional to common on the desert, mostly in the creosote bush and shadscale associations.

Kochia californica Watson. Rare on the subalkaline clay flats at Rosamond Dry Lake. Reported from the alkali flats near Bakersfield by W. L. Jepson (A Flora of California, 1:443). (*Kochia americana* var. *californica*.)

Monolepis Nuttalliana (Schultes) Greene. Occasional, all associations to the yellow pine forests, usually growing in winter-wet often subalkaline places.

Nitrophila occidentalis (Nuttall) Moquin-Tandon. Occasional, usually in small colonies in moist alkaline soil, in the desolate waste lands of the Adobe Road district southwest of Arvin, on the subalkaline flats at Isabella, and at the north end of Rosamond Dry Lake.

Salicornia depressa Standley. According to Herbert L. Mason (A Flora of the Marshes of California, p 467), the plant is known from Monolith (probably no longer present there because of the drying of Proctor Lake). (*S. europaea*.)

Salicornia subterminalis Parish. Glasswort. Colonies are scarce in moist places on the alkali plains in the Adobe Road district south and west of Arvin.

Salsola Kali Linnaeus var. **tenuifolia** Tausch. Russian-thistle, and in much of Kern County, tumbleweed. Common in arid light or disturbed soils. It is especially common on sandy plains on the desert, in the valley, and in arid farmlands in the Temblor Range. Russian-thistle is occasional through the yellow pine forest (often locally common in the Jeffrey pine forest, especially in disturbed places). The plant has an unusually broad altitude range throughout California; in 1959 the writer observed it at sea level at the San Francisco International Airport and collected it at 10,400 feet elevation in the White Mountains of Inyo County (T5812). Native of Eurasia.

Russian-thistle is a winter host for beet leaf hopper, necessitating widespread aerial spraying, and is a most troublesome tumbleweed in the fall and winter in the desert and valley towns and along highways and fences. It is often a severe pest in cultivated farmlands and its pollen is sometimes a cause of hay fever. However, the young plants are first-rate feed for livestock and rain soaked winter skeletons of mature plants are also eaten with relish, often when no other natural feed is available. In dry years with late rains, Russian-thistle densely covers thousands of acres in the western San Joaquin Valley and Indian Wells Valley on the desert; this growth either eliminates or greatly reduces the summer and fall dust storms that follow dry winters. Thus, its value as an addition to the native flora is, in mildest terms, controversial.

Suaeda fruticosa (Linnaeus) Forskål. Common and often abundant over great areas of alkaline plains in the valley and about Koehn Lake on the desert; less common around the dry lakes at Edwards Air Force Base. (*S. Moquini.*)

Suaeda Torreyana Watson. Occasional about the alkali sinks on the desert and in the valley. The pubescent var. *ramosissima* (Standley) Munz is perhaps more common in the valley west to Carrisa Plains in San Luis Obispo County. The degree of pubescence varies greatly and the segregation of the variety is of dubious merit.

AMARANTHACEAE. Amaranth Family

Amaranthus albus Linnaeus. Tumble weed. An often abundant weed in summer fallowed fields in the Temblor Range that is also common in the valley. It is less frequent to the chaparral in the mountains and in the farmlands about Cantil and Rosamond. Tumble weed was abundant in the sand of the dry bed of Lake Isabella in 1961. Native of tropical America. (*A. graecizans* of authors, not Linnaeus.)

Amaranthus californicus (Moquin-Tandon) Watson. California amaranth. Occasional in wet soil (or in recently dried mud) in canal bottoms, etc., from Buena Vista Lake to Buttonwillow, mostly in the Buena Vista Slough. It is rare in dessicated vernal pools in the Douglas oak woodland in the Greenhorn Range and is common in loose sand on the flood plain above Lake Isabella, where it hybridizes with A. albus (T6486, 10,215). In the great drouth of 1961 when the lake was almost dry, *A. californicus* and *A. albus* grew densely in the dry granitic sand of the lake bed; these were the only plants that were at all common in this very "open" community.

Amaranthus graecizans Linnaeus. Prostrate tumble weed. Common in disturbed places and plowed grain fields in the Temblor Range and the valley; less common to the chaparral in the mountains. (*A. blitoides* Watson).

Amaranthus Palmeri Watson. Palmer careless weed. Although apparently a relatively recent arrival in the San Joaquin Valley, Palmer careless weed is a widespread pest of cotton fields and is common in waste places. It is occasional, mostly as a transient, through the Douglas oak woodland. Native of the southwestern United States and Mexico.

Amaranthus Powellii Watson. Known from a waste place at Kernville (Howell 38,911, 25 September 1962). Although a weed here, apparently it is native in California.

Amaranthus retroflexus Linnaeus. Careless weed, rough pigweed. A widespread cropland and waste place weed in the valley, where it is the leading pest of cotton fields. It is scarce in sandy irrigated fields in South Fork Valley and a rare waif in the other mountains and in the Temblor Range. Native of tropical America.

Nyctaginaceae. Four O'clock Family

Abronia pogonantha Heimerl. Sand verbena. A common and colorful plant, with pleasing lemon-scented flowers, of sandy places in the desert; occasional in the valley (mostly in sandy places, such as Sand Ridge, Belridge Plains, and Point of Rocks).

Abronia turbinata Torrey. Rare in loose brown clay in Sand Canyon northeast of Monolith (T7277). According to Philip A. Munz (A California Flora, p. 396), it ranges from Mt. Pinos in Kern County northeast through the mountains.

Abronia villosa Watson. Desert sand verbena. A scattered colony of a few plants grows in the shadscale scrub on the sandy plains along County Line Road east of Ridgecrest (T11,912).

Mirabilis Bigelovii Gray var. **aspera** (Greene) Munz. Occasional in rocky places in the desert ranges, in the sand dunes around the desert dry lakes, and on the arid slopes of the mountains west to South Fork Valley. (*M. laevis* var. *aspera.*)

Mirabilis Bigelovii var. **retrorsa** (Heller) Munz. The common form on the desert, growing mostly on slopes and in rocky canyons but occasionally on well-drained sand flats and even in semi-stabilized dunes. (*M. laevis* var. *retrorsa*).

Mirabilis Froebellii (Behr) Greene. Wagon wheels, devil's root. Occasional on the northwest flank of Mt. Abel northwest to Grocer Grade and through the San Emigdio Range, mostly below the chaparral. Although the plants are sometimes locally common, colonies are scarce in the foothills bordering the east side of the valley.

Mirabilis laevis (Bentham) Curran. Wishbone bush. Common in the central Temblor Range, usually growing in rocky places in the chaparral or the arid Douglas oak woodland, in the foothills around the valley, mostly below the woody Upper Sonoran associations, and rare along the edge of the desert at the mouth of Cow Heaven Canyon on the west side of Indian Wells Valley.

Illecebraceae. Knotwort Family

Herniaria cinerea DeCandolle. Locally common, usually in bare or beaten places, from the foothills of the mountains to the Douglas oak woodland, and at the mouth of Middlewater Canyon in the Temblor Range. Native of southern Europe.

AIZOACEAE. Carpet Weed Family

Glinus lotoides Linnaeus. Rare in summer in drying heavy soil in the Buena Vista Slough and the Buena Vista Lake bed. Native of Europe.

Mollugo verticillata Linnaeus. Indian chick weed. Occasional in wet places in the valley, especially in drying canal bottoms, and on moist banks along Kern River above Lake Isabella. Native of the Old World.

Sesuvium verrucosum Rafinesque. Sea purslane. Common in low alkali places in the central valley. The bright purple sepals are most attractive in contrast with the grayish barren soil where the plant usually grows. (*S. sessile.*)

Trianthema Portulacastrum Linnaeus. Horse purslane. A recent immigrant that is now a widespread weed of cottonfields, usually in light subalkaline soils (cf. Leafl. West. Bot., 9:145, 1961). Native from Imperial Valley to tropical Mexico.

PORTULACACEAE. Purslane Family

Calandrinia ciliata (Ruiz Lopez & Pavon) DeCandolle var. **Menziesii** (Hooker) Macbride. Red maids. Common in good often heavy soils in the cismontane Upper Sonoran grassland and in moist soils in the valley; scarce on the desert (Soledad Mountain, T4331). (*C. caulescens.*)

Calyptridium monandrum Nuttall. Sand cress. Occasional, usually in bare places around shrubs, in the Temblor Range; common in the mountains to the yellow pine forest, especially in the more arid regions; occasional in the arid shrub association; scarce in favorable years in the creosote bush association (Soledad Mountain, T4313).

Calyptridium Parryi Gray. Known from Griffins, Mt. Pinos, in Ventura County (Elmer 3948, POM). The plant undoubtedly occurs in Kern County in the same region.

Calyptridium umbellatum (Torrey) Greene. Pussy paws. Common in gravelly or disturbed soil in the yellow pine forest, particularly at higher elevations. (*Spraguea umbellata.*)

Lewisia rediviva Pursh. Bitter root. Scarce in oak woodland in the Woody-Glennville region.

Lewisia rediviva var. **minor** (Rydberg) Munz. Lesser bitter root. Common in gravel and gravelly loam on the summits of Sawmill Mountain and Mt. Pinos.

Montia Chamissoi (Ledebour) Durand & Jackson. Scarce in very wet meadows or in shallow creeks at Evans Flat and Shirley Meadows in the Greenhorn Range, along Mill Creek on Breckenridge Mountain, along the fork of Cannell Creek at Pine Flat and upper Bartolas Creek on the Kern Plateau, and rare in cienegas on Mt. Pinos.

Montia fontana Linnaeus. Water purslane. Rare in sunny wet places in the Upper Sonoran associations in the Greenhorn Range (Delano Road 4 miles west of Woody, T11,782; Cedar Creek near the Rock Corral, Farnsworth 470, T11,796, Wagy Creek, T10,962).

Montia gypsophiloides (Fischer & Meyer) Howell. Colonies are scarce in shaded humus in the central Temblor Range (Black Canyon, T201; summit of Associated Canyon, T1981).

Montia perfoliata (Donn) Howell. Miner's lettuce, Indian lettuce. Occasional in the valley and very rare on the desert in wet years (recently dried rain pool 14.9 miles north of Mojave, T6773, 10 April 1962). It is common in all the other associations to the borders of the yellow pine forest. Colonies are especially common and often dense in humus under Douglas oaks. Miner's lettuce has a pleasingly piquant taste, and was used by the pioneers for salad.

The plant occurs in many forms; the more significant of those that have been named and occur in Kern County are forma *parviflora* (Douglas) J. T. Howell, which is occasional in dry shallow sandy soils in the shade of chaparral in the central Temblor Range and in moist shaded loam in the Douglas oak woodland along Cedar Creek in the Greenhorn Range; var. *depressa* (Gray) Jepson, a bright red form that is rare at the west end of Tollgate Ridge in the northeastern Tehachapi Mountains (T12,326) and 2.5 miles north of Greenhorn Pass (T12,402); and var. *utahense* (Rydberg) Munz, which is rare in damp places in talus slides in the El Paso Range (Last Chance Canyon, T11,869) and on the northwest slope of Government Peak in the Rand Mountains (T12,021).

Montia spathulata (Douglas) Howell. Primarily a plant of the Coast Ranges from San Luis Obispo County north, it is scarce although widespread in Kern County in the woody Upper Sonoran associations, usually growing in chaparral or shin oak brush.

Montia spathulata var. **exigua** (Torrey & Gray) Robinson. Scarce in moist soil in Mountain Meadow in the red fir forest on the east slope of Sunday Peak (T8641). (*M. exigua.*)

Montia spathulata var. **viridis** A. Davidson. A small colony of this distinctive green variety grows in heavy shade along Peel Creek on the west slope of Sunday Peak (T5219). (*M. exigua* var. *viridis.*)

Portulaca oleracea Linnaeus. Purslane. A common weed of moist soils in the valley and the Temblor Range; occasional in the mountain valleys. Native of Europe.

CARYOPHYLLACEAE. Pink Family

Arenaria californica (Gray) Brewer. California sandwort. Occasional in the Douglas oak woodland in the Greenhorn Range, usually growing in poor coarse granitic soil; less frequent south to Tejon Canyon.

Arenaria Douglasii Fenzl. Douglas sandwort. Occasional to scarce in the upper Sonoran grasslands through the chaparral and in the Jeffrey pine forest at Tecuya Ridge in the Mt. Pinos region.

Arenaria macradenia Watson var. **macradenia.** Desert sandwort. Scarce along the granite outcrops at the south end of the Kern Plateau (Howell & True 41,824, T8662).

Arenaria macradenia var. **arcuifolia** Maguire. Rare in rocky places at the lower levels of the yellow pine forest in the Piute Mountains (Harris Grade near Landers Meadow, T7443) and on the south side of Black Mountain in the Greenhorn Range (4,700 feet elevation, Howell 38,256).

Arenaria macradenia ssp. **Ferrisiae** Abrams. Occasional, sometimes locally common, in the Jeffrey pine forest in the Piute Mountains, in the pinyon woodland on the east slope of Morris Peak in the southern Sierra Nevada, and in the arid Douglas oak woodland on the east slope of the Greenhorn Range.

Arenaria macradenia var. **Parishiorum** Robinson. Rare on the ridge along the east summit of Tunis Canyon in the Tehachapi Mountains (T4600), an isolated occurrence of this distinctive variety that ranges from Inyo County to Arizona and New Mexico.

Arenaria pusilla Watson. Dwarf sandwort. Known from the east base of the Greenhorn Range (Hardham 3601).

Cerastium glomeratum Thuillier. Mouse-ear. Occasional in moist places in the Douglas oak woodland in the mountains. Native of Europe. (*C. viscosum* in part.)

Loeflingia pusilla Curran. Colonies are scarce in favorable years in the dunes at the south end of Rogers Dry Lake (T10,714) and the northeast end of Buckhorn Dry Lake (T10,777). Both *L. pusilla* and *L. squarrosa* grow in the dunes around these desert dry lakes; *L. pusilla* is more common in stiffer more alkaline soils. Intergrades between the two are common; further study may reveal that *L. pusilla,* long known only from the type collection, is merely an extreme variation of *L. squarrosa.* The type was collected between Tehachapi and Girard (now Marcel) Siding, probably somewhere near the summit of Tehachapi Canyon, by Katharine Layne Curran in June, 1884.

Loeflingia squarrosa Nuttall. Colonies are scarce in rather barren sandy places in the cismontane Upper Sonoran associations and in the semi-stabilized massive dunes around Buckhorn Dry Lake on the desert.

Sagina apetala Arduino var. **barbata** Fenzl. Known from Tillie Creek, 2 miles west of Lake Isabella (Hardham 3622); along Cedar Creek below the narrows on the Farnsworth Ranch (Farnsworth 544); and from 3 miles northwest of Keene at 2,500 elevation (Howell 32,769). Native of Europe.

Sagina saginoides (Linnaeus) Karsten var. **hesperia** Fernald. Common in wet places in the yellow pine forest in the Greenhorn Range and on Breckenridge Mountain; rare in the cienega at the head of Bitter Creek on Mt. Pinos.

Silene antirrhina Linnaeus. Sleepy catchfly. Scarce in the Douglas oak woodland on Gighas Ridge in the Tehachapi Mountains (T6257).

Silene californica Durand. Indian pink. A plant with colorful red flowers that is occasional, mostly in the chaparral, in the mountains south to the Mt. Abel region.

Silene gallica Linnaeus. Windmill pink. A fairly common weed on grassy slopes and around rock outcrops in the Douglas oak woodland from Woody and Glennville north to the Tulare County line. It is a town weed at Kernville (J. T. Howell, 21 May 1962). Native of Europe.

Silene Lemmonii Watson. Occasional in the ponderosa pine forest in the Greenhorn Range, scarce at the higher elevations of the Douglas oak woodland.

Silene montana Watson ssp. **bernardina** (Watson) Hitchcock & Maguire. Occasional about the summit of Mt. Pinos and in the yellow pine forest on Breckenridge Mountain. It extends north in Tulare County to The Needles on the Western Divide and to Brush Mountain on the Kern Plateau; the Kern County and southern Sierra occurrence in Tulare County is a range extension for this plant previously known from the east slope of the Sierra Nevada in Inyo and Mono counties. (*S. bernardina.*)

Silene montana var. **sierrae** Hitchcock & Maguire. Scarce in the yellow pine forest south to Evans Flat in the Greenhorn Range, on Breckenridge Mountain, the southwestern Kern Plateau, and rare on the north slope of Tehachapi Peak.

Silene verecunda Watson ssp. **platyota** (Watson) Hitchcock & Maguire. Occasional around granite blocks and outcrops in the southern Kern Plateau, rare at Lookout Point in the Piute Mountains, and on Black Mountain in the Greenhorn Range.

Spergularia atrosperma R. P. Rossbach. Scarce at the alkali playas 9 miles north of Lost Hills (T887); rare in Red Rock Canyon (T7094). (*S. diandra.*)

Spergularia Bocconii (Scheele) Foucaud. Abundant in subalkaline soil along the north shore of Castac Lake near Lebec (T9737, 11,097).

Spergularia macrotheca (Hornemann) Heynhold var. **leucantha** (Greene) Robinson. Scarce in moist alkali soil in the valley. It grows luxuriantly in warm mud around Scovern Hot Springs near Isabella.

Spergularia marina (Linnaeus) Grisebach. Salt marsh spurrey. Occasional in moist alkaline soil in the valley, the Temblor Range, and at Rogers and Rosamond dry lakes on the desert. The poorly marked var. *tenuis* (Greene) R. P. Rossbach is rare in the valley. (*S. salina.*)

Spergularia rubra (Linnaeus) J. & C. Presl. Purple sand spurrey. A dense colony grows in dry soil at the borders of Shirley Meadows. It is a ranchground weed at the Farnsworth Ranch near Glennville. Native of Europe.

Stellaria crispa Chamisso & Schlechtendal. Rare in wet soil in heavy shade at Mountain Meadow on the east slope of Sunday Peak (T7758, 6465).

Stellaria Jamesiana Torrey. Occasional on shaded north slopes in the yellow pine forest in the Greenhorn Range and the Piute Mountains; rare near the summit of Breckenridge Mountain.

Stellaria longipes Goldie. Common in wet soil in sun along the fork of Cannell Creek near Pine Flat and upper Bartolas Creek on the Kern Plateau; in shallow water in Mill Creek in the yellow pine forest on Breckenridge Mountain.

Stellaria media (Linnaeus) Villars. Comon chickweed. Often common in the Upper Sonoran grassland in the east-side foothills, extending to the Douglas oak woodland and the chaparral. It is a common weed of moist places and gardens in the valley and the Temblor Range. Native of Eurasia.

Stellaria nitens Nuttall. Shiny chickweed. Common in the Upper Sonoran grassland in the foothills of the mountains, extending to the dryish meadows in the yellow pine forest; scarce in the San Emigdio and Temblor ranges.

Stellaria sitchana Steudel var. **Bongardiana** (Fernald) Hultén. Occasional at Shirley Meadows (T4134) and Greenhorn Mountain Park (T8611), apparently a southern range extension from Mariposa County.

Ceratophyllaceae. Hornwort Family

Ceratophyllum demersum Linnaeus. Hornwort. A dense colony grows in a deep pool behind a wier on Poso Creek at the Henry Bowen ranch grounds in Lynns Valley (T9967); once collected in Bakersfield in a ditch at what was then the north edge of town (Coville 1242 in 1891), and at Buttonwillow (Evermann in 1914).

Ranunculaceae. Crowfoot Family

Aconitum columbianum Nuttall. Monkshood. Rare in the cienega on the east slope of Sunday Peak below Mountain Meadow (T6396). Charlotte Smith reports (personal communication) it also occurs in similar places on the west slope (between Greenhorn Summit Store and the Tulare County line, C. N. Smith 711, JEPS). The plant is uncommonly beautiful but is extremely poisonous to livestock.

Aquilegia formosa Fischer var. **pauciflora** (Greene) Bootham. Occasional in moist shaded places in the mountains.

Aquilegia formosa var. **truncata** (Fischer & Meyer) Baker. Crimson columbine. Occasional in the yellow pine forest in the Greenhorn Range; less common south to the Mt. Abel region. (*A. truncata.*)

Clematis ligusticifolia Nuttall. Virgin's bower. Scarce, usually in moist places, in the Temblor Range and in the Douglas oak woodland through the lower elevations of the yellow pine forest in the mountains; east to the pinyon woodland along the South Fork of Kern River.

Delphinium gracilentum Greene forma **versicolor** Ewan. Known from Richbar in Kern Canyon (Benson 8692, POM).

Delphinium gypsophilum Ewan. Gypsum larkspur. Common in the Temblor Range, mostly in light soil in the Upper Sonoran grassland on the east slope; occasional in the valley to the arid McKittrick and Elk Hills, southeast to the western part of the San Emigdio Range; generally scarce in the low east-side foothills but locally common in light dry soils in the Poso Hills and the Greenhorn foothills. The plant is exceptionally drouth resistant; according to Harlan Lewis and Carl Epling (Evolution 6:256), the roots can survive for as long as 8 years without even forming rosette leaves. Away from the arid climate where it is native, gypsum larkspur is extremely susceptible to mildew; this is probably a factor in limiting its distribution to the arid slopes around the San Joaquin Valley.

The plant is the leading livestock poisoner of the Temblor Range; so great is the hazard that many areas have to be grazed with special attention and timing to avoid serious losses.

Delphinium Hanseni (Greene) Greene. Occasional in the Douglas oak woodland and the Upper Sonoran grassland south to the Tehachapi Mountains. It extends through the yellow pine forest at rocky places and is a characteristic plant of the highest summit granite crests. (*D. Hanseni* ssp. *kernense.*)

Delphinium hesperium Gray ssp. **pallescens** (Ewan) Lewis & Epling. Pallid western larkspur. Scarce in the Douglas oak woodland and the Upper Sonoran grassland south to Poso Hills.

Delphinium inopinum (Jepson) Lewis & Epling. Scarce at the higher elevations of Piute Mountain and Mt. Pinos, often along exposed rocky ridgetops. (Endemic to these places and Kern River Canyon in Sequoia National Park in Tulare County.) (*D. Parishii* var. *inopinum.*)

Delphinium Parishii Gray. Parish larkspur. The only larkspur of the Mojave Desert in Kern County, where it is an attractive plant with light blue flowers, occasional from the creosote bush association to the pinyon woodland. (*D. amabile.*)

Delphinium Parishii ssp. **purpureum** Lewis & Epling. Scarce in the pinyon woodland in the Mt. Pinos region east to Cuddy Valley. The type was collected along the Cuddy Valley Road 0.1 mile from the junction with the Cuddy Canyon Road (Harlan Lewis & David Dunn 478, 5 June 1943).

Delphinium Parryi Gray ssp. **seditiosum** (Jepson) Ewan. Blue larkspur. Common in heavy soils on north slopes in the Upper Sonoran associations from the Temblor Range south to the pinyon woodland in the Mt. Abel region. (*D. hesperium* var. *seditiosum*).

Delphinium patens Bentham ssp. **Greenei** (Eastwood) Ewan. Occasional from the chaparral to the yellow pine forest in the mountains south to Tejon and El Paso canyons.

Delphinium patens ssp. **montanum** (Munz) Ewan. Occasional in the Jeffrey pine forest in the Mt. Pinos region.

Delphinium Purpusii Brandegee. Kern Canyon larkspur, Purpus larkspur, pink larkspur. A scarce and distinctive endemic, occurring from Roads End in Tulare County to the vicinity of Democrat Hot Springs in Kern Canyon; east to Erskine Canyon at the north end of the Piute Mountains; on the east slope of the Greenhorn Range below Wagy Flat, and along Poso Creek near Poso Mine. Kern Canyon larkspur usually grows in steep shaded moist rock outcrops. It is avidly eaten by deer and only in protected places is it beyond their reach. The plant is easily distinguished; it is the only rose-pink larkspur native to North America. The type was collected at Erskine Creek (Joseph Anton Purpus 5015 in 1897).

Delphinium recurvatum Greene. Alkali larkspur. Occasional, in wet years relatively common, in the alkali sinks and plains in the valley south to Goose Lake; common around Soda Lake in San Luis Obispo County. In contrast to its occurrence at subalkaline places in these very arid regions, it is scarce in wet heavy blue clay (decomposed serpentine) in the Douglas oak woodland on Parkfield Grade in southwestern Fresno County (T1950). (*D. hesperium* var. *recurvatum.*)

Isopyrum occidentale Hooker & Arnott. Colonies are scarce but widespread from the west slope of the Greenhorn Range through the mountains southwest to Black Bob Canyon in the San Emigdio Range, usually growing in moist shaded places in chaparral or Douglas oak woodland.

Myosurus minimus Linnaeus. Mouse tail. Rare in wet places in the valley (alkali playas 10 miles north of Lost Hills, T1002, 910).

Myosurus minimus var. **filiformis** Greene. Known from the shade of boulders on a hillside near Onyx (Fay A. MacFadden 13,547, 13 April 1935) and from Cedar Creek at the Rock Corral (Farnsworth 572).

Ranunculus aquatilis Linnaeus var. **capillaceus** (Thuillier) DeCandolle. Water buttercup. Common in pools and quiet water through the mountains to Cuddy Valley, sometimes growing in large dense colonies.

Ranunculus aquatilis var. **hispidulus** Drew. A small colony grew in a rapidly drying small artificial pool bed 0.5 mile west of White Oak Lodge in the Tejon Pass region in August, 1963 (T8858).

Ranunculus californicus Bentham. California buttercup. A colony grows in friable clay at a spring about 2 miles north of the mouth of Alamo Canyon at the southeast end of the Tehachapi Mountains (T6948) and another at Water Creek on the southwest slope of Bear Mountain (T9660).

Ranunculus canus Bentham var. **ludovicianus** (Greene) L. Benson. Common in the Douglas oak woodland south to the Tehachapi Mountains; rare in the San Emigdio Range. The type, collected at Tehachapi by Katharine Layne Curran in 1884, was apparently destroyed in the San Francisco fire of 1906. Lyman Benson (Amer. Mid. Nat. 40:68) has designated a collection made by Katharine Curran Brandegee at Tehachapi on 16 May 1905 as a neotype.

Ranunculus Cymbalaria Pursh var. **saximontanus** Fernald. Common in sunny wet places or in water in the mountains, mostly in Douglas oak woodland but sometimes extending to the yellow pine forest.

Ranunculus hebecarpus Hooker & Arnott. Scarce in the mountains south to Tejon Canyon, mostly in shin oak brush but also in chaparral and in the Douglas oak woodland.

Ranunculus occidentalis Nuttall var. **Eisenii** (Kellogg) Gray. Common in Douglas oak woodland in the mountains south to Lebec; occasional in the yellow pine forest.

Thalictrum polycarpum (Torrey) Watson. Meadow rue. Occasional in shade, mostly in Douglas oak woodland, in the Greenhorn Range and on the south side of Kern Canyon.

BERBERIDACEAE. Barberry Family

Berberis dictyota Jepson. Barberry. Colonies, often of a few plants or even only of one shrub, are scarce but widely distributed through the Greenhorn Range from the middle elevations of the Douglas oak woodland to the lower levels of the ponderosa pine forest. At Tollgate Lookout in the northeastern Tehachapi Mountains, a dense colony of about three acres grows on a broad open sunny high slope. (*B. californica.*)

LAURACEAE. Laurel Family

Umbellularia californica (Hooker & Arnott) Nuttall. California bay. Colonies are occasional just below the yellow pine forest on the west slope of Breckenridge Mountain, in shaded canyons on the south side of Kern Canyon, and in shaded moist canyons on Rattlesnake Grade in the Douglas oak woodland in the Greenhorn Range. California bays grows well in cultivation and makes a neat round-headed tree. The leaves can be used in cooking; the amount given in recipes for bay leaves should be divided by two. Early settlers tied branches in chicken houses to discourage lice.

PAPAVERACEAE. Poppy Family

Argemone munita Durand & Hilgard. Prickly poppy. Occasional in dry places in the Temblor Range and from Mt. Abel northeast to the desert slopes and ranges. Forms intermediate between ssp. *munita* and ssp. *rotundata* are common from Mt. Abel north through the Temblor Range. (*A. platyceras.)*

Argemone munita ssp. **rotundata** (Rydberg) G. Ownbey. Occasional in the arid Douglas oak woodland on the east slope of the Greenhorn Range.

Canbya candida Parry. A delicate miniature white poppy that is rare in colonies in sandy soil on the west borders of Indian Wells Valley, the sand dunes between Buckhorn and Rosamond dry lakes, and west in South Fork Valley to Lake Isabella. An inconspicuous plant that perhaps is overlooked.

Dendromecon rigida Bentham. Bush poppy, chaparral poppy. Local in the chaparral at Geringer Grade at the southeast end of Piute Mountain (T3065). Charlotte Smith reports a colony on the ridge 0.5 mile south of the old Greenhorn School on Rattlesnake Grade in the Greenhorn Range (C. N. Smith 1033, JEPS). Mrs. Smith writes (personal communication) that in 1948 the fire-following colony was a large one, growing among small bushes of *Ceanothus cuneatus* and making the hillside yellow with bloom. By 1961, the *Ceanothus* had become a thicket and only about a dozen plants of *Dendromecon* survived.

Eschscholzia caespitosa Bentham. Common in the Greenhorn Range and on the north slope of Breckenridge Mountain, growing in Douglas oak woodland and chaparral; rare, east to Havilah.

Eschscholzia caespitosa ssp. **kernensis** (Wolf) Munz. Tejon Ranch poppy. Endemic to the low east-side foothills, where it is normally scarce although it sometimes grows in dense colonies in wet years, usually occurring in stiff clay. Often it furnishes the only spots of bright color on the otherwise dry bleak clay slopes between Chanac and Tejon canyons, where it grows at least as depauperate plants in all but the driest of years. It is rare north to the mesas east of Bakersfield and at Dry Bog Knoll at the head of Adobe Canyon in the Greenhorn Range foothills. Although the variety probably developed by isolation because of the increasing aridity of the upper San Joaquin Valley, a close relationship to *E. caespitosa* is questionable and the plant is in need of critical study. Unless the bur-like seeds are collected, it is sometimes difficult to distinguish from glabrous forms of *E. Lemmonii,* which reaches its northeastern limits in the same region. The type was collected in the Tejon Hills, 2 miles northwest of the Tejon Ranch Headquarters (Carl B. Wolf 6358, 15 April 1935).

Eschscholzia californica Chamisso. California poppy. Common and colorful in all associations except the most arid open plains on the desert and in the yellow pine and red fir forests. The plant is extremely variable, ranging from the delicate annuals of dry years on the valley plains to the robust heavily rooted perennials of deep soils in the mountains. The plants bloom the first year and Stanton A. Cook (Evolution, 16:278) has demonstrated that some of these are obligate annuals (probably all of our valley plants). Regretably, he did not elucidate characters that would make it possible to distinguish them from drouth-stricken young perennials but he does seem to imply that these annual poppies are conspecific with *E. mexicana* Greene that occurs from southern California and Baja California east to Texas. A twice-blooming form is common in sandy soils in the Upper Sonoran associations in the northern Temblor Range; this is apparently var. *peninsularis* (Greene) Munz and not var. *crocea* (Bentham) Jepson, which the author called it in his Flora of the Temblor Range (Wasm. Journ. Biol., 14:233).

John Thomas Howell has paid an eloquent tribute to this our state flower:

> "No poet has yet sung the full beauty of our poppy, no painter has successfully portrayed the satiny sheen of its lustrous petals, no scientist has satisfactorily diagnosed the vagaries of its variations and adaptability. In its abundance, this colorful plant should not be slighted: cherish it and be ever thankful that so rare a plant is common."

While Howell was writing of the plant as it occurred in Marin County, he might well have been contemplating the broad flaming beds that cover in spectacular fashion the southeast slopes of the Tehachapi Mountains along Antelope Valley. At the other extreme, scattered plants on the plains west of Lost Hills in a dry year consist of a small inconspicuous tuft of leaves topped by one large flower; truly Cinderella dressed for the ball!

Eschscholzia glyptosperma Greene. In favorable years, a colony grows along County Line Road in the shadscale scrub on the plains east of Ridgecrest (T11,827) and it is occasional on sandy vacant lots in Ridgecrest (Meng 279). These are the local western limits for this desert poppy; it is common to the east on the plains and mesas at the head of Salt Wells (Poison) Canyon in San Bernardino County.

Eschscholzia hypecoides Bentham. Colonies that grow only in favorable years are scarce in brown shale gravel slides in Cedar Canyon in the Temblor Range (Alec Cook Rocks, T13,023; Cedar Canyon, 6 miles south of Bitterwater Road, T13,022).

Eschscholzia Lemmonii Greene. Lemmon's poppy. Common, often in dense colonies in poor light soils, on dry slopes in the Temblor Range, the plains on the west side of the valley, in the San Emigdio Range, and rare in the east-side foothills north to Comanche Point. In the latter region, glabrous forms of *E. Lemmonii* can be distinguished only with difficulty from *E. caespitosa* ssp. *kernensis,* particularly the depauperate plants of dry years (the diagnostic character of pubescent foliage given for *E. Lemmonii* is often not valid).

Eschscholzia Lobbii Greene. Frying pans. Occasional in small colonies in the Douglas oak woodland in the foothills around Woody and scarce in the Upper Sonoran grassland in the low hills west of Granite Station, usually growing in deep well-watered soils. According to Wallace R. Ernst (personal communication), our plants approach *E. pulchella* as proposed by Greene.

Eschscholzia minutiflora Watson. Small-flowered poppy. Occasional on the desert plains (the common poppy of that region), west to Weldon and the east slope of the Piute Mountains. It has also been collected in Lockwood Valley in Ventura County and at an indefinite Cuyama Valley station in the Ozena region. Further collections of adequate material are needed to establish with more certainty the Cuyama Valley occurrence of the plant. The completely disjunct and non-desert region of high rainfall around Poso in San Luis Obispo County has been included in the range of the plant (Theodore Mosquin, Madroño, 16:91) on the basis of what almost certainly must be a collection with incorrect locality data or of a most unusual waif (Armstrong 1112, UC). (Mosquin's inclusion of all of the Temblor Range in his map of the range of *E. minutiflora* seems to have no foundation in fact.)

Eschscholzia minutiflora var. **rutaefolia** (Greene) Jepson. Scarce on a slope of limestone talus 9.4 miles east of Onyx on the Walker Pass Highway (T7051A), where it grows with typical *E. minutiflora.* Although usually considered a form of *E. minutiflora* of no significance, the variety has distinctive slender stems and ternate-biternate foliage with broadly cuneiform lobes that readily set it apart. The type locality is given, perhaps in a highly general fashion, as Havilah (Townshend Stith Brandegee, 15 May 1891). In addition to Havilah and the north end of the Piute Mountains, the variety is also known from the east base of the San Bernardino Mountains.

Eschscholzia Parishii Greene. Parish poppy. Readily distinguishable from the other common desert poppy, *E. minutiflora,* by its large petals, Parish poppy is occasional in the Rand Mountains and infrequent from the east end of the El Paso Range to Indian Wells Valley and Salt Wells (Poison) Canyon in nearby San Bernardino County. This large flowered form of *E. Parishii* is a biotype peculiar to this region; the typical biotype grows in the southern Mojave and Colorado desert. (*E. minutiflora* var. *Parishii.*)

Eschscholzia procera Greene. Kernville poppy. A distinctive plant of the sand flats at the north limits of Kernville (Howell 37,132, T8568, 11,370). It is the largest species in the *E. californica* group, the plants becoming almost bushy with stems to three feet long. Kernville poppy has a long woody tap root and garden cultures (T9217, 9548, 10,644) have established that, unlike other California poppies, it has little drouth resistance. Thus, it is probably limited to these sand flats by the need for ample subsurface moisture. The plant is notable for the flowers that occur both with and without torus rims; usually, however, these are lacking. The type was collected near Kernville (Townshend Stith Brandegee, 13 May 1891). (*E. californica* in very small part.)

Hesperomecon linearis (Bentham) Greene. Occasional from the higher plains of the valley through the chaparral, usually growing in small dense colonies on north slopes. Plants from Kern Canyon and from scattered stations in the Greenhorn Range have yellow outer petals and white inner ones. This form, considered to be of minor morphological significance, is *H. pulchellum* Greene. (*Meconella linearis.*)

Meconella californica Torrey. Uncommon in the Woody region, growing in small dense colonies in good soil on shaded or protected canyons and flats in the Douglas oak woodland. (*M. oregana* var. *californica.*)

Papaver hybridum Linnaeus. Point of Rocks poppy. A European poppy known in the United States from the Point of Rocks (where it was first collected by Eben McMillan in 1950) and Antelope Valley south to the New Well Wash and Blackwells Corner in western Kern County and from extreme eastern San Luis Obispo County at the Sinsheimer Flats in the Choice Valley Hills. The plant is rare, but is becoming more common; in favorable places such as light soils in abandoned farming fields or even in planted wheat fields in poor soil it is well established and sometimes locally common. This charming little red-flowered poppy is a welcome addition to the immigrant flora of the arid western San Joaquin Valley. (Cf. Leafl. West. Cot., 6:112, 10:19, Wasm. Journ. Biol., 14:234.) (*Papaver apulum* Tenore var. *micranthum* (Boreau) Fedde of California references.)

Platystemon californicus Bentham. Cream cups. Common in all cismontane associations below the yellow pine forest. It is local on the western borders of the desert and in good soils above 4,000 feet elevation in the Rand and El Paso mountains. A form with shaggy pubescence has been described as var. *crinitus* by Greene; this is most common in the transmontane regions although it is occasional in the Sierra Nevada foothills south to the Piute and Tehachapi mountains and in Kelso Valley at the east base of the Piute Mountains. Although it intergrades completely with the typical variety and W. L. Jepson stated correctly that it "merely represents a terminus in a line of variation," the extreme forms are quite distinct. The type of var. *crinitus* was collected at Tehachapi by Edward L. Greene in 1889.

Stylomecon heterophylla (Bentham) G. Taylor. Wind poppy. Occasional from the cismontane foothills to the edge of the valley, growing from the Upper Sonoran grassland to the chaparral. Wind poppy is especially common in the chaparral after burns. (*Papaver heterophyllum.*)

Stylomecon heterophylla var. **micropetala** Hoover. Scarce in the Upper Sonoran subshrub association. It grows on north slopes, usually only in good years, in the low foothills along the west side of the valley south to Maricopa (and to Baja California); rare in Elk Hills. The plant is distinguished by its small petals. (Cf. R. F. Hoover. Leafl. West. Bot., 10:344, 1966.)

FUMARIACEAE. Fumitory Family

Dicentra chrysantha (Hooker & Arnott) Walpers. Golden ear drops. Occasional in the mountains from the chaparral to the lower levels of the yellow pine forest, usually growing around rocky places.

Dicentra pauciflora Watson. Small-flowered bleeding heart. Rare near a rock outcrop on the Evans Flat Road near the Wagy Flat Road junction one mile east of Evans Flat (C. N. Smith 654, JEPS). Mrs. Smith has observed this colony for many years and reports it rarely blooms. (She has never seen the plant elsewhere in the Greenhorn Range.) This station in the lower levels of the southern limits of the Sierran ponderosa pine forest is decidedly anomolous; it is typically a plant of the subalpine or the alpine zones and is next known far to the north at Mt. Moses and Mineral King in Tulare County. It is not known north of the Kaweah colonies in the Sierra Nevada, but occurs in the Marble Mountains of northern California.

CRUCIFERAE. Mustard Family

Arabidopsis Thaliana (Linnaeus) Heynhold. A small colony grows in shaded dryish loam on the south side of Petitt Meadow in the ponderosa pine forest on the west slope of the Greenhorn Range (T10,984); another grows in a moist hollow on the French Ranch about five miles southwest of the summit of Eugene Grade (C. N. Smith 1384, JEPS). Naturalized from Europe. (*Arabis Thaliana.*)

Arabis glabra (Linnaeus) Bernhardi. Tower mustard. Occasional in the Douglas oak woodland from the Glennville region south to Poso Flat; scarce through the mountains to the southwest (Squirrel Meadow on Breckenridge Mountain, Bear Valley, Ft. Tejon, Tecuya Ridge, Iris Point on Mt. Pinos).

Arabis platysperma Gray. Rock cress. A typical plant of the Sierra Nevada that reaches its southern limits on the north slopes of Sunday Peak and along the ridge southeast of Pine Flat on the Kern Plateau.

Arabis pulchra Jones. Occasional in the pinyon woodland and arid shrub association on the east slope of the mountains southwest to Ballinger Canyon in upper Cuyama Valley, often growing through low shrubs on arid slopes.

Arabis repanda Watson. Scarce in the ponderosa pine forest in the Greenhorn Range; rare on the ridge southwest of Little Cannell Meadow on the Kern Plateau; and very rare on the north slope of Tehachapi Peak. W. L. Jepson (Flora of California, 2:63) cites Mt. Pinos (sans collector or collection).

Arabis sparsiflora Nuttall var. **arcuata** (Nuttall) Rollins. Occasional in rocky places in the Douglas oak woodland in the Kern Canyon region. (*A. Holboelii* var. *arcuata.*)

Athysanus pusillus (Hooker) Greene. A delicate annual common in shade in the cismontane Upper Sonoran associations.

Barbarea orthoceras Ledebour. Winter cress. Scarce in wet or moist places in the mountains from the Douglas oak woodland to the yellow pine forests.

Brassica geniculata (Desfontaines) J. Ball. Mediterranean mustard, summer mustard. A common weed of cismontane fields and waste places that is a scarce recent roadside adventive on the desert, mostly in low places. This perennial is the most drouth resistant of the California weedy mustards. It is eagerly grazed in summer by livestock; this prevents it from becoming even more common. Native of Europe.

Brassica Kaber (DeCandolle) L. C. Wheeler. Charlock, wild mustard. A widespread and often troublesome winter and spring weed in the valley that is becoming more common in the dry farm wheat fields in the northern Temblor Range. The young plants make excellent stewed greens, not unlike rather pungent spinach. Native of the Old World, probably Eurasia.

Capsella Bursa-pastoris (Linnaeus) Medicus. Shepherd's purse. A common weed of gardens, low areas, beaten places, etc., that is also common in the valley and lower foothills, becoming less frequent through the cismontane Upper Sonoran associations. Native of Europe.

Cardaria Draba (Linnaeus) Desvaux. Hoary cress. A colony has grown for many years at the Davies Ranch one mile south of Annette in the Temblor Range (T1112). Another grows in low alkali soil along Rosamond Boulevard just west of Rosamond Dry Lake (T10,853). Native of Europe. (*Lepidium Draba*).

Cardaria pubescens (C. A. Meyer) Jarmolenko var. **elongata** Rollins. White top. Occasional in farmed or disturbed areas in the arid mountains, from Cuddy Valley northeast to the Inyo County line. It is locally common and troublesome in farmlands near Weldon, in Walker Basin, and in the Tehachapi region. Probably a native of Asia.

Caulanthus amplexicaulis Watson. Known from Seymour Creek, elevation 6,400 feet, Mt. Pinos, Ventura County (Munz 6980, POM). It probably also occurs in Kern County in that region, where county lines are difficult to ascertain. (*Streptanthus amplexicaulis.*)

Caulanthus californicus (Watson) Payson. Rare and growing only in wet years on the alkali plains north of Semitropic (11.5 miles north of U. S. 466 on the Corcoran Highway, T4167). It is also known in light soil in Rancheria Gulch in the Upper Sonoran grassland of the low Greenhorn foothills (C. N. Smith 53, JEPS). (*Streptanthus californicus.*)

Caulanthus Cooperi (Watson) Payson. Occasional in sandy soil on the desert, often growing under creosote bushes; infrequent through the arid shrub association. (*Thelypodium Cooperi.*)

Caulanthus Coulteri Watson. Common in the cismontane Upper Sonoran associations in the mountains; rare on the east slopes, approaching the desert near the mouth of Jawbone Canyon. (*Streptanthus Coulteri.*)

Caulanthus inflatus Watson. Desert candle, squaw cabbage. Common in the creosote bush association, where it is one of the more striking and widely recognized spring flowers; occasional in arid places in the Tehachapi Mountains and around the southeast end of the valley; occasional on arid slopes in the Temblor Range and in the upper Cuyama Valley. Rare in Rancheria Gulch in the Greenhorn foothills (C. N. Smith 255, 288, JEPS). The Indians are said to have eaten the succulent stems. (*Streptanthus inflatus.*)

Caulanthus Lemmonii Watson. Lemmon's jewel flower. Occasional in the Upper Sonoran grassland in poor soil or fine clay in the northern Temblor Range. A disjunct colony grows in friable clay at Dry Bog Knoll at the head of Adobe Canyon in the foothills of the Greenhorn Range. (*Caulanthus Coulteri* var. *Lemmonii, Streptanthus Lemmonii.*)

Dentaria californica Nuttall. California toothwort. Uncommon but widespread in moist heavily shaded ravines and deep swales in the Douglas oak woodland in the mountains south to Palos Altos Canyon northeast of Lebec. (*D. integrifolia* var. *californica.*)

Descurainia pinnata (Walter) Britton ssp. **glabra** (Wooton & Standley) Detling. Occasional in light soils on the desert, locally abundant in South Fork Valley, and scarce about Bakersfield. (*Sisymbrium pinnatum* in part.)

Descurainia pinnata ssp. **Menziesii** (DeCandolle) Detling. Tansy mustard. A common weed of all associations below the yellow pine forest, especially in grasslands and heavily grazed or disturbed places (it is not readily grazed by cattle or sheep). Tansy mustard is occasional and sometimes even locally common in the Jeffrey pine forest, especially in the Tehachapi Mountains (a scattered colony even grows around the low granite outcrops at the summit of Bear Mountain). It is scarce on the desert. (*Sisymbrium pinnatum* in part.)

Descurainia Richardsonii (Sweet) O. E. Schulz ssp. **viscosa** (Rydberg) Detling. Scarce (but recurrent) on the north slope of Tehachapi Peak, growing in full sun along a fire guard at 6,200 feet elevation in the Jeffrey pine forest (T7235).

Descurainia Sophia (Linnaeus) Webb. Occasional, sometimes locally common, on the desert west to Monolith and Weldon. It is especially common in recently disturbed sandy soil. The plant is rare in the valley, mostly in the region from Bakersfield to Buena Vista Lake, where it is becoming more common. At Edwards Air Force Base it grows densely in very light sandy soil in a field of about 20 acres that was once irrigated farmland but was abandoned about 12 years ago (the old Pancho Barnes place). Native of Europe.

Dithyrea californica Harvey. Spectacle pod. Common in sandy soils on the desert, often growing under creosote bushes. It is not particularly drouth resistant and is widespread only in years of good rainfall.

Draba verna Linnaeus. Whitlow-grass. Known from a colony near the shipping corral on the west slope of Breckenridge Mountain in Douglas oak woodland at 3,200 feet elevation (C. N. Smith 1072, JEPS), from the Greenhorn Mountains near the summit of Eugene Grade (C. N. Smith 1444, JEPS), and from colonies above and below the Lumreau Creek narrows, also in Douglas oak woodland (Farnsworth 414, 664, 676). Native of Europe.

Eruca sativa Hill. This vile smelling and sometimes troublesome crucifer grew as a waif two miles east of Lost Hills along Highway U. S. 466 in 1952; it did not persist. Native of Europe.

Erysimum argillosum (Greene) Rydberg. Rare in granite sand around massive granite slabs and blocks on the arid rim of the Kern Plateau southwest of Pine Flat (T8651). This wall flower of Nevada, Arizona, and New Mexico is otherwise known in California from isolated stations in Inyo and Mono counties.

Erysimum capitatum (Douglas) Greene. Wild wall flower. Occasional but widespread in the Douglas oak woodland through the yellow pine forest in the mountains; less common on the east slope in the pinyon woodland and the arid shrub association. (*E. asperum.*)

Erysimum capitatum var. **Bealianum** (Jepson) G. Rossbach. Alkali wall flower. Occasional, usually in colonies, in sandy subalkaline places from the San Bernardino County line near Kramer (Four Corners) west to Rosamond.

Erysimum capitatum var. **stellatum** J. T. Howell. Occasional in good soil on open slopes in the Glennville region. The plant has orange flowers that soon become reddish brown or maroon and the hairs of the leaves, flowers, and silques are three to six parted. The type was collected at Glennville (John Thomas Howell 5153, May 1930). (Cf. Madroño, 2:18, 1931.)

Erysimum moniliforme Eastwood. Yellow wall flower. Common, usually in light soil, on grassy often arid slopes in the Temblor and San Emigdio ranges, less common southwest to Ballinger Canyon in upper Cuyama Valley. Although many botanists consider it conspecific with *E. capitatum,* apart from the early blooming and ecological preference for light soils on arid slopes, the characters of the slender pods somewhat constricted between the seeds, the entire leaves, and the pale yellow to yellow flowers seem sufficient to set it apart from that species.

Erysimum perenne (Watson) Abrams. Perennial wall flower. A plant of the high Sierra that is rare along the crest of Piute Mountain (Breedlove 3930, T8912) and at the summit of Mt. Pinos (T6330).

Heterodraba unilateralis (Jones) Greene. Common in the Upper Sonoran associations; rare in wet years in the valley.

Hutchinsia procumbens (Linnaeus) Desvaux. Occasional in the alkali sinks in the valley and the southern Temblor Range, west to Soda Lake, usually growing in small dense colonies, often in the shade of salt bushes. It is rare on the desert near Rosamond and on the subalkaline flats near Isabella. (*Capsella procumbens.*)

Idahoa scapigera (Hooker) Nelson & Macbride. A small colony grows in a wet mossy seep on rocks along Cedar Creek about 0.75 mile above the mouth of Fulton Creek (Farnsworth 255) and another on a damp shady gravelly slope near Cedar Creek just above the Rock Corral (Farnsworth 453). This is a remarkable southern range extension from Sierra County and the Tahoe region; thus the plant's Sierra Nevada occurrence is at the northern and southern limits of that range. (*Platypspermum scapigerum.*)

Lepidium campestre (Linnaeus) R. Brown. Known in moist soil in low places in Bear Valley (T9676) and in Cummings Valley (B. K. Easley, 26 June 1964). Native of Europe.

Lepidium densiflorum Schrader. Occasional in disturbed often moist places in the Greenhorn Range; especially common between Shirley Meadows and Greenhorn Pass. It is occasional in moist meadows in Lynns Valley and was common along the old Kern River channel in the dry bed of Lake Isabella in 1961. The plant occurred as a transient at the Temblor Ranch grounds in 1954 (T1345).

Lepidium dictyotum Gray. Pepper-grass. Common in winter-wet usually alkali soil in the valley and on the desert; infrequent in similar places in the Temblor Range.

Lepidium dictyotum var. **acutidens** Gray. This distinctive variant grows in dense colonies in the range of the typical species, often with or near it. It is less common than the species in the valley and on the western edge of the desert but is more common in the Temblor Range. (*L. oxycarpum* var. *acutidens.*)

Lepidium flavum Torrey. Yellow pepper-grass. A characteristic yellow-flowered annual of sandy soils and gravelly slopes in the creosote bush association. It furnishes bright splashes of yellow for the desert mosaic of spring color in the recently all-too-rare wet years.

Lepidium Fremontii Watson. Desert alyssum. A common and characteristic subshrub of the creosote bush association of the high desert and the arid shrub association in the neighboring canyons.

Lepidium lasiocarpum Nuttall. Colonies are scarce in sandy and gravelly canyon bottoms in the El Paso Range and on the alkali flats north of Koehn Lake.

Lepidium nitidum Nuttall. Common pepper-grass. Common and often abundant in the lower foothills and the valley; occasional to the chaparral and in the creosote bush association on the desert. Pepper-grass grows rapidly after the first rains; in times of drouth, mature plants less than an inch tall have ample seed. The seeds are extremely nutritious and this accounts for the strange sight of a fat sheep in dry years on what appears to be plains devoid of forage. The seeds are also important food for birds and small mammals. Many people enjoy their sharp taste; some local cooks add them to soups, sauces, and salads.

Lepidium nitidum var. **Howellii** C. L. Hitchcock. Occasional in colonies, usually in sandy places, in the creosote bush association. The type was collected near Mojave (Amos Arthur Heller 7760, 24 April 1905).

Lepidium nitidum var. **insigne** Greene. This large-fruited variety is often abundant in heavy soils on north slopes in the Tejon Hills and the lower slopes of the San Emigdio Range. A dense colony grows in friable (dry bog) clay in Bitterwater Canyon at the San Luis Obispo County line and another on the flats at the head of Messa Ridge in the Temblor Range.

Lepidium nitidum var. **oreganum** (Howell) C. L. Hitchcock. Scarce in winter-wet often alkali soil in the Temblor Range; known from the valley at Goose Lake (T1726B).

Lepidium perfoliatum Linnaeus. Shield cress. Colonies are rare in the alkali sinks east of Rosamond, at the north end of Rosamond Lake, and at the south end of Kelso Valley. Native of Europe.

Lepidium strictum (Watson) Rattan. Known from Glennville (Howell 5145, May, 1930). Native of South America.

Lepidium virginicum Linnaeus var. **pubescens** (Greene) C. L. Hitchcock. Common in disturbed places about Glennville; an uncommon roadside and ranchground weed in the Greenhorn foothills west to Woody and Granite Station; occasional in dryish soil at Little Cannell Meadow on the Kern Plateau; local in the cienegas at the head of Tejon Canoyn.

Matthiola bicornis DeCandolle. Night-scented stock. Known as a roadside weed from the Old Town Road near Tehachapi (L. B. Krauter, 17 June 1964, Calif. Dep't. Agr. Herb.). Native of Greece.

Nasturtium officinale R. Brown. Water cress. Common in wet sunny places in the cismontane Upper Sonoran associations, especially in the Temblor and San Emigdio ranges; infrequent in the Jeffrey pine forest; scarce in desert canyons (Indian Wells Canyon, Budgett 35; Alphie Springs in a fork of Jawbone Canyon, T9327). It grows most luxuriantly in sunny subalkaline seeps and creeks. The plant is an excellent salad green and garnish for roasts. Native of Eurasia. (*Radicula Nasturtium-aquaticum.*)

Raphanus sativus Linnaeus. Wild radish. A widespread weed of irrigated fields, waste places, etc., in the valley. Native of Europe.

Rorippa curvisiliqua (Hooker) Bessey. Western yellow cress. Common in disturbed often moist soil in the Shirley Meadows region and at Tiger Flat in the Greenhorn Range. It was common along the old Kern River channel in the dry Isabella Lake bed in 1961. (*Radicula curvisiliqua.*)

Rorippa islandica (Oeder) Borbás var. **occidentalis** (Watson) Butters & Abbe. A common weed of wet places in the valley, mostly along canals, spring overflow areas, etc. (*Radicula palustris.*)

Sisymbrium altissimum Linnaeus. Tumble mustard. Common, mostly in waste places and farmed fields in the San Emigdio Range, the nearby extreme southern Temblor Range, and the Tehachapi Mountains. Tumble mustard is an occasional weed in the oak woodland in the Greenhorn Range and is scarce on the desert, mostly as a roadside weed east to Boron. Native of Europe.

Sisymbrium Irio Linnaeus. London rocket. A common and widespread winter weed of the valley that is scarce in the Temblor Range and rare on the desert east to Boron, mostly as a roadside weed, and in gardens at China Lake. London rocket has become much more common in the San Joaquin Valley in the last 15 years. Native of Europe.

Sisymbrium orientale Linnaeus. Oriental mustard. A roadside weed that is becoming rather common in the northern Temblor Range and in the valley; scarce in Caliente Canyon; rare on the desert (Highway U. S. 466 at the San Bernardino County line, T9308). Native of Europe.

Stanleya pinnata (Pursh) Britton. Golden prince's plume. Occasional in the desert, where it is colorful and sometimes locally common on slopes and in canyons. It occurs sparingly west to Cuyama Valley and is occasional in the Temblor Range north to Chico Martinez Canyon and along the western edge of the valley near Maricopa. The plant concentrates selenium and is occasionally a hazard to cattle.

Streptanthella longirostris (Watson) Rydberg. Occasional on the desert, usually growing in light sandy often wind blown soil. (*Thelypodium longirostres.*)

Streptanthella longirostris var. **derelicta** J. T. Howell. Known from a recently graded lot on Burrles Street in Ridgecrest (Meng 245B in 1966). This is an extension of range for this plant of the Colorado and extreme southern Mojave deserts.

Streptanthus cordatus Nuttall var. **piutensis** J. T. Howell. Piute jewel flower. An extensive colony occurs at the north end of Piute Mountain, occupying much the same area as the Bodfish Piute cypress grove, where it grows in various soils from shattered metamorphic rock to fine red clay. This highly local endemic is scarce in metamorphic gravel and gravelly loam in the isolated and inaccessible mountains at the head of Jawbone and Pine Tree canyons (Champagne Spring on the northwest shoulder of Cache Peak, T12,355; Sweetwater Ridge southeast of Cache Peak, T12,373). The typical variety of this distinctive jewel flower ranges from Inyo

County north along the east slope of the Sierra Nevada and in the White Mountains, eastward to the Rocky Mountains.

The type of Piute jewel flower was collected one mile above Bodfish Gap (Dennis E. Breedlove 3840, 14 July 1962). Other collections from the type locality include Bacigalupi 4536, Hardham 3641, Raven & Lewis 9348, and T3621, 8893, and 11,269.

Streptanthus Farnsworthianus J. T. Howell. Evalyn's jewel flower. Known from three small colonies at the head of Lumreau Creek at 4,600 feet elevation at the upper limits of the Douglas oak woodland in the Greenhorn Range, where it grows in decomposed slate on a broad sunny slope. (Evalyn L. Farnsworth 477, 13 May 1965, is the type for foliage and flowers, her No. 182, 21 June 1964, the type for fruit. Other collections by Mrs. Farnsworth from the type and the two nearby localities are her Nos. 575, 638, 640, 641, and 661. These collections give an unusually complete record for the plant; they include all phases of growth from seedlings to fully ripened seed.)

This remarkable plant, notable for its diverse foliage with purple cauline leaves and curiously bipinnatifid bright green basal leaves is also known from Rush Creek Mill and near Trimmer Springs in Fresno County and above Three Rivers in Tulare County.

Thelypodium integrifolium (Nuttall) Endlicher. W. L. Jepson's Flora of California (2:36) and more recent manuals include the Tehachapi Mountains in the range of this crucifer of wet places, apparently based on a collection from Tehachapi by Edward L. Greene which served as the type for *T. affine* Greene, a synonym. This needs verification with a collection to furnish a definite modern record. However, with the reclamation for farming and settlement and other drying of wet places in Tehachapi Valley, searches for it may continue to be fruitless.

Thelypodium lasiophyllum (Hooker & Arnott) Greene. California mustard. Common from the Lower Sonoran grassland to the chaparral and in the creosote bush association, especially in disturbed light soils. The form with entire leaves common in the El Paso and Rand mountains is var. *utahense* (Rydberg) Jepson.

Thelypodium Lemmonii Greene. Occasional in the Upper Sonoran grassland in the Temblor Range, mostly in light soils. Isolated colonies occur in friable (dry bog) clay on Pleito Ridge in the San Emigdio Range (T11,025) and on Dry Bog Knoll in Adobe Canyon in the foothills of the Greenhorn Range (C. N. Smith 307, JEPS; T10,601).

Thysanocarpus curvipes Hooker. Fringe pod. Common from the borders of the valley through the chaparral. The plants are highly variable. Var. *longistylis* Jepson is a distinctive form with a persistent style 1.3 to 2 mm. long; it is occasional on shaded slopes in the Greenhorn Range.

Thysanocarpus laciniatus Nuttall. Occasional from the edge of the valley through the chaparral, scarce in canyons and on slopes in the desert.

A form with wings with rays is var. *crenatus* (Nuttall) Brewer; it occurs at random with the species. Var. *Hitchcockii* Munz, distinguished by its silicles with clavate hairs, is occasional in the Rand and El Paso mountains and rare in metamorphic rock at Soledad Mountain and on the arid granite slopes at the head of Spanish Needle Creek northwest of Walker Pass.

Tropidocarpum gracile Hooker. Common in the valley and surrounding cismontane Upper Sonoran associations, particularly in the grasslands; occasional in the creosote bush association. Var. *dubium* (Davidson) Jepson is common with the species, sometimes replacing it on the high plains bordering the Temblor Range.

CAPPARIDACEAE. Caper Family

Cleomella obtusifolia Torrey & Frémont var. **pubescens** Nelson. Stink weed. Occasional in the dunes around the Muroc dry lakes; scarce in Indian Wells Valley.

Cleomella parviflora Gray. W. L. Jepson (Flora of California, 2:12) cites a collection from Rosamond (Davy 2247). A diligent search for the plant in the Rosamond region has been without success and it possibly no longer grows there.

Isomeris arborea Nuttall. Bladder pod. Occasional on desert washes, slopes, and canyons in the arid shrub association.

Isomeris arborea var. **angustata** Parish. Desert bladder pod. Scarce in canyons and flood plains in the deep desert at the east end of the El Paso Mountains. Fine plants can be observed along the highway between Ridgecrest and Randsburg.

Isomeris arborea var. **globosa** Coville. Valley bladder pod. Common on the arid slopes bordering the valley, occasional on the high plains at the edge of the valley and in the Elk Hills. The plant is often the dominant shrub in the Upper Sonoran subshrub association. It grows in extensive colonies on the steep slopes from Rose Station to Grapevine Canyon and on many of the lower east slopes of the Tehachapi Mountains. There the plants are never browsed by livestock; according to Roy V. Parker (oral communication) cattle will starve before eating them. In the Temblor Range, where the variety appears to be morphologically identical and there are no significant edaphic differences in its habitat, they commonly browse bladder pod freely and beneficially. Thus, there is an inexplicable difference in the taste of the plants in the two regions (both are equally and strongly unappetizing to human taste). The type was collected at Caliente Creek near Caliente (Frederick Vernon Coville 1107, 24 June 1891).

Wislizenia refracta Engelmann. Jackass-clover. Occasional in colonies in the alkali sinks in the valley. The flowers are favorites of honey bees. (*W. californica.*)

CRASSULACEAE. Stone-crop Family

Dudleya cymosa (Lemaire) Britton & Rose. Common in the mountains south to Lebec, usually growing around rocky places in the woody Upper Sonoran associations. (*Cotyledon nevadensis*).

Dudleya lanceolata (Nuttall) Britton & Rose. Live-forever. Scarce in rock outcrops in the Temblor Range; occasional northeast through the mountains to the Tehachapi Mountains. Given garden culture with ample watering, both of these species grow robustly and little resemble their wild form. The leaves remain green and fleshy all year and the plant develops long stout caudices. This establishes that in moist soil the summer dormancy with shrivelling leaves common in the field is not obligatory. (*Cotyledon lanceolata.*)

Parvisedum Congdoni (Eastwood) Clausen. Occasional in dense colonies on moist granite sheets in the Douglas oak woodland in the Greenhorn Range. (*Sedella pumila.*)

Sedum spathulifolium Hooker ssp. **anomalum** (Britton) Clausen & Uhl. Pacific stonecrop. A colorful yellow-flowered succulent known from a dense colony on a shaded rocky bank in El Paso Canyon (0.5 mile above the Meadows Place, elevation 3,500 feet, T6160) and another near the summit of Black Bob Canyon in the San Emigdio Range (T8330). Both of these grow in very local white fir groves in rocky canyon bottoms.

Tillaea aquatica Linnaeus. Water pigmy weed. A tiny plant of pool beds and mud flats known from the (old) Kern County Airport (Benson 2975, 23 February 1932, POM).

Tillaea erecta Hooker & Arnott. Sand pigmy weed. Colonies are widespread in the valley and in the surrounding ranges through the Douglas oak woodland, usually growing in bare places.

SAXIFRAGACEAE. Saxifrage Family

Heuchera caespitosa Eastwood. San Emigdio alum root. Known with certainty only from the type collection, made in San Emigdio Canyon by Mrs. William Jasper in May, 1895.

Heuchera rubescens Torrey var. **alpicola** Jepson. Occasional around rocks at Lookout Point and along the main crest of Piute Mountain.

Lithophragma Bolanderi Gray. Woodland star. Occasional in shade in the yellow pine forest and scarce in the woody Upper Sonoran associations, south to the Tehachapi Mountains. (*Lithophragma scabrellum,* and *L. scabrellum* var. *Peirsonii* in part.)

Lithophragma Cymbalaria Torrey & Gray. Woodland star. Occasional in the Douglas oak woodland and chaparral in the Temblor Range.

Lithophragma parviflorum (Hooker) Nuttall. Occasional to common in shade, often in humus, from the woody Upper Sonoran associations to the lower levels of the yellow pine forest in the mountains; scarce in the San Emigdio and Temblor ranges. (*L. tripartitum.*)

Ribes californicum Hooker & Arnott. California gooseberry. The only Kern County colony of this coastal gooseberry grows at Joe Messa Spring near the summit of Messa Canyon in the Temblor Range (T746, 1709).

Ribes cereum Douglas. Squaw currant. Occasional from the pinyon woodland through the yellow pine forest, where it is especially common around exposed summits and ridgetops.

Ribes malvaceum Smith. A currant of the Coast Ranges that is rare in Douglas oak woodland at the head of Palos Altos Canyon in the southwestern Tehachapi Mountains (T9538).

Ribes Menziesii Pursh var. **ixoderme** Quick. Rare in the Douglas oak woodland on the east slope of Blue Mountain in the Greenhorn Range (T9417) and in Caliente Canyon at 2,300 feet elevation (T9538).

Ribes nevadense Kellogg. Sierra currant. Occasional from the chaparral to the red fir forest, often common in the yellow pine forests. The fruits are used for exceptionally good jelly.

Ribes quercetorum Green. Chaparral gooseberry. A common and characteristic shrub of the Douglas oak woodland, often forming large dense colonies. It is scarce in the desert-facing canyons of the Tehachapi Mountains; in Cameron Canyon it grows with *Yucca brevifolia* var. *Herbertii, Opuntia basilaris,* and *Quercus lobata* on a slope with an unusual complex of desert and cismontane vegetation types.

The dense colonies afford haven for small game, particularly cottontail rabbits and California quail. In the Greenhorn Range the ground around them is often barren from rabbits feeding close to protection. These bare places are where some of the rarest plants of the region grow; among them are *Mimulus pictus, M. viscidus, M. Johnstonii,* and *Scribneria Bolanderi.*

Ribes Roezlii Regel. Sierra gooseberry. Common, especially on north slopes in the yellow pine forest; occasional in the Douglas oak woodland of the mountains and in the deeper desert-facing canyons of the pinyon woodland. Usually the colonies are small but in the Tehachapi Mountains east of Keene they frequently densely cover an acre or more on north slopes at the highest elevations of the Douglas oak woodland at such places as the heads of Tollgate and Stevenson canyons.

Ribes velutinum Greene var. **glanduliferum** (Heller) Jepson. Occasional in the Jeffrey pine forest in the Mt. Pinos region (it grows on the exposed summit plateau of Mt. Pinos) and the Tehachapi and Piute Mountains; scarce at the higher elevations of the Douglas oak woodland.

Ribes viscosissimum Pursh. A small colony grows under *Castanopsis sempervirens* on the north slope of Sunday Peak (T12,382), the southern limit for the species.

Saxifraga californica Greene. Occasional from the Douglas oak woodland through the shin oak brush; scarce in the yellow pine forest and in the Temblor Range. (*S. virginiensis* var. *californica.*)

PLATANACEAE. Sycamore Family

Platanus racemosa Nuttall. Sycamore. A common and picturesque tree of stream banks in the mountains below the yellow pine forest, extending along Kern River to Bakersfield.

ROSACEAE. Rose Family

Adenostoma fasciculatum Hooker & Arnott. Chamise. Infrequent in Kern County, colonies of only a few plants are scattered and scarce in chaparral and Douglas oak woodland from Caliente and Back Canyons south to the east slope of the extreme southwestern Tehachapi Mountains. Elsewhere in California, chamise is a common shrub of the chaparral association, often occurring in almost pure stands over hundreds of acres. There are locally dense and extensive colonies in Kern Canyon in the Roads End region of Tulare County, well to the north of the Kern County line.

Alchemilla occidentalis Nuttall. Western lady's mantle. A miniature member of the rose family, this little plant is common in small dense colonies in bare places in the Douglas oak woodland in the Glennville-Woody region; colonies are occasional south to Tejon Canyon. Under magnification, the fruits closely resemble miniature rose hips, an interesting illustration of the family relationship of plants that superficially seem to have little in common. (*A. arvensis.*)

Amelanchier pallida Greene. Service berry. Scattered colonies grow along Catskin Creek near its head in the Tehachapi Mountains, on the ridges east of Keene, and an isolated colony grows in Water Canyon in the Douglas oak woodland at the north base of Tehachapi Peak. Service berry is occasional in the Greenhorn Range, growing at widely separated stations from the Douglas oak woodland to the lower levels of the ponderosa pine forest. A specimen with bright pink flowers grows in a damp ravine two miles west of Cedar Creek along the Glennville highway (Meng 112).

Cercocarpus betuloides Nuttall. Mountain mahogany. Common in the chaparral and occasional in the Douglas oak woodland in the mountains. Mountain mahogany is an important browse plant for livestock and deer. Evalyn Farnsworth states (oral communication) that seedlings are often abundant following brush burns in the Greenhorn Range and that the plants respond well to vigorous browsing when relatively young.

Cercocarpus intricatus Watson. Occasional in the sparse Jeffrey pine forest in the Bartolas Country of the southern Kern Plateau (Howell & True 41,813). Plants at the summit of the steep west wall of the canyon of the South Fork of Kern River appear to be intermediate with *C. ledifolius,* which occurs a few miles to the north in Tulare County (T12,179).

Cercocarpus ledifolius Nuttall. Gnarled old shrubs are local and scarce along the east fork of upper San Emigdio Creek and on nearby Tecuya Ridge, growing in pinyon woodland. Two young shrubs grow on a slope southeast of Mil Potreros.

Chamaebatia foliolosa Bentham. Mountain misery. Occasional, often in extensive colonies, in the ponderosa pine forest south to Cedar Creek in the Greenhorn Range, the southern limits for this common plant of the Sierran forests.

Coleogyne ramosissima Torrey. Black bush. Common in the desert-facing slopes of the mountains in the arid shrub association; rare in the creosote bush association in the Rand Mountains and on the plains north of Red Rock Canyon.

Geum macrophyllum Willdenow. Avens. Scarce in Browns Meadow in the Piute Mountains (T3196, Breedlove 3981).

Holodiscus Boursieri (Carriérre) Rehder. Cream bush. Occasional at high altitudes in the yellow pine forest, mostly around exposed summits. (*H. discolor* in part.)

Photinia arbutifolia (Aiton) Lindley. Toyon. A colony of about 25 senescent shrubs was discovered in 1961 by Donald J. Bedell in Dons Canyon, a fork of upper Cedar Canyon, in the California juniper association of the Temblor Range (Bedell 74-4, 65-1, T9215, 11,681). The decadent state of this colony is typical of the isolated stations for coastal plants in the Temblor Range and is evidence of the severe drouths of the past 40 years and less directly perhaps evidence of the drying of the climate over the past several hundred years.

Most of the shrubs are infested by the large and interesting fungus, *Gymnosporangium exiguum* Kern, a rust known in California from this colony, from Sobaba Hot Springs in Riverside County, and at the Santa Barbara Botanic Garden. Its alternate hosts are members of the Rosaceae and the genus *Juniperus*. The fungus is common from Texas into Mexico where it usually infests *Crataegus* and *Juniperus*.

The large toyon on the slope just south of the old hotel at Lebec which appears spontaneous was planted. (*Heteromeles arbutifolia.*)

Potentilla biennis Greene. Occasional in wet places in the mountains in the Upper Sonoran associations, scarce through the yellow pine forest.

Potentilla Bolanderi (Gray) Greene var. **Parryi** (Watson) Munz & Johnston. Known from Cuddy Valley (Hoffmann, 13 June 1927, SBM); also known from nearby Lockwood Valley in Ventura County (Hoffmann, 4 June 1930, SBM). (*P. Parryi, Horkelia Bolanderi* var. *Parryi.*)

Potentilla glandulosa Lindley. Common in wet places, mostly in the yellow pine forest. Ssp. *nevadensis* (Watson) Keck occurs in the Greenhorn Range and at the cienega at the head of Bitter Creek on Mt. Pinos; ssp. *reflexa* (Greene) Keck also grows in the Greenhorn Range and on Mt. Pinos;

and the more distinctive Sierran ssp. *Hansenii* (Greene) Keck reaches its southern limits at an isolated station at Browns Meadow in the Piute Mountains (T3194).

Potentilla gracilis Douglas ssp. **Nuttallii** (Lehmann) Keck. Cinquefoil. Occasional in moist to dryish places, usually near the borders of meadows in the Jeffrey pine forests; less common in the adjoining pinyon woodlands.

Potentilla millegrana Engelmann. Scarce in moist black clay in the cienegas in the yellow pine forest at the head of Tejon Creek (T8842). (*P. rivalis* var. *millegrana.*)

Potentilla pectinisecta Rydberg. Common in moist soil in Cuddy Valley (T6297); scarce at Mil Potreros in the pinyon woodland in the San Emigdio Range (T5288) and at the north end of French Meadow in the Jeffrey pine forest in the Piute Mountains (C. N. Smith 879, JEPS).

Potentilla purpurascens (Watson) Greene. Occasional in moist places and in dry meadows on the Kern Plateau and Breckenridge Mountain. It grows in an unusually robust form at Grouse and French meadows in the Piute Mountains. (*Ivesia purpurascens.*)

Potentilla santolinoides (Gray) Greene. Occasional about the summits of Mt. Pinos and Sawmill mountains; scarce, usually about rock outcrops or in decomposed granite, in the yellow pine forest in the Tehachapi Mountains (Cache Peak), the Greenhorn Range, and on the Kern Plateau. (*Ivesia santolinoides.*)

Potentilla saxosa Lemmon ssp. **sierrae** Munz. Known from 12 miles northeast of Weldon in crevices in a cliff shaded by *Pinus Sabiniana* (Hardham 2110) (apparently the mouth of Chimney Creek Canyon) and from the rim of the Kern Plateau east of Fay Creek Canyon (Howell & True 41,808). A southwestern extension of range for a plant of the east base of the Sierra Nevada and the White Mountains in Inyo County.

Prunus Andersonii Gray. Desert peach. Scarce in desert-facing canyons in the pinyon woodland and the arid shrub association (Horse Canyon, Short Canyon). A colony grows on the sand flats along Canebrake Creek just east of Walker Pass Station in the pinyon woodland on the west slope of Walker Pass, and another at the mouth of Bath House Canyon near Kernville. The shrub is a colorful mound of deep pink during its brief blooming season.

Prunus demissa (Nuttall) Walpers. Choke cherry. Occasional in the mountains from the Douglas oak woodland through the Jeffrey pine forest; scarce in the San Emigdio Range. An isolated colony grows at Rock Spring in McLean Canyon in the Temblor Range. Although choke cherry is sometimes poisonous to livestock in other regions, it has never been a problem in Kern County. (*P. virginica* var. *demissa*).

Prunus emarginata (Douglas) Walpers. Bitter cherry. Common at higher altitudes in the mountains south through the Tehachapi Mountains, often growing in dense colonies along the summit ridges, occasionally descending to the chaparral and the Douglas oak woodland.

Prunus fasciculata (Torrey) Gray. Desert almond. Common in the arid shrub association in upper Jawbone and Pine Tree canyons; occasional in the other desert-facing slopes and canyons. Scarce in light sandy soil at the base of a hot steep slope of granite blocks 1.6 mile east of Onyx (Eastwood & Howell 7621, T10,939).

Prunus ilicifolia (Nuttall) Walpers. Holly-leaf cherry. Occasional in chaparral on the northwest flank of Mt. Abel. An extensive colony grows in an unusual arboreal form in Temblor Canyon in the Temblor Range where it is a handsome small round-headed tree. This is possibly the adventitious result of browsing of all but the rapidly growing main leader of the young shrubs; however, a genetic difference may be involved.

Prunus subcordata Bentham. Sierra plum, western plum. Occasional in the Greenhorn Range, scarce south to Black Bob Canyon in the San Emigdio Range, mostly in places with better than average moisture in the chaparral and shin oak brush.

Shrubs with stout spines and pubescent ovaries are occasional on sunny slopes with good soil in the Glennville region; these may be hybrids with *P. Andersonii.*

Purshia glandulosa Curran. Desert bitter brush. Occasional, often locally common, on the mountain slopes in the arid shrub association and the pinyon woodland from the eastern Sierra Nevada southwest to Dry Canyon in the Mt. Pinos region in Ventura County. In the southeastern part of the Kern Plateau in Kern and Tulare counties, *P. glandulosa* is common through the pinyon woodland and *P. tridentata* reaches its southern limits in the Jeffrey and lodgepole forests. While the ranges of the two apparently do not overlap, they are only narrowly separated in the region along the South Fork of the Kern River and hybrids are not uncommon. As far south as Kelso Canyon at the east base of the Piute Mountains, *P. glandulosa* seems to reflect introgression with genes of *P. tridentata.*

Exceptionally well developed shrubs of desert bitter brush grow in the deep gravel on the steep slopes on the northeast side of Piute Mountain along Harris Grade, the densely branched plants often as much as six feet tall and nine feet across. It is locally important as a browse plant for cattle and deer; seedlings are eaten so completely that in many areas young plants are rare. The type was collected on the Mojave side of Tehachapi Pass [near Cameron] (Katharine Layne Curran in June, 1884).

Purshia tridentata (Pursh) DeCandolle. Antelope brush, bitter brush. Reaches its southern limits in the Bartolas Country in the arid Jeffrey pine forest of the southern Kern Plateau, where it is scarce (Howell & True 41,822, T12,175). Antelope brush becomes common a few miles to the north in the Jeffrey and lodgepole pine forests of the Kern Plateau in Tulare County.

Rosa Aldersonii Greene. Occasional in creek beds, around springs, and in washes with subsurface moisture in the east-side foothills south to Ft. Tejon. It often grows in dense colonies that give shelter to quail, rabbits, and other wild life. (*R. californica* in part.)

Rosa californica Chamisso & Schlechtendal. California wild rose. Scarce, although often growing in dense colonies, in the San Emigdio Range (Neasons Flat, the potrero in upper San Emigdio Canyon, Ft. Tejon). It also occurs at the east end of Bear Valley in the Tehachapi Mountains.

Rosa Eglanteria Linnaeus. Sweet briar. Two plants, not certainly spontaneous, grow on the Kern River flood plain 3.4 miles south of Kernville (T6547). Native of Europe. (*R. rubiginosa.*)

Rosa pinetorum Heller. Scarce in moist places in the ponderosa pine and red fir forests in the Greenhorn Range southwest to Basket Peak and in the Jeffrey pine forest of the southwestern Kern Plateau.

Rosa Woodsii Lindley var. **ultramontana** (Watson) Jepson. Interior wild rose. Occasional in moist sunny places in the lower Jeffrey pine forest and the upper levels of the pinyon woodland from the Lamont Meadow region of the eastern Sierra Nevada and from Kern River below Kernville southwest through the mountains to the Mt. Pinos region. (*R. californica* var. *ultramontana.*)

Rubus glaucifolius Kellogg. Scarce in the Jeffrey pine forest on the north end of Piute Mountain (near Saddle Springs, C. N. Smith 858, JEPS; summit road, 2 miles south of Saddle Springs, T8905; 7.7 miles above Bodfish Gap, T10,175).

Rubus leucodermis Douglas. Western raspberry. A dense colony grows on a shady moist slope in Tejon Canyon at 2,750 feet elevation. It is rare in the Greenhorn Range along Alder Creek just above Alder Creek Public Camp (T11,385) at the lowest level of the ponderosa pine forest.

Rubus parviflorus Nuttall. Thimble berry. Scarce in moist often shaded places in the ponderosa pine forest in the Greenhorn Range. According to W. L. Jepson (Flora of California, 2:169), thimble berry was collected in the Tehachapi Mountains by Edward L. Greene.

Rubus procerus P. J. Mueller. Himalaya berry. A rare escape on the sandy flats along Kern River between Lake Isabella and Kernville and around old canals and ditches in the Arvin-Greenfield region. Native of Europe.

Rubus ursinus Chamisso & Schlechtendal. Wild blackberry. Scarce in the Douglas oak woodland in the Greenhorn Range south to Kern Canyon and along Kern River near Kernville.

Rubus vitifolius Chamisso & Schlechtendal var. **titanus** (Bailey) Bailey. Bramble, California blackberry. Occasionally spontaneous in the valley, especially along old canals in the Weedpatch and Arvin regions. ***Rubus***

vitifolius is native to California coastal regions; our plants originated from horticultural forms of this native species.

It has recently been discovered in Tulare County that grape leaf-hopper, a serious pest of vineyards, can be controlled by plantings of *Rubus* spp., which are hosts for the leaf-hopper parasite *Anagrus.* As little as 0.1 acre of *Rubus* to 40 acres of grapes is effective. Preliminary observations show that mound building species such as *R. ursinus* and *R. vitifolius* var. *titanus* are most effective. (Frederik L. Jensen. U. C. Agr. Ext. Ser. Grape Notes, 19 March 1964).

LEGUMINOSAE. Pea Family

Alhagi Camelorum Fischer. Camel thorn. Once common and obnoxious at Lerdo, Bellevue, Rosedale, and Old River, this persistent pest has been rigorously controlled and possibly has finally been eradicated. Native of Asia Minor.

Astragalus acutirostris Watson. Keel beak. An occasional to common annual of mostly sandy places in the creosote bush association.

Astragalus asymmetricus Sheldon. Horse loco. Common in the Annette Hills, horse loco reaches its southern limits at the summit of Drake Ridge in the central Temblor Range. It is cumulatively toxic to livestock, especially to horses, which become addicted to the plant. The process is remarkably similar to narcotic addiction in man, and if allowed to run its course, the animal dies from malnutrition, deterioration of the central nervous system, and nuerological collapse. Plants in the northern Temblor Range are often infested by an unidentified gall wasp; study of this insect may lead to biological control of horse loco. Meanwhile, despite serious efforts at control on most ranches for decades, the plant is still locally plentiful. (*A. leucophyllus.*)

Astragalus didymocarpus Hooker & Arnott. Dwarf loco, and locally and erroneously, wild vetch. Common in the grasslands, particularly on the high plains on the west side of the San Joaquin Valley where it is sometimes dense in years with early rainfall. It is occasional to the Douglas oak woodland and in the desert creosote bush association. Dwarf loco is not toxic and is excellent forage for livestock.

Astragalus didymocarpus var. **dispermus** (Gray) Jepson. Scarce in sandy places in the creosote bush association. (*A. dispermus.*)

Astragalus Douglasii (Torrey & Gray) Gray. Douglas loco. Occasional and sometimes locally common in the Temblor Range and the mountains, reaching its northeastern limits at Spanish Needle Creek northwest of Walker Pass. The robust bushy plants with many stems that grow in the region from Lebec to Tejon Pass have been called var. *tejonensis* Jones (the type was collected in the hills near Tejon Pass by Cyrus G. Pringle, 13 May 1882). While not clearly separable from the prostrate form of the typical variety, these leafy plants with their ascending to upright stems are locally hazardous to livestock, especially horses. The plant grows in regions where *A. lentiginosus* var. *idriensis* also commonly occurs and apparently is responsible for the livestock losses sometimes attributed to the latter.

Astragalus Gambellianus Sheldon. Gambel loco. Occasional in the Douglas oak woodland in the Greenhorn Range; rare, south to the southwestern Tehachapi Mountains. It usually grows in thin rocky or gravelly soil in the Upper Sonoran grassland and Douglas oak woodland. (*A. nigrescens.*)

Astragalus Hornii Gray. Sheep loco. Once common but now scarce in the valley, growing in moist usually subalkaline soil. The plant was formerly a great hazard to livestock but rigorous control and conversion of much of its habitat to cultivated farming land has very nearly eliminated sheep loco from most of its original range. The type locality, given as Ft. Tejon, was probably a station somewhere at the head of the San Joaquin Valley rather than in the immediate neighborhood of the fort. (George Henry Horn in 1863.)

Astragalus Layneae Greene. Colonies are occasional, usually in wind-drifted light sand, in the creosote bush association. The plant's subterranean stems are well adapted to the shifting sands where it grows. Colonies are most common around Randsburg and in Indian Wells Valley; a particularly fine one grows in the massive slope of sand below the old Yellow Aster Mine in the Rand Mountains.

Astragalus lentiginosus Douglas var. **albifolius** Jones. Known from an alkali flat near Muroc [now Edwards] (Hoffmann, 20 May 1929, SBM). If the plant still grows in the Muroc region, it is extremely rare.

Astragalus lentiginosus var. **idriensis** Jones. Tehachapi loco. Occasional from the Tehachapi region southwest to Mt. Abel; scarce along the summit of the Temblor Range (and at scattered stations north to the Mt. Hamilton Range). It grows most frequently in thin gravelly soils, often in shale, in the Douglas oak woodland, rarely ascending to the Jeffrey pine forest. The plant has striking metallic purple-blue flowers. (*A. lentiginosus* var. *caesariatus, A. lentiginosus* var. *tehatchapiensis.*)

Astragalus lentiginosus var. **nigricalycis** Jones. Black loco. Common, often abundant, in the valley grassland; occasional from the southern Temblor Mountains to the southern Carrisa Plains and Cuyama Valley. It is one of the first of the valley plants to bloom, often as early as late January. Black loco furnishes excellent forage for livestock; its toxicity, if any, is only minor. Seeds are often dispersed by the digestive processes of cattle and modern local cattle transportation by truck has resulted in its adventive appearance in numerous nearby places where it does not persist (Glennville, Woody, northern Temblor Range at the Dorothy Twisselmann Ranch, probably the occurrence in the Santa Lucia Range, etc.). The type was collected at Bakersfield (Alice Eastwood, 24 March 1893). It was apparently destroyed in the San Francisco fire of 1906 but a photograph survives at the Pomona College Herbarium. Rupert Barneby (Leafl. West. Bot., 4:127) has designated a collection made by Eastwood on 8 March 1893 in San Emigdio Canyon as a neotype.

Astragalus lentiginosus var. **variabilis** Barneby. Desert loco. Common in the creosote bush association and occasional to the pinyon woodland on the desert-facing slopes of the mountains. It reaches a record altitude of 5,600 feet elevation in Horse Canyon on Kiahvah (Scodie) Mountain (T8177). The plant is scarce and probably a relatively recent introduction along the east side of the valley, from the Bakersfield region north through the Poso Hills and near Shafter. A perennial that blooms the first year while quite young, it sometimes behaves as an annual on the desert plains in times of drouth (where it is usually the first plant to bloom in the spring). In the creosote bush association desert loco has prostrate stems; in the pinyon woodland it is tufted and has silvery leaves. Rupert Barneby writes (Atlas of North American Astragalus, 2:949) that it tends to intergrade with var. *Fremontii* on the east slope of the Sierra Nevada north of Walker Pass.

Astragalus macrodon (Hooker & Arnott) Gray. Salinas Valley rattle weed. A Salinas Valley endemic that is rare in Kern County along the central summit of the Temblor Range, where it grows in thin soil over shale and sandstone.

Astragalus oxyphysus Gray. Diablo loco. Common in the valley, especially on the high plains, ascending to the Douglas oak woodland at Granite Station and Keene; southeast to Cuyama Valley and the Carrisa Plains. The plant is highly susceptible to mildew, which may be a factor in restricting its distribution to regions of relatively low humidity (although it is not uncommon in the well developed Douglas oak woodland of the Castle Mountain region of Monterey and Fresno counties). Diablo loco is highly toxic to livestock and has been much controlled and in some areas eliminated.

Astragalus pachypus Greene. A bushy astragalus with exceptionally handsome large white flowers that is occasional in the pinyon woodland, chaparral, and Douglas oak woodland from Quatal Canyon in upper Cuyama Valley northeast to Caliente Canyon and in rocky canyons in the desert ranges. The type was collected at Bealville in the Tehachapi Mountains (Katharine Layne Curran, June, 1884).

Astragalus Purshii Douglas var. **tinctus** Jones. Woolly pod. Common in rocky and gravelly soils from the Sierra Nevada north of Walker Pass and the Piute Mountains southwest to Mt. Pinos, growing in the arid shrub association and the pinyon and Douglas oak woodland through the Jeffrey pine forest. It is common about the summit of Mt. Pinos. The only colony known from the Greenhorn Range grows in the Douglas oak woodland on the north flank of Pine Mountain, one mile from the summit (C. N. Smith 1440, 1446, 1452, 1462, JEPS). (*A. Purshii* var. *longilobus.*)

Astragalus subvestitus (Jepson) Barneby. A small but dense colony grows in metamorphic talus around the massive outcrops at Lookout Point in the Piute Mountains. This is very probably the type locality, given as Erskine Creek (Joseph Anton Purpus 5098 in 1897). The plant is a rare endemic, known from here and from isolated colonies at Monache Meadow and Bald Mountain (T11,467) on the Kern Plateau in Tulare County.

Astragalus trichopodus (Nuttall) Gray var. **phoxus** Jones. A small dense colony grows on the hillside and adjoining freeway cut just south of the old inn at Lebec (T10,861, Hildreth 50, DAV, Jepson 12,423, JEPS). (*A. Antiselli* var. *phoxus.*)

Astragalus Whitneyi Gray. Common on the summits of Mt. Pinos and Sawmill Mountain; rare along the crest of Piute Mountain. These are southwestern extensions of range of this Astragalus of the east slope and central crest of the Sierra Nevada from Inyo County to Alpine County, where it ascends to 12,000 feet elevation.

Cassia armata Watson. Desert senna. Colonies are occasional in the higher creosote bush and the arid shrub associations, mostly in gravelly soils in desert canyons, where it is sometimes locally common and colorful.

Cercis occidentalis Torrey. Western red bud. A small colony grows at the base of the slope of granite blocks 2.4 miles east of Onyx (T4499, 6690, Meng 270, Epling & Anderson in 1931, DS). Tony Gasbarro reports (personal communication) it also occurs in similar places in nearby South Fork Canyon. These colonies in this very arid region are an interesting southeastern range extension from King George Mountain in the Sierra Nevada foothills of Tulare County.

Dalea Fremontii Torrey var. **Saundersii** (Parish) Munz. Indigo bush. Occasional on slopes and canyons in the El Paso and Rand mountains, scarce on the high plains at the edge of the desert. (*D. Saundersii.*)

Glycyrrhiza glabra Linnaeus. Common licorice. A long-established colony grows in a moist waste place near Goose Lake (T2076). Native of Eurasia.

Glycyrrhiza lepidota Pursh. Wild licorice. Scarce at wet places in the Temblor Range, possibly introduced here by the pioneers as a medicinal herb.

Glycyrrhiza lepidota var. **glutinosa** (Nuttall) Watson. Widespread in wet places in the valley; occasional in the mountains through the Douglas oak woodland. An extensive colony grows around Mesquite Springs on the desert at the southeastern base of the El Paso Range.

Hoffmannseggia densiflora Bentham. Scattered colonies grow in low moist subalkaline soil near the Rosedale Ranch in the Rosedale region (Snow Road, 0.25 mile west of the Kern County Land Company's Rosedale Ranch, T4796). Other colonies have been reported from near Famosa and Delano. Native from southeastern California to Texas and Mexico.

Lathyrus latifolius Linnaeus. Everlasting pea. Escaped and apparently established around the Greenhorn Mountain Park headquarters (Howell 38,764, T9011). Native of Europe.

Lathyrus laetiflorus Greene. A single plant grows in Black Bob Canyon in the San Emigdio Range at 3,250 feet elevation (T8356) and another in Studhorse Canyon, a fork of Caliente Canyon (T9528), both in shrubby areas in Douglas oak woodland. These isolated plants are a surprising northern range extension from the San Gabriel and Santa Monica mountains.

Lathyrus pauciflorus Fernald ssp. **Brownii** (Eastwood) Piper. Occasional in the Jeffrey pine forest in the Piute Mountains.

Lathyrus sulphureus Brewer. Occasional in the ponderosa pine forest and at the upper levels of the Douglas oak woodland in the Greenhorn Range; scarce in the shin oak brush at the north end of Piute Mountain.

Lotus argophyllus (Gray) Greene. Occasional in the mountains south to Lebec, growing from the pinyon and Douglas oak woodlands to the yellow pine forest. The many lax stems with their silvery leaves form beautiful mounds that are especially attractive around rock outcrops. Var. *decorus* (Johnston) Ottley is scarce around the granite blocks and outcrops on the steep hot slopes at the head of Spanish Needle Creek northwest of Walker Pass.

Lotus corniculatus Linnaeus. Birdsfoot trefoil. A common pasture plant widely escaped in moist places in the valley. It is well established in the dense growth on the moist flats around Castac Lake where the plant was seeded about 1950. Native of Europe.

Lotus crassifolius (Bentham) Greene. Broad leaf lotus. Occasional in the yellow pine forest, mostly in small dense colonies in disturbed often moist soils.

Lotus Davidsonii Greene. Occasional from the chaparral to the Jeffrey pine forest in the Mt. Pinos region where it reaches its northern limits on the northwest flank of Mt. Abel.

Lotus grandiflorus (Bentham) Greene. Large-flowered lotus. Occasional in the mountains, mostly in good soil in shin oak brush, chaparral, and Douglas oak woodland; rare in the yellow pine forest.

Lotus humistratus Greene. Common from the valley grasslands through the Douglas oak woodland; occasional from the desert plains through the pinyon woodland.

Lotus micranthus Bentham. Scarce in disturbed places in the Douglas oak woodland in the Greenhorn Range from the Tulare County line south to Blue Mountain, the southern limits for the species.

Lotus nevadensis Greene. Sierra lotus. Occasional to common in the yellow pine forest, ascending to the highest elevations. (*L. Douglasii* var. *nevadensis.*)

Lotus oblongifolius (Bentham) Greene. Scarce in the chaparral; common in the yellow pine and red fir forests; rare on the desert-facing slopes in the pinyon woodland. It usually grows in wet often shaded places.

Lotus procumbens (Greene) Greene. Scarce from the Kern Plateau and Mt. Owen southwest to Mt. Pinos, from the pinyon and Douglas oak woodlands to the Jeffrey pine forest. Rare in the Greenhorn Range in the Douglas oak woodland (Fulton Creek, 0.25 mile above Cedar Creek, Farnsworth 154). (*L. leucophyllus.*)

Lotus procumbens var. **Jepsoni** (Ottley) Ottley. Occasional in granitic sand between massive granite blocks and on sandy slopes on the southern Kern Plateau. (*L. leucophyllus* var. *Jepsoni.*)

Lotus Purshianus (Bentham) Clements & Clements. Spanish clover, mustang clover. Common, sometimes abundant, in disturbed sandy soil, from the oak woodland to the lower levels of the yellow pine forest in the mountains; occasional in moist or recently dried soil in the valley, especially in waste places. The plant is considered of secondary value for forage. (*L. americanus.*)

Lotus salsuginosus Greene. Scarce on gravelly slopes in the Lower Sonoran grassland in the San Emigdio foothills (Wheeler Ridge, T8075).

Lotus salsuginosus Greene var. **brevivexillus** Ottley. Scarce in canyons in the El Paso Range (Last Chance Canyon, T4360).

Lotus scoparius (Nuttall) Ottley. Deer lotus. Common in the cismontane shrubby Upper Sonoran associations, especially in disturbed places, such as road banks. The plant does not seem to survive prolonged heavy grazing.

Lotus scoparius var. **brevialatus** Ottley. Scarce on transmontane slopes where it replaces the typical variety. It is also the common form on the slopes of the Piute Mountains south of South Fork Valley west to Bodfish Gap and Havilah. An anomalous local occurrence of this variety of arid regions is in the well developed Douglas oak woodland at Oak Flat on the southwest flank of the Greenhorn Range (Myrick 1117).

Lotus strigosus (Nuttall) Greene. Strigose calf lotus. Common in the cismontane associations below the yellow pine forest, east to Kelso Canyon (where it grows in an unusually robust form on the divide between Kelso Canyon and Kelso Valley). It is especially common in the Upper Sonoran grasslands.

Lotus strigosus var. **hirtellus** (Greene) Ottley. Occasional from the desert-facing slopes of the mountains west to the east slope of the Greenhorn Range and the west slope of Piute Mountain, growing in the arid shrub association, the pinyon woodland, and the arid phases of the Douglas oak woodland.

Lotus subpinnatus Lagasca y Segura. Calf lotus. Common, sometimes abundant, in all cismontane associations below the yellow pine forest, especially in the Upper and Lower Sonoran grasslands, where it makes an important contribution to the forage.

Lotus tomentellus Greene. Hairy calf lotus. Scarce in the desert canyons of the Sierra Nevada north of Walker Pass and on the plains on the broad west side of Indian Wells Valley south to the El Paso Range. It grows most often on sandy canyon bottoms.

Lupinus adsurgens E. Drew var. **undulatus** C. P. Smith. Occasional in the pinyon woodland and on dry slopes in the yellow pine forest, mostly from the northern Tehachapi Mountains to Inyo County; scarce on dry slopes in the Jeffrey pine forest in the Piute Mountains; rare on the east slope of the Greenhorn Range. Plants from the west slope of Walker Pass were described by Alice Eastwood as *L. salticola* (type: Alice Eastwood & John Thomas Howell 9634, 16 May 1941); these seem to be completely synonymous with *L. adsurgens* var. *undulatus.*

Lupinus albicaulis Douglas. Rare at Alder Creek in the Greenhorn Range, where it grows in the Kellogg oak woods just below the ponderosa pine forest (T3584).

Lupinus albicaulis var. **shastensis** (Heller) C. P. Smith. Occasional on sandy flats in the Douglas oak woodland on the west slope of the Greenhorn Range. It is scarce in a very different habitat in the granite rocks and sand along the rim of the Kern Plateau southwest of Pine Flat.

Lupinus albifrons Bentham. Bush lupine, silver lupine. Common and colorful, usually in colonies, in the Temblor Range south to the Mt. Abel region; infrequent in the foothills of the mountains, east to Erskine Canyon. Plants of var. *eminens* (Greene) C. P. Smith grow from the Temblor Range south to Ballinger Canyon in the upper Cuyama Valley in situations (such as steep banks) that protect it from browsing and attendant breakage; these would seem to be no more than large robust individuals of the typical variety.

Lupinus Andersonii Watson. Rare in the Jeffrey pine forest on Mt. Pinos and Tecuya Ridge.

Lupinus arbustus Douglas. A colony grows at the summit of Shirley Peak in the Greenhorn Range (Howell 38,518, T8613), a southern range extension from the Sierra Nevada of Mariposa County. (*L. laxiflorus.*)

Lupinus Benthamii Heller. Bentham lupine. A common and colorful spring annual that often blooms in great masses on the east side of the valley and the adjoining foothills through the Douglas oak woodland. It is particularly abundant in the Granite Station region and is one of the common spring flowers of the Sand Ridge Wildflower Preserve near Edison.

Lupinus bicolor Lindley. Ground lupine, calf lupine. Common, in one form or another, throughout much of the county, excepting on the higher mountains and the desert plains. Ssp. *marginatus* Dunn is common in sandy soil from Lebec northeast to Kern Canyon, extending occasionally to the west side of the desert. It is most frequent in the pinyon woodland and is occasional in the Douglas oak woodland to the west and in the arid shrub association to the east. Ssp. *microphyllus* (Watson) Dunn is occasional from the Upper Sonoran grassland through the Douglas oak woodland in the east-side foothills, extending to the southeast end of the Piute Mountains. Ssp. *Pipersmithii* (Heller) Dunn is common in all the Upper Sonoran associations, extending to the borders of the ponderosa pine forest in the Greenhorn Range; it is especially common and frequently abundant in sandy adobe soil in the northern Temblor Range. Finally, ssp. *tridentatus* (Eastwood) Dunn is common in all associations to the borders of the yellow pine forest; it is especially common in the Upper Sonoran grassland.

Lupinus Breweri Gray. Occasional in sand around granite outcrops and in sandy openings in the red fir forest in the Sunday Peak region.

Lupinus Breweri var. **bryoides** C. P. Smith. A characteristic matted plant of the exposed summits of Mt. Pinos and Sawmill Mountain, growing in extensive mat-like colonies in gravelly soil. (These plants have been segregated as *L. tegeticulatus* by Alice Eastwood; cf. Leafl. West. Bot., 2:251.)

Lupinus Breweri var. **grandiflorus** C. P. Smith. Rare in colonies in the Jeffrey pine forest along the crest of Piute Mountain and on the ridges at the south end of the Kern Plateau.

Lupinus concinnus Agardh var. **Orcuttii** (Watson) C. P. Smith. Occasional in sandy or gravelly soils on the desert-facing slopes of the mountains, southeast to the Jeffrey pine forest on Tecuya Ridge, west to the east slope of the Greenhorn Range. It is infrequent in the creosote bush association and in the desert ranges and sometimes occurs as a waif along Kern River west to Buena Vista Lake.

Lupinus confertus Kellogg. A distinctive lupine that is occasional in the higher elevations of the Jeffrey pine forest on Mt. Pinos and on sandy flats and about sagebrush meadows on the southern Kern Plateau. It occurs as a waif along Kern River in the Kernville region. Plants on the open Mt. Pinos summit plateau have deep purple-blue flowers that are especially colorful.

Lupinus densiflorus Bentham var. **lacteus** (Kellogg) C. P. Smith. A white-flowered annual lupine that is occasional in the Upper Sonoran associations, mostly in the Douglas oak woodland, south through the Tehachapi Mountains; less frequently to the Mt. Pinos region. An attractive form with clear shell-pink flowers grows on sandy flats in hills south of Glennville (Farnsworth 680, 682, 691). The type was collected at Oak Creek, 14 miles from Tejon Pass [west of Mojave] by Samuel Brannan.

Lupinus elatus Johnston. Occasional on sage brush flats in the Mil Potreros region of the San Emigdio Range (T6283) and in Cuddy Valley

(T6299); known from Browns Meadow on Piute Mountain (Breedlove 3979).

Lupinus excubitus Jones. Common on the arid desert-facing slopes of the mountains; east to the El Paso Range; west to the arid east slopes of the Piute Mountains and the Greenhorn Range, and southeast to the Mt. Pinos region. In the desert canyons large colonies grow on the steep often barren talus slides; here the plant with its rich royal blue flowers is one of the more spectacular sights of a desert spring. Particularly fine colonies grow in Horse Canyon on the east slope of Kiahvah Mountain and in Nine Mile Canyon in extreme southern Inyo County.

Lupinus excubitus var. **austromontanus** (Heller) C. P. Smith. An herbaceous form that most commonly grows in the Jeffrey pine forest from the southern Kern Plateau, the Piute Mountains, and Breckenridge Mountain south through the Tehachapi Mountains. However, an extensive colony grows in Upper Sonoran grassland at the summit of Oak Creek Pass between Mojave and Tehachapi. The type was collected at Tehachapi (Amos Arthur Heller 7825 in 1905).

Lupinus formosus Greene. Beautiful lupine. Common in the Douglas oak woodland, chaparral, and pinyon woodland through the yellow pine forest in the mountains. The plants are highly variable, especially those of the Tehachapi Mountains, and the species is in need of study. Var. *robustus* C. P. Smith is an herbaceous form that grows in loose sand in Walker Basin.

Lupinus Grayi (Watson) Watson. Occasional in dry places in the yellow pine forest in the Greenhorn Range, the Piute Mountains, and on the Kern Plateau.

Lupinus horizontalis Heller. Sunset lupine. Rare in the Tejon Hills (high slope on the north side of the mouth of Tejon Canyon, T8032; 3 miles northwest of the old Tejon Ranch headquarters, T6881); known in 1952, a year with an exceptionally good spring, as a rarity on the Antelope Plains southwest of Blackwells Corner (T273), and in 1951 at the foot of the Elkhorn Hills Road west of Maricopa (Bacigalupi 3218, JEPS); once rare and probably now extinct on the sandy plains in the Arvin region that have been converted to farm lands. Although apparently restricted to the high plains and low hills about the head of the San Joaquin Valley, this rare lupine is not fully understood and critical field and herbarium study of it and related species may considerably extend our knowledge of its range. The type was collected at Sunset [near Arvin] (Amos Arthur Heller 7725, April, 1905).

Lupinus horizontalis var. **platypetalus** C. P. Smith. Common on the desert, mostly on sandy plains in the creosote bush association. According to Philip A. Munz (A California Flora, p. 809), it occurs west to the Greenhorn Range.

Lupinus latifolius Agardh var. **columbianus** (Heller) C. P. Smith. Occasional in wet soil in the meadows in the Jeffrey pine forest in the Piute

Mountains and at Calf Creek in the ponderosa pine forest in the Greenhorn Range.

Lupinus latifolius var. **Parishii** C. P. Smith. A dense colony grows in a small wet meadow 1.7 mile east of the summit of Mt. Pinos (T3145).

Lupinus nanus Douglas ssp. **Menkerae** (C. P. Smith) Dunn. Sky lupine. Common, often abundant and colorful, in sandy soils in the valley, especially on the plains on the west side. In wet years, sky lupine colors the broad head of the valley a bright blue which is familiar to travelers emerging from Grapevine Canyon on the Ridge Route from Los Angeles. The once broad fields around Arvin have largely been replaced by irrigated farmlands. The variety is endemic to the upper San Joaquin Valley; the type was collected at Bakersfield (Amos Arthur Heller 7588 in April, 1905).

Lupinus odoratus Heller. A beautiful lupine with rich deep blue flowers that is occasional from the desert west to the east slope of the Piute Mountains and in South Fork Valley west to Weldon. It usually grows in small scattered colonies in sandy soil.

Lupinus polyphyllus Lindley var. **superbus** (Heller) Munz. A characteristic plant of wet places in the yellow pine and red fir forests of the Greenhorn Range, the southwestern Kern Plateau, and at French Meadow in the Piute Mountains.

Lupinus ruber Heller. Rare in the Tehachapi Mountains, mostly in friable (dry bog) clay. The type was collected along the railroad tracks just west of Tehachapi (Amos Arthur Heller 7827 in 1905). (*L. microcarpus* var. *ruber.*)

Lupinus Shockleyi Watson. Scarce in favorable years in the shadscale scrub along County Line Road on the sandy plains east of Ridgecrest (T11,829). W. L. Jepson (Flora of California, 2:281) reports a collection by Katharine Brandegee from Tehachapi.

Lupinus subvexus C. P. Smith. Purple lupine. Widespread and often common in the valley and surrounding foothills, the Temblor Range, and the Cuyama Valley region. Rare on the desert (Last Chance Canyon, El Paso Range, Peñalosa 881). (*L. microcarpus.*)

Lupinus succulentus Douglas. Scarce in heavy often disturbed soil in the northern Temblor Range and in the hills around the valley (sometimes, as at Wheeler Ridge, locally common). Appearing early in the order of plant succession, it is often common on fresh roadcuts, fills, etc.

Lupinus vallicola Heller. Rare in the Douglas oak woodland south to Caliente Canyon.

Lupinus vallicola ssp. **apricus** (Greene) Dunn. David Dunn reports (Aliso 3:167) it is occasional in the Douglas oak woodland south to Pastoria Creek in the Tehachapi Mountains.

Medicago hispida Gaertner. Bur-clover. Widespread and often abundant in the grasslands, growing densely in heavy soils. It is occasional in the Douglas oak woodland. Bur-clover is a highly favored forage plant. Var. *confinis* (Koch) Burnat, a form in which the pods are without prickles, is rare with the typical variety (Carneros Rocks, T3333; Kern Hot Springs, T12,210). Native of southern Europe.

Medicago lupulina Linnaeus. Black medic. Widespread but infrequent in cismontane moist places; rare in dryish meadows in the Jeffrey pine forest. Native of Europe.

Medicago sativa Linnaeus. Alfalfa. Frequently spontaneous and perhaps very locally established in the cismontane associations below the yellow pine forest; occasional along roads and in dry washes. A surprisingly drouth-resistant plant of great forage value that would seem to justify research to develop strains that could withstand grazing. Native of the Old World.

Melilotus albus Desrousseaux. White sweet-clover. A widespread and common cismontane weed of moist soil. It reaches the desert's edge at the Onyx Ranches line camp in Freeman Canyon. Native of Eurasia.

Melilotus indicus (Linnaeus) Allioni. Indian sweet-clover. A scarce but widespread cismontane weed growing in moist soils below the yellow pine forest, most frequent in the valley. Native of Eurasia.

Melilotus officinalis (Linnaeus) Lamarck. Yellow sweet-clover. Scarce in the often dense weedy growth in wet places in South Fork Valley. Native of Eurasia.

Parkinsonia aculeata Linnaeus. Horse bean, Mexican palo verde (not the palo verde of the southern California deserts). A handsome large summer-blooming shrub known from one spontaneous mature plant, now gone, in a dry wash near Maricopa. It is to be expected near extensive freeway and garden plantings in the valley. Native from Arizona and Florida to South America.

Pisum sativum Linnaeus. Garden pea. Often spontaneous near fields where it has been planted in the valley but never persistent.

Poinciana Gilliesii Hooker. Bird-of-paradise tree. An occasional escape in the valley and the Temblor Range, especially about cities and towns. It is outlawed as a noxious plant by city ordinance in Delano because of the erroneous belief that it is a common cause of hay fever. Contrary to widespread opinion, apparently the plant is not poisonous but the green seed pods are a severe irritant to the digestive tract of animals and man. Native of South America. (*Caesalpinia Gilliesii.*)

Prosopis juliflora (Swartz) DeCandolle var. **Torreyana** L. Benson. Mesquite. Common on hundreds of acres of alkali plains east and northeast of Buena Vista Lake; less common throughout the slough country north to Tulare Lake. An isolated colony grows along Poso Creek above the Round Mountain Road in the Poso Hills. Occasional on the desert, where it grows at springs or places with subsurface moisture (especially common in the dunes south of Koehn Lake). (*P. juliflora* var. *glandulosa.*)

Psoralea californica Watson. A colony grows in friable (dry bog) clay at the head of Quatal Canyon and another in very fine clay at the northwest end of Piute Mountain just east of Bodfish Gap. W. L. Jepson (Flora of California, 2:34) cites a collection by T. S. Brandegee from Kern Hot Springs.

Robinia Pseudoacacia Linnaeus. Black locust. Widely planted and frequently spontaneous (Buena Vista Slough north of Tupman, Bakersfield, Hart Park, etc.). Native of the eastern United States.

Sphaerophysa salsula (Pallas) DeCandolle. Austrian pea weed. A colony which has been eradicated was discovered in 1960 about 2 miles northwest of Old River (T. C. Fuller 4232). The station on which the Munz listing is based (A California Flora, p. 888) grew along Highway U. S. 99 just south of Bakersfiield (Gander 7573 in 1939, SD) and is no longer present. Thomas C. Fuller says (personal communication) that Austrian pea weed can become an extremely troublesome pest that should be rigorously controlled should it again appear in the county. Native from the Caspian Sea to China.

Trifolium albopurpureum Torrey & Gray. Rancheria clover. Scarce in the Lower Sonoran grassland; a common and important forage plant of the Upper Sonoran grassland; occasional to the edge of the yellow pine forest.

Trifolium ciliolatum Bentham. Tree clover. Occasional, sometimes common, in the Douglas oak woodland in the Greenhorn Range and the Tehachapi Mountains, particularly at higher elevations; rare in the San Emigdio Range (Pleito Canyon near Neasons Flat).

Trifolium cyathiferum Lindley. Known from moist ground on Mt. Pinos, Ventura County (Hoffmann, 12 June 1927, SBM), apparently near the Kern County line.

Trifolium depauperatum Desvaux. Occasional in colonies in moist alkaline places in the valley and the Temblor Range.

Trifolium fragiferum Linnaeus. Salina strain of strawberry clover. Roy V. Parker reports (personal communication) that Salina clover is locally established along El Paso Creek at the old Tejon Ranch headquarters. Native of Europe.

Trifolium gracilentum Torrey & Gray. Pin-point clover. Common in the cismontane grasslands, mostly below the Douglas oak woodland, and from the desert ranges to the pinyon woodland. Var. *inconspicuum* Fernald occurs occasionaly with the typical variety; it is doubtfully more than a depauperate form.

Trifolium hirtum Allioni. Seeded in 1952 and 1953 with scant success at the University Test Plot near Lost Hills. A few plants grew spontaneously in 1954 and 1955 (T1985). None has since been found. Native of Europe.

Trifolium microcephalum Pursh. Widespread but scarce from the valley to the yellow pine forest. It grows in a variety of habitats, usually in small dense colonies, from wind-drifted sand at Point of Rocks on the arid west side of the valley to moist soil along streamlets in the Jeffrey pine forest on the Kern Plateau.

Trifolium monanthum Gray. Occasional in wet places in the yellow pine and red fir forests, often growing in dense colonies.

Trifolium obtusiflorum Hooker. Clammy clover, creek clover. Occasional in wet places in Kern Valley. Rare and probably transient along Kern River in the valley (Enos Road near Stockdale, T2868).

Trifolium olivaceum Greene. Scarce on grassy slopes on the hills between Lumreau and Cedar creeks in the Douglas oak woodland in the Greenhorn Range (Farnsworth 435A).

Trifolium pratense Linnaeus. Red clover. Well established in the moist meadows in Lynns Valley; often spontaneous in regions where it is planted in pastures in the valley; a street weed in Mojave. Native of Eurasia.

Trifolium repens Linnaeus. White clover. An occasional escape in moist places near plantings in irrigated pastures in the valley. Native of Europe.

Trifolium spinulosum Douglas. Occasional in moist places from the pinyon woodland through the yellow pine forest.

Trifolium stenophyllum Nuttall. A small dense colony grows along an alkali sink 7 miles southwest of Wasco on Kimberlina Road (T4252). W. L. Jepson (A Flora of California, 2:290) cites a collection from Caliente (Heller 7620).

Trifolium tridentatum Lindley. Tom-cat clover. Common and often abundant in the Upper and Lower Sonoran grasslands and on open slopes in the Douglas oak woodland; occasional from the chaparral to the borders of the yellow pine forests. Tom-cat clover is a very important forage plant, adapted to a wide variety of soils. It grows best in wet years, particularly those with early rains.

Trifolium variegatum Nuttall. White-tip clover. Occasional in wet places in the Douglas oak woodland and common in similar places from the higher chaparral through the yellow pine forest.

Trifolium variegatum var. **melananthum** Greene. W. L. Jepson (A Flora of California, 2:295) cites a collection from Tehachapi (Heller 7821).

Trifolium variegatum var. **pauciflorum** (Nuttall) McDermott. Occasional in usually shady wet meadows in the yellow pine forest in the Greenhorn Range and the Piute Mountains, often forming broad colonies. It is rare in the very arid Douglas oak woodland along Cannell Creek east of River Kern.

Trifolium Wormskioldii Lehmann. Cow clover. Common in wet often subalkaline soils in the pinyon woodland and yellow pine forest, occasional in the Douglas oak woodland, from Inyo County southwest to Mil Potreros, west to the east slope of the Greenhorn Range (rare in wet places in the Douglas oak woodland on the west slope.) (*T. involucratum.*)

Vicia americana Muhlenberg ssp. **oregana** (Nuttall) Abrams. American vetch. Occasional in the Mt. Pinos region and the Tehachapi Mountains; rare, north to the Greenhorn Range, mostly in the yellow pine forest and Douglas oak woodland.

Vicia californica Greene. California vetch. Occasional on open slopes from the Douglas oak woodland to the yellow pine forest in the Greenhorn Range, on Breckenridge Mountain, and at the north end of Piute Mountain; scarce south to the Tehachapi Mountains.

Vicia exigua Nuttall. Colonies grow in friable (dry bog) clay at the head of Bitterwater Creek in the Pinole Hills in extreme western Kern County and at Dry Bog Knoll at the head of Adobe Canyon in the Greenhorn foothills.

Vicia Faba Linnaeus. Horse bean. Rarely spontaneous about plantings in the valley farm lands (Lerdo Road, 0.5 mile east of Spicer City Gin, T4211). Native of Eurasia.

Vicia villosa Roth. Hairy vetch. Scarce along a ditch at the Poso Creek bridge one mile west of Glennville (Meng 146, 13 June 1965). Native of Europe.

GERANIACEAE. Geranium Family

Erodium Botrys (Cavanilles) Bertoloni. Broad-leaf filaree. Common in the valley, especially on the east side, and in the Greenhorn foothills below the Douglas oak woodland; occasional in the Temblor Range. It is most frequent in colonies in bare often rather heavy soil in such places as those where cattle have been fed, in fireguards, graded roadsides, etc. The broad completely prostrate basal leaves tend to discourage competitive seedlings. Native of the Mediterranean region.

Erodium cicutarium (Linnaeus) L'Héritier. Red-stem filaree. Widespread and often abundant below the Douglas oak woodland; locally common on dry sunny slopes from the Douglas oak woodland to the edge of the yellow pine forest; common on the desert, where it is one of the few annuals that succeed in dry years.

Red-stem filaree is most common in regions with less than 12 inches rainfall; when regions with this amount are reached in the mountains, red-stem yields to the much less desirable white-stem filaree (*E. moschatum*). However, in the good soils on the high rounded slopes of the Tehachapi Mountains, it is common to at least 6,400 feet elevation (summit of Purdy Ridge). Here red-stem grows with such montane plants as *Artemisia tridentata, Eriogonum Kennedyi* var. *austromontanum,* and *E. Wrightii* var *subscaposum.* These slopes and ridgetops are quite cold in winter, often with semi-persistent snow, but are warmed by the late spring and summer thermal winds that flow over them when the San Joaquin Valley temperatures soar in the mornings. Thus, it is late when red-stem (and other annuals usually found at far lower elevations) makes its rapid growth, usually blooming in June, fully two or even three months after the species has dried in the valley far below.

The plant grows with little rain, is exceptionally nutritious, and thrives even with moderately heavy grazing. In the Lower Sonoran grassland and in much of the Upper Sonoran grassland it is the primary forage plant. It is also of great importance to wild life. Both the green plants and the seeds are leading sources of food for the large bird population of the region. Kangaroo rats and other rodents depend on the seed for much of their basic diet. They, in turn, are the primary food of the large population of several species of hawks and owls, and of kit foxes, coyotes, and rattlesnakes. Finally, the plant is the chief food of several species of ants, one abundant species of grasshopper, and many other insects (which in turn are food for reptiles, mammals, and birds, and so on through the biological complex). Thus, in the arid Lower Sonoran grassland of the upper San Joaquin Valley a great deal of both the region's wild life and its human economy are dependent on a single plant species, and that species one that is not native.

Although very common in much of California, red-stem filaree is an immigrant from the Mediterranean region. It became widespread and well established very early in Kern County; contrary to the opinion of many plant ecologists, long before grazing could have been a factor in its widespread occurrence. While the country was still a wilderness, John C. Frémont wrote on 11 April 1844 (a dry year) of the region between present day Porterville and Bakersfield:

"Instead of grass, the whole face of the country is covered with *Erodium cicutarium,* here only two or three inches high" (Narrative of the Exploring Expedition to the Rocky Mountains in the Year 1842 and Northern California in the Years 1843–44, p. 365).

The plant is superbly adapted to light disturbed soils and probably spread rapidly during the first drouth cycle in the interior after its introduction from the coastal mission region.

Grazing early in the season tends to benefit red-stem as well as many native annuals at the expense of the taller weedy plants, particularly wild oats, ripgut, the foxtails, and other introduced grasses. It has remarkable ability to develop depauperate seed-bearing plants in dry years; plants only 1.5 inches tall produce viable seeds.

Erodium macrophyllum Hooker & Arnott. Round-leaved filaree. Occasional, usually growing in heavy soil, often in friable (dry bog) clay, in the Temblor Range, the foothills of the mountains east to Tehachapi, at

Mexican Mine in the extreme southwestern Tehachapi Mountains along the northwest side of the desert Antelope Valley, and at Dry Bog Knoll at the head of Adobe Canyon in the Greenhorn foothills.

Erodium moschatum (Linnaeus) L'Héritier. White-stem filaree. Common in the foothills of the mountains, extending through the Douglas oak woodland; scarce in the San Emigdio and Temblor ranges. It is not nearly as nutritious as red-stem filaree; some livestock people consider it of only secondary value. Native of the Mediterranean region.

Erodium obtusiplicatum (Maire, Weiller & Wilczek) J. T. Howell. Common, often abundant, on the plains on the east side of the valley; occasional but becoming more common in the hills around the valley. Native of North Africa.

Erodium texanum Gray. Texas filaree, desert hornbill. Texas filaree reaches its northwestern limits at the mouth of Last Chance Canyon in the El Paso Range where a highly local colony grows on a gravelly desert fan (T4441). Apparently it is not otherwise known from the Mojave Desert northwest of the Calico Mountains.

Geranium californicum G. N. & F. F. Jones. Common in moist meadows in the Jeffrey pine forest on the Kern Plateau, Breckenridge Mountain, and in the Piute Mountains.

Geranium carolinianum Linnaeus. Carolina cranesbill. Scarce in sunny moist soil along Cedar Creek at the Rock Corral (Farnsworth 565).

Geranium dissectum Linnaeus. Cranesbill. Scarce in meadows at the Henry Bowen Ranch in Lynns Valley. Native of Europe.

LIMNANTHACEAE. Meadow Foam Family

Limnanthes montana Jepson. Meadow foam. Occasional in colonies in moist sunny places in the Douglas oak woodland in the Greenhorn Range, often growing densely in very shallow moist soil over granite sheets.

OXALIDACEAE. Oxalis Family

Oxalis corniculata Linnaeus. Yellow sorrel. A common garden weed, especially in lawns, that is well naturalized along Lumreau Creek just above Cedar Creek in the Douglas oak woodland. Native of Europe.

Oxalis pes-capri Linnaeus. Cape oxalis, Bermuda-buttercup. Well naturalized on rocks on the canyon-side at the power house 4.2 miles west of Richbar in Kern Canyon (T5934). Native of South Africa.

LINACEAE. Flax Family

Linum Lewisii Pursh. Blue flax. Rare on gravelly slopes in the Jeffrey pine forest on Mt. Pinos and Tecuya Ridge (*L. perenne* ssp. *Lewisii.*)

Linum micranthum Gray. Dwarf flax. Occasional in red clay in the thin chaparral at the north end of Piute Mountain.

Zygophyllaceae. Caltrops Family

Larrea divaricata Cavanilles. Creosote bush. The widespread and dominant shrub of the desert, covering tens of thousands of acres. It occupies almost all of the Kern County portion of the Mojave Desert that has rainfall of less than six inches and soil without shallow hardpan and that is not subalkaline. Three plants grow west of the mountains in Sand Canyon just above Poso Creek and an isolated plant is said to have once grown on the Caliente Canyon flood plain. A single celebrated plant, locally known as the Dead Man's Bush, grows on the lonely Antelope Plains near Point of Rocks on the west side of the valley (T315, 1387). The local legend that the plant grew from seed in the clothing of a dead man found under the plant in the 1880's, thought to be an unknown traveller from the desert, seems to have been fairly well discredited by recent research by local historians. Ernest M. Still (oral communication) recalls the shrub was fully grown when he picknicked there as a small boy in 1900 when less than 20 years had elapsed since the skeleton had been found. This hardly seems long enough for a creosote bush to reach maturity. The plant was first reported in the botanical literature by a pioneer Annette rancher, Abram A. Still, whose letter of 7 February 1898 to Willis Linn Jepson was published in Erythea (6:38, 1898). Mr. Still enquired whether the dead man of "12 or 15 years ago" might have been poisoned by eating some of the leaves and added "or the seed may have been in his pocket and grew the year after he died." At any rate, the plant is of botanical interest in that it is possibly the only fully mature specimen whose minimum age is known; it is now at least 80 years old.

Although the shrub occupies a vast expanse in the North American deserts, these out-of-range plants are remarkably few. Factors not completely understood rather rigidly limit its distribution and the borders of the creosote bush association are abrupt; Mary Austin aptly wrote "if you have any doubt, know that the desert begins with the creosote."

Willis Linn Jepson (A Flora of California, 2:439) gives an excellent account of its role in the Indian and pioneer white economy of the desert and its geographical distribution. He reports the plant has been used for firewood, burning as well when green as when dry, and that the root crown, which can be lifted with a crowbar, burns especially well. Indians used gum from the leaves for resin for pottery mending and water-proofing baskets. Teamsters made a decotion from the leaves for treating collar sores of draft animals. Other authors have written of various medicinal uses for the plant; these seem to be the standard early day "cure alls" made from any odorous or vile-tasting plant.

Creosote bush is never browsed, although isolated plants, such as the Dead Man's Bush, can suffer severely from rubbing by livestock, especially by bulls. It does not seem to survive attempts at garden culture away from the regions where it is native (unlike many desert shrubs that thrive under such care and often little resemble their native form). (*L. tridentata* var. *glutinosa.*)

Tribulus cistoides Linnaeus. The California record reported by Duncan M. Porter (Contrib. Gray Herb., 192:127) and by the writer (Leafl. West. Bot., 10:55) was based on a collection labelled "Kern River Valley, 7 miles below Hobo Hot Springs, Jesse D. Skoss, 7 May 1948" and distributed by the University of California at Davis. Painstaking investigation by John Tucker (personal communication) has revealed that this collection was actually made by A. S. Crafts in Santo Domingo, probably at La Romana. (The labelling confusion was in no way the fault of either Dr. Skoss or Dr. Crafts.) Thus, the species should be deleted from both the Kern and California floras.

Tribulus terrestris Linnaeus. Puncture vine. A common noxious weed of fields, bare places, roadsides, etc., in all the cismontane associations below the chaparral. Troublesome everywhere, it is especially a problem in melon fields. Native of the Old World.

Zygophyllum Fabago Linnaeus var. **brachycarpum** Boissier. According to Philip A. Munz (A California Flora, p. 158), it is established near Rosamond and Muroc [now Edwards]. Alice Eastwood quotes a letter (Leafl. West. Bot., 1:15, 1932) from Mrs. Susan W. Hutchinson that states "it has been growing for some time just outside of a ranch near the dry lake in the Muroc region." It is possible that the conversion of Muroc to Edwards Air Force Base and the rapid growth of both residential areas and irrigated farmlands around Rosamond has eliminated this Old World weed.

SIMARUBACEAE. Ailanthus Family

Ailanthus altissima (Miller) Swingle. Ailanthus, tree-of-heaven. A frequent escape (that is occasionally adventive far from the nearest colony) in the foothills, in South Fork Valley, the Temblor Range, and about Bakersfield. It is especially well naturalized at Long Tom Mine in the Greenhorn foothills and in Havilah Canyon. Straggly forlorn often solitary trees are sometimes the only marker of old homestead sites in the Temblor and San Emigdio ranges.

Ailanthus is declared a noxious weed by city ordinance in Delano, because of a presumed hay fever hazard. Although often despised for its weedy nature, it is a handsome tree that is drouth resistant, neat, and free from diseases and insect infestation and grows easily in many situations in which very few other trees succeed.

A native of Asia, introduced in California in the early mining days by the Chinese (who probably brought it to Havilah Canyon and Long Tom Mine in the gold boom of the 1850's and 60's). (*A. glandulosa.*)

EUPHORBIACEAE. Spurge Family

Croton californicus Mueller-Argovensis var. **mohavensis** Ferguson.

Mojave croton. Scarce in sandy places in the desert and on flood plains along the east side of the valley from Sand Ridge to Tejon Canyon.

Eremocarpus setigerus (Hooker) Bentham. Turkey mullein. Common in all the cismontane associations to the edge of the yellow pine forest; especially common in summer fallowed fields and sandy soils in the Upper Sonoran grassland. It is occasional on sandy plains and on roadsides in the desert. The seeds are a favorite autumn food of doves, and in the southern Temblor Range, of the introduced chukar partridge.

Euphorbia albomarginata Torrey & Gray. Rattlesnake weed. Common from the creosote bush association through the pinyon woodland, southwest to the Mt. Pinos region; west to Kernville and Erskine Canyon. It usually grows in loose light soils.

Euphorbia Lathyrus Linnaeus. Caper-spurge, gopher plant. A rare escape that has been persistent for many years at the head of Big Canyon on the Ernest Still Ranch near Annette in the northern Temblor Range (T6045 in 1961). Native of Europe.

Euphorbia maculata Linnaeus. Spotted spurge. A common garden weed, often troublesome in lawns (Bakersfield, Kernville, China Lake), that is naturalized along Cedar Creek just below the mouth of Fulton Creek. Native of the eastern United States. (*E. supina* of many California references.)

Euphorbia ocellata Durand & Hilgard. Yerba golondrina. Common in the cismontane Upper Sonoran associations, especially in the Temblor Range; less common in the valley. An extremely drouth tolerant summer annual that is successful in regions almost and often entirely devoid of summer annuals. The type was collected at Poso Creek by Adolphus L. Heermann in 1853.

Euphorbia Peplus Linnaeus. Petty spurge. Occasional in moist soil in gardens in Bakersfield and probably the other cities and towns.

Euphorbia serpyllifolia Persoon. Occasional in disturbed places in the Jeffrey pine forest in the Piute and Tehachapi mountains.

Euphorbia spathulata Lamarck. Scarce in ultra-fine red clay at the north end of Piute Mountain and in friable (dry bog) clay deposits in the northern Temblor Range. (*E. dictyosperma.*)

Euphorbia vallis-mortae (Millspaugh) J. T. Howell. Death Valley euphorbia. Rare in sandy places in desert canyons, from Red Rock Canyon north to Short Canyon (in the Sierra Nevada north of Walker Pass), west to Kelso Valley and the west slope of Walker Pass. The type was collected a few kilometers north of Indian Wells (Frederick Vernon Coville and Frederick Funston 1008, 21 June 1891). Although the type is attributed to both of these members of that important desert scientific expedition, it was actually collected by Coville; Funston was returning to the San Joaquin

Valley by way of Mammoth Pass when the collection was made (Coville, F. V. Botany of the Death Valley Expedition, pp. 7–10).

Ricinus communis Linnaeus. Castor bean. Frequently spontaneous in moist soils in the valley and well established as a stream bank plant in lower Kern Canyon. All parts of the plant, especially the seeds, are extremely poisonous. Native of Asia and Africa.

Stillingia paucidentata Watson. Occasional on the desert northwest to Dove Spring Canyon and Indian Wells Valley, usually growing in drifted sand or in very light soil. (*S. linearifolia* var. *paucidentata.*)

CALLITRICHACEAE. Water-starwort Family

Callitriche heterophylla Pursh var. **Bolanderi** (Hegelmaier) Fassett. Rare in Lake Ming in 1961 (T6643). Rhodora, 53:177 cites a collection from [old] Kernville, a station now flooded by Lake Isabella (in which the plant has not yet been found).

Callitriche marginata Torrey. Common on dessicating pools and flats in Lynns Valley.

MELIACEAE. Mahogany Family

Melia Azedarach Linnaeus var. **umbraculiformis** Berckmans. Texas umbrella tree, china berry tree. Occasionally spontaneous in waste places in the valley where it is often persistent around old or abandoned dwellings. Native of southwest Asia.

ANACARDIACEAE. Sumac Family

Rhus diversiloba Torrey & Gray. Poison-oak. Occasional in rocky places in the Douglas oak woodland and chaparral in the mountains, scarce in the Temblor Range. The Temblor plants are rigid shrubs with coriaceous leaves; those of the moist shaded swales along Kern Canyon are lax and almost vine-like. Although containing a volatile oil that is a severe skin irritant to many people, cattle eagerly browse the plant, probably destroying many seedlings.

Rhus trilobata Nuttall. Squaw bush. Rare in the Douglas oak woodland in the Greenhorn foothills (3 miles east of Woody, C. N. Smith 37, JEPS).

Schinus Molle Linnaeus. Pepper tree. Commonly planted in the valley and at lower elevations elsewhere, pepper trees are occasionally spontaneous, especially along the river in lower Kern Canyon and around Bakersfield. Once established, it is extremely drouth tolerant (although damaged by cold at temperatures of less than about 25° F.). Mature trees have survived at an abandoned oil pumping station on the arid plains at North Belridge since 1950 without apparent drouth damage; the region has had a mean annual rainfall of less than four inches during this time. Native of Peru.

ACERACEAE. Maple Family

Acer macrophyllum Pursh. Big-leaf maple. A small colony of this large beautiful tree grows on a flat with subsurface moisture along El Paso Creek in the Tehachapi Mountains and another grows near the summit of Black Bob Canyon in the San Emigdio Range.

Acer Negundo Linnaeus ssp. **californicum** (Torrey & Gray) Wesmael. California box elder. Occasional along canyon bottoms and creeks from Lebec to El Paso Canyon. A small colony grows in upper Jawbone Canyon just above the Wilderness Ranch. It is a handsome round-headed tree well adapted to cultivation that is common in the plantings at Hart Memorial Park near Bakersfield.

HIPPOCASTANACEAE. Buckeye Family

Aesculus californica (Spach) Nuttall. California buckeye. Common in the mountains southwest to Lebec, often forming a woodland with Douglas oak and digger pine. California buckeye is most common in Kern Canyon, where its spring blossoming is an attractive feature of the steep canyon-sides. It is rare in Pleito Canyon in the Emigdio Range. The flowers are poisonous to honey bees.

RHAMNACEAE. Buckthorn Family

Ceanothus cordulatus Kellogg. Snow brush. Common in the high mountains, mostly in exposed rocky places, often around rugged summits. Browsed plants develop stout thorns capable of seriously puncturing the unwary. The dense colonies are often favorite hiding places of mule tail deer.

Ceanothus cuneatus (Hooker) Nuttall. Buck brush. A common element of the chaparral in the mountains that is scarce on high ridges in the central Temblor Range. Excessive fire control has resulted in many senescent and some dead colonies. When these eventually do burn, what would have naturally been a relatively harmless or even beneficial fire becomes a holocaust. Seeds of this fire-adapted shrub sprout prolifically after burns.

Ceanothus diversifolius Kellogg. Pine-mat. A prostrate ceanothus with deep purple-blue flowers that is scarce in and just below the ponderosa pine forest on the west slope of the Greenhorn Range south to Cedar Creek.

Ceanothus integerrimus Hooker & Arnott var. **californicus** (Kellogg) G. T. Benson. Deer brush. Scarce in a dense growth of shrubby canyon live oaks near the summit of Tejon Canyon (T8870) and rare on a bank in the ponderosa pine forest just south of Greenhorn Pass (T12,411).

Ceanothus integerrimus var. **puberulus** (Greene) Abrams. Deer brush. Occasional in the Jeffrey pine forest in the Mt. Pinos region.

Ceanothus leucodermis Greene. Chaparral white thorn. Occasional, often locally common, (especially in old burns), in the chaparral in the mountains. (*C. divaricatus.*)

Ceanothus vestitus Greene. Common in the chaparral in the mountains, mostly on the east slopes where it replaces *C. cuneatus,* but extending southwest to Mt. Abel and west to Kern Canyon. In the Cache Peak region of the extreme northeastern Tehachapi Mountains, *C. vestitus, C. cuneatus,* and *Quercus turbinella* var. *californica* form an interesting and unusual chaparral on the steep arid slopes of the region. The type was collected at the borders of the pine forest on the mountains near Tehachapi (Edward L. Greene, 23 June 1889). (*Ceanothus Greggii* var. *vestitus.*)

Rhamnus californica Eschscholtz ssp. **cuspidata** (Greene) C. B. Wolf. Coffee berry. Occasional from the Kern Plateau through the arid slopes of the mountains southwest to the Tehachapi Mountains. It grows from the pinyon woodland to the Jeffrey pine forest, where it ascends to 7,500 feet elevation on the southern Kern Plateau. (*R. californica* var. *viridula.*)

Rhamnus californica ssp. **tomentella** (Bentham) C. B. Wolf. Occasional in the chaparral in the mountains. Although the range of these subspecies overlap and intermediates are common, ssp. *cuspidata* tends to occur in the more xeric regions. However, this is not always true; ssp. *cuspidata* grows at Keene, a region of relatively high rainfall; ssp. *tomentella* is rare at the extremely arid head of Nine Mile Canyon in southern Inyo County.

Rhamnus crocea Nuttall. Buckthorn, red berry. Rare in Douglas oak woodland in the central Temblor Range (Messa Ridge, Cedar Canyon).

Rhamnus ilicifolia Kellogg. Holly-leaf berry. Occasional in the chaparral in the mountains and in the central Temblor Range; scarce in the sparse Jeffrey pine forest on the southern Kern Plateau, occurring at elevations of more than 7,000 feet. Both *R. crocea* and *R. ilicifolia* are excellent evergreen garden shrubs. The berries make a tart beverage, sometimes used by the pioneers as a summer cooler. Venturesome modern ranchers find it a pleasantly astringent addition to drinks with a gin base. (*R. crocea* ssp. *ilicifolia.*)

VITACEAE. Grape Family

Parthenocissus incertus Fritsch. Virginia creeper. Occasional in lower Kern Canyon in dense colonies in moist places along the highway. While these plants are doubtfully spontaneous, efforts to discover their origin have been without success. The species is native as far west as Arizona but the related species, *P. quinquefolia,* is much more commonly planted in gardens, so the status of the Kern Canyon colonies as native or naturalized remains in doubt. Until the answer is known, it is best to assume they were planted.

Vitis californica Bentham. Wild grape. A dense and impressive colony of huge vines that covers the steep banks of the canyon bottom and festoons the Douglas and canyon live oaks grows in El Paso Canyon. Wild grape was also once abundant in Grapevine Canyon (and was the source of the name "The Grapevine" originally used for the old Ridge Road to Los Angeles) but these colonies are now nearly eradicated by freeway construction. The plant is scarce along Kern River about one mile below the China Grade bridge in Bakersfield.

Vitis Girdiana Munson. Desert grape. Several dense colonies grow in Grapevine Canyon in the southern Sierra Nevada north of Walker Pass. These furnish cover and food for large flocks of California valley quail.

Vitis vinifera Linnaeus. Domestic grape. A dense spontaneous colony (possibly of clonal origin) grows along an old canal on Sunset Boulevard just west of Edison Road in the Weedpatch district (T5439). Native of Europe.

Malvaceae. Mallow Family

Althaea rosea (Linnaeus) Cavanilles. Garden hollyhock. A common escape, especially in the Tehachapi region, possibly sparingly naturalized about ranchgrounds, vacant lots, and waste places. Native of China.

Eremalche exilis (Gray) Greene. Desert mallow. Common in sandy places on the desert. A form with a broader and more abruptly pointed calyx than is typical has been tentatively assigned to this species by Thomas H. Kearney; it is rare in moist soil at the Buena Vista Canal 2.6 miles east of Lost Hills (T1883). (*Malvastrum exile.*)

Eremalche kernensis C. B. Wolf. Kern mallow. An endemic that is highly local on the Belridge Plain and in the Soda Lake region in San Luis Obispo County. The Kern plants have predominantly white flowers; those at Soda Lake and the nearby Panorama Hills vary from white to lavender-blue. The species rests on what have proven to be inconstant characters of pubescence and petal color and Thomas H. Kearney implied it might well be reduced to a subspecies of *E. Parryi* (Madroño, 13:242). He suggested the possibility that it may have originated as a hybrid between that species and the desert *E. exilis.* The type was collected 7 miles northwest of McKittrick on the Lost Hills Road [along Salt Creek] (Carl B. Wolf 8413, 11 April 1937).

Eremalche Parryi (Greene) Greene. Parry mallow; at Belridge, farewell-to-spring. Common to abundant in light soils in the valley, especially on the west-side plains; scarce in the Temblor Range through the Douglas oak woodland and in the San Emigdio Range through the pinyon woodland. Parry mallow is colorful in late spring on the Belridge Plains, blooming when most of the spring annuals of the region have faded, and in favorable years forming great drifts of lavender-pink. These displays are particularly impressive on the wind-blown sand around Middlewater Pump Station. (*Malvastrum Parryi.*)

Eremalche rotundifolia (Gray) Greene. Desert five-spot. Rare in metamorphic talus in Last Chance and El Paso Canyons in the El Paso Range. These are western outposts for a plant common in the desert ranges to the east. (*Malvastrum rotundifolium.*)

Gossypium hirsutum Linnaeus. Cotton. Widely planted in the valley and occasionally spontaneous but nowhere even sparingly established. Probably a native of tropical America.

Malacothamnus marrubioides (Durand & Hilgard) Greene. Although attributed to the Tehachapi Mountains by Philip A. Munz (A California Flora, p. 126) and by Thomas H. Kearney (Leafl. West. Bot., 6:131), these citations were apparently based on collections from the Liebre Mountains of Los Angeles County (along the old Ridge Route) that were erroneously labeled Kern County. Its range north to the type locality at Millerton in Fresno County (now under Millerton Lake) and its reported occurrence on the Ventura County slopes of Mt. Pinos strongly suggest that this very rare mallow does grow in our region. However, Kearney (*loc. cit.*) notes of the Millerton collection: ". . . the species has not been collected since so far to the north" so the northern distribution of this shrub of the mountains of Los Angeles and Ventura counties remains in doubt. (*Sphaeralcea densiflora.*)

Malacothamnus orbiculatus (Greene) Greene. Tehachapi mallow. Occasional, usually in disturbed rocky places, from Mt. Pinos along the desert-facing slopes of the Tehachapi Mountains to the Piute Mountains and the southern Sierra Nevada. While ascending to higher elevations than any other species of *Malacothamnus* (at 7,800 feet elevation in bulldozed slate near the summit of Chimney Peak in southeastern Tulare County and at 8,800 feet elevation at Onion Meadow in Inyo County), it also occurs on the arid saltbush plains west and south of Belridge at elevations of only 700 feet. On the eastern slope of the Temblor Range it is locally abundant following brush burns, becoming rare in the prolonged absence of fire. The plant also grows as a fire-follower in the Cache Peak region east of Caliente Canyon. The type was collected in the mountains south of Tehachapi by Edward L. Greene in 1889.

Malva neglecta Wallroth. Cheese weed. An occasional weed of gardens, fields, etc., everywhere except the desert and the higher mountains. Native of Eurasia. (*M. rotundifolia.*)

Malva parviflora Linnaeus. Bull mallow. A widespread weed of all associations below the yellow pine forests except on the desert, where it is occasional about towns. It often grows densely for many years at old cattle and sheep camps. Native of Eurasia.

Sida hederacea (Douglas) Torrey. Alkali mallow. Common in the valley, the Temblor Range, and South Fork Valley, growing in light usually subalkaline soils, where it spreads from rootstocks and forms dense colonies.

Sidalcea malvaeflora (DeCandolle) Gray ssp. **sparsifolia** C. L. Hitchcock var. **sparsifolia.** Checker bloom. Common on grassy north slopes and at winter-wet swales and moist meadows in the Douglas oak woodland in the mountains south to Mt. Abel. It is especially common in the Tehachapi Mountains. The type was collected one mile south of Ft. Tejon (C. Leo Hitchcock 19,546, 29 May 1952).

Sidalcea malvaeflora ssp. **sparsifolia** var. **uliginosa** C. L. Hitchcock. Kern Hot Springs checker bloom. Endemic to the wet subalkaline soil at and below Kern Hot Springs and at the subalkaline springs about 2 miles north of Weldon. The type was collected at Kern Hot Springs (C. Leo Hitchcock 19,543, 29 May 1952).

Sidalcea ranunculacea Greene. Greenhorn checker bloom. A distinctive checker bloom that is occasional in moist places in the yellow pine and red fir forests in the Greenhorn Range. It is endemic to the southern extension of the Great Western Divide north to the Kaweah River drainage.

Sphaeralcea ambigua Gray. Desert mallow, apricot mallow. Occasional in Indian Wells Valley and the neighboring Sierran canyons north of Walker Pass.

Sphaeralcea ambigua var. **monticola** Kearney. Desert mallow, apricot mallow. Occasional and colorful in the creosote bush association. It is especially common in the Boron region.

Sterculiaceae. Cacao Family

Fremontodendron californicum (Torrey) Coville. Fremontia, flannel bush, slippery elm (a misnomer that happily is being less commonly used). Common in chaparral and the pinyon woodland in the mountains, east to the desert-facing canyons. The large deep yellow flowers make it a colorful and often spectacular shrub when in bloom. The colonies on the arid slopes of granite blocks east of Weldon are particularly striking. Unusually large plants grow in the Frazier Park region. Carl B. Wolf reports (Rancho Santa Ana Bot. Garden Occ. Papers, 1:67, 1938) that one of these is probably the largest specimen known. It is 23 feet tall, 30 feet across, and has 4 main trunks whose diameters are 16, 16, 12, and 18 inches (Frazier Park Road, 2.6 miles west of the Ridge Route, Cuddy Canyon, C. B. Wolf 6993, 23 May 1935, RSA). It is disappointing that such a colorful shrub has had scant success in cultivation. The form that has been called *Fremontia californica* var. *integra* M. Harvey, distinguished by its entire leaves, is common on the west slope of Breckenridge Mountain, at scattered stations in Kern Canyon, and on Blue Mountain in the Greenhorn Range. (*Fremontia californica.*)

Guttiferae. St. Johns-wort Family

Hypericum anagalloides Chamisso & Schlechtendal. Tinker's penny. Occasional in wet places in the Greenhorn Range from the higher elevations of the Douglas oak woodland to the ponderosa pine forest.

Hypericum perforatum Linnaeus. Klamath weed. A small dense colony of this pernicious weed grows in a wet ditch one mile south of Kernville on the road to Weldon (Howell 38,927, 25 September 1962, T10,222, 16 September 1964). Native of the Old World.

Hypericum Scouleri Hooker. St. Johns-wort. Common and colorful in wet places in the mountains, mostly in the yellow pine forest. (*H. formosum* var. *Scouleri.*)

ELATINACEAE. Waterwort Family

Bergia texana (Hooker) Seubert. Rare in muddy places in the valley.

Elatine brachysperma Gray. Very rare around a vernal pool near Fugitt Road in Lynns Valley (T10,895 in 1965).

FRANKENIACEAE. Frankenia Family

Frankenia grandifolia Chamisso & Schlechtendal var. **campestris** Gray. Alkali heath. A common and characteristic plant of alkali sinks in the valley, desert, and the Temblor Range.

TAMARICACEAE. Tamarisk Family

Tamarix aphylla (Linnaeus) Karsten. Desert-pine, salt-cedar, athel. Although widely planted in the valley and on the desert for its drouth and wind resistance and tolerance of subalkaline soil, desert-pine is rarely spontaneous. Several definitely spontaneous trees grow in the Poso Creek bed near Blankenship Road north of Wasco (T5596). Old plantings are often all that remain at abandoned homesteads, camps, oil installations, etc.; these often appear adventive. Cuttings grow easily when stuck in damp or wet soil; these frequently seem part of the natural growth around springs, creeks, and other wet places. Native of Asia and Africa.

Tamarix pentandra Pallas. Occasional in wet places in the valley, mostly in alkali waste land. A single young shrub grows at Castac Lake near Lebec. Seedlings were abundant in 1964 on freshly flooded portions of the Kern National Wildlife Refuge. Native of the Mediterranean region.

Tamarix tetrandra Pallas. French tamarisk, garden tamarisk. A huge shrub that is a mass of pink during its early spring blooming. It is common at moist places on the desert and occasional in the northern Temblor Range. French tamarisk is widespread over many acres at the south end of Rosamond Dry Lake and on the dreary alkali flats that border it on the west and south. Native of the Mediterranean region. (*T. aphylla.*)

RESEDACEAE. Mignonette Family

Oligomeris linifolia (Vahl) J. F. Macbride. A scattered colony grows in moist bare soil around Mesquite Springs just east of the mouth of Mesquite Canyon at the south base of the El Paso Range (T11,889). A second much smaller colony occurs in dry subalkaline semi-compacted gravel near the spring along the Garlock Fault about 1.5 mile east of the mouth of Last Chance Canyon (T11,987).

VIOLACEAE. Violet Family

Viola aurea Kellogg ssp. **mohavensis** Baker & Clausen. Mojave violet. Rare on the east slope of the Tehachapi Mountains, southwest to Mt. Pinos. An isolated colony grows in thin white shale in a broad swale in the Douglas oak woodland one mile north of the McKittrick Highway along the summit of the Temblor Range (T1975). The type was collected at Tehachapi (Alice Eastwood 3229, 13 May 1913) [probably actually in the yellow pine forest on the mountain slopes in the vicinity of Tehachapi].

Viola Douglasii Steudel. Douglas violet. Colonies are scarce in heavy soil in Douglas oak woodland in the Glennville-Woody region. It is rare south to Keene and Tehachapi Valley (Benson 3488, POM).

Viola glabella Nuttall. Stream violet. Reaches its southern limits with small colonies along Sandy Creek on the west slope of Sunday Peak (T5242) and at Little Poso Creek just below Pettit Meadow (T8542). Both colonies grow in moist soil in deep shade.

Viola Macloskeyi Lloyd. White violet. Occasional in wet places in the ponderosa pine and red fir forests in the northern Greenhorn Range. A small colony grows along Cedar Creek at the Rock Corral in the Douglas oak woodland in the Greenhorn Range, far below the plant's typical range (T11,790). It is known from moist soil at 8,500 feet elevation on Mt. Pinos (Hoffmann, 19 July 1927, SBM). (*V. blanda.*)

Viola purpurea Kellogg. Occasional in the Jeffrey pine forest from the Piute Mountains southwest to Mt. Pinos, usually growing in gravelly open places.

Viola purpurea ssp. **xerophyta** Baker & Clausen. Occasional in sandy places at high altitudes in the yellow pine forest from Black Mountain in the Greenhorn Range southwest to the Mt. Pinos region.

Viola quercetorum Baker & Clausen. Mountain violet. Common in the Douglas oak woodland; rare at the lower levels of the yellow pine forest. The type was collected 2 miles west of Glennville (David Daniels Keck and Jens Christian Clausen 3186).

Viola Sheltonii Torrey. Small colonies grow in heavily shaded loam with much humus at Shirley Meadows and Shirley Peak (Howell 38,408, T8616) and on the north slope of Basket Peak (T8551) in the ponderosa pine forest in the Greenhorn Range.

Loasaceae. Loasa Family

Eucnide urens (Gray) Parry. Rock nettle. Rare in rocky canyons in the El Paso Range, said to occur in lower Jawbone Canyon.

Mentzelia affinis Greene. Valley stickseed. Common in sandy soils in the valley and at lower elevations in the Temblor Range, growing mostly in the Lower Sonoran grassland.

Mentzelia albicaulis Douglas. Small-flowered blazing star. Common in the creosote bush association west to the pinyon woodland on the desert-facing slopes of the mountains.

Mentzelia congesta (Nuttall) Torrey & Gray. Rare on gravelly slopes in the pinyon woodland on the arid east slopes of the Sierra Nevada north of Walker Pass (northeast flank of Morris Peak, T4537, 7161) and around a granite outcrop on the east fork of upper Bartolas Creek in the thin Jeffrey pine forest of the southern Kern Plateau (T12,484).

Mentzelia dispersa Watson. Yellow-flower. Common, usually in dense colonies in light often gypseous soils, from the Temblor Range south to Quatal Canyon, east through the San Emigdio Range to Wheeler Ridge; rare on arid slopes in the southwestern Tehachapi Mountains and at Adobe Canyon in the Greenhorn foothills. Although it grows in all the Upper Sonoran associations, it is most common in the Upper Sonoran grassland. Yellow-flower is a grain field pest in light soils in the northern Temblor Range.

Mentzelia gracilenta Torrey & Gray. Common in the upper Sonoran grasslands and the pinyon woodland from the Mt. Abel region through the Tehachapi Mountains.

Mentzelia involucrata Watson. Scarce in canyons in the desert ranges and on the desert-facing slopes of the mountains, growing in talus or fractured rock.

Mentzelia laevicaulis (Douglas) Torrey & Gray. Blazing star. A striking plant that is scarce in the woody Upper Sonoran associations, often in sandy washes, at scattered stations from Poso Mine in the Greenhorn Range southwest to Frazier Park, San Emigdio Canyon, and the Mt. Abel region.

Mentzelia montana Davidson. Common around rock outcrops and on gravelly or sandy slopes through the yellow pine forest; occasional to the pinyon woodland; scarce in the Douglas oak woodland in the Greenhorn Range.

Mentzelia nitens Greene var. **eremophila** (Jepson) Darlington. Shining stickweed. Occasional in colonies in the creosote bush and shadscale scrub associations, growing most often on flood plains or on gravelly slopes, especially on talus, in the desert canyons (*M. nitens, M. Lindleyi* var. *nitens.*)

Mentzelia pectinata Kellogg. San Joaquin blazing star. Common on dry slopes in very light often loose soils in the hills about the San Joaquin Valley. The bright orange petals with their coppery bases make a colorful showing on these arid slopes near the end of the wildflower season. (*M. gracilenta* var. *pectinata.*)

Mentzelia tricuspis Gray var. **brevicornuta** Johnston. Known from the mud hills in upper Red Rock Canyon (Howell 4947).

Mentzelia Veatchiana Kellogg. Common through the Upper Sonoran associations in the mountains, southwest to Cuyama Valley; east through the creosote bush association. (*M. gracilenta* var. *Veatchiana.*)

Petalonyx Thurberi Gray. Sandpaper plant. Rare, usually on rocky canyon sides, in Red Rock and Jawbone canyons and at Salisbury Mine in the Rand Mountains. It is also rare in the arid digger and pinyon pine woodland on the steep rocky walls of Chimney Creek Canyon north of South Fork Valley (Pittman 25).

DATISCACEAE. Datisca Family

Datisca glomerata (Presl) Baillon. Durango root. Occasional in moist canyon bottoms and along streams in the Douglas oak woodland from the Tehachapi Mountains north through the Greenhorn Range; rare along Cannell Creek east of River Kern. (The Cannell Creek plants are the most robust in the county, some having arching stems more than six feet long.) Durango root is insidiously poisonous to livestock; its cumulative effects, usually developing over a long time, are often mistaken for severe Vitamin A deficiency. Ranchers in the Glennville region are attempting to control and eventually eliminate the plant.

CACTACEAE. Cactus Family

Echinocactus polycephalus Engelmann & Bigelow. Cotton-top cactus. A colony of about 75 plants of all ages, some having as many as 15 well-developed stems, grows in rhyolite on the steep south slope of Galileo Hill, a volcanic butte on the broad desert plains south of the west end of the Rand Mountains. A scattered colony occurs on the low nameless buttes a few miles northwest of Galileo Hill and colonies probably grow in other rhyolite exposures in that region. Cotton-top is very rare on the rocky walls of lower Last Chance Canyon in the El Paso Range and is said to have once occurred in the Rand Mountains. (Apparently both it and Mojave fish hook were formerly more common in the desert ranges. Neither of these rare cacti should be collected for gardens; the first is readily available in nurseries, the second rarely succeeds in cultivation. Such collecting for gardens and curiosities is probably the chief cause of the depletion of the species.)

Echinocactus polyancistrus Engelmann & Bigelow. Mojave fish hook, hermit cactus. Scarce in the El Paso Range, growing most frequently in basalt on Black Mountain. According to reports, it once grew in the Rand Mountains. Unlike the colonial cotton-top, Mojave fish hook tends to occur as isolated individuals (hence the common name "hermit cactus"), often growing through low protective shrubs such as *Eriogonum fasciculatum* ssp. *polifolium.*

Opuntia basilaris Engelmann & Bigelow. Beaver-tail. Scarce in rocky places in the El Paso and Rand mountains, west through the pinyon woodland; rare on the east side of the valley north to Woody (and possibly to Three Rivers in Tulare County). (At least some of the plants along the borders of the San Joaquin Valley are adventive.) It is sometimes locally common on the north slope of Kern Canyon, at Caliente, and in the arid Douglas oak woodland on the east slope of the Greenhorn Range. Var. *ramosa* Parish, a form with the joints branched above, is common on Delonegha and Tucker ridges on the north side of Kern Canyon and is scarce in Kelso Canyon.

Opuntia Treleasei Coulter. Bakersfield cactus. Endemic to the mesas east of Bakersfield, south to Comanche Point. Bakersfield cactus forms dense almost impenetrable colonies, probably from broken joints which root readily (although seed also germinates easily). These colonies were once common on the low hills between Kern River and Caliente Wash and in other nearby areas but most of the great colonies have been destroyed by farming and by the rapid suburban developments of recent years. The best of those that survive are along Kern River below the Kern Front Oil Field, in the hills near Cottonwood Creek at the west base of Breckenridge Mountain, and in Caliente Wash and the adjoining Sand Ridge. The Caliente Wash-Sand Ridge occurrence of Bakersfield cactus is one of the notable cactus fields of California. Densely covering much of the broad wash and the adjoining steep scarp of Sand Ridge, the colony is nearly four miles long and as much as one half mile across. Even this is only a remnant; the west end is abruptly terminated by intensive farming. Fortunately, a significant sample of this magnificent colony is within the boundaries of the Nature Conservancy Sand Ridge Wildflower Preserve.

Even here the cactus is not entirely free of hazards. The region was once intensively grazed; this is no longer true. However deleterious such grazing may have been to the native annuals and especially perennials, it was unquestionably beneficial to the cactus. Control of rank weedy immigrants may be an unexpected challenge at the Wild Flower Preserve. Further problems may come from the highly developed farming that borders the Preserve. Accidental sprayings, invasive cropland pests, and other incidents of agricultural technology are more than speculative possibilities.

Thus, there is still the possibility that the plant may become extinct. Fortunately, it starts easily from cuttings (pads) and grows in a variety of soils in cultivation. Widespread use in gardens is desirable as one means of ensuring preservation of the species. Meanwhile, the success of the various movements to have it named the official Kern County flower would be helpful in bringing the precarious state of this interesting and rare plant to public attention and ultimate adequate protection.

Bakersfield cactus has a further distinction: it was one of the first Kern County plants mentioned in the historical record. In April, 1844, Charles Preuss, Frémont's mapmaker, wrote in his journal of the low hills between Kern River and Caliente Canyon:

> "The hilly country is completely bleak, without any vegetation except a beautiful species of cactus whose magnificent red blossoms grace this sad, sandy desert in a strange manner" (Exploring with Frémont, p. 123).

The type was collected at Caliente by William Trelease in 1892. Trelease (1857–1945) was Director of the Missouri Botanical Garden and later Professor of Botany at the University of Illinois. The details of his botanical visit to Kern County in 1892 are lacking. (*O. basilaris* var. *Treleasei.*)

Opuntia echinocarpa Engelmann & Bigelow. Silver cholla. Occasional on the high desert; uncommon west to Kernville (although the finest colony in the county grows on the slopes at the broad mouth of Fay Canyon north of Weldon). Although cacti are often thought of as being among the most drouth resistant of plants, numerous colonies of silver cholla have died during the extended drouths of the last 20 years.

ELAEAGNACEAE. Oleaster Family

Shepherdia argentea Nuttall. Buffalo berry. A small colony of very old plants grow in damp soil in full sun at Mil Potreros in the San Emigdio Range (T5286). This plant also occurs in the Cuyama Valley in Santa Barbara County.

LYTHRACEAE. Loose-strife Family

Ammannia coccinea Rottboell. Common in wet places in the valley; an especially common rice field weed.

Lythrum californicum Torrey & Gray. California loose-strife. Common in wet places in the valley; occasional in the surrounding foothills; local along Kern River in Kern Valley.

Lythrum Hyssopifolia Linnaeus. Very rare in wet soil in Lynns Valley at the Henry Bowen Ranch (T9969) and along Poso Creek 1.25 mile above Poso Mine (C. N. Smith 1281, JEPS).

PUNICACEAE. Pomegranate Family

Punica Granatum Linnaeus. Pomegranate. Seedlings are occasional in the Delano region and scarce at other places where the tree has been planted. Native of southern Asia.

MYRTACEAE. Myrtle Family

Eucalyptus rostrata Schlechtendal. About 50 trees are spontaneous in a waste area in an oil field just north of Gordons Ferry along Kern River northeast of Bakersfield. The tree is occasionally spontaneous elsewhere (Democrat Hot Springs in Kern Canyon; along El Paso Creek about 1.5 mile below the old Tejon Ranch headquarters). *E. polyanthemos* Schauer (silver dollar tree) was once much planted at camps, etc., in the valley. It is probably never spontaneous although isolated survivors often appear to be. *E. rudis* Endlicher, which grows at a spring along the Kern Canyon highway near Richbar, was probably planted. All are natives of Australia. According to reports, the genus was much planted in Kern County in the 1870's to dry swamps for malaria control (The Versatile Eucalyptus. San Francisco Chronicle, p. 11, 20 November 1961).

ONAGRACEAE. Evening-primrose Family

Boisduvalia densiflora (Lindley) Watson. Occasional in wet places in the cismontane Upper Sonoran associations in the mountains south to the Tehachapi Mountains; rare at Grouse Meadow in the Jeffrey pine forest in the Piute Mountains.

Boisduvalia glabella (Nuttall) Walpers. Rare in wet places, vernal poolpeds, etc., in the valley and the surrounding foothills. (*B. glabella* var. *campestris.*)

Boisduvalia stricta (Gray) Greene. Scarce in moist places in the Greenhorn Range, mostly in the Douglas oak woodland.

Camissonia Boothii (Douglas) Raven ssp. **decorticans** (Hooker & Arnott) Raven. Widespread in the arid cismontane Upper Sonoran associations and the higher elevations of the Lower Sonoran grassland, usually growing in small dense colonies on gravel slides and roadbanks or in poor often gypseous soil. (*Oenothera decorticans.*)

Camissonia Boothii ssp. **desertorum** (Munz) Raven. Bottle cleaner. Occasional on the desert, mostly in bare gravelly or sandy soil, often on banks. It is infrequent west to the east slope of the Greenhorn Range. (*Oenothera decorticans* var. *desertorum.*)

Camissonia californica (Nuttall) Raven. Scarce in sandy places, around rocks, or in gravel slides in the cismontane Upper Sonoran associations, mostly in the Upper Sonoran subshrub association around the head of the valley. (*Oenothera leptocarpa.*)

Camissonia campestris (Greene) Raven. Sun cups. Common in light or sandy soils in all associations to the borders of the yellow pine forest. The plant is abundant in colorful colonies in wind-blown soils in the valley and on the desert. (*Oenothera dentata* var. *campestris.*)

Camissonia claviformis (Torrey & Frémont) Raven. Common on the desert, especially in wind-drifted or other loose light sandy soil. (*Oenothera clavaeformis.*)

Camissonia contorta (Douglas) Kearney. This hexaploid species is common in the Upper and Lower Sonoran associations, usually growing in sandy or disturbed gravelly soils, especially in creek beds, along dry washes, etc. It is occasional to the yellow pine forest and pinyon woodland, reaching its southern limits in Kern County. (*Oenothera contorta* in part.)

Camissonia dentata (Cavanilles) Reiche. A tetraploid species common in light soils in the cismontane Upper Sonoran associatons. Peter H. Raven will describe a second tetraploid species of this complex that is erect, with narrow, entire leaves, which is endemic to arid slopes in the Kern River drainage from Hobo Hot Springs to the vicinity of Weldon. (*Oenothera contorta* in part.)

Camissonia graciliflora (Hooker & Arnott) Raven. Scarce in small colonies from the Upper Sonoran grassland to the Douglas oak woodland and the chaparral. Although the plant sometimes grows in dense stands of other annuals, it usually occurs in rather bare poor gravelly soil. (*Oenothera graciliflora.*)

Camissonia hirtella (Greene) Raven. Scarce in the mountains south to Lebec, growing in disturbed soil from the higher elevations of the chaparral to the Jeffrey pine forest and the borders of the ponderosa pine forest. (*Oenothera micrantha* var. *Jonesii.*)

Camissonia kernensis (Munz) Raven. Kern sun-cups. Endemic to the arid shrub association and the pinyon woodland on the eastern slope of the mountains from Grapevine Canyon (north of Walker Pass) south to the Red Rock Canyon region and west to the vicinity of Claraville in the Piute Mountains and near Lake Isabella in South Fork Valley. It is rare in the El Paso Range east to the El Paso peaks. The plant is most common in the desert-facing canyons with deep slopes of sand and gravel; often it is locally abundant in these places. The type was collected in Freeman Canyon east of Walker Pass (Frank Peirson 8822, May, 1930). (*Oenothera kernensis.*)

The taxon known as *Oenothera kernensis* ssp. *Gilmanii* (Munz) Munz, (*Oe. campestris* var. *Gilmanii, Oe. kernensis* ssp. *mojavensis.*) A second subspecies of this species, is occasional in sandy soils in the creosote bush association.

Camissonia micrantha (Hornemann) Raven. Scarce in bare fine sandy soil around the picturesque sandstone outcrops at Alec Cook Rocks in Cedar Canyon in the Temblor Range (T1194). (*Oenothera micrantha.*)

Camissonia pallida (Abrams) Raven. Pallid sun-cups. Scarce but widespread on the desert, extending to the pinyon woodland on the eastern slopes of the mountains. (*Oenothera micrantha* var. *exfoliata, Oe. Abramsii.*)

Camissonia Palmeri (Watson) Raven. Colonies of usually only a few scattered plants are scarce in sandy soil or on clay flats in the creosote bush association; rare, west to the vicinity of Tejon Pass; very rare in the inner south Coast Ranges to San Luis Obispo County. (*Oenothera Palmeri.*)

Camissonia parvula (Nuttall) Raven. Common in the pinyon woodland from the Walker Pass region west to the Piute Mountains and Erskine Creek. (*Oenothera contorta* var. *pubens.*)

Camissonia refracta (Watson) Raven. Scarce on sandy flats in the creosote bush association. (*Oenothera refracta.*)

Circaea alpina Linnaeus var. **pacifica** (Ascherson & Magnus) Jones. Enchanter's nightshade. Rare on a shaded slope near Shirley Meadows (Howell 38,777). (*C. pacifica.*)

Clarkia cylindrica (Jepson) Lewis & Lewis. Farewell-to-spring. Common from the Upper Sonoran grassland through the Douglas oak woodland, mostly on north slopes where it blooms in dense colorful colonies in late spring. These colonies are especially widespread and conspicuous on broad slopes in the Upper Sonoran grassland in the Temblor Range. (*Godetia Bottae* var. *cylindrica.*)

Clarkia Dudleyana (Abrams) Macbride. Occasional in small colonies in rocky places in the Douglas oak woodland along the east-side foothills south to Kern Canyon. (*Godetia Dudleyana.*)

Clarkia exilis Lewis & Vasek. A narrow endemic of the Sierra Nevada foothills of Tulare and Kern counties that is scarce in rocky places in the Douglas oak woodland south to Kern Canyon. The type was collected one mile from [above] the mouth of Kern Canyon (Harlan Lewis, Margaret Ensign Lewis, and Frank Vasek 908, 19 April 1952).

Clarkia purpurea (Curtis) Nelson & Macbride ssp. **quadrivulnera** (Douglas) Lewis & Lewis. Common, often growing in dense colonies on north slopes in the Upper Sonoran grassland and in broad openings in the Douglas oak woodland. (*Godetia purpurea* var. *quadrivulnera.*)

Clarkia purpurea ssp. **viminea** (Douglas) Lewis & Lewis. A form with unusually large petals grows in small colonies that are scarce on the high slopes in the Douglas oak woodland on the north slope of the Tehachapi Mountains south of Tejon Canyon. (*Godetia viminea.*)

Clarkia rhomboidea Douglas. Occasional in chaparral and the higher elevations of the Douglas oak woodland and common in the yellow pine forest in the mountains south to the Tehachapi Mountains.

Clarkia speciosa Lewis & Lewis ssp. **polyantha** Lewis & Lewis. Occasional south to the Tehachapi Mountains, mostly growing in chaparral and shin oak brush.

Clarkia tembloriensis Vasek. Temblor clarkia. Occasional in the Temblor Range, growing in colonies on well-drained north slopes or in shady places in the Upper Sonoran subshrub and California juniper associations, usually in decomposed shale or shale talus. The plant is restricted to arid slopes along the west side of the San Joaquin Valley from the Temblor Range to Alameda County. According to Frank C. Vasek (The evolution of *Clarkia unguiculata* derivatives adapted to relatively xeric climates. Evolution, 18:26–42) *C. tembloriensis* evolved from *C. unguiculata* in response to an increasingly arid climate. Vasek's extensive field and laboratory studies of this species suggest similar processes are involved in the speciation of other members of the xerophytic San Joaquin flora. The type was collected at Carneros Rocks [about 1 mile west of the rocks at the foot of Messa Ridge] (Harlan Lewis 1028, 15 April 1956).

Clarkia unguiculata Lindley. Occasional to common in chaparral and Douglas oak woodland south to the San Emigdio Range and the Mt. Pinos region. It grows mostly in rocky places on north slopes in more mesic habitats than the closely related *C. tembloriensis* of the valley west-side. These species furnish an interesting example of floristic divergence in regions with only modest differences in rainfall. (*C. elegans.*)

Clarkia Xantiana Gray. Occasional, usually in rocky places in the Douglas oak woodland, from the east slope of the Greenhorn Range south through the Tehachapi Mountains. The type was collected at Ft. Tejon (Xantus de Vesey 31, 1857–58).

Epilobium adenocaulon Haussknecht var. **holosericeum** (Trelease) Munz. Scarce on shaded moist banks along Kern River below Lake Ming (T6600).

Epilobium adenocaulon var. **Parishii** (Trelease) Munz. Widespread in wet places from the Upper Sonoran associations to the yellow pine forest; rare in the Temblor Range (Holland Spring, McMillan 36–2 in 1957).

Epilobium brevistylum Barbey. Scarce in moist places in the yellow pine forest.

Epilobium glaberrimum Barbey. Scarce in sunny wet soil at Shirley Meadows (Howell 38,400, T8627) and near Greenhorn Mountain Park headquarters (Howell 38,455).

Epilobium oregonense Haussknecht. Scarce at Shirley Meadows (Howell 38,616; 38,796) and at a spring in the Bartolas Country on the Kern Plateau (Howell & True 41,803).

Epilobium paniculatum Nuttall. Willow-herb. Common in dry soil in the Upper Sonoran associations, occasional through the yellow pine forest. The form which has been called var. laevicaule (Rydberg) Munz is occasional with the species, usually growing in somewhat moister habitats.

Gayophytum decipiens Lewis & Szweykowski. A colony grows in disturbed sand on a sagebrush flat above Pine Flat near the southwest rim of the Kern Plateau (T9885).

Gayophytum diffusum Torrey & Gray ssp. **parviflorum** Lewis & Szweykowski. Common in the yellow pine forest, often in disturbed places. (*G. lasiospermum, G. lasiospermum* var. *Hoffmannii*).

Gayophytum eriospermum Coville. Common in the yellow pine forest in the Greenhorn Range and on Breckenridge Mountain. (*G. lasiospermum* var. *eriospermum.*)

Gayophytum heterozygum Lewis & Szweykowski. Common in the yellow pine and red fir forests in the Greenhorn Range, the Piute Mountains, and on Breckenridge Mountain. (*G. diffusum* var. *villosum.*)

Gayophytum humile Jussieu. Scarce in wet places in the red fir and ponderosa pine forests in the Greenhorn Range (1.5 mile south of Greenhorn Summit, Howell 38,416). (*G. Nuttallii.*)

Gayophytum racemosum Torrey & Gray. Scarce in wet to dryish soil in the Greenhorn Range, on Piute Mountain, and at the cienega at the head of Bitter Creek on Mt. Pinos. (*G. caesium.*)

Heterogaura heterandra (Torrey) Coville. Rather common from the Douglas oak woodland through the lower elevations of the yellow pine forest south to the Tehachapi Mountains, mostly on shaded slopes. (*Gaura heterandra.*)

Ludwigia palustris (Linnaeus) Elliott. Marsh-purslane. Scarce along Kern River east of Wofford; rare in Kern River in Kern Canyon.

Ludwigia peploides (Kunth) Raven. Yellow water-weed. Occasional in ponds, sloughs, and permanently muddy places in the valley and at the intake pools above Lake Isabella. An attractive plant that often floats in lush green masses on the deeper pools. (*Jussiaea californica, J. repens* var. *peploides.*)

Oenothera deltoides Torrey & Frémont. Desert primrose. Colonies are scarce but widespread in light sandy soil in the creosote bush association; sometimes locally common in the sand drifts around the alkali sinks.

Oenothera deltoides ssp. **cognata** (Jepson) Klein. Devil's lantern, and in the western San Joaquin Valley, sand lily. Scarce but widespread in loose soil, often wind-drifted sand, in the valley.

Oenothera Hookeri Torrey & Gray ssp. **angustifolia** (Gates) Munz. Evening primrose. Occasional at wet places in the mountains, mostly in the Douglas oak woodland. Ssp. *grisea* (Bartlett) Munz grows in the mineralized soil below Kern Hot Springs in Bath House Canyon near Kernville. (Howell 40,110). The tall well-branched plants that grow at Richbar in Kern Canyon (T6507) are ssp. *venusta* (Bartlett) Munz.

Oenothera primiveris Gray. Yellow sun cups. Rare in low sandy places on the desert (Red Rock Canyon, the alkali plains near Boron, roadside in Fremont Valley, sand flats near Inyokern).

Zauschneria californica Presl. Humming-bird fuchsia. Occasional from the woody cismontane Upper Sonoran associations to the lower levels of the yellow pine forest in the mountains and the Temblor Range, usually growing around rocks. The plants are pollinated by humming birds. The brilliant red flowers in late summer and fall give a note of bright color to otherwise sere landscapes. The plant does well in cultivation and is an excellent addition to the rock garden.

Zauschneria latifolia (Hooker) Greene. Occasional in the yellow pine forest, mostly in rocky places (*Z. californica* ssp. *latifolia.*)

Zauschneria viscosa Moxley. Occasional around rock outcrops at high altitudes in the Jeffrey pine forest on the Kern Plateau, in the Piute Mountains, and on Breckenridge Mountain. (*Z. californica* ssp. *latifolia.*)

HALORAGIDACEAE. Water-milfoil Family

Myriophyllum hippuroides Nuttall. Parrot's feather. Once known from Kern River near Hart Park (Benson 8025, POM).

UMBELLIFERAE. Parsley Family

Angelica lineariloba Gray var. **Culbertsonii** Jepson. Occasional in rocky places from the Kern Plateau and Sunday Peak south to the Piute Mountains and the summit of Breckenridge Mountain, often growing on exposed ridgetops. The plant, easily distinguished by its broad incised leaflets, was long known only from the Little Kern River in Tulare County. However, it is not uncommon in the yellow pine, lodgepole, and red fir forests of the southern Sierra Nevada in Tulare and Kern counties. (*A. lineariloba.*)

Apiastrum angustifolium Nuttall. Mock celery. Known from two widely separated colonies in fine clay; one is at the head of McClean Canyon in the Temblor Range (T1179) and the other on the hill east of Bodfish Gap at the north end of Piute Mountain (T10,946).

Apium graveolens Linnaeus. Celery. Occasionally naturalized in wet places (Lavrock Canyon and Temblor Creek in the Temblor Range; forks of San Emigdio Canyon and the lower potrero in San Emigdio Canyon; south side of Castac Lake; Kern River near Hart Park; Kelso Creek 9 miles south of Weldon). Native of Europe.

Berula erecta (Hudson) Coville. Water parsnip. Occasional in wet sunny often subalkaline places in the mountains below the yellow pine forest southeast to San Emigdio Canyon.

Bowlesia incana Ruiz & Pavon. Occasional, usually in shade, in the cismontane Upper Sonoran associations. (*B. lobata.*)

Caucalis microcarpa Hooker & Arnott. Occasional in the cismontane Upper Sonoran associations, usually growing in colonies on shaded slopes.

Cicuta Douglasii (DeCandolle) Coulter & Rose. Water hemlock; locally, erroneously, and somewhat dangerously, wild parsnip. Occasional in wet places in the mountains, mostly in Douglas oak woodland (El Paso Creek; Angel Creek near Glennville; Kern River at Kernville); less common in the yellow pine forest (Greenhorn Mountain Park; French Meadow in the Piute Mountains). This extremely poisonous plant (said to be the most toxic native North American plant) with its highly virulent tuberous roots was once much more common, but persistent control has greatly reduced its occurrence. However, it has been the cause of death of cattle in the Tehachapi Mountains as recently as 1962. Human fatalities were not uncommon in early times; Mary McDermott Doyle (A Child Went Forth, p. 357) wrote of attempting to treat Greek mine laborers in the Owens Valley who mistook the roots for parsnips. Mary Austin (Earth Horizon, p. 207) tells of its use for suicide by two Indian girls who had been kidnapped and assaulted by young white toughs at Bishop: "The Indian girls closed their part of the episode by eating wild parsnip root—the convulsions induced by that bane being mercifully shorter than the sufferings already endured." Marcia Rittenhouse Wynn (Pioneer Family of Whiskey Flat, p. 96) wrote that the children of pioneers in the Glennville region were taught never to touch the plant, so they used it to play "poison" with their dolls. (She reports the rag dolls usually recovered; if not, they were given elaborate funerals.)

Because of the superficial similarity of poisonous and non-poisonous species of the parsley family, the untrained should be wary in eating any wild member of the family.

Conium maculatum Linnaeus. Poison-hemlock. Common in dense colonies along Poso Creek in Lynns Valley and at Angel Creek at Glennville; occasional in small colonies at scattered stations elsewhere (Greenhorn Mountain Park; near Poso Mine; Lebec Road 1.6 mile north of Lebec). Although poisonous, especially the root, there are no records of local livestock fatalities and it is not considered a problem. Native of Europe.

Cymopterus panamintensis Coulter & Rose. A plant of the Death Valley region that reaches its extreme southwestern limits with isolated colonies at the summit of Mesquite Canyon in the El Paso Range, where about 30 plants grow in basalt talus on an open canyonside (T4407, 10,729), and in a metamorphic gravel slide at Salisbury Mine on the northwest flank of Government Peak in the Rand Mountains (T12,020).

Daucus Carota Linnaeus. Common carrot. A small colony, which seems well established, grows along Angel Creek at Glennville (T6389). It is also known from a roadside colony on Highway U. S. 46 west of Wasco (T8947 in 1963). Colonies, often transient, are occasional in similar places. Native of Europe.

Daucus pusillus Michaux. Rattlesnake weed. Scarce in usually small dense colonies on dry slopes and in disturbed places in the cismontane Upper Sonoran associations, most frequently in chaparral. It was widely used by the pioneers as a poultice for infections and for rattlesnake bites. Evalyn Farnsworth (oral communication) has found it effective for these purposes and for promoting healing of cuts and wounds.

Eryngium spinosepalum Mathias. Occasional from the Upper Sonoran grassland to the Douglas oak woodland in the Greenhorn foothills east to Lynns Valley and Poso Flat, usually growing in swales, roadside low places, and other situations where water stands intermittently in winter. (*E. globosum, E. Vaseyi* var. *globosum*.)

Eryngium Vaseyi Coulter & Rose. Coyote-thistle. Common in low places, especially in pool beds, in the Temblor Range; scarce in similar places in the valley.

Foeniculum vulgare Miller. Sweet fennel, anise. Although a common weed in much of California, sweet fennel is a rare adventive in Kern County. A single plant grew along Highway U. S. 99 at the Ft. Tejon overpass in 1964 (T9767) and a roadside colony of three plants grows at the Isabella alkali flats (T10,072). Native of the Mediterranean region.

Hdyrocotyle verticillata Thunberg. Known from small colonies in a seep between Lake Ming and Kern River (T6613); at Erskine Creek in the arid Douglas oak woodland at the limestone cliffs 2.8 miles south of Isabella (T8843); and along Poso Creek near Poso Mine (C. N. Smith 1430).

Lomatium californicum (Nuttall) Mathias & Constance. Hog fennel (generally applied indiscriminately to *Lomatium* species). Scarce in the mountains at scattered stations (San Emigdio Canyon just above the forks, T3769; Tehachapi Mountain Park, T7241; crest of Piute Mountain, T8904). (*Leptotaenia californica.*)

Lomatium caruifolium (Hooker & Arnott) Coulter & Rose. Alkali parsnip. Scarce in the grasslands in the low hills west of Woody (Delano Road, 9.5 miles west of Woody, T4245).

Lomatium dasycarpum (Hooker & Arnott) Coulter & Rose. Scarce in rocky places in the central Temblor Range (Upper Messa Ridge, T366; Drake Rocks above Drake Ridge, T1156).

Lomatium dasycarpum ssp. **tomentosum** (Bentham) Theobald. Occasional in the Douglas oak woodland in the Glennville-Woody region; scarce south to the Tehachapi Mountains. (*L. tomentosum.*)

Lomatium dissectum (Nuttall) Mathias & Constance var. **multifidum** (Nuttall) Mathias & Constance. Scarce in the mountains from the chaparral and Douglas oak woodland to the lower levels of the yellow pine forest. (*Leptotaenia dissecta* var. *multifida.*)

Lomatium macrocarpum (Hooker & Arnott) Coulter & Rose. Scarce but widespread, sometimes locally common, in the cismontane Upper Sonoran associations from Breckenridge Mountain and the Piute Mountains to Mt. Abel and in the central Temblor Range.

Lomatium mohavense Coulter & Rose. Desert parsley. Occasional from the desert floor to the pinyon woodland in the mountains bordering the desert; scarce in the Jeffrey pine forest. The flowers on the desert are purplish black; those in the neighboring mountains yellow (except at the summit of Mt. Abel where the flowers are purplish black).

Lomatium nevadense (Watson) Coulter & Rose var. **Parishii** (Coulter & Rose) Jepson. Scarce on the east slopes of the Sierra Nevada from the arid shrub association to the Jeffrey pine forest, west to the Piute Mountains and the dryish meadows on the Kern Plateau. The plant is also known from Cuddy Valley (Hoffmann, 13 June 1927, SBM).

Lomatium nudicaule (Pursh) Coulter & Rose. Pestle parsnip. Willis Linn Jepson (A California Flora, 2:646) cites Hall 6294 in 1905 from Ft. Tejon.

Lomatium utriculatum (Nuttall) Coulter & Rose. Wild parsley. Common in the cismontane Upper Sonoran associations, especially the grassland; east to the Joshua tree groves in Kelso Valley, ascending to the borders of the Jeffrey pine forest in the Tehachapi Mountains. It is one of the earliest plants to bloom, often in January in the warmer regions.

Lomatium Vaseyi (Coulter & Rose) Coulter & Rose. Known from a grassy hill near Caliente (Hoffmann, 28 April 1930, SBM).

Oenanthe sarmentosa Presl. Occasional in wet places in the Douglas oak woodland south to Cedar Creek in the Glennville region.

Oreonana Clementis (Jones) Jepson. According to J. T. Howell and G. H. True (Kern Plateau Botany, 1966. Ms.) it grows on the rim of the Kern Plateau east of Fay Creek Canyon at 7,200 feet elevation. In southernmost Tulare County, Howell and True collected the plant on the Cherry Hill Road 3 miles south of the Cannell Meadow intersection (41,775) and it grows on the ridge between South Dry Lake and Rattlesnake Meadow (T12,490B).

Osmorhiza brachypoda Torrey. Sweet cicely. Common from the Douglas oak woodland through the yellow pine forest in the mountains.

Osmorhiza chilensis Hooker & Arnott. A colony grows in the small dense incense-cedar woods just above Shirley Meadow (Howell 38,485, T7894, 8623, 11,542); another grows in thick shaded humus near the summit of nearby Shirley Peak (T8616). (*O. nuda.*)

Oxypolis occidentalis Coulter & Rose. Scarce in wet places in the yellow pine and red fir forests in the Greenhorn Range and the Piute Mountains.

Perideridia Bolanderi (Gray) Nelson & Macbride. Scarce in deep shade at a spring between massive granite boulders on the Wagy Flat Road 7.6 miles east of the summit of the Greenhorn Range (upper limits of the Douglas oak woodland at 4,100 feet elevation, T4476). (*Eulophus Bolanderi.*)

Perideridia Parishii (Coulter & Rose) Nelson & Macbride. Common in wet meadows in the yellow pine forest. (*Eulophus Parishii.*)

Perideridia Pringlei (Coulter & Rose) Nelson & Macbride. Common in the Douglas oak woodland and chaparral, rare in the pinyon woodland and the Jeffrey pine forest (summit of Tehachapi Peak), southwest to the upper Cuyama Valley, east to Sand Canyon near Monolith. The plant usually grows in heavy clay. (*Eulophus Pringlei.*)

Sanicula bipinnata Hooker & Arnott. Poison sanicle. Occasional in the Upper Sonoran grassland and the Douglas oak woodland from the Temblor Range through the mountains to Caliente Canyon. There is no local evidence that this plant with its fetid odor is actually poisonous; the writer once observed an extensive colony eaten by two Brahma steers which suffered no ill effects.

Sanicula bipinnatifida Douglas. Purple sanicle. Occasional in the cismontane Upper Sonoran associations, mostly in heavy soils in the grassland.

Sanicula crassicaulis Poeppig. Gambel weed. Willis Linn Jepson (Madroño, 1:111) cites Davy 2363 from Ft. Tejon.

Sanicula graveolens Poeppig. Common in the yellow pine forest in the Greenhorn Range; less common in the mountains southwest to Tecuya Ridge. (*S. nevadensis.*)

Sanicula tuberosa Torrey. Tuberous sanicle, turkey sanicle. Occasional in the Greenhorn foothills, mostly in the Douglas oak woodland.

Sphenosciadium capitellatum Gray. Button parsley, white heads. Occasional and colorful in wet sunny meadows and moist places in the high mountains south to Mt. Pinos.

Tauschia Hartwegii (Gray) Macbride. Occasional in shaded canyons in the foothills south to Kern Canyon; rare southwest to Tunis Ridge in the southwestern Tehachapi Mountains. (It is locally common under canyon live oaks on the north slopes in the Tollgate Ridge region of the northeastern Tehachapi Mountains.) (*Velaea Hartwegii*).

Tauschia Parishii (Coulter & Rose) Macbride. Occasional on the desert-facing slopes of the mountains west to the Piute Mountains and Breckenridge Mountain, southwest to Brush Mountain in the Mt. Abel region; scarce in granite sand on the arid slopes in the Jeffrey pine forest on the southern Kern Plateau. (*Velaea Parishii.*)

GARRYACEAE. Silk-tassel Family

Garrya flavescens Watson var. **pallida** (Eastwood) Bacigalupi. Silk-tassel bush. Occasional in shin oak brush and common in chaparral from the east slope (rare on the west slope) of the Greenhorn Range southwest through the mountains to the Mt. Abel region. It is scarce in the arid Jeffrey pine forest on the southern Kern Plateau at elevations of more than 7,000 feet. Vigorous new growth appears at the base of burned shrubs; these shoots are much browsed by livestock and deer. (*Garrya pallida.*)

CORNACEAE. Dogwood Family

Cornus Nuttallii Audubon. Western dogwood. A tree of great beauty when in bloom that is scarce along Sandy Creek on the west slope of Sunday Peak (T5230) and near Spout Spring along the highway just west of

Greenhorn Summit (T3823, 8471). (In 1963 one of the trees at Sandy Creek had severely shredded bark where bears had sharpened their claws.) The southern Sierran limit for the species is a tree reported by Charlotte Smith (personal communication) that grows near Mill Creek on the west slope of Basket Peak. In 1958 the late Louis Edmunds told the author this magnificent flowering tree almost never succeeds in cultivation, even in the favorable climate of the San Francisco Bay region.

Cornus glabrata Bentham. Scarce along Lumreau Creek at the Milk Ranch in the Greenhorn Range (Farnsworth 698, 756).

Cornus stolonifera Michaux. Creek dogwood. Common in the cienega below Mountain Meadow on the east slope of Sunday Peak; rare, mostly in the yellow pine forest, southwest to the head of Black Bob Canyon in the San Emigdio Range.

LENNOACEAE. Lennoa Family

Pholisma arenarium Nuttall. Rare in widely separated colonies in loose sand on the desert: in the arid shrub association in Bird Spring Canyon west of Ricardo (T3540); in the creosote bush association at Homestead on the west side of Indian Wells Valley (Mark Kerr, 29 May 1938); in Red Rock Canyon (Cecil Hart, 10 June 1926); and at Koehn Dry Lake (Benson 3514, POM).

PYROLACEAE. Indian Pipe Family

Pterospora andromedea Nuttall. Pine drops. A saprophyte that is occasional in shaded pine duff in the yellow pine and red fir forests.

Pyrola aphylla Smith. Scarlet shin-leaf. A bright scarlet plant that is rare in shaded humus in the ponderosa pine forest near the summit of the west slope of Basket Peak in the Greenhorn Range (Farnsworth 647). (*P. picta* forma *aphylla.*)

Pyrola picta Smith ssp. **dentata** (Smith) Piper. White-veined shin-leaf. Scarce in the ponderosa pine forest north of Greenhorn Pass; rare in the arid Jeffrey pine forest on the southern Kern Plateau. (*P. dentata.*)

Sarcodes sanguinea Torrey. Snow plant. Occasional, sometimes locally common, in the yellow pine forest, especially on Mt. Abel, Mt. Pinos, and Sawmill Mountain. Under Kern County Ordinance no. 250, it is illegal to collect this bright red saprophyte for any purpose.

ERICACEAE. Heather Family

Arctostaphylos glauca Lindley. Big-berried manzanita. Occasional, often in large colonies, from the San Emigdio Range and Mil Potreros to Canyon de los Gatos Montes on the southeast slope of the Tehachapi Mountains. Big-berried manzanita is an important element of the chaparral in these regions; sometimes scattered plants or small colonies also occur in the pinyon woodland and the Jeffrey pine forest. A colony of large shrubs, some fully 12 feet tall, grows in the Douglas oak woodland near the head of Bear Trap Canyon in the Tehachapi Mountains.

Arctostaphylos glauca var. **puberula** J. T. Howell. Common in the limited chaparral of the central Temblor Range. An especially fine stand occurs among the sandstone outcrops on upper Messa Ridge.

Arctostaphylos Mariposa Dudley. Mariposa manzanita. Common in the yellow pine and red fir forests south to the Piute Mountains.

Arctostaphylos nevadensis Gray. Pine-mat manzanita. This prostrate manzanita typical of the Sierran forests reaches its southern limits with numerous colonies in the red fir forest on the north slope of Sunday Peak (Howell 38,841, T3861, 12,389, C. N. Smith 730, JEPS). The manzanitas growing against the granite outcrops are one of the distinctive features of the summit.

Arctostaphylos Parryana Lemmon. Parry manzanita. Occasional from the Douglas oak woodland and the chaparral to the yellow pine forest from the Mt. Abel region through the Tehachapi Mountains northeast to the Cache Peak region. Parry manzanita is never densely colonial and often occurs as isolated individuals. The type was collected 4 miles west of Keene [actually, south, cf. W. L. Jepson, Madroño, 1:89] (John Gill Lemmon, 25 June and 5 August 1888).

Arctostaphylos patula Greene. Green-leaf manzanita. Common in the yellow pine forest in the Greenhorn Range, the Piute Mountains, and on the Kern Plateau. The bright green leaves and rich reddish brown bark add attractive summer color to the mountain slopes, especially where the plants grow around gray granite outcrops.

Arctostaphylos pungens Humboldt, Bonpland & Kunth. Mexican manzanita. Occasional from the chaparral through the Jeffrey pine forest in the Mt. Pinos region and along the east slope of the Tehachapi Mountains west of Antelope Valley.

Arctostaphylos viscida Parry. White-leaf manzanita. Scarce in the higher elevations of the Douglas oak woodland and rare in the ponderosa pine forest in the Greenhorn Range. An isolated colony grows in chaparral on Tollgate Ridge in the northeastern Tehachapi Mountains (T12,281).

Rhododendron occidentale (Torrey & Gray) Gray. Western rhododendron. Charlotte N. Smith reports (personal communication) that a colony grows at Cane Springs and another along the creek about a mile northeast of the springs in the ponderosa pine forest in the Greenhorn Range. Apparently it is not otherwise known in the Sierra Nevada south of Fresno County.

PRIMULACEAE. Primrose Family

Anagallis arvensis Linnaeus. Pimpernel. A common garden weed, especially at residences in the foothills. Native of Europe.

Androsace elongata Linnaeus ssp. **acuta** (Greene) Robbins. Rare, usually in small dense colonies, from the cismontane grasslands through the Douglas oak woodland. (*A. occidentalis* var. *acuta.*)

Dodecatheon Clevelandii Greene ssp. **patulum** (Greene) H. J. Thompson. Shooting star. Occasional from the Douglas oak woodland through the shin oak brush in the Greenhorn Range, usually growing in good often heavy soil. (*D. patulum.*)

Dodecatheon Hansenii (Greene) H. J. Thompson. Hansen's shooting star. Occasional from the Douglas oak woodland to the edge of the ponderosa pine forest on the west slope of the Greenhorn Range. (*D. Hendersonii* var. *Hansenii.*)

Oleaceae. Olive Family

Forestiera neomexicana Gray. Desert olive, delia. Occasional in moist places in canyon bottoms from the desert ranges through the arid mountain slopes to the Temblor Range; west to Kernville. The plants often form dense impenetrable thickets that are of great value for wild life protection, particularly for California valley quail. Desert-olive grows well in cultivation and is especially adaptable for espalier treatment.

Fraxinus latifolia Bentham. Oregon ash. Occasional in Kern Canyon just above Kernville (and on to the north in Tulare County) and in the Greenhorn Range to Wagy Flat; south to Caliente Canyon. The species freely intergrades with *F. velutina* Torrey var. *coriacea* (Watson) Rehder. Virtually all of our trees are intermediate in one or more characters and it is included here in *F. latifolia* somewhat arbitrarily. Philip A. Munz and J. D. Laudermilk (Aliso, 2:49) state that such introgression is not uncommon in southern California. (*F. oregana.*)

Gentianaceae. Gentian Family

Centaurium exaltatum (Grisebach) Wight. Colonies are scarce in moist sunny often subalkaline places in the Douglas oak woodland in Caliente Canyon and about Kernville.

Centaurium venustum (Gray) Robinson ssp. **Abramsii** Munz. Canchalagua. Scarce in wet and recently dried places in the Douglas oak woodland in the Greenhorn Range and at Cannell Creek on the east side of Kern Canyon 1.5 mile southeast of River Kern.

Frasera neglecta Hall. Mt. Pinos green-gentian. Scarce in the Jeffrey pine forest at Mt. Pinos. (*Swertia neglecta.*)

Frasera tubulosa Coville. Green-gentian. Occasional in the yellow pine forest in the Greenhorn Range, the Piute Mountains, and on the Kern Plateau, usually growing in sandy rather dry open areas. (*Swertia tubulosa.*)

Gentiana simplex Gray. Hiker's gentian. A scattered colony grows in wet soil at Shirley Meadows in the Greenhorn Range (T3914) and another in a small meadow on the northwest slope of Sunday Peak (C. N. Smith 1012, JEPS). Shirley Meadows has recently been devastated for winter sports development; hiker's gentian and other typical Sierran meadow plants may have been eliminated.

APOCYNACEAE. Dogbane Family

Apocynum cannabinum Linnaeus var. **glaberrimum** A. DeCandolle. Indian hemp. Occasional at low places on the plains south and west of Bakersfield from Greenfield to Greeley; local along Cedar and Lumreau creeks at the Farnsworth Ranch near Glennville.

Apocynum medium Greene var. **floribundum** (Greene) Woodson. Western dogbane. Scarce on north slopes in the yellow pine forest in the Greenhorn Range; rare at the Lumreau Creek falls in the Douglas oak woodland far below its typical range (Farnsworth 726). Edward L. Greene collected the type in the Tehachapi Mountains in June, 1889; the locality is given as "the mountains west of the Mojave Desert in Kern County."

Vinca major Linnaeus. Periwinkle; locally, myrtle. Occasionally escaped in moist waste places near gardens; perhaps sparingly established around the Stockdale Country Club near Bakersfield. Native of Europe.

ASCLEPIADACEAE. Milkweed Family

Asclepias californica Greene. California milkweed. An attractive red-flowered milkweed that is occasional from the Douglas oak woodland to the Jeffrey pine forest in the mountains. The type was collected at Ft. Tejon (Xantus de Vesey, 1857–58).

Asclepias cordifolia (Bentham) Jepson. Scarce from the chaparral to the yellow pine forest in the Greenhorn Range, the Piute Mountains, and on Breckenridge Mountain.

Asclepias eriocarpa Bentham. Woolly milkweed. Occasional from the Temblor Range through the mountains to the Piute Mountains and Breckenridge Mountain, rare in the Greenhorn Range, growing from the Upper Sonoran associations to the yellow pine forest.

Asclepias erosa Torrey. Occasional in the east-side foothills southwest to Ballinger Canyon and the northwest flank of Mt. Abel, growing mostly in the Upper Sonoran grassland, occasionally in the Douglas oak woodland. It is rare and perhaps adventive on the desert (along the road to California City; at Green Acres Station in Indian Wells Valley).

Asclepias fascicularis Decaisne. Mexican milkweed, narrow-leaved milkweed. Common, often in moist places, from the valley through the chaparral; rare in disturbed places in the yellow pine forest. A form with decidedly atypical lanceolate-ovate leaves grows in disturbed gravel at

Shirley Peak (Howell 38,867). In the early days in times of severe drouth, hungry sheep were frequently killed by eating this unpalatable milkweed when all other forage was gone. Jacob C. Martens says (oral communication) that it is still hazardous to sheep, even with good native pasture, in late summer when other green plants are lacking. Roy V. Parker (personal communication) reports that a valuable bull in the Caliente region that had ample feed mysteriously died in 1961. An autopsy showed a large ingestion of Mexican milkweed was the cause of death; apparently a plant that is ordinarily repugnant to cattle was fatally attractive to this bull. Such bovine vagaries in taste are not uncommon and plants "rarely eaten" cause more losses than are commonly believed.

Oddly enough, the young pods are edible. Mary Austin wrote (The Flock, p. 82) that they were cooked like string beans by Basque sheepherders. (*A. mexicana.*)

Asclepias vestita Hooker & Arnott. Occasional along the foothills of the Tehachapi Mountains and the bordering east-side valley plains; rare around sandstone outcrops at the central summit of the Temblor Range; southwest to upper Cuyama Valley.

CONVOLVULACEAE. Morning-glory Family

Convolvulus arvensis Linnaeus. Wild morning-glory, field bindweed. A common, troublesome, and persistent weed in irrigated fields, gardens, and moist places in the valley that is occasional along roadsides in the cismontane foothills. It is becoming a problem in grain fields in heavy soil in the northern Temblor Range and is rare in moist places about towns on the desert. Wild morning-glory is rarely found in pasture lands as it cannot withstand prolonged grazing by livestock. Native of Europe.

Convolvulus fulcratus (Gray) Greene var. **deltoides** (Greene) Jepson. Endemic from the Piute Mountains to the Mt. Pinos region, where it is occasional in the Douglas oak and pinyon woodland and scarce in the Jeffrey pine forest. Particularly fine colonies grow on the broad sand flats at Neasons Flat in the pinyon woodland of the San Emigdio Range. The type was collected at Tehachapi (Edward L. Greene, 22 June 1889).

Convolvulus longipes Watson. Occasional in the mountains, mostly on arid slopes in chaparral and around rocky places in the Douglas oak and pinyon woodlands.

Convolvulus malacophyllus Greene ssp. **pedicellatus** (Jepson) Abrams. Occasional in the pinyon woodland and scarce in the Jeffrey pine forest in the San Emigdio Range and the Mt. Pinos region.

Convolvulus simulans L. M. Perry. An extensive colony grows in the ultra-fine clay at Dry Bog Knoll at the head of Adobe Canyon in the Greenhorn foothills (C. N. Smith 308, JEPS, 1078, T10,597). It occurs in similar soil in the Pinole Hills in the Temblor Range in San Luis Obispo County very near the Kern County line (C. F. Smith 2705). (*C. pentapetaloides.*)

Convolvulus tomentellus Greene. Greenhorn morning-glory. Endemic from the Greenhorn Range near Greenhorn Pass south to the Tehachapi Mountains, usually growing in dryish places, on banks, etc. It is most common in the Jeffrey pine forest in the Piute Mountains. The type was collected in the Greenhorn Mountains, 6-7,000 feet elevation [possibly the colony along the summit road just north of Greenhorn Pass] (Edward Palmer 55, 12 June 1888).

Cressa truxillensis Humboldt, Bonpland & Kunth var. **vallicola** (Heller) Munz. Cressa, alkali weed. Common in subalkaline to alkaline often heavy soils in the valley, Temblor Range, and around the desert dry lakes. (*C. cretica.*)

Cuscuta californica Hooker & Arnott. California dodder. Occasional, usually on *Eriogonum* spp., especially *E. fasciculatum* subspecies, in all the cismontane Upper Sonoran associations. In the annual grasslands, sometimes this brilliant orange parasite will begin life on the annuals that grow and mature early, such as red-stemmed filaree, and, as these dry, grow onto the later annual wild buckwheats. This allows a much longer growing season for the dodder than it would have had from either host alone.

Cuscuta campestris Yuncker. Field dodder. A parasite on *Aster exilis* at Lake Woolloomes (near Delano) (T8958) and on *Asclepias* on the east slope of Breckenridge Mountain (T3177). (*C. arvensis* in part.)

Cuscuta Ceanothi Behr. Canyon dodder. Occasional from the chaparral to the yellow pine forest, mostly on *Eriodictyon, Artemisia,* and *Solidago.* (*C. subinclusa.*)

Cuscuta denticulata Engelmann. Scarce on shrubby Compositae on the desert slopes and canyons, west to Onyx.

Cuscuta salina Engelmann. Marsh dodder, alkali dodder. Common, often abundant, on *Suaeda* spp. in alkali sinks in the valley. It also grows on *Atriplex semibaccata* in the Cottonwood Road district near Greenfield and on *Conyza* in South Fork Valley.

Cuscuta Suksdorfii Yuncker var. **subpedicellata** Yuncker. Rare on *Gayophytum* and *Allophyllum* on the slope above Pine Flat on the trail to Little Cannell Meadow in the arid Jeffrey pine forest on the southern Kern Plateau (T9882).

Ipomoea hederacea (Linnaeus) Jacquin. A rare cotton field weed at Semitropic in 1959 (T5598) and scarce in a field of gyp corn 1.5 mile west of Wasco in 1965 (T11,555). Native of the southeastern United States.

Ipomoea purpurea (Linnaeus) Roth. Garden morning-glory. A common vineyard weed in the Delano region and a frequent escape about gardens in cities and towns. Native of tropical America.

POLEMONIACEAE. Phlox Family

Allophyllum divaricatum (Nuttall) A. & V. Grant. A small colony grows under canyon live oaks on a shady flat at 3,500 feet elevation in upper El Paso Canyon in the Tehachapi Mountains. According to the range map of Verne and Alva Grant (Aliso 3:103), it apparently also occurs in the San Emigdio Range. (*Gilia divaricatum.*)

Allophyllum gilioides (Bentham) A. & V. Grant. Common in the cismontane Upper Sonoran associations, especially in the chaparral and at higher elevations in the Douglas oak woodland; rare in the Temblor Range. (*Gilia gilioides.*)

Allophyllum integrifolium (Brand) A. & V. Grant. Occasional on north slopes in the ponderosa pine forest in the Greenhorn Range, often in disturbed places. According to Verne and Alva Grant's distribution map (Aliso 3:103) it apparently also occurs in the San Emigdio Range. (*Gilia gilioides* var. *integrifolia.*)

Allophyllum violaceum (Heller) A. & V. Grant. Scarce in the yellow pine and red fir forests. (*Gilia violacea.*)

Collomia grandiflora Douglas. Scarce in the mountains from the higher elevations of the Douglas oak woodland to the yellow pine forest. Its occurrence is sporadic and unpredictable but the plant is most common in old disturbed places in the yellow pine forest, often growing in colonies of very few plants or even as isolated individuals.

Collomia heterophylla Douglas. W. L. Jepson (A California Flora, 3:169) cites a collection by Brandegee from [old] Kernville. The nearest presently known stations are far to the north at Beach and Little Troy meadows on the central Kern Plateau in Tulare County.

Collomia linearis Nuttall. Known from Wagy Creek at the Wagy Flat Road (Hardham 2558) in the Greenhorn Range and from Toad Springs in the pinyon woodland at the head of Quatal Canyon in the Mt. Abel region (Pollard, 16 July 1949).

Eriastrum densifolium (Bentham) Mason ssp. **austromontanum** (Craig) Mason. Common in sandy places and around rock outcrops in the mountains (except the west slope of the Greenhorn Range), usually growing in the pinyon woodland and the yellow pine forest, less frequently in the chaparral. (*Hugelia densifolia* in part.)

Eriastrum densifolium ssp. **elongatum** (Bentham) Mason. Occasional on dry slopes in the mountains, mostly below the yellow pine forest in the chaparral, pinyon woodland, and the arid shrub association. Ssp. *elongatum* and ssp. *austromontanum* often intergrade in our region and determinations sometimes must be arbitrary. (*Hugelia densifolia* in part.)

Eriastrum densifolium ssp. **mohavense** (Craig) Mason. Scarce on desert slopes and flats from Red Rock Canyon to the southwestern Tehachapi

Mountains where it ascends to the Douglas oak woodland in Bear Trap Canyon. The type was collected between Rosamond and Mojave (Thomas Theodore Craig 1360), very probably in the Rosamond Hills southeast of Soledad Mountain, where several colonies grow in sandy washes. (*Hugelia densifolia* in part.)

Eriastrum eremicum (Jepson) Mason. Rare in gravel and sand in arid canyons in the eastern El Paso Range, at Government Peak in the Rand Mountains (Meng 336), and in Short Canyon north of Walker Pass. (*Hugelia eremica.*)

Eriastrum Hooveri (Jepson) Munz. Hoover star-flower. Scarce around sandstone outcrops or on shale banks in the Temblor Range (Cedar Canyon forks, T4466; Alec Cook Rocks, T1243; Messa Ridge, Annetta Carter in 1956, UC). It is rare in the valley, where apparently it grows only in wet years (between Lost Hills and Semitropic, Hoover 1801, JEPS; Oildale, Hoover 4081, JEPS). The type was collected at Shafter (Robert F. Hoover 1846). (*Hugelia Hooveri.*)

Eriastrum pluriflorum (Heller) Mason. Star flower. Common in the valley and the Temblor Range, occasional in the mountains to the edge of the yellow pine forest. The plant, with its rich blue flowers that bloom quite late in the spring when all the other colorful annuals are dry, thrives in almost incredibly xeric places such as the hot barren white shale slopes on the southwest side of the valley. The type was collected at Sunset [near Arvin] (Amos Arthur Heller 7734 in 1905). (*Hugelia Brauntonii.*)

Eriastrum pluriflorum ssp. **Sherman-Hoytae** (Craig) Mason. Desert star flower. Scarce in the arid shrub association and the pinyon woodland on the slopes bordering the desert; occasional in the dunes around the Muroc dry lakes, east to the Kramer Hills.

Eriastrum sapphirinum (Eastwood) Mason ssp. **ambiguum** (Jones) Mason. Occasional in dunes and wind-modified light soil from the creosote bush association west to the pinyon woodland in the Walker Pass region, where it usually grows in sand or on gravelly slopes. (*Hugelia virgata* var. *sapphirina.*)

Eriastrum sparsiflorum (Eastwood) Mason. Occasional on the arid slopes of the mountains southwest to the Mt. Abel region; west to Breckenridge Mountain and the east slope of the Greenhorn Range. Although occasional in the arid shrub association and the chaparral, it is most common in the pinyon woodland and the Jeffrey pine forest. The plant is rare in the Douglas oak woodland on the west slope of the Greenhorn Range. (*Gilia sparsiflora.*)

Gilia achilleaefolia Bentham ssp. **multicaulis** (Bentham) V. & A. Grant. Colonies are rare in sandy-adobe soil in the Upper Sonoran grassland in the northern Temblor Range. (*G. multicaulis.*)

Gilia aliquanta A. & V. Grant. Scarce on sandy flats and gravelly slopes in the desert-facing canyons of the mountains, from Red Rock Canyon north to Short Canyon.

Gilia brecciarum Jones. Occasional on the high desert, from the higher elevations of the creosote bush association through the pinyon woodland; west to the Piute Mountains, southwest to Cuyama Valley. The plant is rare in the Jeffrey pine forest at Mt. Pinos and in the Piute Mountains.

Gilia brecciarum ssp. **argusana** A. & V. Grant. Occasional in the creosote bush association in the El Paso and Rand mountains, west to Dove Spring in the extreme southern Sierra Nevada.

Gilia brecciarum ssp. **neglecta** A. & V. Grant. Occasional along the border of the desert from Jawbone Canyon north, west to Kelso and South Fork valleys. The type was collected 7 miles east of Isabella in Kern [South Fork] Valley (Verne and Alva Grant 9109B, 22 April 1952).

Gilia cana (Jones) Heller ssp. **speciosa** (Jepson) A. & V. Grant. Endemic and scarce from Red Rock Canyon northwest in the desert-facing canyons of the southern Sierra Nevada to Nine Mile Canyon in Inyo County. The type was collected in Red Rock Canyon by Allison Krames. (*G. tenuiflora* var. *speciosa.*)

Gilia capillaris Kellogg. Occasional in the Jeffrey pine forest on Mt. Pinos.

Gilia capitata Sims ssp. **abrotanifolia** (Nuttall) V. Grant. Globe gilia. Occasional from the cismontane Upper Sonoran associations to the yellow pine forest, sometimes locally common in loose clay on gravel banks in the Douglas oak woodland in the mountains; rare, east to Spanish Needle Creek northwest of Walker Pass. (*G. abrotanifolia.*)

Gilia clivorum (Jepson) V. Grant. Occasional in sandy washes and gravelly slopes from the northern Temblor Range south to Cuyama Valley.

Gilia filiformis Parry. A species of the deep desert ranges that is very rare in metamorphic talus just above the mouth of Last Chance Canyon in the El Paso Range (T11,857).

Gilia hutchinsifolia Rydberg. Scarce on a sandy wash in Red Rock Canyon (Thorne 31,784, RSA).

Gilia interior (Mason & A. Grant) A. Grant. Common in the Upper Sonoran associations in the mountains east to the arid shrub association in the desert-facing canyons; southwest to Cuyama Valley, north on the west slope of the Temblor Range in San Luis Obispo County to the Red Hills near Shandon. The plants from Cuyama Valley to Red Hills have been called *G. inconspicua* (Smith) Sweet ssp. *austroocidentalis* A. & V. Grant.

Gilia inyoensis Johnston. Inyo gilia. Rare in loose soil in an old fire guard on the northwest slope of Black Mountain in the Greenhorn Range (Howell 38,567, T8594). This is a highly local western range extension of a plant known from the South Fork of Kern River in Tulare County to Rock Creek in Mono County. (*G. campanulata* var. *breviuscula.*)

Gilia jacens A. & V. Grant. Endemic and scarce from Liebre Mountain in Los Angeles County and Oso Canyon in the southeastern Tehachapi Mountains northwest through the Mt. Pinos region to Ybarra Canyon in the Temblor Range. It usually grows in sandy or gravelly soils in the arid Upper Sonoran associations. The type was collected at the Kern-Ventura County line between Stauffer Junction and Mt. Pinos (Verne Grant 9980, 28 April 1957).

Gilia latiflora (Gray) Gray. Common, often in colorful colonies, on the dunes about the Muroc dry lakes and in the drifted sand in the Boron region.

Gilia latiflora (Gray) Gray ssp. **cuyamensis** A. & V. Grant. Local, mostly in sandy places, in the pinyon woodland and the chaparral from upper Cuyama Valley to the northwest flank of Mt. Abel.

Gilia latiflora ssp. **Davyi** (Milliken) A. & V. Grant. Common in the creosote bush association on the sandy plains west of Soledad Mountain; occasional southwest through the Tehachapi Mountains to Cuyama Valley; rare, north along the west side of the Temblor Range to Cholame in San Luis Obispo County. It usually grows in scattered colonies on sandy washes and flats. (*G. tenuiflora* var. *Davyi.*)

Gilia latiflora ssp. **elongata** A. & V. Grant. Scarce on the sandy plains on the west side of Indian Wells Valley; occasional in dense colonies on the sandy slopes from the east end of the El Paso Range to Searles Station in San Bernardino County.

Gilia latiflora ssp. **excellens** (Brand) A. & V. Grant. Common, often abundant, on the desert plains west to the east slope of the Piute Mountains. In wet years, the flowering plants form a great blanket of blue on the sandy plains of the creosote bush association while its haunting fragrance permeates the air; at such times it is difficult to recall how bleak and barren the region is in most months and, indeed, even in most years. Exceptionally robust and well developed plants grow on the steep slopes of granite sand along Harris grade on the east side of the Piute Mountains. The type was collected near Cameron (Katharine Layne Curran in June, 1884).

Gilia latifolia Watson. Rare in gravel in canyons in the El Paso Range and on their outflows (Last Chance Canyon, T4364, Iron Canyon, T11,892, flood plain of Mesquite Canyon near Garlock, T5051).

Gilia leptantha Parish ssp. **pinetorum** A. & V. Grant. Occasional, sometimes locally common, in the Mt. Pinos region, mostly in the Jeffrey pine forest; rare at the summit of Tehachapi Peak (T11,148). The subspecies is endemic to these places.

Gilia leptantha ssp. **Purpusii** (Milliken) A. & V. Grant. Common in the yellow pine forest south to the Piute Mountains and Breckenridge Mountain, usually growing in colonies in sunny forest openings where the plant is often the only common colorful annual. (*G. tenuiflora* var. *Purpusii.*)

Gilia leptantha ssp. **vivida** A. & V. Grant. Verne and Alva Grant (Aliso 3:239) cite a collection from the Tehachapi Mountains (Tehachapi Peak, W. R. Dudley 319), where it is apparently rare in the pine woods. The plant is otherwise known from the San Gabriel Mountains of Los Angeles County.

Gilia leptomeria Gray. Scarce in sandy places on the desert (dunes southwest of Koehn Lake, T5054).

Gilia malior A. Day. Scarce in sandy places in the creosote bush association, southwest through the arid Upper Sonoran associations to the Temblor Range. The type was collected at the buttes south of Mojave (Verne & Alva Grant 10,053, 14 March 1958).

Gilia micromeria Gray. Scarce in the creosote bush association in the desert-facing canyons (mouth of Short Canyon, T7163; Jawbone Canyon, T3542).

Gilia minor A. & V. Grant. Occasional on the desert, the Temblor Range and the bordering high plains of the San Joaquin Valley, and the upper Cuyama Valley; scarce through the creosote bush association.

Gilia ochroleuca Jones. Widespread in sandy soil on the desert, extending west to the arid Douglas oak woodland in Erskine Canyon in the Piute Mountains.

Gilia ochroleuca ssp. **bizonata** A. & V. Grant. Occasional from the Upper Sonoran grassland to the yellow pine forest in the mountains and the Temblor Range. The type was collected at the Ventura-Kern County line west of Stauffer Junction on the Mt. Pinos road (Verne & Alva Grant 16,040, 14 June 1951).

Gilia scopulorum Jones. A deep desert species that is very rare on a steep slope of rocks and gravel in semi-shade at 2,500 feet elevation in the creosote bush association in Last Chance Canyon (T11,871).

Gilia sinuata Douglas. Scarce in sandy soil on the plains and slopes at the western edge of the desert.

Gilia stellata Heller. Scarce in the El Paso Range and on the east side of Indian Wells Valley.

Gilia tenuiflora Bentham. Dense colorful colonies grow in bare sandy places in the chaparral on Messa Ridge in the Temblor Range (T1962, 1159), a southeastern range extension for this Salinas Valley plant.

Gilia tenuiflora ssp. **amplifaucalis** A. & V. Grant. This subspecies of the upper Salinas Valley reaches its southern limits in Choice Valley in the northern Temblor Range where a colony grows in Kern County in sandy-adobe soil 3/8 mile north of the Choice Valley School (T6040).

Gilia tetrabreccia A. & V. Grant. Common in the Jeffrey pine forest in the Mt. Pinos region; occasional in the pinyon woodland at the south end of the Piute Mountains, and in sagebrush on the high slopes in the southwestern Tehachapi Mountains.

Gilia transmontana (Mason & Grant) A. & V. Grant. Scarce in the creosote bush association in the Rand and El Paso mountains.

Gilia tricolor Bentham ssp. **diffusa** (Congdon) Mason & A. Grant. Bird's eye gilia. Common, often abundant, from the valley plains through the east-side foothills. It is occasional in the Douglas oak woodland in the mountains, an association in which it does not occur in the Temblor Range. This is an interesting contrast with the large number of valley plants that grow in the Temblor Range but not in the foothills of the mountains. Bird's eye gilia is often abundant in subalkaline soils; the area now farmed between Semitropic and Wasco was once noted for its spring show of hundreds of acres of this colorful plant. Great colonies still survive on the alkali waste lands between Semitropic and Alpaugh in Tulare County.

Ipomopsis aggregata (Pursh) V. Grant ssp. **Bridgesii** (Gray) V. & A. Grant. Scarlet gilia, sky rocket. Rare in the thin Jeffrey pine forest on the southwestern Kern Plateau, usually growing around granite outcrops. (*Gilia aggregata.*)

Langloisia flaviflora Davidson. Scarce in sandy places in the El Paso Range. (*L. Schottii* in part.)

Langloisia Matthewsii (Gray) Greene. Desert calico. Occasional on the desert, mostly on sandy flats; local west to South Fork Valley and the east slope of the Piute Mountains.

Langloisia Schottii (Torrey) Greene. Common in the creosote bush association; rare in shallow sand around sandstone outcrops in the Temblor Range (Alec Cook Rocks, lower Messa Ridge).

Leptodactylon pungens (Torrey) Rydberg var. **Hallii** (Parish) Mason. Colorful colonies grow around the rocks at the summits of Breckenridge Mountain, Piute Peak, Mt. Pinos, Sawmill Mountain, and Mt. Abel and around granite outcrops and blocks on the southern Kern Plateau.

Leptodactylon pungens ssp. **pulchriflorum** (Brand) Mason. An attractive matted plant of exposed ridges at Lookout Point and along the crest of Piute Mountain, around the rimrock at the southern end of the Kern Plateau, and on Mt. Owen.

Linanthus aureus (Nuttall) Greene. Golden gilia. Common, often in broad bright yellow colonies, from the desert dunes to the pinyon woodland in the bordering mountains, rare in the Jeffrey pine forest (Tecuya Ridge, T5309); west to Ballinger Canyon in the upper Cuyama Valley.

Linanthus bicolor (Nuttall) Greene. Occasional in the ranges around the valley from the Upper Sonoran grassland through the chaparral and in the pinyon woodland in the Mt. Pinos region, where it is scarce in the lower levels of the Jeffrey pine forest.

Linanthus Bigelovii (Gray) Greene. Rare in sandy soil in the desert, west to the east slope of Walker Pass and on the subalkaline flats east of Weldon. Most *Linanthus* species are colonial, many densely so; this usually grows in colonies of only four or five plants and often as isolated individuals.

Linanthus ciliatus (Bentham) Greene. Common, southwest through the mountains to the San Emigdio Range. It grows from the Douglas oak woodland through the lower elevations of the yellow pine forest, often in dense colonies on open slopes in good soil.

Linanthus dichotomus Bentham. Evening snow. Common in the valley and the Temblor Range and through the Douglas oak woodland in the mountains; local in the creosote bush association on the west side of the desert. The plant usually grows in dense drifts, the pleasingly fragrant flowers opening in the late afternoon. In the northern Temblor Range, evening snow and California poppies often grow together on north slopes in sandy adobe. These slopes are bright golden orange until late afternoon; then the poppies close, the evening snow opens, and in about an hour the golden slope becomes snowy white.

Linanthus filipes (Bentham) Greene. A brightly colored little Linanthus that is scarce in small dense colonies in heavy soil in the Greenhorn foothills, usually growing in the Douglas oak woodland but occasionally in the Upper Sonoran grassland.

Linanthus liniflorus (Bentham) Greene ssp. **pharnaceoides** (Bentham) Mason. Common in the valley and from the Temblor Range south to the San Emigdio Range and the north flank of Mt. Abel, growing mostly in the Lower and Upper Sonoran grasslands, often in dense colonies in light soil.

Linanthus montanus (Greene) Greene. Mustang-clover. Common in the Greenhorn Range and on the north slope of Breckenridge Mountain. It grows, often in dense drifts, from the Douglas oak woodland to the chaparral and shin oak brush, and occasionally in the ponderosa pine forest.

Linanthus nudatus Greene. Common from the Douglas oak woodland to the lower elevations of the yellow pine forest, south to Brites Valley in the Tehachapi Mountains; rare, southwest to Lebec. When Willis Linn Jepson (A Flora of California, 3:217) described this species as *L. Nashianus* (based on C. N. Smith 209, a collection from the Davis Guard Station in the Greenhorn Range) he was apparently confused as to the identity of Greene's plant, an unusual lapse by California's most meticulous botanist.

Linanthus Nuttalii (Gray) Greene. A very attractive shrublet that grows around rock outcrops from Saddle Springs along the crest of Piute Mountain south to the massive outcrops at Piute Peak. (*Leptodactylon Nuttallii.*)

Linanthus Parryae (Gray) Greene. Parry linanthus. Common from the creosote bush association west to the arid slopes of the San Emigdio Range, Cuyama Valley, and the east slope of the Greenhorn Range. The flowers on the desert are usually white; those in the surrounding mountains are generally blue.

Linanthus parviflorus (Bentham) Greene. Common in dense colonies in the shrubby Upper Sonoran associations from the Temblor Range south to the Mt. Abel region. A disjunct colony with lavender-blue flowers grows in granite sand in the very arid Douglas oak woodland at the steep head of Spanish Needle Creek northwest of Walker Pass (T10,900).

Linanthus pygmaeus (Brand) J. T. Howell. Occasional in small dense colonies in the chaparral on Ross and Messa ridges in the Temblor Range, usually growing in shallow sandy soil in the shade of sandstone outcrops. According to Philip A. Munz (A California Flora, p. 509), it also occurs in the Tehachapi Mountains. (*Gilia pusilla.*)

Linanthus serrulatus Greene. Known from the Douglas oak woodland in the hills near Glennville at 3,200 feet elevation (Howell 5132).

Navarretia divaricata (Torrey) Greene. Dense colonies grow in dry disturbed loamy soil at the borders of Shirley Meadows. It is also known from 1.5 mile southeast of Greenhorn Pass (Howell 38,421). Some native plants benefit from even the most egregious of man's devastating activities; *N. divaricata* has been particularly abundant since Shirley Meadows was virtually destroyed in 1963 for the construction of a toboggan slide.

Navarretia intertexta (Bentham) Hooker. A scattered colony grows in the bed of the vernal pool one mile southwest of French Ranch (near Oak Flat) in the Greenhorn Range (T11,551, C. N. Smith 1390, JEPS, 1393, JEPS, Myrick 1114, SBBG; all in 1965). The plant is scarce in moist heavy black soil below a spring just south of the Bernard Ranch about four miles below the summit of Eugene Grade (C. N. Smith 1463, JEPS).

Navarretia mitracarpa Greene. Occasional in heavy soils in the Temblor Range south to the summit of Black Canyon.

Navarretia nigellaeformis Greene. Charlotte N. Smith reports (oral communication) colonies with plentiful plants in good years are occasional in heavy soil in Long Tom Gulch between Long Tom Mine and the Hugh Smith Ranch in the Douglas oak woodland in the Greenhorn Range (C. N. Smith 1181, 1183).

Navarretia peninsularis Greene. A plant primarily of the southern California mountains that is known from Mt. Pinos (Hall 6615, UC) and is abundant in disturbed loam in the meadow at Breckenridge Public Camp in the yellow pine forest on the north slope of Breckenridge Mountain (T11,241). (*N. divaricata* var. *peninsularis.*)

Navarretia pubescens (Bentham) Hooker & Arnott. Colonies are scarce in heavy soil in the Douglas oak woodland in the Greenhorn Range.

Navarretia setiloba Greene. A rare endemic of heavy soils from Grasshopper Flat in the southwestern Tehachapi Mountains north to a station near Jack Ranch in extreme southern Tulare (or possibly extreme northern Kern) County. Other colonies are known from Poso Station, in Rancheria Gulch in the low Greenhorn foothills, and between Walker Basin and Hackberry Canyon (a fork of Caliente Canyon). The type was collected on the west slope of the divide between Bodfish and Havilah (Frederick Vernon Coville & Frederick Funston 1049, 23 June 1891).

Phlox diffusa Bentham ssp. **subcarinata** Wherry. White phlox. A matted plant of exposed high elevations that grows in colorful drifts on the broad open summit plateau of Mt. Pinos. Colonies are scarce in the rocks along the crest of Piute Mountain. (*Phlox Douglasii* in part.)

Phlox gracilis (Hooker) Greene. Slender phlox. Widespread and common in the cismontane Upper Sonoran grassland; occasional to the yellow pine forest, where it usually grows in disturbed places. (*Microsteris gracilis.*)

Polemonium micranthum Bentham. Known from near Tehachapi (Eastwood & Howell, 6 May 1941.)

Hydrophyllaceae. Waterleaf Family

Emmenanthe penduliflora Bentham. Whispering bells. Occasional in the cismontane woody Upper Sonoran associations, especially in disturbed places in the chaparral; scarce in canyons and on slopes in the desert. It is often abundant following fires.

Emmenanthe penduliflora var. **rosea** Brand. A pink-flowered form of disjunct distribution in the inner south Coast Ranges that is known in Kern County from the slopes west of Cuddy Valley (Dudley & Lamb 4503, DS). The type was collected at nearby Stauffer in Ventura County (Harvey Monroe Hall 6446 in 1905). Although W. L. Jepson states (A Flora of California, 3:283) that this is apparently the common form on Mt. Pinos, all of the colonies observed by the writer are the typical yellow-flowered variety.

Eriodictyon californicum (Hooker & Arnott) Torrey. Yerba santa. Common in the chaparral south to Breckenridge Mountain and the north end of Piute Mountain, especially in old disturbed places in rocky soil.

Eriodictyon crassifolium Bentham var. **denudatum** Abrams. Occasional in the pinyon woodland in the San Emigdio Range and the Mt. Pinos region.

Eriodictyon crassifolium var. **nigrescens** Brand. A southern California variety that reaches its northern limits from the summit of Grapevine Canyon through Castac Valley and Cuddy Canyon to Frazier Park.

Eriodictyon tomentosum Bentham. Woolly yerba santa. Two small colonies grow in chaparral on Ross Ridge in the Temblor Range, the southeastern limits for the species. (*E. crassifolium* var. *niveum.*)

Eucrypta chrysanthemifolia (Bentham) Greene. Common in shade in the cismontane Upper Sonoran associations, often forming dense colonies under Douglas oaks. (*Ellisia chrysanthemifolia.*)

Eucrypta chrysanthemifolia var. **bipinnatifida** (Torrey) Constance. Scarce, mostly in the shade of rocks, from the desert ranges west to South Fork Valley.

Eucrypta micrantha (Torrey) Heller. Scarce in small colonies in the shade of metamorphic outcrops in Last Chance Canyon in the El Paso Range; rare on the northwest flank of Government Peak in the Rand Mountains. (*Ellisia micrantha.*)

Hesperochiron pumilus (Grisebach) Porter. Known from west of Cummings Valley in the Tehachapi Mountains (Yuassy 336, POM). It is also known from a spring near Lockwood Valley in extreme northern Ventura County (Hoffmann, 5 June 1930, SBM).

Hydrophyllum occidentale (Watson) Gray. Waterleaf. Occasional in shade or on north slopes in the yellow pine forest south to the Tehachapi Mountains (where it sometimes occurs at the higher elevations of the Douglas oak woodland).

Lemmonia californica Gray. Occasional to scarce in disturbed soils in the cismontane Upper Sonoran associations; scarce in the yellow pine forest, usually in dry soil around meadows.

Nama demissum Gray var. **deserti** Brand. Occasional on the desert, west to Kelso Canyon, usually growing in colonies on sandy flats or washes.

Nemophila maculata Bentham. Mountain five-spot. Occasional in the Greenhorn Range, growing in colorful colonies on open slopes in good soil in the ponderosa pine forest; scarce in the chaparral and shin oak brush.

Nemophila Menziesii Hooker & Arnott. Baby-blue-eyes. A common and most attractive spring annual in sandy soil in the cismontane Upper Sonoran associations. Scattered colonies occur east to the desert-facing canyons of the mountains. The flowers are a bright blue of uncommon intensity; Alice Eastwood (Zoe 4:145) described the blossoms at San Emigdio in 1893 as "bits of the sky drawn down to the earth."

Nemophila Menziesii ssp. **integrifolia** (Parish) Munz. Very rare in heavy clay at the summit of the central Temblor Range (high hill above the summit of McLean Canyon, T425).

Nemophila parviflora Douglas. Occasional in the Douglas oak woodland and chaparral in the Greenhorn Range, usually growing in small but dense colonies in good soil on north slopes.

Nemophila parviflora var. **quercifolia** (Eastwood) Chandler. Occasional from the Douglas oak woodland through the yellow pine forest south through the Tehachapi Mountains, usually growing at the base of rock outcrops.

Nemophila pedunculata Douglas. Occasional in the cismontane Upper Sonoran shrubby associations; scarce on the desert-facing slopes of the mountains. The pedicels recurve sharply as the fruiting capsules mature, often effectively burying the seed in the light soils in which the plant usually grows.

Nemophila pulchella Eastwood. Occasional in dense colorful bright violet-blue colonies on north slopes from the Douglas oak woodland to the yellow pine forest in the Greenhorn Range and on Breckenridge Mountain.

Nemophila pulchella var. **Fremontii** (Elmer) Constance. Occasional on north slopes in the Douglas oak woodland in the Glennville-Woody region; scarce east to Onyx and southwest to the San Emigdio Range.

Nemophila pulchella var. **gracilis** (Eastwood) Constance. Scarce in colonies in the shade of shrubs, on shady banks, etc., in the Douglas oak woodland in the Woody-Glennville region.

Nemophila spatulata Coville. Scarce in damp usually shaded places, often on north slopes, from the shin oak brush to the yellow pine forest.

Phacelia affinis Gray. Scarce in the pinyon woodland in the Mt. Pinos region; rare in the Douglas oak woodland in the Tehachapi Mountains (Water Canyon on the north slope of Tehachapi Peak, elevation 5,500 feet, T7247).

Phacelia austromontana J. T. Howell. Occasional in metamorphic sand in the small openings in the dense brush on the steep exposed west slope of Mt. Pinos (T6319).

Phacelia bicolor Torrey. Rare in the creosote bush and arid shrub associations in the Indian Wells Valley region (Indian Wells Valley, Kirby 721; Short Canyon, Kirby 1279).

Phacelia cicutaria Greene. Caterpillar phacelia. Common, especially around rock outcrops, in the Upper Sonoran associations in the mountains; east to Cactus Butte and Robbers Roost on the borders of the desert. (*P. hispida* var. *cicutaria.*)

Phacelia ciliata Bentham. Blue-flower, valley phacelia. Abundant in the Temblor and San Emigdio ranges, in the valley in good years, and occasional in the foothills of the mountains. It grows in a variety of soils but is especially dense in heavy or fine clay on open slopes, frequently coloring large areas a rich blue. These often form spectacular patterns with the bright yellow of equally dense colonies of *Monolopia lanceolata* that grow in lighter soils. Blue-flower is a weed in dry-farm grainfields in the Temblor Range, where it is often abundant in late winter and early spring in summer-fallowed fields that have not been weeded.

Phacelia Congdonii Greene. Occasional in colonies from the chaparral to the yellow pine forest south to the Tehachapi Mountains. It grows most often in fractured rock or in gravel in sunny places.

Phacelia crenulata Torrey. An extensive colony grows in metamorphic talus in Iron Canyon in the El Paso Range (T11,981, T11,899). It is common in favorable years in the desert ranges to the east. The colorful bright lavender flowers belie the vile smell of the foliage.

Phacelia curvipes Torrey. Common from higher elevations in the chaparral and the pinyon woodland to the Jeffrey pine forest from Breckenridge Mountain and the Piute Mountains south to Mt. Pinos.

Phacelia Davidsonii Gray. Occasional in the Jeffrey pine forest on Mt. Pinos and in the chaparral north to the Piute Mountains. The type was collected in Kern County by George Davidson. (*P. curvipes* var. *macrantha.*)

Phacelia distans Bentham. Fern-phacelia. Common in all the cismontane associations to the chaparral; infrequent in the shin oak brush and the lower elevations of the yellow pine forest. Colonies are occasional and sometimes locally common under shrubs on the desert. The plant is highly variable; the colonies on the flood plains and fans along the El Paso Range are var. *australis* Brand (mouth of Bond Buyer Canyon, T6775; east fork of Red Rock Canyon, T11,906).

Phacelia Douglasii (Bentham) Torrey. Douglas phacelia. Colonies are common in the Lower Sonoran grassland and occasional in all other cismontane associations below the yellow pine forest.

Phacelia egena (Brand) J. T. Howell. Rock phacelia. Occasional but widespread in many associations, from the Upper Sonoran grassland in the Temblor Range to the highest elevations of the mountains, having the greatest ecological range of any of our phacelias. It usually grows around dry sunny outcrops but is occasional on open gravelly slopes in the Temblor and San Emigdio ranges.

Phacelia Fremontii Torrey. Frémont's phacelia. Common in all the upper Sonoran associations, occasional to the yellow pine forest and in the Lower Sonoran grassland and the creosote bush association. The plant is common in favorable years and its flowers especially vivid on the semi-stabilized dunes around the Muroc dry lakes and on the sandy plains east of Ridgecrest. The form with small flowers that grows at higher elevations, especially in the pinyon woodland, is P. Hallii Brand.

Phacelia humilis Torrey & Gray var. **Dudleyi** J. T. Howell. Tehachapi phacelia. A rare variety that is known from the Jeffrey pine forest in the

Tehachapi Mountains and the head of Mill Creek on the north slope of Breckenridge Mountain. In both places this distinctive variety is very local in dense colonies in disturbed or loose loam on north slopes (bank of the fire trail on the north slope of Tehachapi Peak at 6,700 feet elevation, T7233, 11,124; northwest slope of the summit ridge of Breckenridge Mountain, elevation 7,340 feet, T11,218). Anstruther Davidson wrote (Muhlenbergia, 4:65–68, 1908) that the plant "literally covered" the slopes of Cummings Mountain in many places in June, 1907.

Var. *Dudleyi* is distinguished from the typical variety by its longer stamens, larger seeds, and much longer calyx lobes, and perhaps merits specific recognition. The type was collected in the vicinity of Bisses Station in the Tehachapi Mountains (William Russel Dudley 338, 27 June 1895).

Phacelia mohavensis Gray. Mojave phacelia. This inappropriately named little phacelia of the Sierra Nevada of Kern and Tulare counties and of the San Gabriel and San Bernardino mountains is scarce in usually dense colonies in sunny moist places in the yellow pine forest in the Greenhorn Range, the Kern Plateau, Breckenridge Mountain, and the Piute Mountains.

Phacelia mutabilis Greene. Occasional in dense colonies in wet places in the ponderosa pine forest in the Greenhorn Range. The colony at Shirley Meadows is particularly impressive.

Phacelia Nashiana Jepson. Charlotte's phacelia. A phacelia with stunning deep blue flowers with white throats that is rare in talus on the slopes of the steep arid desert-facing canyons from Jawbone Canyon north to Nine Mile Canyon in southernmost Inyo County. It grows on slopes of deep decomposed granite gravel or sand in the arid shrub association, reaching the pinyon woodland on the east slope of Morris Peak and at the type locality. Charlotte's phacelia is rare in metamorphic talus near the mouth of Last Chance Canyon in the El Paso Range; in 1966 a few waifs grew on the broad fan below that desert canyon (Meng 214).

The type was collected in Nine Mile Canyon, 6,000–6,250 feet elevation (Charlotte Nash Smith 160, 5 May 1940). Mrs. Smith writes (personal communication) of the type collection: "In 1935, Mr. V. L. Haynes, who was road foreman, found a plant of this in Jawbone Canyon, and brought it to me at Cantil. As it was so lovely, I pressed it and kept it. (I had never heard or dreamed of such a thing as collecting, that there could be any desire by the botanists for specimens, or had I any idea how to press and dry them.) I didn't see the plant again until 1940, when I found the Nine Mile colony, the only one I ever saw on the desert."

Phacelia pachyphylla Gray. Scarce in small but sometimes dense colonies in compacted gravel, in metamorphic talus, and on sesmotite banks in canyons in the El Paso Range. Philip A. Munz (A California Flora, p. 541) attributes the plant to alkaline flats; this is true of none of the Kern County colonies.

Phacelia Purpusii Brandegee. Purpus phacelia. Although usually scarce in the yellow pine forest in the Greenhorn Range and on the southern Kern

Plateau, in some years it is common in dense colonies in disturbed places, especially on slopes that have recently been subjected to the rigors of logging operations or other places with severely disturbed loamy soil.

Phacelia ramosissima Douglas var. **suffrutescens** Parry. Common from the woody Upper Sonoran associations to the yellow pine forest, east to the large rock outcrops on the borders of the desert. The plant grows at virtually every rock outcrop in the mountains and is common in other protected places. Var. *valida* Peck grows at Lookout Point (T7460) and at the summit of Harris Grade east of Landers Meadow (T7441) in the Jeffrey pine forest in the Piute Mountains.

Phacelia rotundifolia Torrey. Round-leaved phacelia. Scarce in shaded cracks that are probably moist in winter in the large metamorphic outcrops on the east side of the mouth of Last Chance Canyon in the El Paso Range (T11,849, 11,864).

Phacelia stimulans Eastwood. Scarce from the pinyon and Douglas oak woodlands to the yellow pine forest in the Tehachapi Mountains and the Mt. Pinos region, usually growing in sunny dry rocky or gravelly places. (*P. imbricata* in part.)

Phacelia tanacetifolia Bentham. Caterpillar phacelia, fiddleneck phacelia. Common in all associations below the yellow pine forest, especially the valley, the Temblor Range, and the desert.

Phacelia vallis-mortae J. Voss. Death Valley phacelia. Common around shrubs on the desert, especially in the arid shrub association.

Phacelia vallis-mortae var. **heliophila** (Macbride) J. Voss. Rare in shale or gravel on slopes in the Douglas oak woodland around the head of the valley. The type was collected at Sunset [near Arvin] (Amos Arthur Heller 7730 in 1905).

Pholistoma auritum (Lindley) Lilja. Fiesta flower. Occasional in shade in the Douglas oak woodland and shin oak brush south through the Tehachapi Mountains, often growing around rocks. (*Nemophila aurita, Ellisia aurita.*)

Pholistoma membranaceum (Bentham) Constance. White fiesta flower. Common in shade in all the cismontane Upper Sonoran associations, often growing in dense masses, rare under shrubs in desert canyons. (*Ellisia membranacea.*)

Tricardia Watsonii Torrey. Occasional on the desert, usually growing under or through shrubs.

Turricula Parryi (Gray) Macbride. Rare in the chaparral in the mountains north to the Walker Pass region and on the east slope of the Greenhorn Range. It is most common following fires or in recently graded or bulldozed places. (*Nama Parryi.*)

BORAGINACEAE. Borage Family

Allocarya acanthocarpa Piper. Adobe allocarya. Common in winter-wet often alkaline soils in the valley and the Temblor Range. The type was collected at Caliente by Townshend Stith Brandegee. (*Plagiobothrys acanthocarpus.*)

Allocarya bracteata Howell. Common in the Glennville region and in Bear Valley in the Tehachapi Mountains, growing in poolbeds, often in dense colonies. (*Plagiobothrys bracteatus.*)

Allocarya hispidula Greene. Occasional in meadows in the yellow pine forest in the mountains south to Saddle Springs at the north end of Piute Mountain and on the north slope of Breckenridge Mountain. (*Plagiobothrys hispidulus.*)

Allocarya leptoclada Greene. Occasional in the valley and the Temblor Range in vernal poolbeds that are often quite alkaline; rare in the alkali sinks in the Rosamond region. (*Plagiobothrys leptocladus.*)

Amsinckia Douglasiana A. DeCandolle. Fireweed, fiddleneck. Abundant in the Temblor Range; common in the valley; occasional through the Douglas oak woodland in the Tehachapi Mountains, where it grows at elevations as high as 6,400 feet on the high slopes south of Tejon Canyon. Fireweed is one of the spectacular sights in a good spring in the Temblor Range, where it often colors great expanses of broad slopes brilliant orange. However, the light soils of these slopes accumulate minerals in periods of extended drouth. When these drouth cycles are followed by a winter of unusually heavy rainfall, the lush growth of *Amsinckia* has a high nitrate content, causing severe losses of cattle (1952, 1961). The local common name, fireweed, may have originated from the fiery color of the masses of plants in bloom, or it may have been less sympathetically inspired by the violent itching caused by haying or harvesting in fields infested by the plant.

Amsinckia intermedia Fischer & Meyer. Common, often abundant, in sunny places with light soils in all the cismontane associations below the yellow pine forest; occasional, sometimes locally common, on the desert. The cismontane colonies are often very heavily infested by a mildew that seems to have no serious effect on the plant's growth or fruiting.

Amsinckia intermedia var. **californica** (Suksdorf) Jepson & Hoover. Occasional, sometimes in fairly dense colonies on broad slopes, in the east-side foothills south to the west slope of the Tehachapi Mountains. The type was collected at Caliente by Katharine Curran Brandegee.

Amsinckia intermedia var. **Eastwoodae** (Macbride) Jepson. The common form in the Douglas oak woodland in the southern Sierra Nevada. Unlike the typical variety, it is not densely colonial, often occurring as scattered plants on grassy slopes or as small colonies around sunny rock outcrops.

Amsinckia Menziesii (Lehmann) Nelson & Macbride. Scarce to rare in the Temblor Range, usually occurring as scattered or isolated individuals in dense stands of other species. (*A. Helleri, A. retrorsa.*)

Amsinckia tessellata Gray. Common in the desert; occasional to the pinyon woodland (although sometimes locally common on slopes with light soil); abundant in dense colonies on the high arid slopes of the San Emigdio Range and on white shale slopes in the Temblor Range, otherwise scarce in the hills around the valley.

Amsinckia vernicosa Hooker & Arnott. Scarce in the Temblor Range in usually dense colonies in light often gypseous soils; sometimes locally common on steep slopes of white shale gravel.

Coldenia Nuttallii Hooker. Scarce in loose sand in low places in the creosote bush association; rare in South Fork Valley (dunes at the southwest end of Rogers Dry Lake, T10,820; dunes at the south end of Koehn Lake, T5056; east fork of Red Rock Canyon, T3532; South Fork Valley, 1.6 mile east of Onyx, T10,941).

Coldenia plicata (Torrey) Coville. Rare in deep sand in the dunes at the south end of Rogers Dry Lake (T10,821) and at the northeast end of Buckhorn Dry Lake (T10,695, 10,842).

Cryptantha affinis (Gray) Greene. Common in the ponderosa pine and red fir forests in the Greenhorn Range.

Cryptantha angustifolia (Torrey) Greene. Scattered plants grow in the wind-blown sand in the shadscale scrub along County Line Road on the plains east of Ridgecrest, the western limit for this common species of the deep desert (T11,834).

Cryptantha barbigera (Gray) Greene. Occasional in the creosote bush association and scarce in the arid shrub association and the pinyon woodland, southwest to the Mt. Pinos region, west to Lake Isabella.

Cryptantha circumscissa (Hooker & Arnott) Johnston. Common on the desert, west to the Jeffrey pine forest; occasional in the Upper Sonoran associations on the west slope of the mountains and in the Temblor Range. It usually grows in light sandy soils. (*Greeneocharis circumscissa.*)

Cryptantha circumscissa var. **hispida** (Macbride) Johnston. Scarce in sandy soil from South Fork Valley near Weldon to the Jeffrey pine forest in the Piute Mountains and on the Kern Plateau.

Cryptantha decipiens (Jones) Heller. Common in the creosote bush association on the sandy plain along Silver Queen Road northeast of Soledad Mountain (T4342); rare at Alphie Springs in a fork of Jawbone Canyon (T9332) and in the east fork of Red Rock Canyon (T11,905). W. L. Jepson (A Flora of California, 3:342) cites a collection by Katharine Brandegee from Kernville.

Cryptantha decipiens var. **corollata** Johnston. Ivan M. Johnston (Contr. Gray Herb. 74:61) cites a collection from Ft. Tejon (Xantus de Vesey 85 in 1857–58). (*C. corollata.*)

Cryptantha dumetorum Greene. Scarce on the desert, usually growing under creosote bushes, sometimes through other shrubs. Although strictly a desert plant, it only succeeds in years with better than average rainfall. The type was collected at Tehachapi Pass [near Cameron northwest of Mojave] (Katharine Layne Curran in 1884). (*C. intermedia var. dumetorum.*)

Cryptantha echinella Greene. The most common cryptantha of the mountains, growing from the Douglas oak and pinyon woodlands to the highest elevations. It is scarce in the Upper Sonoran grassland at Middle Water Spring in the low foothills west of Woody.

Cryptantha flaccida (Douglas) Greene. Common from the Douglas oak woodland through the chaparral in the mountains and the Temblor Range.

Cryptantha incana Greene. Known from the yellow pine forest on the northwest slope of Black Mountain in the Greenhorn Range (Howell 38,545).

Cryptantha intermedia (Gray) Greene. Common in the Upper Sonoran associations from the Temblor Range south to the Mt. Abel region.

Cryptantha micrantha (Torrey) Johnston. Common in sandy soil on the desert; rare in sand at Little Cannell Meadow in the Jeffrey pine forest on the Kern Plateau.

Cryptantha micrantha var. **lepida** (Gray) Johnston. Scarce in the yellow pine forest in the Greenhorn Range (northwest slope of Black Mountain, Howell 38,565), at Mace Meadow in the Piute Mountains (T3054), and, according to Philip A. Munz (A California Flora, p. 567), at Mt. Pinos.

Cryptantha microstachys (Greene) Greene. Tejon cryptantha. Scarce in the chaparral and Douglas oak woodland in the mountains south to Lebec. The type was collected at Ft. Tejon (Xantus de Vesey 84 in 1857–58).

Cryptantha mohavensis (Greene) Greene. Scarce along the desert-facing slopes of the mountains from Tecuya Ridge northeast to Inyo County; west in the Tehachapi Mountains to the east slope of Bear Mountain. The type was collected in lower Tehachapi Pass [below Cameron, northwest of Mojave] by Katharine Layne Curran in 1884.

Cryptantha muricata (Hooker & Arnott) Nelson & Macbride. Occasional to common in the cismontane Upper Sonoran associations in the mountains and the Temblor Range, scarce in the Jeffrey pine forest. (*C. muriculata.*)

Cryptantha muricata var. **denticulata** (Greene) Johnston. Known from the yellow pine forest along the crest of Piute Mountain (Breedlove 3853) and from Black Mountain in the Greenhorn Range (Howell 38,251).

Cryptantha muricata var. **Jonesii** (Gray) Johnston. Rare in the pinyon woodland on Harris Grade 4.4 miles east of Landers Meadow on the east slope of the Piute Mountains (T7445). W. L. Jepson (A California Flora, 3:338) cites a collection by T. S. Brandegee from Kernville. (*C. vitrea.*)

Cryptantha nemaclada Greene. Scarce in light or sandy soils from the Tehachapi Mountains through the Mt. Pinos region to upper Cuyama Valley and Reyes Station at the summit of Maricopa Grade; rare in the Temblor Range in ultra-fine friable clay (dry bog adobe).

Cryptantha nevadensis Nelson & Kennedy. Common on the desert, usually growing in loose sand under shrubs; rare west to the east slope of the Greenhorn Range. (*C. barbigera* var. *inops.*)

Cryptantha nevadensis var. **rigida** Johnston. Common in the shrubby Upper Sonoran associations in the Temblor Range; occasional south and east through the mountains to the desert ranges where it is scarce. It is rare in good years in the Lower and Upper Sonoran grasslands. Plants in the east-side foothills approach *C. intermedia* (0.25 mile east of the mouth of Kern Canyon, T5954); those in the Tehachapi region are intermediate with *C. barbigera* (old Tehachapi Road, 2.5 miles southeast of Keene, T7166).

Cryptantha oxygona (Gray) Greene. Common in the creosote bush association, especially at its higher elevations; occasional in the desert-facing slopes and canyons of the mountains, west to Bodfish and southwest to the Mt. Pinos region in the Upper Sonoran associations; occasional in the Temblor Range.

Cryptantha pterocarya (Torrey) Greene. Common in canyons and on slopes and flood plains in the desert; occasional west to the pinyon woodland. Known from Messa Ridge in the Temblor Range (Hardham in 1955).

Cryptantha pterocarya var. **cycloptera** (Greene) Macbride. Rare in the mountains bordering the desert from Spanish Needle Creek northwest of Walker Pass to the south end of the Piute Mountains [head of Spanish Needle Creek, T10,920; west of Walker Pass summit at 5,200 feet elevation (growing with the typical variety), Howell 37,244A; south fork of upper Back Canyon, T9502].

Cryptantha pterocarya var. **Purpusii** Jepson. Rare at the head of Indian Wells Canyon (T7153) and at 4,800 feet elevation on the west slope of Walker Pass (T10,934). Both stations are at the upper levels of the arid shrub association.

Cryptantha similis Mathew & Raven. Scarce in sandy places in the creosote bush association, where it usually grows with *C. circumscissa* (from which it is easily distinguished by its larger corollas). (*C. circumscissa* in part; cf. Mathew, Kunjamma & Peter H. Raven, Madroño, 16:169).

Cryptantha simulans Greene. Common from the higher elevations of the Douglas oak and pinyon woodlands through the Jeffrey pine forest in the mountains south to Mt. Pinos.

Cryptantha sparsiflora (Greene) Greene. Scarce in the chaparral and Douglas oak woodland on the east slope of the Greenhorn Range and along Erskine Creek in the Piute Mountains.

Cryptantha Torreyana (Gray) Greene. Scarce from the shin oak brush to the ponderosa pine forest in the Greenhorn Range and in the arid Jeffrey pine forest on the southern Kern Plateau, usually growing in disturbed soil.

Cryptantha utahensis (Gray) Greene. Rare in Last Chance Canyon in the El Paso Range, where it grows in gravelly soil along a ridgetop at 2,800 feet elevation (T4426), and in compacted metamorphic gravel on the northwest slope of Government Peak in the Rand Mountains (T12,030). An anomalous station for this desert species is in the well developed Douglas oak woodland of the southwest flank of the Greenhorn Range, where it grows on a steep graded roadbank on Rattlesnake Grade 0.5 mile north of the Record Ranch buildings at 4,500 feet elevation on the north slope of Oak Flat Peak (C. N. Smith 1189 in 1963, JEPS).

Cynoglossum occidentale Gray. Hound's tongue. Scarce in the ponderosa pine forest south to Black Mountain in the Greenhorn Range; rare on the southwestern rim of the Kern Plateau.

Hackelia velutina (Piper) Johnston. Stick-seed. Rare in the red fir forest on the north slope of Sunday Peak (T4642). (*Lappula velutina.*)

Heliotropium curassavicum Linnaeus var. **oculatum** (Heller) Johnston. Common in winter-wet often subalkaline low places, sandy washes, etc., in all associations below the chaparral. It is especially common on canal banks in the valley where it often forms dense colonies. Although primarily a plant of moist soil, once established it can survive and even thrive in dry bare places. What would appear to var. *curassavicum,* with glabrous leaves that are not succluent, grows in a colony mixed with var. *oculatum* at the mouth of Red Rock Canyon (T6469) (about two-thirds of the plants are this form and one-third var. *oculatum*).

Myosotis laxa Lehmann. A rare weed in California that is occasional along the banks of Kern River from Kernville to Lake Isabella and in the dense growth in the wet mineralized soil at nearby Kern Hot Springs. Native of the eastern and northwestern United States, possibly as far south as Plumas and Humboldt counties.

Pectocarya heterocarpa (Johnston) Johnston. Scarce in all the cismontane associatons below the yellow pine forest and on the sandy desert plains. The species is doubtfully distinct from the closely related *P. penicillata.*

Pectocarya linearis DeCandolle var. **ferocula** Johnston. Common in sandy places in all associations below the yellow pine forest.

Pectocarya penicillata (Hooker & Arnott) A. DeCandolle. Common in sandy soils in all associations below the yellow pine forest, especially in the Upper and Lower Sonoran grasslands.

Pectocarya pusilla (A. DeCandolle) Gray. Scarce in the chaparral and shin oak brush in the Greenhorn Range.

Pectocarya recurvata Johnston. Occasional in rocky or gravelly places in the desert ranges, mostly on canyonsides or around the desert buttes, west to the arid Douglas oak woodland at the head of Spanish Needle Creek (northeast of South Fork Valley).

Pectocarya setosa Gray. Common in rocky or gravelly places on the desert, west to Kernville; scarce in the California juniper association in the Temblor Range.

Plagiobothrys arizonicus (Gray) Greene. Arizona pop-corn flower. Common in the foothills around the valley (to the yellow pine forest on Breckenridge and Bear Mountain) and in the desert canyons. It is occasional on the desert plains and in the Lower Sonoran grassland in the valley. All species of *Plagiobothrys* are charged with purple dye; this is especially true of *P. arizonicus.*

Plagiobothrys canescens Bentham. Valley pop-corn flower. Common in the Lower and Upper Sonoran grasslands and on open slopes in the Douglas oak woodland; scarce on the plains at the western edge of the desert. The large seeds are nutritious. Plants such as this make a contribution to the unusual quality of the forage of the San Joaquin Valley and the surrounding foothills that is frequently overlooked by people concerned with range management.

Plagiobothrys infectivus Johnston. An extensive colony grows on a broad slope of white shale at the summit of Ybarra Canyon in the Temblor Range (T1972).

Plagiobothrys Jonesii Gray. Rare in metamorphic talus at the base of a steep canyonside in Last Chance Canyon in the El Paso Range (T4363).

Plagiobothrys nothofulvus (Gray) Gray. Pop-corn flower. A common and characteristic annual of the Douglas oak woodland, usually growing in shade on north slopes.

Plagiobothrys tenellus (Nuttall) Gray. Slender pop-corn flower. Occasional to common in the cismontane Upper Sonoran associations, usually growing on shaded slopes in the Douglas oak woodland.

Plagiobothrys Torreyi (Gray) Gray. Colonies are scarce at the lower elevations of the ponderosa pine forest in the Greenhorn Range and on the north slope of Breckenridge Mountain.

VERBENACEAE. Verbena Family

Lippia nodiflora Michaux var. **rosea** (D. Don) Munz. Lippia. Occasional in wet places in the valley, where the dense spreading colonies are helpful in preventing erosion on canal banks. Lippia is sometimes used as a ground cover but many people find it objectionable because it attracts bees, often in large numbers. Native of South America.

Verbena Abramsii Moldenke. Abrams verbena. Common in the Douglas oak woodland in the Greenhorn Range; rare in vernal poolbeds in the northern Temblor Range. (*V. lasiostachys* in part.)

Verbena bracteata Lagasca y Segura & Rodriquez. Bracted verbena. Common in wet soil, on dessicating flats, and in vernal poolbeds in the valley and the Temblor Range; occasional around Lake Isabella. It is abundant in wet soil and shallow water at the Kern National Wildlife Refuge.

Verbena lasiostachys Link. California vervain, western verbena. Common in the Douglas oak woodland in the Greenhorn Range; occasional through the mountains south to the Mil Potreros region in the San Emigdio Range. (*V. prostrata.*)

Verbena robusta Greene. A small colony has grown since at least 1956 in mud in full sun at the edge of a small pool on McGovern Grade one mile east of Annette in the Temblor Range (T1535). Although this plant is native in California, Ernest M. Still, the owner of the ranch where the pool is located, says (oral communication) it is relatively recent there and that he has never observed it elsewhere in the Annette Hills (which he has known intimately since the early 1900's).

Verbena tenuisecta Briquet. Moss verbena. A most attractive plant occasionally escaped and at least persistent in vacant lots, waste places, etc., about Bakersfield and at the Antelope Ranch grounds on the valley west-side. Native from Georgia to Louisiana.

Vitex Agnus-castus Linnaeus. Butterfly bush. A single robust obviously spontaneous shrub grows in a dry canal bottom in the alkali sink ⅜ mile east of Cottonwood Road on Weedpatch Loop in the Weedpatch region (T7409). The shrub in colorful full bloom in July seems incongruous and almost unreal on the hot dreary alkali plain. Other spontaneous plants away from habitations can be expected, as butterfly bush volunteers freely when planted in Kern County gardens. Native of southern Europe.

LABIATAE. Mint Family

Agastache urticifolia (Bentham) Kuntze. Rare in the ponderosa pine forest in the Greenhorn Range and on the north slope of Breckenridge Mountain. It has grown as a waif at the forks of Lumreau and Cedar creeks (Farnsworth 653) and along Lumreau Creek in the Elsie Casad field (Farnsworth 716) in the Douglas oak woodland on the western slope of the Greenhorn Range.

Lamium amplexicaule Linnaeus. Giraffe's head, henbit. A common ranchground and garden weed in the Glennville region and in Bakersfield. It is occasional in the moist meadows in Lynns Valley, and is scarce but becoming more common in moist swales in the Douglas oak woodland in the hills south of Glennville (Farnsworth 666). Native of Europe.

Lycopus americanus Muhlenberg. Water horehound. A fairly common streambank plant along Kern River northeast of Bakersfield. Small colonies grow along Cedar Creek south of Glennville, along Kern River at the Borel Intake east of Wofford, and at Sycamore Creek 0.5 mile east of Arvin.

Marrubium vulgare Linnaeus. Horehound. Common in waste places, at old habitations, ranchgrounds, etc., below the yellow pine forest; rare in similar places on the desert (although it is well established in the Rosamond-Willow Springs region). Horehound is naturalized on the broad wind-blown sand slopes on the northwest side of the desert Antelope Valley where it is the only common herbaceous perennial in the creosote bush association. It was much planted by early settlers for medicinal uses, mostly for a syrup for respiratory ailments. Native of Europe.

Mentha arvensis Linnaeus. Marsh-mint, field penny-royal. Scarce at scattered localities in the Greenhorn Range (Lynns Valley at the Henry Bowen Ranch, Angel Creek near Glennville, Lumreau Creek falls, Cedar Creek at the Rock Corral) and along Kern River south of Kernville and at Hart Park. The pubescence of the plant is extremely variable; varieties based on this factor seem to be of little merit, at least as to California plants. Some races of the plant are apparently native, others may have been introduced.

Mentha citrata Ehrhart. Bergamot mint. Once collected at a swampy flat just below old Kernville (C. N. Smith 012, JEPS), a station now inundated by Lake Isabella. Native of Europe.

Mentha rotundifolia (Linnaeus) Hudson. Apple-mint. Known from Cedar Creek near the Rock Corral in the Greenhorn Range (Farnsworth 307). A rare adventive in California, apple-mint also grows in Mariannas Canyon in the Temblor Range in San Luis Obispo County not far from the Kern County line. Native of Europe.

Mentha spicata Linnaeus. Spearmint, garden mint. Widespread and often well established at wet places in the mountains, mostly in the Douglas oak woodland (El Paso Creek, Cedar Creek, Caliente Creek, Lynns Valley) and along Kern Island Canal south of Bakersfield. Most naturalized colonies are heavily grazed by deer and rarely succeed in blooming or setting seed. The plant was once a standard garden herb used for mint jelly, as a garnish, and as a flavoring for various drinks. Native of Europe.

Monardella Breweri Gray. Colonies are rare in the California juniper phase of the Douglas oak woodland in the Temblor Range and in the pinyon woodland in the Mt. Abel region.

Monardella candicans Bentham. Occasional in colonies on open slopes in the Douglas oak woodland in the Greenhorn Range foothills south to Cedar and Lumreau creeks, the southern limits for the species.

Monardella exilis (Gray) Greene. Desert penny-royal. Scarce in the arid shrub association and the pinyon woodland from Kelso Canyon to Weldon. Scattered plants grow on the dunes at Buckhorn Dry Lake. (*M. candicans* var. *exilis.*)

Monardella lanceolata Gray. Scarce from the chaparral to the edge of the yellow pine forest in the mountains southwest to Cuddy Valley.

Monardella linoides Gray. Occasional from the chaparral to the edge of the yellow pine forest south to Breckenridge Mountain and the Piute Mountains.

Monardella linoides ssp. **oblonga** (Greene) Abrams. Occasional in the Jeffrey pine forest and the pinyon woodland from Lookout Point in the Piute Mountains south to Ventura County in the Mt. Pinos region. The type was collected in the mountains south of Tehachapi by Edward L. Greene in 1889.

Monardella odoratissima Bentham. ssp. **parvifolia** (Greene) Epling. Mountain penny-royal. Occasional in the ponderosa pine forest in the Greenhorn Range and in the Jeffrey pine forest near the summit of Breckenridge Mountain.

Pogogyne Douglasii Bentham. Occasional, often in dense colonies, around poolbeds in the Glennville region, especially in Lynns Valley.

Salazaria mexicana Torrey. Bladder sage. A typical shrub of the desert canyons and slopes. This very xerophytic shrub is often leafless most of the year.

Salvia carduacea Bentham. Thistle-sage. Common in all the associations below the chaparral, usually growing in light or well drained soils. The large flowers of clear lavender blue in whorls on relatively tall stems make it one of the showiest of the spring annuals.

Salvia Columbariae Bentham. Chia. Common in all associations below the yellow pine forest. The seeds, which were a staple in the diet of the Piute Indians, are exceptionally nutritious; a tablespoonful a day is said to be an adequate emergency diet.

Salvia Dorrii (Kellogg) Abrams. Blue sage. Occasional, often in colonies, in the arid shrub association and pinyon woodland south to Jawbone Canyon and Kelso Valley, west to South Fork Valley, and east to Mesquite Canyon in the desert El Paso Range. It is rare southwest to the slopes below Tejon Pass on the southeastern slopes of the Tehachapi Mountains. Blue sage is an important browse plant for deer and livestock. (*S. carnosa* var. *pilosa.*)

Salvia leucophylla Greene. Purple sage. This sage of the southern California coast has been reported from 1.25 mile northwest of Blue Point, Tejon Ranch (Nordstrom 561, RSA) and from the ridge road near Lebec (Jones in 1924, POM). Although the point directly across Castac Valley from Lebec is locally known as "Blue Point," 1.25 mile northwest of it would be in Castac Valley, almost at Fort Tejon. Thus, the use of "Blue Point" as a reference for the locality and the unlikely possibility that purple sage ever grew in Castac Valley suggest that this collection, and perhaps also that of Jones, came from the nearby Liebre Mountains of Los Angeles County, where the plant is relatively common.

Salvia mellifera Greene. Black sage. Occasional in the chaparral in the Temblor Range. Black sage responds well to fire. It was becoming rare in the central Temblor Range before the controlled burns of recent years. Colonies of vigorous young plants have replaced the senescent ones that were burned.

Scutellaria Austinae Eastwood. Austin skull cap. Scarce from the higher elevations of the Douglas oak woodland to the yellow pine forest in the Greenhorn Range and at Erskine Creek and Weldon Meadow in the Piute Mountains, growing in sunny rather moist places. It is rare in the Tehachapi Mountains (ridge above Rising Canyon, T11,086).

Scutellaria Bolanderi Gray. Sierra skull cap. Scarce along Cedar Creek from the mouth of Fulton Creek to the Rock Corral and along Lumreau Creek in the Douglas oak woodland in the Greenhorn Range.

Scutellaria tuberosa Bentham. Dannie's skull cap. Known from the south side of Wildcat Gulch, about 2 miles east of Woody (C. N. Smith 341, JEPS) and from the slopes above Buck Springs west of Woody (Farnsworth 787).

Stachys albens Gray. White hedge-nettle. Widespread at wet places in all the cismontane associations through the yellow pine forest, approaching the desert at Oak Creek in the Tehachapi Mountains. The type was collected at Ft. Tejon (Xantus de Vesey 80 in 1857–58).

Trichostema austromontanum Lewis. Occasional in meadows and disturbed places in the Piute Mountains, at Squirrel Meadow on Breckenridge Mountain, and on the southern Kern Plateau northeast to Kennedy Meadows on the South Fork of Kern River in Tulare County. According to Harlan Lewis (Brittonia 12:93), the plant is of hybrid origin. One parent, *T. oblongum,* grows at Wagy Flat and along Cedar Creek in the nearby Greenhorn Range; the other, *T. micranthum,* reaches its northwestern limits in the distant San Bernardino Mountains. The isolation of *T. austromontanum* is probably due to geographical separation of the parents by increasing post-Pleistocene aridity.

Trichostema lanceolatum Bentham. Turpentine weed, vinegar weed, and among beekeepers, blue curl. Common in sandy soil in the Upper Sonoran grassland and the Douglas oak woodland, especially in the dry-farm grainfields in the Temblor Range. The plant is of great importance to beekeepers; its copious nectar makes honey of exceptional quality. Pioneer cowboys crushed the leaves in their bed rolls to discourage fleas.

Trichostema oblongum Bentham. Common in moist usually shaded soil along Lumreau and lower Cedar creeks in the Douglas oak woodland on the west slope of the Greenhorn Range; rare along Wagy Creek near Wagy Flat on the east slope (Hardham 2559). These are the southern limits for this Sierran species.

Trichostema ovatum Curran. San Joaquin turpentine weed. Endemic to the Lower Sonoran grassland of the San Joaquin Valley south of Fresno County. San Joaquin turpentine weed grows only in the east half of the valley in Kern County, reaching its western limits between Semitropic and Lost Hills. Although it is usually rare, colonies are rather common in the infrequent years with late spring rains. In such favorable seasons, what are apparently transient colonies grow around vernal poolbeds as high as the Douglas oak woodland in the Greenhorn Range (Oak Flat, T11,552 in 1965). An interesting occurrence of this San Joaquin Valley native was the dense colony that grew on a recently graded vacant lot on Oxford Street just north of 11th Avenue in downtown Delano in 1959 (T5567). The plant was recently discovered on a sandy flat near the mouth of Ballinger Canyon in the Upper Cuyama Valley in Santa Barbara County, where it is possibly an introduction (Chandler 934, 19 September 1962).

Unlike many California plants of restricted range, San Joaquin turpentine weed is in no danger of extinction; in favorable years it thrives not only in graded vacant lots but in loose soil in vineyards, along service roads in rose farms, on low levees in cotton fields, and in other places with disturbed light soils. Thus, it is the exceptional rare plant that is probably more common now than it was in ancient times. The type was collected along the railroad south of Sumner [East Bakersfield] by Katharine Layne Curran in 1884.

SOLANACEAE. Nightshade Family

Datura meteloides A. DeCandolle. Tolguacha, and locally, Jimson weed. Common on washes, sandy flats, waste places, roadsides, etc., everywhere below the yellow pine forest (a colony grows at the borders of the forest on Rattlesnake Grade in the Greenhorn Range). The plant was a source of narcotics used in initiation rites by the California Indians. All *Datura* species have narcotic properties; those of *D. Stramonium* were known to the English soldiers at Jamestown as early as 1676, which accounts for the origin of the common name used in the west for the similar *D. meteloides.* An old Texas cowboy song, "Riding through the Jimson Weed," apparently obliquely refers to its use as a drug on the southwestern frontier. The press has recently reported that it is being used in the Santa Barbara and Ventura regions by adolescents involved in the current juvenile enthusiasm for hallucigens. Such use is fraught with hazard as all parts of the plants are toxic and the margin between a narcotic and a fatal dose is narrow.

Although rarely eaten by mature cattle, calves are occasionally killed by it. The plant is sometimes grown for its large flaring white flowers that are particularly impressive in gardens lighted at night.

According to W. L. Jepson (A California Flora, 3:457), tolguacha (a name of probable Aztec origin) was introduced in California from Mexico, almost certainly before the American occupation of the state, probably by the Indians prior to the Spanish exploration.

Datura Stramonium Linnaeus var. **Tatula** (Linnaeus) Torrey. Thorn-apple, in the eastern United States, Jimson weed. Occasional along Kern River near the Stockdale Country Club and at waste places in the Old River region; rare along Kern River in Kern Valley. This, too, is a highly poisonous plant. It is the source of the drug, stramonium, a standard pharmaceutical for controlling blood clotting and for immobilizing eyes after surgery. Native of tropical America.

Lycium Andersonii Gray. Box thorn. Scarce in Ballinger Canyon in the upper Cuyama Valley in Kern and Santa Barbara counties and in the low hills along the San Andreas Fault scarp west of Elkhorn Plain in extreme southeastern San Luis Obispo County.

Lycium Andersonii var. **deserticola** (C. L. Hitchcock) Jepson. Occasional in the desert-facing canyons of the Tehachapi Mountains and the southern Sierra Nevada, west to the rugged point northeast of Lake Isabella (Howell 37,148); rare on the dunes north and northwest of Buckhorn Dry Lake. An isolated colony grows in the steep rocky Poso Creek Gorge below Poso Mine (C. N. Smith 182, JEPS) on the western slope of the Greenhorn Range.

Lycium Cooperi Gray. Common in the dunes around the Muroc dry lakes; occasional on the desert and the desert-facing slopes of the mountains west to Kelso Canyon, Cameron, and the arid slopes north of South Fork Valley; scarce in the Elk and McKittrick hills; rare (one plant) on the Belridge Plain in the valley. The densely branched shrubs are excellent cover for birds and small mammals.

Lycopersicon esculentum Miller. Tomato. Occasionally spontaneous in wet places in the valley, especially on weedy canal banks. It is possibly somewhat persistent at some stations. Native of South America.

Nicotiana attenuata Torrey. Mountain Indian tobacco. Common in disturbed places in the yellow pine forest; occasional in the Douglas oak woodland and chaparral. The plant is poisonous.

Nicotiana Bigelovii (Torrey) Watson. Indian tobacco. Common in vernal pool beds, often in dense colonies, and occasional in waste places in the Temblor Range and the foothills around the valley; rare in low places in the valley. The plant is very poisonous and sometimes causes livestock losses.

Even inhalation of the smoke from burning dried plants is said to be hazardous for humans.

Nicotiana glauca Graham. Tree tobacco. Common in creeks, waste places, etc., in all the associations below the Douglas oak woodland, especially in the valley and the low east-side foothills. This attractive large shrub is poisonous and sometimes causes livestock losses. Native of South America.

Physalis acutifolia (Miers) Sandwith. Ground cherry. Scarce in moist soils from east of Buena Vista Lake to the Buttonwillow region. Native from Los Angeles County to Texas and Mexico. (*P. Wrightii.*)

Physalis lanceifolia Nees. Lance-leaved ground cherry. A common weed of moist soils in the valley; known as a waif along Cedar Creek at the Rock Corral in the Douglas oak woodland in the Greenhorn Range (Farnsworth 650) and at the Temblor Ranch grounds. The fruits are edible and are a staple food in Mexico. In California they are sometimes sold canned as "tomatillos" in specialty stores. Apparently a native of tropical America, possibly north to Texas and Arizona.

Solanum californicum Dunal. California nightshade. An attractive herbaceous perennial that is occasional in the woody Upper Sonoran associations from the Temblor Range south to the San Emigdio Range. The plant is relished by livestock and deer. (*Solanum umbelliferum* var. *incanum.*)

Solanum elaeagnifolium Cavanilles. Bull nettle, and about Bakersfield, Noriega weed. A common and sometimes troublesome weed of waste places and farm lands in the valley and around Mojave; scarce elsewhere on the desert. The local common name derives from the weed's early appearance at the old Noriega Ranch along Kern River near Bakersfield. Native from the central United States to Mexico.

Solanum nodiflorum Jacquin. Mock black nightshade. A very common cismontane weed of moist soils, particularly in the valley and at lower elevations in the mountains and the Temblor Range. The plant is often confused with the Old World black nightshade, *S. nigrum*, which is apparently rare in California. Contrary to common belief, the fruits of *Solanum nodiflorum* are not poisonous and there seems to be no evidence that any part of the plant is toxic. Native of tropical America.

Solanum sarachoides Sendtner. Known as a rare weed in a cotton field near Poso Creek at Whistler Road, northeast of Wasco (T8973); as a scarce ranchground weed at the Farnsworth Ranch near Glennville (Farnsworth 244); and from a single plant that grew in 1965 along the South Fork of the Kern River near Weldon (T10,942). It will be interesting to see if this currently rare but apparently widespread weed becomes common in the county. Native of Brazil.

Solanum triflorum Nuttall. A plant of the transmontane regions that is known from the Douglas oak woodland 3 miles south of the head of Eugene Grade in the Greenhorn Range (C. N. Smith 1049, JEPS).

Solanum tuberosum Linnaeus. Potato. A leading valley crop, the potato is often spontaneous but never persistent around fields where it has been planted. Potatoes also occasionally grow as garden and waste place weeds. Native of South America, but introduced in the United States as horticultural forms from Europe.

Solanum Xanti Gray. Chaparral nightshade, purple nightshade. Occasional from the pinyon woodland and the chaparral through the yellow pine forest in the mountains southeast to Kelso Valley. The type was collected at Ft. Tejon (Xantus de Vesey in 1857–58).

Solanum Xanti var. **intermedium** Parish. This glandular variety is occasional in colonies that overlap the range of the other varieties, occurring in the Douglas oak woodland from Castaic Valley to Cuddy Canyon in the Mt. Pinos region, at Purdy Ridge and Tehachapi Mountain Park in the Tehachapi Mountains, and at the summit of Tweedy Creek and on Sweetwater Ridge in the mountains at the head of Caliente Canyon. It is the common form in the Jeffrey pine and lodgepole forest of the Kern Plateau in Kern and Tulare counties.

Solanum Xanti var. **montanum** Munz. An herbaceous form perhaps more common than the typical subspecies from the Tehachapi Mountains through the Greenhorn Range. It is rare in the pinyon woodland in Quatal Canyon in the Mt. Abel region. When in bloom, the plants become beautiful mounds of lavender-purple.

SCROPHULARIACEAE. Figwort Family

Antirrhinum cornutum Bentham var. **leptaleum** (Gray) Munz. Known from Rancheria Creek in the chaparral on the southwest slope of Piute Mountain (road to Piute Mountain, 4.6 miles north of the Caliente Creek Road junction, Robert Hull 175, JEPS).

Bacopa Eisenii (Kellogg) Pennell. Water-buttercup. Common in pools, lakes, rice fields, and other quiet water in the valley. The plant is avidly eaten by migrating water fowl in the fall. In 1958 in early October several acres of the drainage pool at Buena Vista Lake were densely covered with water-buttercup; by mid-November it had been completely consumed. (*B. rotundifolia.*)

Castilleja Applegatei Fernald. Scarce in the Jeffrey pine forest from the Mt. Pinos region northeast to the Kern Plateau, east to Champagne Spring on the north slope of Cache Peak. It often grows at exposed rock outcrops at the highest elevations. (*C. pinetorum.*)

Castilleja Breweri Fernald. Scarce in metamorphic rock along the crest of Piute Mountain (C. N. Smith 861, JEPS, T8914).

Castilleja chromosa A. Nelson. Desert paint brush. Occasional in desert canyons where the bright red flowers and bracts are colorful against the tan or gray rocks. (*C. angustifolia.*)

Castilleja disticha Eastwood. Occasional in rocky places at the higher elevations of the Jeffrey pine forest in the Piute Mountains.

Castilleja foliolosa Hooker & Arnott. A colony of several plants grows in shale along a long-abandoned road bed just within Kern County on the east slope of Cottonwood Pass in the Diablo Ranges (T13,106). Colonies are probably occasional in rocky places in that region in the extreme northwest corner of the county.

Castilleja Jepsoni. Bacigalupi & Heckard. Occasional from the Douglas oak woodland to the Jeffrey pine forest (except on the Kern Plateau and in the ponderosa pine forest in the Greenhorn Range) in the mountains and the Temblor Range, east to Walker Pass, northeast to Spanish Needle Creek. The plant is usually associated with *Eriogonum* species, especially *E. fasciculatum* ssp. *polifolium* and ssp. *foliolosum*, which it possibly parasitizes. In Cummings Valley it grows with *Artemisia tridentata.* (Cf. Leafl. West. Bot., 10:281, 1966.)

Castilleja linariaefolia Bentham. Known from the sand banks at the upper end of Claraville Meadow in the Piute Mountains (C. N. Smith 839, JEPS) and along Kern River just above Kernville (T9034).

Castilleja miniata Douglas. Known from between the Greenhorn Summit Store and the Tulare County line (C. N. Smith 710, JEPS).

Castilleja plagiotoma Gray. Rare in adobe in the Upper Sonoran subshrub association in the central Temblor Range (Ripley & Barneby 3277). An interesting colony that is well out of range, both geographically and ecologically, grows in sagebrush at the borders of Landers Meadow in the Jeffrey pine forest in the Piute Mountains (T7451).

Castilleja Roseana Eastwood. Occasional at higher elevations in the Jeffrey pine forest in the Mt. Pinos region, east to nearby Tecuya Ridge; scarce on Tehachapi Peak and Bear Mountain.

Castilleja stenantha Gray. Scarce in wet places from the Douglas oak woodland through the chaparral in the mountains south to Lebec.

Collinsia bartsiaefolia Bentham var. **Davidsonii** (Parish) Newsom. Occasional in colonies on north slopes in the cismontane Upper Sonoran associations. In favorable years, colonies of vigorous plants with intensely colored flowers grace the massive semistabilized dunes around Buckhorn Dry Lake (T10,680). The hot desert floor is in startling contrast to the shaded north slopes where the plant generally grows.

Collinsia callosa Parish. Occasional in small colonies in the Jeffrey pine forest and the pinyon woodland and scarce in the higher elevations of the Douglas oak woodland from the summit of Quatal Canyon in the Mt. Abel region northeast to the Piute Mountains.

Collinsia Childii Parry. Colonies are scarce in shade, often in disturbed places, from the shin oak brush to the highest elevations in the mountains. It is most common in the yellow pine forest.

Collinsia heterophylla Buist. Chinese houses, innocence. Occasional in colonies on shaded north slopes in the cismontane Upper Sonoran associations, rare to the yellow pine forest. (*C. bicolor.*)

Collinsia parviflora Douglas. Scarce in moist sand along Cedar Creek above the Rock Corral in the Douglas oak woodland in the Greenhorn Range (Farnsworth 442).

Collinsia tinctoria Hartweg. Rare in the Douglas oak woodland and chaparral on the west slope of the Greenhorn Range and the north slope of Breckenridge Mountain. On the Kern Plateau, a small colony grows near a moist seep in the open Jeffrey pine forest on the trail from Pine Flat to Little Cannell Meadow and it is rare along upper Bartolas Creek.

Collinsia Torreyi Gray. A small colony grows in moist granitic sand on a slope southeast of Pine Flat near the trail to Little Cannell Meadow in the arid Jeffrey pine forest on the southwestern Kern Plateau (T9858), a typical example of the Sierran plants that reach their southern limits on the terminus of the plateau in Kern County.

Collinsia Wrightii Watson. Blue-eyed Mary. Common in the yellow pine forest in the Greenhorn Range and the Mt. Pinos region, mostly at high altitudes. The type was collected in the Greenhorn Mountains, 12 miles from Kernville (William Greenwood Wright 4, 12 June 1888). Wright, an entomologist and operator of a planing mill at San Bernardino, accompanied Edward Palmer on the first of his two ill-fated and poorly planned attempts in 1888 to reach Mt. Whitney from Kernville. Their Greenhorn collections are a fortuitous addition to the Kern flora; the Greenhorn trip was made when they learned on their arrival in Kernville that snow blocked the way to the Mt. Whitney region. (*C. Torreyi* var. *Wrightii.*)

Cordylanthus compactus Pennell. Bird's beak. Occasional, usually in heavy soil, in the Douglas oak woodland and scarce in the pinyon woodland and yellow pine forest in the mountains south to Mt. Pinos region. (*C. rigidus* ssp. *brevibracteatus.*)

Cordylanthus Ferrisianus Pennell. Occasional in small dense colonies in the ponderosa pine forest in the Greenhorn Range; rare on the rim of the Kern Plateau at the head of Fay Creek Canyon (Howell & True 41,825).

Cordylanthus hispidus Pennell. Known from the alkali sinks 15 miles southeast of Bakersfield (Bauer 8, 22 October 1927, JEPS). The collection served as the type for *C. mollis* Gray var. *viridis* Jepson. Lawrence R. Heckard writes (personal communication) that the plant needs further collecting and careful study before its status can be evaluated. However, the alkali sinks southeast of Bakersfield have been drastically altered in the last 40 years and this interesting plant may no longer be present.

Cordylanthus Nevinii Gray. Scarce along rocky crests and ridges at high elevations in the Piute Mountains (Saddle Springs, T3141, 8902, crest of Piute Mountain, C. N. Smith 860, JEPS) and at Mt. Pinos (Hoffmann, 19 July 1927, SBM). It is also known from Reyes Peak in Ventura County (Chandler 3202, SBBG).

Cordylanthus rigidus (Bentham) Jepson. Mule weed. Occasional in the Douglas oak woodland and chaparral in the Temblor Range. It usually grows in small rather dense colonies in heavy soils. The flowering stems are often eaten by deer.

Diplacus calycinus Eastwood. Bush monkey flower. Common in rocky places in the Upper Sonoran associations, reaching the edge of the desert at the massive rock exposures in the eastern Sierran canyons. The massed plants growing in cracks and fissures in rock outcrops are very colorful, particularly in contrast with the light buff sandstone at such places as Carneros Rocks in the Temblor Range. (*Mimulus longiflorus* ssp. *calycinus.*)

Gratiola ebracteata Bentham. Rare in a pool bed on a broad open slope (Water Hole Flat) on the Farnsworth Ranch between Cedar Creek and the Basket Pass Road (Farnsworth 514).

Howelliella ovata (Eastwood) Rothmaler. A scattered colony of only a few plants of this rare and distinctive plant of the south inner Coast Ranges grew on the Agua Media Canyon flood plain on the Antelope Plains in 1962, the only San Joaquin Valley record for the species (Bedell 74–2). The only other Kern County record is a single plant that grew in shale gravel in the Cedar Canyon Creek bed near the canyon forks in 1955 (T2232). Although it seems obvious that the seeds for both were carried by flood waters, the plant has not yet been found in the high central Temblor Range. (*Antirrhinum ovatum.*)

Limosella aquatica Linnaeus. Northern mud-wort. Scarce in a drying pool bed in Bear Valley (T9686 in 1964).

Linaria dalmatica (Linnaeus) Miller. Dalmatian toad flax. Locally established in the pinyon woodland around the town of Frazier Park, slowly spreading eastward in creek beds and on roadsides to Castac Valley. Although it had been present here for many years and seemed to present no problem, the Kern County Board of Supervisors declared it a noxious weed in 1964 and the County Department of Agriculture is attempting to eradicate the plant. Apparently it is considered a serious range invader in more mesic regions; this seems highly unlikely in the arid pinyon woodland at Frazier Park. Native of the Mediterranean region.

Lindernia anagallidea (Michaux) Pennell. Scarce in wet places in the valley and along the Poso Creek drainage system from the low foothills to the Douglas oak woodland; occasional in wet sand along Kern River above Lake Isabella.

Mimulus. The genus *Mimulus,* with at least 77 species in California and about 150 in the world, is remarkably well represented by the 26 species that occur in Kern County. Possibly no other area of similar size has such a concentration of monkey flowers as this region where those of the coastal regions, the mountains, and the desert meet. The plants occupy diverse habitats, from those of *M. Bigelovii,* common and colorful in favorable years on the desert dunes and gravelly slopes, to that of *M. primuloides,* that grows in very wet soils and even shallow water in the high mountain meadows. None occurs in the Lower Sonoran grassland of the San Joaquin Valley and only two grow in the Temblor Range. On the other hand, no less than 18 species are known from the Greenhorn Range. Although some species are fairly common in the plant associations in which they occur, only one, common monkey flower (*M. guttatus*), with its stunning yellow flowers, is widespread; it grows at almost all moist places in the county that are not strongly alkaline.

One group forms an important part of the "gooseberry community" of the east-side foothills. These annual species, which include *Mimulus pictus, M. Congdonii, M. Douglasii, M. Kelloggii,* and *M. viscidus,* are poor competitors with the spring annuals of the foothills. However, they are unpalatable to rabbits. Thus, these monkey flowers grow most often near colonies of chaparral gooseberries that afford protection for cottontail rabbits. The rabbits graze heavily near the sheltering gooseberries, effectively eliminating plants that compete with the monkey flowers.

Taxonomically, the last word is far from having been written on the specific limits and relationships within the genus. This is especially true in inner central California of the group of annuals that have sessile flowers. However, their taxonomic complexity is no barrier to the enjoyment of their bright cheerful flowers, often with intricate patterns of sometimes flamboyant colors.

Mimulus androsaceus Curran. Tehachapi monkey flower, and at Granite Station, gramaphone plant. Scarce, mostly in shin oak brush and chaparral, sometimes on sunny slopes in the Douglas oak woodland and the yellow pine forest, south to the Tejon Pass region. The local common name refers to the resemblance of the slender corolla tube with its flaring lobes to the speakers of early phonographs. The type was collected at Tehachapi (Katharine Layne Curran, June, 1884).

Mimulus barbatus Greene. This colorful bicolored mimulus of Inyo and Tulare counties is scarce in the arid Jeffrey pine forest on the southern Kern Plateau, growing in colonies in moist granitic sand (Pine Flat, T11,334; in a seep near the trail from Pine Flat to Little Cannell Meadow, T9861A; along upper Bartolas Creek, T12,462).

Mimulus Bigelovii (Gray) Gray. Desert monkey flower. Scattered colonies grow in favorable years in gravelly and sandy soils on the desert, usually in canyons or on slopes of compacted gravel. Extensive and unusually colorful colonies of robust plants occur in good years on the semi-stabilized sand dunes at Buckhorn Dry Lake.

Mimulus Bolanderi Gray. Scarce in disturbed soil at Mil Potreros in the

San Emigdio Range. According to Philip A. Munz (A California Flora, p. 620), it also occurs in the Tehachapi Mountains.

Mimulus Breweri (Greene) Coville. Dwarf money flower. Rare in small dense colonies in moist soil in the yellow pine forest in the Greenhorn Range and on the north slope of Breckenridge Mountain. It is known from a spring in the Bartolas Country on the Kern Plateau (Howell & True 41,801) and from Mt. Pinos at 8,200 feet elevation (Hoffmann, 18 June 1927, SBM).

Mimulus cardinalis Douglas. Scarlet monkey flower. A handsome plant that is occasional in wet places in the mountains from the Douglas oak woodland to the yellow pine forest.

Mimulus Congdonii Robinson. Scarce in the Douglas oak woodland on the west slope of the Greenhorn Range, where it usually grows in bare areas around colonies of chaparral gooseberry (Schwab Place, 3 miles east of Woody, T9230, C. N. Smith 259, JEPS; Edna Williams place on Nightingale Creek, T10,595; Farnsworth Ranch 2 miles south of Glennville, Farnsworth 380; Pine Mountain at 2,800 feet elevation, C. N. Smith 322, JEPS).

Mimulus constrictus Pennell. Common in colonies, often in disturbed places, in the yellow pine forest; occasional in the chaparral. (*M. viscidus* ssp. *constrictus, M. subsecundus* var. *constrictus.*)

Mimulus Douglasii (Bentham) Gray. Known from the Douglas oak woodland in the foothills 2.5 to 3 miles east of Woody (C. N. Smith 261, JEPS) and from Lynns Valley (C. N. Smith 57, JEPS).

Mimulus Dudleyi Grant. Granite monkey flower. Occasional in moist cracks in granite outcrops in the Greenhorn foothills; scarce south to the mouth of El Paso Canyon, and east to Kelso Canyon and the head of Short Canyon in the Sierra Nevada north of Walker Pass.

Mimulus floribundus Lindley. Scarce in wet soil in the yellow pine forest in the Greenhorn Range, the Bartolas Country of the Kern Plateau, and at French Meadow in the Piute Mountains. It is known as a rare waif along streams in the Douglas oak woodland in the Greenhorn Range and along Kern River near Buena Vista Lake.

Mimulus Fremontii (Bentham) Gray. Frémont monkey flower. Common in the chaparral and pinyon woodland from the Temblor Range south to Mt. Abel. The colonies growing in bare shallow sand over sandstone in the openings in the chaparral in the central Temblor Range are especially colorful.

Mimulus guttatus Fischer. Common monkey flower. Common in wet places in all associations. The large yellow flowers with their scarlet markings are uncommonly beautiful; because of the plant's ubiquitous occurrence, this beauty is often overlooked. As might be expected of a plant which grows in moist soil in virtually all of California below 10,000 feet elevation, common monkey flower is highly variable and grows in many forms. Some have been named but these seem tc reflect no clear-cut morphological entities. *M. microphyllus* Bentham is one of the more distinct; this depauperate form grows in moist cracks in granite in Kern Canyon and on the west slope of the Greenhorn Range. Another is *M. nasutus* Greene, an annual form that is occasional in poolbeds in the Temblor Range and the low foothills on the east side of the valley.

Mimulus inconspicuus Gray. A small colony grows along Cedar Creek near the Rock Corral (Farnsworth 593) and another at the Mill Creek crossing on the west slope of Basket Peak (C. N. Smith 530, JEPS), both in the Douglas oak woodland.

Mimulus Johnstonii Grant. Known from loose sliding soil in a road cut 3.4 miles west of Frazier Park (Bacigalupi 4257, 19 June 1953, JEPS). This colony is apparently no longer extant but the species is fairly common in disturbed gravel in nearby Ventura County, especially on the lower slopes of Frazier Mountain.

Mimulus Kelloggii (Curran) Curran. Scarce around chaparral gooseberry colonies in the Douglas oak woodland in the Greenhorn Range (foot of Pine Mountain, C. N. Smith 323, JEPS; Edna Williams place on Nightingale Creek, T10,593).

Mimulus montioides Gray. Common in moist places in the yellow pine forest south to the Piute Mountains and Breckenridge Mountain. (*M. discolor.*)

Mimulus moschatus Lindley. Musk flower. Common in wet places in the yellow pine forest in the Greenhorn Range and occasional in similar places in the other mountains.

Mimulus Palmeri Gray. Occasional around seeps and springs on the desert; scarce from the Douglas oak woodland through the yellow pine forest south to the Tehachapi Mountains; rare on the east side of the valley. Especially fine colonies grow along the short ephemeral stream in Red Rock Canyon and at the spring at the old Cudahy Camp in Last Chance Canyon.

Mimulus Parishii Greene. Rare in damp soil along Cottonwood Creek at Geringer Grade in the Douglas oak woodland on the southeast end of Piute Mountain (T7513). This plant of the San Bernardino Mountains and the southern California desert ranges is also rare at Marques Spring in the pinyon woodland on the Kern Plateau in southeastern Tulare County (T7866).

Mimulus pictus (Curran) Gray. Calico flower. Endemic to the foothills from Tejon Canyon north to Porterville, this colorful mimulus is occasional in the Douglas oak woodland in the Greenhorn Range and scarce to the south, usually growing in colonies in bare ground around chaparral gooseberry patches or granite outcrops. (The ground around the gooseberry patches is grazed heavily by cottontail rabbits; calico flower apparently cannot compete well with other annuals but seems to be unpalatable to rabbits.) The fitting common name was given the plant more than one hundred years ago by William Higgins, a pioneer settler in the Woody region, because of the resemblance of the variegated maroon and white corolla lobes to a fine piece of that once popular fabric. The type was collected at Keene by Katharine Layne Curran in June, 1884.

Mimulus pilosus (Bentham) Watson. Occasional but widespread in wet places in the valley and the mountains.

Mimulus primuloides Bentham. Meadow monkey flower. Scarce in wet meadows in the northern Greenhorn Range, at Little Cannell Meadow on the Kern Plateau, and at Saddle Springs on Piute Mountain.

Mimulus primuloides var. **pilosellus** (Greene) Smiley. Creeping meadow monkey flower. Common in wet meadows in the Greenhorn Range and at the head of Bitter Creek on Mt. Pinos.

Mimulus tricolor Hartweg. Known from west of Delano (Mrs. Stella Meadows, POM). (The most likely station would probably be in the wet subalkaline sinks along Garces Road.)

Mimulus viscidus Congdon. Rare around chaparral gooseberry colonies in the Douglas oak woodland in the Greenhorn Range (the Schwab Place along Howling Gulch, about 3 miles east of Woody, T9229; Cedar Creek near the mouth of Fulton Creek, Farnsworth 583).

Mimulus Whitneyi Gray. Rare in dry open areas in the yellow pine forest in the Greenhorn Range (Shirley Meadows, near the ski run, Howell 38,405, Myrick 1138, SBBG; Windy Gap, T3702). Rimo Bacigalupi suggests (personal communication) that the plants at Shirley Meadows may be strongly introgressed with the genes of *M. constrictus.*

Mohavea breviflora Coville. Rare on metamorphic talus in Last Chance Canyon in the El Paso Range (T4353, 11,852), the southwestern limit for this interesting plant of the ranges of the Death Valley region.

Orthocarpus attenuatus Gray. White owl's clover. Colonies are common from the Lower Sonoran grassland through the Douglas oak woodland. Plants in the east-side foothills have atypically ample inflorescences and exserted styles (T10,526, 10,562).

Orthocarpus erianthus Bentham. Johnnytuck, butter and eggs. Abundant on the alkali plains north of Pond in the valley and in good loamy soil on flats in the Douglas oak woodland in the Poso Flat region of the Greenhorn Range.

Orthocarpus erianthus var. **micranthus** (Gray) Jepson. Occasional in the Upper Sonoran grassland and the Douglas oak woodland in the foothills in the Woody region. This variety grows on sunny slopes with well drained soil rather than on the vernally moist flats preferred by the typical variety.

Orthocarpus hispidus Bentham. Reaches its southern limits in the arid Jeffrey pine forest of the southern Kern Plateau, where a few small dense colonies grow in the sunny swale southeast of Pine Flat (T9875, 11,358) and a small dense cluster of many plants occurs at the head of Bartolas Creek (T12,452).

Orthocarpus linearilobus Bentham. Occasional on grassy slopes in the Douglas oak woodland in the Glennville region.

Orthocarpus purpurascens Bentham. Owl's clover. Common in broad colonies on open hillsides in good usually deep soil in all the cismontane associations through the Douglas oak woodland, often coloring the slopes a warm rich pink.

Orthocarpus purpurascens var. **ornatus** Jepson. Desert's owl's clover. This variety, with its intense colors, is occasional from the desert creosote bush association to the pinyon woodland of the neighboring slopes. It grows in colonies of only a few scattered plants or sometimes as lone individuals.

Pedicularis densiflorus Bentham. Indian warrior. A small colony grows around the granite outcrops above the Poso Creek Narrows in Douglas oak woodland in the Greenhorn Range (T8238, Record 82–1) and another in Back Canyon at the Piute cypress grove at the south end of the Piute Mountains (T9506). Robert Luthey writes (personal communication) that it also occurs on the east slope of the Greenhorn Range below the Tillie Creek crossing of the old Greenhorn Summit road.

Pedicularis semibarbata Gray. Mountain Indian warrior. Occasional in the yellow pine forest.

Penstemon breviflorus Lindley. Small-flowered penstemon. Occasional in the mountains, mostly in the Douglas oak woodland and the chaparral, where it grows around rock outcrops. Its copious supply of nectar makes the plant very attractive to bees.

Penstemon Bridgesii Gray. Scarlet penstemon. Common and showy in the red fir forest and the higher elevations of the yellow pine forest, often growing around exposed summit rock outcrops.

Penstemon caesius Gray. A characteristic plant of rocky places at high elevations in the yellow pine and red fir forests that reaches its southern limits at Sunday Peak and Black Mountain in the Greenhorn Range and on the rocky southwestern rim of the Kern Plateau.

Penstemon centranthifolius Bentham. Scarlet bugler. Occasional in the Upper Sonoran shrubby associations from the Temblor Range southeast to Ft. Tejon. The plant does well in cultivation and its tall scarlet spikes make a colorful addition to the garden in late spring.

Penstemon Grinnellii Eastwood ssp. **scrophularioides** (Jones) Munz. Occasional in the arid Upper Sonoran associations from Ballinger Canyon in the Mt. Abel region northeast through the mountain canyons to Inyo County. Most of our penstemons have flowers of uncommon beauty; the large flesh-colored bells of this one always inspire springtime superlatives from visitors to the arid canyons where it grows. (*P. Palmeri* in part.)

Penstemon heterophyllus Lindley. Chaparral penstemon. Occasional in the chaparral and Douglas oak woodland in the Temblor Range.

Penstemon incertus Brandegee. Common in the Walker Pass region, west in South Fork Valley to Onyx; scarce on Breckenridge Mountain, along the edge of Indian Wells Valley, and in the desert canyons south to Pine Tree Canyon. It usually grows in the pinyon woodland and the arid shrub association but occurs in the chaparral on Breckenridge Mountain and at the upper levels of the creosote bush association in the desert-facing canyons. An especially fine and extensive colony grows on the sand flats near the junction of the South Fork of the Kern River and Canebrake Creek east of Onyx.

Penstemon labrosus (Gray) Hooker. Scarce in colonies, mostly around rock outcrops, from Iris Point to the summit plateau of Mt. Pinos. The brilliant scarlet flowers, with their strongly deflexed corolla lobes, are a showy addition to the summer flora of the Mt. Pinos summit region.

Penstemon laetus Gray. Occasional on rocky slopes in all associations in the mountains. The type was collected at Ft. Tejon (Xantus de Vesey, 1857–58).

Penstemon Newberryi Gray. Pride-of-the-mountains. Colonies grow at the summit of Sunday Peak and on the north slope of Basket Peak in the Greenhorn Range, around the granite blocks along the southern rim of the Kern Plateau, and at Lookout Point in the Piute Mountains. The vivid deep rose-red flowers make a striking display at these rocky places; the extensive colony on the summit rocks at Sunday Peak is especially colorful.

Penstemon Rothrockii Gray. Rare in the sparse Jeffrey pine forest on the rugged ridges southeast of Pine Flat on the Kern Plateau.

Penstemon speciosus Douglas. Showy penstemon. Occasional in the Jeffrey pine forest on the southern Kern Plateau and in the Piute Mountains. (*P. glaber.*)

Penstemon speciosus var. **piliferus** (Heller) Munz. A distinctive herbaceous form with divergent stems that is occasional on the broad exposed summit of Mt. Pinos and on nearby Sawmill Mountain. On the high summits above Purdy and Lecheria Canyons in the southwestern Tehachapi Mountains it grows with sagebrush or in grassland with a mixture of annual species such as *Amsinckia intermedia* and *Erodium circutarium* that usually grow at elevations two or three thousand feet lower than the penstemon.

Penstemon ternatus Torrey ssp. **septentrionalis** (Munz & Johnston) Keck. Occasional in the Douglas oak woodland in the southwestern end of the

Tehachapi Mountains. The subspecies is endemic to this region and the neighboring Liebre Mountains of Los Angeles County.

Scrophularia californica Chamisso & Schlechtendal var. **floribunda** Greene. Bee plant. Occasional around rock outcrops south to the Tehachapi Mountains, mostly in the Douglas oak woodland.

Scrophularia californica var. **desertorum** Munz. Desert bee plant. Known from Fay Creek at the Fay Ranch, about 5.5 miles north of Weldon (C. N. Smith 952, JEPS).

Verbascum Thapsus Linnaeus. Common mullein. Occasional in the Douglas oak woodland in the Walker Basin region and the Greenhorn Range, where it grows in swales or sandy flats that are vernally moist; on the flats and flood plains along Kern River above Lake Isabella, and in disturbed soil at Evans Flat and Shirley Meadows in the ponderosa pine forest. The size of the plant varies greatly in response to moisture. In 1966, fruiting plants less than a foot tall grew along Cedar Creek in the Greenhorn Range; in the same year, a well watered specimen in the author's garden grew to a height of more than 10 feet. Native of Eurasia.

Verbascum virgatum Stokes. Scarce but persistent in dry sand along the Buena Vista Slough at the Lerdo Road bridge near the Spicer City gin (T489, 1261). Native of Eurasia.

Veronica americana (Rafinesque) Schweinitz. American brooklime. Common in wet places or in shallow water in the mountains south to Mt. Pinos.

Veronica arvensis Linnaeus. Scarce around the vernal pools on the west side of Lynns Valley and on Waterhole Flat south of Glennville in the Douglas oak woodland in the Greenhorn Range.

Veronica peregrina Linnaeus var. **xalapensis** (Humboldt, Bonpland & Kunth) St. John & Warren. Purslane-speedwell. Common in moist soils in the valley, occasional in the surrounding ranges to the Jeffrey pine forest.

Veronica persica Poiret. Persian speedwell. A garden weed at Kernville (Howell 37,083), known in 1957 at the Antelope Ranch on the Valley west-side (T3490). Native of Europe. (*V. Buxbaumii.*)

Veronica serpyllifolia Linnaeus var. **humifusa** (Dickson) Vahl. Colonies grow at Tiger Flat in the Greenhorn Range (T3737) and at a fork of Cannell Creek near Pine Flat on the southwestern Kern Plateau (T8705).

BIGNONIACEAE. Bignonia Family

Catalpa bignonioides Walters. Catalpa. A handsome tree that is an occasional escape at Bakersfield, along Kern River just east of Hart Park, and in moist often subalkaline places in South Fork Valley. Native from Georgia to Florida and Mississippi.

MARTYNIACEAE. Martynia Family

Proboscidea louisianica (Miller) Thellung. Unicorn plant. A scarce but recurrent grainfield weed at the Clarence Still Ranch at Annette in the northern Temblor Range. It is also known from a grassy ravine along Comanche Point Road, 7 miles southeast of Arvin (Breedlove 5447 in 1963). Native of the southern United States. (*Martynia louisianica.*)

OROBANCHACEAE. Broomrape Family

Orobanche californica Chamisso & Schlechtendal var. **Parishii** Jepson. Known from among rocks in a canyon 0.5 mile west of old Ft. Tejon (H. M. Hall 6301, JEPS).

Orobanche corymbosa (Rydberg) Ferris. Rare on a rocky slope along the crest of Piute Mountain (Breedlove 3942, C. N. Smith 864, JEPS; summit road, 3.2 miles south of Saddle Springs, T8907; all probably the same colony). This occurrence is a notable southwestern range extension from the Panamint Mountains of Inyo County. The plant is usually a parasite on *Artemisia tridentata,* but sagebrush does not grow with it on the Piute crest and there is no other obvious host.

Orobanche fasciculata Nuttall. Known in the Greenhorn Range from the Douglas oak woodland at Poso Mine (C. N. Smith 298, JEPS) and from the yellow pine forest at Black Mountain (Howell 38,229, 38,574), where its host is *Eriodictyon californicum.* It is also known from Piute Mountain (Breedlove 3902) and occurs on *Eriogonum fasciculatum* ssp. *polifolium* in the Upper Sonoran subshrub association on Pleito Ridge in the San Emigdio Range (T11,020).

Orobanche Grayana G. Beck var. **Feudgei** Munz. Occasional in the chaparral and pinyon woodland from the east slope of the Greenhorn Range and the Piute Mountains southwest to the Mt. Pinos region. It is rare in the chaparral in the central Temblor Range (Garcia Rocks, C. F. Twisselmann, 23 May 1953; lower Messa Ridge, Hardham 758). The plant is a parasite on *Eriodictyon, Artemisia,* and shrubby species of *Eriogonum.*

Orobanche uniflora Linnaeus ssp. **occidentalis** (Greene) Abrams. Naked broomrape. Distinctive among the broomrapes for its attractive pink flower, naked broomrape is a scarce parasite on *Saxifraga californica* from the Greenhorn Range south to El Paso Canyon. (*O. uniflora* var. *minuta.*)

PLANTAGINACEAE. Plantago Family

Plantago Bigelovii Gray ssp. **californica** (Greene) Bassett. Occasional around alkali sinks in the valley. (*P. Bigelovii.*)

Plantago elongata Pursh ssp. **pentasperma** Bassett. Common in the valley, mostly in alkali sinks. (*P. heterophylla.*)

Plantago Hookeriana Fischer & Meyer var. **californica** (Greene) Poe. Indian-wheat. Common in the Upper Sonoran grassland and occasional in the Lower Sonoran grassland. The seeds are very nutritious for cattle and wild life. (*Plantago erecta* of many California references.)

Plantago insularis Eastwood. Common in light sandy soils in the east-side foothills.

Plantago insularis var. **fastigiata** (Morris) Jepson. Occasional on desert flood plains, fans, and gravelly slopes.

Plantago insularis var. **scariosa** (Morris) Jepson. Scarce around sub-alkaline vernal pools in the valley (Goose Lake, T1716).

Plantago lanceolata Linnaeus. Ribwort, English plantain. A widespread but scarce weed of many cismontane associations below the yellow pine forest. Native of Europe.

Plantago major. Linnaeus. Common plantain. A widespread but scarce weed, mostly in disturbed usually moist soils in the cismontane Upper Sonoran associations. Native of Europe.

RUBIACEAE. Madder Family

Cephalanthus occidentalis Linnaeus var. **californicus** Bentham. Button-willow. Occasional along canals and in wet places from the Buena Vista Lake region north to Buttonwillow; scarce along streams and in moist places in the east-side foothills to about 1,500 feet elevation. The town of Button-willow was named for a large specimen that was once a prominent land-mark for an early trail across the valley. It is now protected as a historical landmark on Buttonwillow Drive about 0.5 mile south of the town. Thus, this valley farming center shares with Palo Alto, Big Pine, and Lone Pine the distinction of being the only California towns of any size named for individual plants. Fortunately, the protected buttonwillow seems safe; the "big pine" and the "lone pine" are gone and the "palo alto" is recurrently threatened by freeways and other developments.

Galium Andrewsii Gray. Needle bedstraw. Occasional from the Upper Sonoran shrubby associations in the Temblor Range south to the Jeffrey pine forest in the Mt. Pinos region; scarce through the mountains to the Tulare County line.

Galium angustifolium Nuttall. Occasional, mostly in rocky places in the chaparral in the Temblor Range and the mountains, occurring in the Jeffrey pine forest at Mt. Pinos.

Galium angustifolium var. **onycense** Dempster. Onyx bedstraw. Scarce on the hot slopes of granite blocks and outcrops from Onyx to Walker Pass Station, northeast to Spanish Needle Creek. It extends to the arid Jeffrey pine forest on the rim of the Kern Plateau east of Fay Creek Canyon (Howell & True 41,811). Onyx bedstraw grows from beneath the rocks, usually on the shaded lower side where the scant soil moisture tends to concentrate. A disjunct colony grows in a markedly more mesic situation on a roadbank on Bald Mountain in the well developed Douglas oak woodland on the west slope of the Greenhorn Range (C. N. Smith 1192, JEPS). The plant is endemic to these places and to isolated stations on the east slope of the San Bernardino Mountains. The type was collected on the hill above the Onyx Ranch (Lauramay Dempster, G. Thomas Robbins, & Rimo Bacigalupi 1015A and 1015B).

Galium Aparine Linnaeus. Goose grass, cleavers. Occasional, often common, on north slopes, mostly in the cismontane grasslands but sometimes locally common to the yellow pine forest. Thought to be introduced from Europe.

Galium bifolium Watson. Occasional in moist soil in the yellow pine forest.

Galium Bolanderi Gray. Occasional from the Douglas oak woodland to the yellow pine forest south to the Tehachapi Mountains. (*G. pubens.* Cf. Dempster, Lauramay T. and G. Ledyard Stebbins, Madroño, 18:106.)

Galium Hallii Munz & Johnston. Occasional around rock outcrops in the yellow pine forest; occurring in the pinyon woodland in Quatal Canyon in the Mt. Abel region.

Galium Matthewsii Gray. Rare in sand at the base of a massive granite canyon wall at the head of Short Canyon in the southern Sierra Nevada north of Walker Pass (T7124, 7125). Lauramay Dempster writes (Brittonia 11:11) that it also grows at Erskine Creek in the Piute Mountains.

Galium Nuttallii Gray. Scarce in the Douglas oak woodland and chaparral on the west slope of the mountains south to Bear Valley.

Galium parisiense Linnaeus. A small dense colony grows near a poolbed on the west side of Lynns Valley along Fugitt Road (T11,387); it is also known from the Fay Ranch northeast of Weldon (C. N. Smith 953, JEPS). Native of Europe.

Galium trifidum Linnaeus var. **subbiflorum** Wiegand. Scarce in wet usually shaded places in the red fir and ponderosa pine forests south to Greenhorn Mountain Park.

Caprifoliaceae. Honeysuckle Family

Lonicera interrupta Bentham. Chaparral honeysuckle. Occasional from the Douglas oak woodland to the yellow pine forest in the Greenhorn Range, on Breckenridge Mountain, and in the Piute Mountains; rare, south to Tweedy Canyon east of Keene. The shrubs are often severely browsed by deer.

Lonicera Johnstonii (Keck) McMinn. Southern honeysuckle. Occasional in the chaparral and the Douglas oak woodland from the Temblor Range southeast through the San Emigdio Range to Ft. Tejon. (*L. subspicata* in part.)

Sambucus caerulea Rafinesque. Blue elderberry, mountain elderberry. Occasional in the yellow pine forest. At the lower elevations, it is not always clearly distinct from *S. mexicana*. (*S. glauca.*)

Sambucus mexicana Presl. Wild elderberry. Occasional to common in moist places in all cismontane Upper Sonoran associations; scarce in the valley along Kern River near Buena Vista Lake. Wild elderberry does well in cultivation, where it can withstand and even benefit from severe pruning, or can be shaped into a small tree. The fruits, which have a taste similar to blue berries, make excellent jelly and wine and are sometimes used for pie. In the wild, they are usually completely eaten by birds; this leads to widespread avian dispersal of the shrub.

Symphoricarpos acutus (Gray) Dieck. Snowberry. Scarce on shaded often moist slopes in the yellow pine forest in the Greenhorn Range and on Breckenridge Mountain. (*S. mollis* var. *acutus.*)

Symphoricarpos Parishii Rydberg. Parish snowberry. Occasional in the Jeffrey pine forest from the Kern Plateau and Mt. Owen to Mt. Pinos, and in the Douglas oak woodland on the high slopes of the southern Tehachapi Mountains. The plant is usually heavily browsed by cattle, sheep, and deer.

Symphoricarpos rivularis Suksdorf. Scarce near the Lumreau Creek Falls at 2,800 feet elevation in the Douglas oak woodland on the west slope of the Greenhorn Range, a disjunct occurrence of a plant of the northern Sierra Nevada and the Coast Ranges from Monterey County north. (*S. albus.*)

CUCURBITACEAE. Gourd Family

Citrullus lanatus (Thunberg) Mansfeld. Water melon. Commonly spontaneous, especially in the valley, but nowhere established or even persistent. Native of tropical and south Africa. (*C. vulgaris.*)

Cucumis Melo Linnaeus. Mush melon, cantaloupe. Occasionally spontaneous in the valley; nowhere persistent. Native of Asia.

Cucurbita foetidissima Humboldt, Bonpland & Kunth. Calabazilla, wild gourd. Occasional in sandy places in the valley through the east-side foothills and on the desert slopes of the Tehachapi Mountains; rare in washes on the west side of the desert. Especially vigorous plants grow on the flood plains east of Lake Isabella in South Fork Valley.

Cucurbita palmata Watson. Coyote melon. Common on cismontane sandy washes and flats, mostly in the Lower Sonoran associations, occasional through the Douglas oak woodland.

Marah fabaceus (Naudin) Greene var. **agrestis** (Greene) Stocking. Man root. Common in the cismontane Upper Sonoran associations, especially in the northern Temblor Range; occasional on the desert east to Iron Canyon in the El Paso Mountains and in the Rosamond Hills. The plants have enormous pulpy roots. (*Echinocystis fabacea.*)

Marah horridus (Congdon) S. Dunn. Occasional in the chaparral from Castac Valley to the Greenhorn Range. (*Echinocystis horrida.*)

Valerianaceae. Valerian Family

Plectritis ciliosa (Greene) Jepson. Common in the Douglas oak woodland in the foothills of Breckenridge Mountain and on the lower slopes of Bear Mountain, and in the chaparral on Blue Mountain in the Greenhorn Range. Colonies are scarce on north slopes in the Upper Sonoran grassland in the northern Temblor Range. (*P. ciliosa* var. *Davyana.*)

Plectritis ciliosa var. **insignis** (Suksdorf) Dempster. Common from the Douglas oak woodland through the shin oak brush; occasional in open areas at the lower elevations of the ponderosa pine forest in the Greenhorn Range.

Plectritis macrocera Torrey & Gray. A small colony grows in brown clay on a shaded north slope in the arid Douglas oak woodland in the south fork of upper Back Canyon (T9499).

Pelectritis macroptera (Suksdorf) Rydberg var. **patelliformis** (Suksdorf) Dyal. Rare in the Kellogg oak woods at the borders of the ponderosa pine forest on the west slope of the Greenhorn Range (9.4 miles east of Glennville on the Greenhorn Summit Road, T3595).

Campanulaceae. Bellflower Family

Githopsis specularioides Nuttall. Colonies are scarce in the Greenhorn Range, where Evalyn Farnsworth reports (personal communication) it is most common following brush burns (slope south of Cedar Creek above the Rock Corral, Farnsworth 458; near the summit of Eugene Grade, C. N. Smith 522, JEPS; foot of Pine Mountain, C. N. Smith 58, JEPS). A colony grows in gravelly soil under an isolated grove of white firs on Purdy Ridge in the Tehachapi Mountains (T11,047).

Heterocodon rariflorum Nuttall. Occasional, sometimes locally common, in shady moist places in the yellow pine forest, rarely descending to the Douglas oak woodland, in the Greenhorn Range and on Breckenridge Mountain. It is also known from the Bartolas Country of the southern Kern Plateau (Howell & True 41,802).

Lobeliaceae. Lobelia Family

Downingia bella Hoover. An isolated colony grows in a small vernal poolbed in the Upper Sonoran grassland on the northwest flank of Mt. Abel in Ventura County 0.1 mile from the Kern County line. The colony is well collected (T2222, Pollard in 1952, Chandler 818, Blakeley 5266, SBBG); the labels often attribute the colony to Kern County. The elevation of the poolbed is 5,310 feet; this is an altitude record for this plant that ranges from Colusa County to Tulare County in the Great Valley.

Downingia cuspidata Greene. Occasional around pools along Fugitt Road on the west side of Lynns Valley (T10,870). (*D. pulchella* var. *arcana.*)

Downingia pulchella (Lindley) Torrey. Monkey face. Locally common around vernal pools and in drying poolbeds in Bear Valley (T9673). Apparently the nearest stations are in Merced and Monterey counties.

Nemacladus capillaris Greene. According to Philip A. Munz (A California Flora, p. 1068), it occurs along the west base of the Sierra Nevada south to Kern County.

Nemacladus glanduliferus Jepson var. **orientalis** McVaugh. Occasional in colonies in sandy washes from the creosote bush association to the pinyon woodland. The plant is sometimes abundant in wind-deposited soil over playa clay around alkali sinks.

Nemacladus gracilis Eastwood. Slender thread-stem. Scarce in sand and gravel in the El Paso Range (east fork of Red Rock Canyon, T3536; Last Chance Canyon, T11,858). G. Thomas Robbins (Aliso 4:141) cites collections from Sunset, Rockpile, and Caliente and states the plant is rarely collected.

Nemacladus interior (Munz) Robbins. Occasional on open slopes in the ponderosa pine forest in the Greenhorn Range and on Breckenridge Mountain. (*N. rigidus* var. *interior.*)

Nemacladus rubescens Greene. Thread-stem. Common in the creosote bush association, mostly in sandy places or in loose light soil.

Nemacladus secundiflorus Robbins. Known from the chaparral on the south slope of Black Mountain on the east slope of the Greenhorn Range (at 3,500 feet elevation, Howell 38,204; 5 miles above the junction of the road to Kernville, Bacigalupi 4538, JEPS).

Nemacladus sigmoideus Robbins. Desert thread-stem. Occasional, often in extensive colonies, on sandy slopes and in canyons on the desert, west to Walker Basin, the Tehachapi Mountains, and the Mt. Pinos region.

Nemacladus Twisselmannii J. T. Howell. Known from only one small colony in granite sand between granite blocks at 7,350 feet elevation along the arid southwestern rim of the Kern Plateau southwest of Pine Flat. (The type and only collection is Ernest Christian Twisselmann 8690, 10 July 1963.) The plant is distinguished from other species of the genus by the pubescent petals, lack of glands, and leafy bract subtending the short pedicel. (Cf. John Thomas Howell. A New California Nemacladus. Leafl. West. Bot., 10:45, 1963.)

At the time of the discovery of this dwarf gray plant, the same color as the granite sand where it grows, the author surmised it would prove to be relatively common on the extensive similar sandy slopes in the then (and still) botanically little known southern Kern Plateau. However, it has not been found in several subsequent searches at the type and other likely localities. In that arid region, it may yet prove to be relatively common in a year of heavy precipitation.

Porterella carnosula (Hooker & Arnott) Torrey. A single plant grew in 1967 in damp sand along Fay Creek just below the Little Cannell Meadow forest road bridge on the Kern Plateau (T13,400).

Compositae. Sunflower Family

Eupatorieae. Eupatory Tribe

Brickellia californica (Torrey & Gray) Gray. Brickell bush. Occasional in the Douglas oak and pinyon woodlands and the chaparral in the mountains south to Mt. Pinos, usually growing around rock outcrops.

Brickellia desertorum Coville. Desert Brickell bush. Scarce in canyons in the El Paso Range, in the desert-facing canyons of the mountains, and those bordering South Fork Valley west to the Erskine Canyon.

Brickellia microphylla (Nuttall) Gray. Scarce in arid rocky places in the Douglas oak woodland from Kern Canyon above Kernville south through the Piute Mountains to Caliente Canyon.

Brickellia multiflora Kellogg. Known from a single plant on a gravelly slope in the arid shrub association just above the creosote bush association in the White Rock fork of Jawbone Canyon, the southwestern limit for this species of the desert ranges northeast to Nevada (T6658).

Brickellia Nevinii Gray. A colony grows on a roadbank 0.5 mile northwest of Valle Vista Public Camp on the ridge leading to Mt. Abel (Chandler 2719, T12,758, 12,565). According to David D. Keck (A California Flora, p. 1269), the plant also occurs in San Emigdio Canyon.

Astereae. Aster Tribe

Acamptopappus sphaerocephalus (Harvey & Gray) Gray var. **hirtellus** Blake. Golden head. A common subshrub of the desert slopes and canyons; occasional in the creosote bush association on the plains.

Aster adscendens Lindley. Occasional in wet meadows in the ponderosa pine forest in the Greenhorn Range (where it occurs as a waif along Cedar Creek in the Douglas oak woodland on the Farnsworth Ranch near Glennville, Farnsworth 276, 291) and in the Jeffrey pine forest on the southwestern Kern Plateau. A colony grows at a seep on the crest of Piute Mountain 1.4 mile south of Saddle Springs (T10,296).

Aster bernardinus Hall. An aster of the mountains of southern California that grows in a scattered colony in dryish soil at the lower end of French Meadow (T8929) and is scarce at Landers Meadow (T10,203) in the Piute Mountains.

Aster chilensis Nees. Occasional in moist places, ditch bottoms, etc., from Ft. Tejon to Tehachapi; a garden escape in the Bakersfield region.

Aster Eatonii (Gray) Hall. Common around the fenced springs in French Meadow in the Jeffrey pine forest in the Piute Mountains (T8931, 10,192, C. N. Smith 873, JEPS). This aster, which ranges from Inyo County to British Columbia, also occurs at Lamont Meadow on the Kern Plateau in the southeastern corner of Tulare County (T10,327). (*A. foliaceus* var. *Eatonii.*)

Aster exilis Elliott. Slender aster. A common weed in moist soil in all associations except those of the highest mountains and the desert.

Aster frondosus (Nuttall) Torrey & Gray. Scarce in moist alkali places in South Fork Valley. A small colony grows around the cement plant's waste water sump at Monolith.

Aster hesperius Gray. Scarce in wet soil at the north end of French Meadow in the Piute Mountains (C. N. Smith 882, JEPS). It is well established at a spring at the head of Big Canyon at the Ernest Still ranch near Annette in the northern Temblor Range (T5909), an obvious escape from the nearby Still Ranch garden where the aster was planted by the late Lelia Penwell Still, possibly as long as 65 years ago. The source of her plants is not known.

Aster intricatus (Gray) Blake. Occasional in alkali sinks in the valley.

Aster pauciflorus Nuttall. Alkali aster. Common on the subalkaline flats at Isabella, especially around Scovern Hot Springs, a western extension of its range from the Amargosa Desert, whence it ranges to Arizona, Texas, and Saskatchewan.

Baccharis Douglasii DeCandolle. A dense colony grows in moist subalkaline soil in Santos Creek in lower Santos Canyon in the Temblor Range (below the old Santos cabin, T821), and another at the potrero in upper San Emigdio Canyon (T11,046).

Baccharis Emoryi Gray. Occasional around Bakersfield, mostly along Kern River. A colony grows in Sand Canyon above Poso Creek in the Poso Hills.

Baccharis glutinosa Persoon. Winter fat. Common at sunny wet places in all the Upper Sonoran associations (rare in the Greenhorn Range); occasional at springs on the desert.

Baccharis sergiloides Gray. Squaw water brush. Occasional at springs in the desert ranges and in the desert-facing canyons of the Sierra Nevada south to upper Jawbone Canyon.

Baccharis viminea DeCandolle. Mule fat. Common in wet places in the Temblor Range, occasional east to Bakersfield; scarce in the Tehachapi Mountains and on the desert. In Kern County, its range completely overlaps that of the desert *B. glutinosa,* although the species rarely grow together. In general, *B. glutinosa* tolerates somewhat more alkaline soil.

Chaetopappa fragilis (Brandegee) Keck. Rare in a disturbed loam at the west end of Weldon Meadow in the Jeffrey pine forest in the Piute Mountains (T7327) and on a sunny bank near Woodard Spring in the ponderosa pine forest on the east slope of the Greenhorn Range (T10,969); very rare in loam under Jeffrey pines at Tollgate Ridge east of Keene (T12,267). The type was collected at Havilah by Townshend Stith Brandegee on May 16, 1891. (Brandegee's use of "Havilah" and "Kernville" for localities for his Kern County collections was probably highly generalized.) (*Pentachaeta fragilis.*)

Chrysopsis Breweri Gray. Scarce in the red fir forest on Sunday Peak, growing most often on the borders of thickets of *Castanopsis sempervirens* (Howell 38,846, T4643, 3956, C. N. Smith 717, JEPS).

Chrysopsis oregona (Nuttall) Gray var. **scaberrima** Gray. Known from the Greenhorn Range 3.5 miles southwest of the summit of Eugene Grade (C. N. Smith 1136, JEPS). Mrs. Smith (personal communication) reports that it also occurs at the foot of Rattlesnake Grade and on the sand flats at the north end of Lynns Valley.

Chrysopsis villosa (Pursh) Nuttall var. **echioides** (Bentham) Gray. Common in the Upper Sonoran associations from the Temblor Range south to the San Emigdio Range and from the Mt. Pinos region east to Lebec; rare in sandy places in the valley east to Panama.

Chrysopsis villosa var. **fastigiata** (Greene) Hall. Common on the sandy flats and washes in the Upper Sonoran associations through the mountains south to Cuddy Valley.

Chrysothamnus nauseosus (Pallas) Britton ssp. **ceruminosus** (Durand & Hilgard) Hall & Clements. Rabbit brush (for all *C. nauseosus* subspecies). The plant is common and autumnally colorful on the sandy slopes on the ridges between Jawbone Canyon and Kelso Valley. It occurs in Kern Canyon from Kernville well into Tulare County and is common in the Lamont Meadow region on the southeastern corner of the Kern Plateau in Tulare County near the Kern County line. According to David D. Keck (A California Flora, p. 1191), it is rare from the Mojave Desert to Tejon Pass. The type was collected at Tejon Pass by Adolphus L. Heermann in September, 1853.

Chrysothamnus nauseosus ssp. **consimilis** (Greene) Hall & Clements.

The common rabbit brush of the pinyon woodland, occasionally extending to the Jeffrey pine forest.

Chrysothamnus nauseosus ssp. **hololeucus** (Gray) Hall & Clements. Common on the desert, even in very arid and often subalkaline places, southwest through the pinyon woodland to Cuyama Valley; west to Kernville (where it is common about the town); rare in the southern Temblor Range (Recruit Canyon, Grocer Grade). Scattered plants grow on the west slope of the Greenhorn Range on ridges in well developed Douglas oak woodland. In 1957, Eben McMillan (41–1) discovered a colony at a graded area where an oil well had been drilled in the La Panza Range in San Luis Obispo County, showing that the plant had no difficulty becoming established in a disturbed area at least 15 miles from its nearest known occurrence.

Chrysothamnus nauseosus ssp. **mohavensis** (Greene) Hall & Clements. Occasional, the distribution disjunct, from the head of Nine Mile Canyon in Inyo County south through the pinyon woodland and the arid shrub association to the western Mojave Desert in Antelope Valley and the Douglas oak woodland at Lebec. It is common around Wofford on the east slope of the Greenhorn Range.

Chrysothamnus nauseosus ssp. **occidentalis** (Greene) Hall. Common and picturesque at high altitudes, often around exposed summits (Breckenridge Mountain, Piute Peak, Tehachapi Peak, and Mt. Abel).

Chrysothamnus nauseosus ssp. **viscosus** Keck. According to David D. Keck (A California Flora, p. 1191), it occurs along the east flank of the Sierra Nevada south to Walker Pass. Despite extensive collecting of *Chrysothamnus* from the Walker Pass region and from the southeastern Sierra Nevada in Tulare County, this subspecies has not been found. This western part of the range given by Keck may be in error.

Chrysothamnus paniculatus (Gray) Hall. Occasional in canyons and on washes and fans on the desert. It is also known from the sparse pinyon woodland in Chimney Creek Gorge northeast of Onyx (Pittman 27). (*Ericameria paniculata.*)

Chrysothamnus Parryi (Gray) Greene ssp. **vulcanicus** (Greene) Hall & Clements. Common in the Jeffrey pine forest in the Piute Mountains, especially on the extensive sagebrush flats; scarce in the Greenhorn Range (south side of Cooks Peak, Howell 38,785; near Tiger Flat, Howell 38,862).

Chrysothamnus teretifolius (Durand & Hilgard) Hall. Occasional at the lower elevations of the desert-facing slopes and canyons of the mountains southwest to the Tehachapi Mountains. The type locality given as the "mountains around Tejon Valley" by Adolphus L. Heermann in 1853 is obscure; it possibly refers to the arid hills at the head of Cottonwood Canyon east of Tejon Pass. (*Ericameria teretifolia.*)

Chrysothamnus viscidiflorus (Hooker) Nuttall. Occasional in the chaparral on the north end of Piute Mountain and common in granite in the

Jeffrey pine forest along the crest of the mountain where it completely intergrades with ssp. *pumilus*. The plant is known from the roadside on the south side of Black Mountain at 3,500 feet elevation in the Greenhorn Range (Howell 38,836) where it is perhaps adventive. The Piute plants are a mass of cheerful yellow in the fall, particularly in a year with ample summer showers.

Chrysothamnus viscidiflorus ssp. **pumilus** (Nuttall) Hall & Clements. Common on the exposed summits of Mt. Pinos, Sawmill Mountain, Mt. Abel, and the crest of Piute Mountain. It grows in autumnally colorful colonies in the rugged fractured slate outcrops at Lookout Point in the Piute Mountains.

Conyza bonariensis (Linnaeus) Cronquist. Scarce in moist places in the valley. Native of South America. (*C. crispa, Erigeron crispus, E. linifolius.*)

Conyza canadensis (Linnaeus) Cronquist. Horseweed. Widespread, often in dense colonies, in cismontane sunny moist places below the yellow pine forest, occasional in irrigated farmlands in the desert. (*Erigeron canadensis* of many California references.)

Conyza Coulteri Gray. Widespread and often abundant in the valley and the farmlands around Cantil on the desert, especially in subalkaline soils. It is scarce in the Temblor Range and rare along Cedar Creek at the Farnsworth Ranch in the Greenhorn Range. The plant's occurrence in the upper San Joaquin Valley is largely concentrated in Kern County and the adjacent parts of Kings and Tulare counties; it is not nearly so common north of Tulare Lake. Although apparently native, with us it almost always behaves as a weed.

Corethrogyne filaginifolia (Hooker & Arnott) Nuttall. Glandular cudweed. Scarce at higher elevations in the Douglas oak woodland from the Greenhorn Range to the Tehachapi Mountains. The subspecific taxa in *Corethrogyne filaginifolia* are confused and the many named varieties are in need of clarification and critical evaluation. The treatment here, for the most part, has followed that of David D. Keck in "A California Flora."

Corethrogyne filaginifolia var. **brevicula** (Greene) Canby. Occasional in the mountains, mostly in the Douglas oak woodland.

Corethrogyne filaginifolia var. **glomerata** Hall. Common from the higher elevations of the Douglas oak woodland to the Jeffrey pine forest south to the Tehachapi Mountains. (Var. *bernardina* in part.)

Corethrogyne filaginifolia var. **Peirsonii** Canby. Occasional about the base of Mt. Pinos from Brush Mountain near Mt. Abel to Castac Valley. According to David D. Keck (A California Flora, p. 1203), it also occurs in the Tehachapi Mountains.

Corethrogyne filaginifolia var. **virgata** (Bentham) Gray. Occasional in the Temblor Range through the low foothills of the mountains to Granite Station. Our plants are not always distinct from var. *glomerata*.

Eastwoodia elegans Brandegee. Eastwoodia. Occasional, often in colonies, on hot dry slopes in the Upper Sonoran subshrub association in the Temblor Range, the lower San Emigdio Range, and the west slope of the Tehachapi Mountains. It is especially common in the very arid southern Temblor Range west of Taft, extending to the southern Carrisa Plains and Cuyama Valley.

Eastwoodia is an indicator for the Upper Sonoran subshrub association, an aggregation of markedly xerophytic plants that occupies a zone shaped like a very long inverted comma on the hot dry slopes on the west side of the San Joaquin Valley from Alameda County south to the head of the valley. While *Haplopappus linearifolius* is by far the most common woody plant in this association, it also grows in many other vegetational complexes; *Eastwoodia* is limited to this local and highly specialized plant community.

The monotypic genus was named for a memorable California botanist, Alice Eastwood, who collected the type in Warthan Canyon near Alcalde in Fresno County southwest of Coalinga. It is fitting that the name of a plant so well adapted to the demands of a harsh and rigorous environment commemorates a pioneer botanical scholar, explorer, and collector who thrived on adversity and challenge.

Erigeron divergens Torrey & Gray. Rare on the slopes around Kern and South Fork valleys, usually growing in cracks in granite (Weldon, Evermann, 19 April 1915; Bull Run Creek, 2 miles west of Kern River, T6360). It is common to the northeast in the Kennedy Meadows and Rock House Basin regions along the South Fork of the Kern River in Tulare County.

Erigeron foliosus Nuttall. Fleabane. Common from the Upper Sonoran associations to the Jeffrey pine forest from Tejon Pass to the Mt. Pinos region.

Erigeron foliosus var. **Covillei** (Greene) Compton. Coville's fleabane. Common in the yellow pine and red fir forests in the Greenhorn Range, the Piute Mountains, and on Breckenridge Mountain.

Erigeron foliosus var. **Hartwegii** (Greene) Jepson. Although far south of its reported southern limits in Monterey and Merced counties, what seems to be this variety grows in the thin Jeffrey pine woods on the ridge above Catskin Meadow in the Tehachapi Mountains (T4563).

Erigeron foliosus var. **stenophyllus** (Nuttall) Gray. Occasional, mostly in the Upper Sonoran associations, in the mountains south to the Tehachapi Mountains.

Erigeron multiceps Greene. The type collection came from "gravelly spots near river banks, north fork of Kern River near Kernville, Kern or Tulare County, California, Edward Palmer and William Greenwood Wright, June 7–15, 1888" (Arthur Cronquist. Revision of the North American Species of *Erigeron*. Brittonia, 6:247, 1947).

The plant is known only from the type collection, despite attempts by many botanists to rediscover it. An erigeron collected 7 miles above [old] Kernville by J. T. Howell (5036, 13 May 1930) has the perennial root of *E. multiceps* but the pubescence of *E. divergens* and plants at Kennedy Meadows on the South Fork of the Kern River in Tulare County have the appressed pubescence of *E. multiceps* but are not perennial (T4844). Cronquist (*loc. cit.*) relates the species closely to *E. divergens,* and until it is recollected, its specific status must be maintained only with reservations. Although Roxanna Ferris (An Illustrated Flora of the Pacific States, 4:360) limits it to Tulare County, the record of Palmer's visit of June, 1888 to Kernville strongly suggests that all of his collections were made within Kern County (Rogers McVaugh. Edward Palmer: Plant Explorer of the American West, pp. 96, 236–37).

Erigeron philadelphicus Linnaeus. Philadelphia daisy. Scarce in moist sunny soil along Cedar Creek from just below the mouth of Fulton Creek to the Rock Corral in the Douglas oak woodland in the Greenhorn Range.

Grindelia camporum Greene. Gum weed. Common in waste places in the valley, scarce in the surrounding hills. It is known as a waif on the desert along Highway 14, 10 miles north of Mojave (Myrick 1281 in 1965, SBBG).

Grindelia camporum var. **australis** Steyermark. A distinctive form common in Tehachapi, Brites, Cummings, and South Fork valleys (south to Antelope Valley in Los Angeles County).

Grindelia procera Greene. Common in low places in the valley, particularly along the Buena Vista Slough. Known as a waif on the desert (roadside near Mojave, Hoffmann, 1 October 1931, SBM).

Gutierrezia bracteata Abrams. Match weed. Common on the west side of the valley, in the Temblor Range south to Cuyama Valley, and on the lower slopes of the San Emigdio Range; scarce in the east-side foothills. (*G. californica* of authors in large part.)

Gutierrezia microcephala (DeCandolle) Gray. Occasional in the shadscale scrub and creosote bush associations south to Fremont Valley. (*G. lucida.*)

Gutierrezia Sarothrae (Pursh) Britton & Rusby. The plant occurs in colonies in the Rosamond Hills and around Willow Springs and Cactus Butte, usually growing on gravelly slopes.

Haplopappus acradenius (Greene) Blake. Scarce in sandy soil or on stabilized dunes from southwest of Rosamond east to Rogers Dry Lake. (*Isocoma acradenia.*)

Haplopappus acradenius ssp. **bracteosus** (Greene) Hall. Pale-leaf golden bush. Common from the Temblor Range south to Apache Saddle in the Mt. Abel region; thence northeast on the arid slopes of the mountains to South

Fork Valley, where it is sometimes abundant in light sandy soils in abandoned farming fields. The plant is occasional in waste places in the valley and on the low foothills at the west base of the Greenhorn Range. Pale-leaf golden bush is a widespread and characteristic subshrub of the Upper Sonoran subshrub association that extends through the other arid Upper Sonoran associations.

Haplopappus arborescens (Gray) Hall. Golden fleece. Occasional in the dense chaparral on Blue Mountain west of Glennville; scarce in the chaparral on the north end of Piute Mountain and on the east slope of the Greenhorn Range. (*Ericameria arborescens.*)

Haplopappus Bloomeri Gray. Known from the rim of the Kern Plateau east of Fay Creek Canyon (Howell & True 41,837), the southern limit for the species. The nearest known stations are at Monache and Summit meadows (Howell 27,344) and at the Tunnel (Howell 25,914) on the northern Kern Plateau. (*Chrysothamnus Bloomeri.*)

Haplopappus Cooperi (Gray) Hall. Occasional, sometimes locally common, from the higher elevations of the creosote bush association to the pinyon woodland on the desert-facing slopes of the mountains, west to Kelso Canyon. (*Ericameria Cooperi.*)

Haplopappus cuneatus Gray. Wedge-leaf golden bush. Occasional around rock outcrops in all the woody Upper Sonoran associations, infrequently to the high arid summits in the Jeffrey pine forest. (The plant is very common at more than 8,000 feet elevation around Lookout Point in the Piute Mountains.) (*Ericameria cuneata.*)

Haplopappus linearifolius DeCandolle. Goldenbush. Common in the Upper Sonoran associations. It is the dominant plant in the Upper Sonoran subshrub association where it grows in often extensive colonies in the Temblor and San Emigdio ranges and locally on the lower slopes of the Tehachapi Mountains. Of no forage value, goldenbush seems to be invasive in good soils that are subjected to prolonged heavy grazing. Controlled burning followed by conservative grazing practices is effective in its control in these places. (*Stenotopsis linearifolius.*)

Haplopappus linearifolius var. **interior** (Coville) Jones. Replaces the typical subspecies in the mountains bordering the desert, west to the east slope of the Greenhorn Range. (*Stenotopsis linearifolius* var. *interior.*)

Haplopappus Palmeri Gray ssp. **pachylepis** Hall. This golden bush of the southern California mountains is rare on a sandy flat at 3,375 feet elevation at the lower limits of the pinyon woodland near the head of Salt Creek Canyon in the San Emigdio Range (T8346). (*Ericameria Palmeri.*)

Haplopappus racemosus (Nuttall) Torrey ssp. **glomeratus** (Nuttall) Hall. Occasional in moist subalkaline soil on the alkali flats below Isabella, in South Fork Valley (Cottage Grove Cemetery), and along Adobe Road near Weedpatch. It grows in good loam at Landers Meadow in the Jeffrey pine forest in the Piute Mountains.

Haplopappus squarrosus Hooker & Arnott ssp. **obtusus** (Greene) Hall. Scarce in colonies around metamorphic outcrops in the Douglas oak woodland in Grapevine and San Emigdio canyons in the San Emigdio Range. An easily accessible colony of this little known subspecies grows at the outcrops just below the drinking fountain along the freeway in Grapevine Canyon. The plant is endemic to Kern County and the interior mountains of Santa Barbara and Ventura counties. The type was collected in San Emigdio Canyon by Alice Eastwood in October 1894. (*Hazardia squarrosa* in part.)

Haplopappus squarrosus ssp. **stenolepis** Hall. Scarce, mostly in decomposed shale, sometimes around sandstone outcrops, on the east slope of Cottonwood Pass in the southernmost extension of the Diablo Ranges in extreme northwestern Kern County and in the central Temblor Range from the summit of Cedar Canyon to the summit of Black Canyon. This subspecies ranges from Parkfield Grade (the type locality) in Fresno and Monterey counties to Cuyama Valley in Santa Barbara County. (*Hazardia squarrosa* in part.)

Haplopappus Whitneyi Gray. Scarce on the ridges near Tiger Flat in the Greenhorn Range, among the granite slabs and blocks along the southwest rim of the Kern Plateau, and on the north end of Piute Mountain near Saddle Springs. Our plants have unusually large involucres. The immature fruits are frequently eaten by larvae, which severely limits seed production. (*Hazardia Whitneyi.*)

Heterotheca grandiflora Nuttall. Telegraph weed. Occasional in the east-side foothills from Poso Creek to Granite Station, mostly along roads; a weed of waste places around Bakersfield. It grew as a waif along the highway between Cameron and Mojave in 1963, a year with unusually heavy summer rains. The plant is native in southern California but probably is an immigrant here.

Lessingia germanorum Chamisso var. **Lemmonii** (Gray) J. T. Howell. Occasional in colonies, usually in sandy soil, from Frazier Park and Cuyama Valley along the arid slopes of the mountains to Inyo County; west to Kernville, east to the desert floor in Indian Wells Valley. (*L. Lemmonii.*)

Lessingia germanorum var. **Peirsonii** J. T. Howell. Occasional in colonies in the mountains south to Frazier Park, growing mostly in the pinyon woodland and arid shrub association but occuring in the Jeffrey pine forest in the Piute Mountains. Colonies are scarce in the creosote bush association on the west side of Indian Wells Valley.

Lessingia germanorum var. **ramulosissima** (A. Nelson) J. T. Howell. Rare in low sandy places in the Rosamond Hills (T6699). J. T. Howell (U. C. Pub. Bot. 16:25) cites a collection from Randsburg (K. Brandegee,

July, 1913, UC). It also occurs on the sandy plains northeast of Boron in San Bernardino County (T9043). (*L. germanorum* var. *Lemmonii* in part.)

Lessingia germanorum var. **tenuipes** J. T. Howell. Occasional in the Upper Sonoran associations, mostly the Douglas oak woodland, from the Temblor Range south to Cuyama Valley; rare in the Douglas oak woodland in the Greenhorn Range (Lumreau Creek on the west slope of Basket Peak, T7711). (*L. germanorum* var. *glandulifera* in part.)

Lessingia germanorum var. **vallicola** J. T. Howell. Common in light soils in the valley and the Temblor Range. (*L. germanorum* var. *glandulifera* in part.)

Lessingia leptoclada Gray. Common in open or disturbed places, especially in the ponderosa pine forest, on the west slope of the Greenhorn Range.

Lessingia nana Gray. Known from Tehachapi Pass (Greene, 9 September 1881), Tehachapi (Barby, DS), and near Bakersfield (Bauer, 10 October 1927, DS). Greene's collection served as the type for his *L. Parryi,* a synonym.

Lessingia nemaclada Greene var. **albiflora** (Eastwood) J. T. Howell. Common in the Temblor Range and the valley, mostly on the west side; rare in Kern Valley (Wofford, J. T. Howell 38,695) and in the Douglas oak woodland in the Greenhorn Range (Lumreau Creek, 0.2 mile below the falls, T11,620). The type was collected at old Rose Station at the head of the San Joaquin Valley (Alice Eastwood, 3 October 1894).

Lessingia tenuis (Gray) Coville. John Thomas Howell writes (personal communication) that the plant is known from Cuddy Canyon (Dudley & Lamb 4476 in 1896, US) and from Cuddy's Ranch near the Ventura County line (Howell 3859). (*L. germanorum* var. *tenuis.*)

Machaeranthera canescens (Pursh) Gray. Rare in light soil in the pinyon woodland on the east slope of Piute Mountain (Harris Grade 5.9 miles east of Landers Meadow, T10,204). It probably occurs on the southeastern Kern Plateau in Kern County where it is known nearby in Tulare County (head of Chimney Creek, T10,314). A discoid form grows at Iris Point on Mt. Pinos in Ventura County near the Kern County line (T12,606). (*Aster canescens.*)

Machaeranthera tortifolia (Gray) Cronquist & Keck. Mojave aster. Occasional on slopes, washes, and fans on the desert, where the large lavender-blue flowers make the plant one of the most attractive spring-blooming perennials. Particularly fine colonies grow on slopes in the Randsburg region. (*Aster tortifolius, A. abatus.*)

Monoptilon bellidiforme Torrey & Gray. Desert star. An interesting dwarf that is scarce on sandy flats in the creosote bush association.

Solidago californica Nuttall. California golden rod. Common in wet sunny places in all the cismontane associations.

Solidago canadensis Linnaeus ssp. **elongata** (Nuttall) Keck. Meadow golden rod. A small colony grows at the fenced spring at the north end of heavily grazed French Meadow in the Piute Mountains (T10,193). (*S. elongata.*)

Solidago occidentalis (Nuttall) Torrey & Gray. Western golden rod. Occasional in all the cismontane associations, mostly below the chaparral. It is especially common along Kern River in the valley.

Solidago spectabilis (D. C. Eaton) Gray. Alkali golden rod. An extensive colony grows in the subalkaline marshy area below Kern Hot Springs near Kernville (Howell 38,929, T9014, 9016), a southwestern extension of its range from Mono and Inyo counties. It occurs north to Oregon.

INULEAE. Everlasting Tribe

Antennaria dimorpha (Nuttall) Torrey & Gray. Scarce in gravel and fractured rock on the Mt. Pinos summit plateau.

Evax acaulis (Kellogg) Greene. Known from the hills near Glennville (Howell 5129, 5159).

Evax caulescens (Bentham) Gray var. **humilis** (Greene) Jepson. Scarce in friable (dry bog) clay at Dry Bog Knoll in the Upper Sonoran grassland at the head of Adobe Canyon in the Greenhorn foothills (C. N. Smith 310, JEPS).

Filago californica Nuttall. Common in the cismontane Upper Sonoran associations, especially in sunny thin sandy-clay soil in the grasslands.

Filago depressa Gray. Colonies are occasional in sandy places in the creosote bush association.

Filago gallica Linnaeus. Occasional in heavy soils in the Douglas oak woodland in the foothills in the Glennville-Woody region; also known from Kernville (Howell 37,134).

Gnaphalium beneolens A. Davidson. Fragrant everlasting. Occasional from the chaparral to the ponderosa pine forest in the Greenhorn Range and in the dense chaparral on the north end of Piute Mountain, usually growing in rocky places.

Gnaphalium californicum DeCandolle. California cudweed. Scarce at the lower elevations in the Douglas oak woodland through the Greenhorn foothills to Kern Canyon; rare and probably adventive on the freeway shoulder at Hart Flat in the Tehachapi Mountains (T11,165). (*G. decurrens* var. *californicum.*)

Gnaphalium chilense Sprengel. Cotton-batting plant. Common in cismontane wet places, mostly below the chaparral. It reaches the edge of the desert on the east slope of the Tehachapi Mountains.

Gnaphalium luteo-album Linnaeus. Occasional in wet often subalkaline places below the yellow pine forest, east to the edge of the desert. Introduced from the Old World.

Gnaphalium microcephalum Nuttall. Occasional around rocky places from the Douglas oak woodland through the ponderosa pine forest in the Greenhorn Range.

Gnaphalium palustre Nuttall. Lowland cudweed. Common, usually in moist soil, in all the cismontane associations. It grows in dry light wind-deposited sand in wet years in the valley.

Gnaphalium thermale E. Nelson. Known from 1.5 mile south of Greenhorn Pass (Howell 38,769). (*G. microcephalum* in part.)

Micropus californicus Fischer & Meyer. Cottonweed. Common in sandy places in the cismontane Upper Sonoran associations, mostly below the chaparral.

Pluchea purpurascens (Swartz) DeCandolle. Marsh fleabane. Common and colorful in a form with rose-pink phyllaries on the borders of the extensive and dense growth of tules (*Scirpus acutus*) in the marsh of thermal-recovery waste water at the forks of Poso Creek and Sand Canyon in the Poso Hills. The more common form with white phyllaries is occasional in small colonies in wet soil on the alkali plains along Adobe Road in the Weedpatch region and along Kern River at Bakersfield. (*Pluchea camphorata.*)

Pluchea sericea (Nuttall) Coville. Arrow weed. Colonies grow on the Caliente Canyon flood plain at the west end of Sand Ridge, along Kern River east and west of Hart Park, and in Grapevine Canyon in the Sierra Nevada north of Walker Pass.

Psilocarphus brevissimus Nuttall. Dwarf woolly heads. Dense colonies grow in the small poolbeds in the Douglas oak woodland in the Greenhorn foothills.

Psilocarphus tenellus Nuttall. Slender wooly heads. Occasional in disturbed or moist places in the ponderosa pine forest in the Greenhorn Range and in bare places in the Douglas oak woodland in Lynns Valley.

Psilocarphus tenellus var. **tenuis** (Eastwood) Cronquist. Occasional in poolbeds, playas, and other wet usually subalkaline places in the valley.

Stylocline filaginea (Gray) Gray. Occasional in arid places in the Upper Sonoran grassland, chaparral, and Douglas oak woodland in the Temblor Range and the foothills of the valley, usually growing in clay or other heavy soil. Var. *depressa* (Gray) Jepson, a form with prostrate stems, grows in friable clay in the Temblor Range.

Stylocline gnaphalioides Nuttall. Rare in bare sand around sandstone outcrops in the central Temblor Range (Alec Cook Rocks, T1991; Ross Ridge, T8375). It is also known from Kernville (Howell 37,129).

Stylocline micropoides Gray. Occasional in rather bare sandy or gravelly places on the desert from the sand dunes around the Muroc dry lakes west to the Soledad Mountain region.

AMBROSIEAE. Ragweed Tribe

Franseria acanthicarpa (Hooker) Coville. Burweed. Common and often troublesome along roadsides, in waste places, about ranchgrounds, etc., throughout the county. It occurs most frequently below the Douglas oak woodland and the arid shrub association but is not uncommon in disturbed sunny places in the yellow pine forest.

Franseria dumosa Gray. Burro brush, bur-sage. Occasional, often common, in the shadscale scrub; occasional in the creosote bush association.

Hymenoclea Salsola Torrey & Gray. White burro brush, cheese brush, bush hops. Common in the shadscale scrub and on sandy and gravelly flats in the arid shrub association; occasional at the higher elevations of the creosote bush association. Occasional in the foothills around the valley from Kern Canyon to the southern Temblor Range and in the Elk and Buena Vista hills, west to Cuyama Valley; in these places it is sometimes locally common on arid shale slopes. The plant is said to be good browse for livestock, although hardly common enough in our region to be of forage significance.

Iva axillaris Pursh var. **pubescens** Gray. Poverty weed. Widespread, often in extensive colonies, in the central Temblor Range (especially on the flats with heavy soil in the Las Yeguas region) and from Frazier Park northeast to Walker Basin and South Fork Valley. It grows most often in clay in dry meadows, sagebrush flats, and low areas, and is very common in Tehachapi Valley. Extensive colonies also grow on the bottom lands and in fields between Weldon and Lake Isabella. Although noxious in some parts of its vast range to Canada and Nebraska, apparently it is not considered particularly troublesome in our region.

Xanthium spinosum Linnaeus. Spanish thistle, spiny clotbur. Colonies are widespread but infrequent in waste places, around springs, etc., mostly below the middle elevations of the Douglas oak woodland. The spiny stems which irritate livestock and sometimes cause accidents for riders of skittish horses and the bur-like fruits which become entangled in hair and wool make it objectionable to ranchers. Thus, it has been much controlled and often locally eradicated.

Xanthium strumarium Linnaeus var. **canadense** (Miller) Torrey & Gray. Cockle bur. Common in moist places in the valley and the Temblor Range, occasional from the east-side foothills to South Fork Valley and Kernville, rapidly becoming established in the recently developed farmlands about Cantil. The young plants are said to be poisonous to livestock but there is no local record of sickness or losses. The bur-like fruits sometimes cause mechanical damage to animals. (*X. canadense, X. italicum.*)

HELIANTHEAE. Sunflower Tribe

Balsamorhiza deltoidea Nuttall. Balsam-root. Occasional from the Mt. Pinos region northeast to Kern Canyon, growing from the chaparral and pinyon woodland to the Jeffrey pine forest.

Balsamorhiza sagittata (Pursh) Nuttall. Known from the west slope of Walker Pass in the pinyon woodland at about 5,200 feet elevation (L. S. Rose 40,302, 10 May 1940).

Bidens cernua Linnaeus. A small colony grows along a canal bank at Cottonwood and DiGiorgio roads in the Greenfield region (T4900).

Bidens frondosa Linnaeus. Beggar ticks. Common along Kern River between Lake Isabella and Kernville and at wet places in Kern Valley; scarce at Hart Park near Bakersfield.

Coreopsis Bigelovii (Gray) Hall. Common in well drained soils in all associations below the yellow pine forest except on the valley floor and the low desert plains.

Coreopsis californica (Nuttall) H. K. Sharsmith. Desert coreopsis. Occasional in sometimes extensive and colorful colonies on the desert, mostly on the sandy plains. In good years the town of Mojave lies in the middle of a deep yellow blanket of several hundred acres of desert coreopsis.

Coreopsis calliopsidea (DeCandolle) Gray. Occasional from the Temblor Range south to Cuyama Valley; scarce on the desert plains and in the valley. It grows most often in clay that is derived from brown shale. The plant's rich yellow flowers borne on relatively tall stems makes it a useful annual for the early spring garden.

Encelia virginensis A. Nelson ssp. **actoni** (Elmer) Keck. Mountain bush sunflower. Occasional, often locally common, on desert slopes and canyons and in the foothills around the valley from Granite Station to Tejon Canyon and in the canyons bordering the upper Cuyama Valley. (*E. actoni.*)

Helianthella californica Gray var. **nevadensis** (Greene) Jepson. William Weber (Amer. Midl. Nat. 48:28) cites a collection from Breckenridge Mountain (Grinnell 271, US) and another from the North Fork of Kern River near Kernville (Palmer in 1888, US).

Helianthus annuus Linnaeus. Common sunflower. Common below the yellow pine forest in waste places, roadsides, etc.; scarce in good years in similar places on the desert. (*H. annuus* ssp. *lenticularis.*)

Helianthus gracilentus Gray. Slender sunflower. An extensive colony grows in recently graded metamorphic gravel and rock at the microwave station at Pescadero Point at the head of Pescadero Canyon at 6,000 feet elevation in chaparral in the southwestern Tehachapi Mountains (T12,547). A small colony grows in fractured metamorphic rock on the road to the point in chaparral and digger pines at 5,100 feet elevation (T12,546). These colonies are eastern range extensions for this sunflower of the Coast Ranges and southern California.

Wyethia invenusta (Greene) W. Weber. Scarce in the lower levels of the yellow pine forest in the Greenhorn Range and at the summit of Breckenridge Mountain. The type was collected in the mountains of Kern County [Greenhorn Range] (Edward Palmer 105, June 5–15, 1888). (*Balsamorhiza invenusta.*)

Wyethia ovata Torrey & Gray. Mule ears. Ocacsional on sunny open slopes, often in disturbed soil, in the yellow pine forest south to the Piute Mountains and Breckenridge Mountain. Rare in the Douglas oak woodland along Cedar Creek on the west slope of the Greenhorn Range (Farnsworth 213).

MADIEAE. Tarweed Tribe

Achyrachaena mollis Schauer. Blow-wives. Common in heavy soils, mostly in the Upper Sonoran grassland, in the Temblor Range and the foothills of the mountains; east to Tehachapi Valley and the southwest end of Antelope Valley.

Blepharizonia plumosa (Kellogg) Greene ssp. **viscida** Keck. Scarce in the Temblor Range south to the summit of Messa Ridge, usually growing in small dense colonies in heavy soil.

Calycadenia hispida (Greene) Greene forma **albiflora** Keck. Abundant in heavy sandy soil in a small abandoned farming field on the low ridge between Brites and Tehachapi valleys (T3994).

Calycadenia spicata (Greene) Greene. Extensive colonies grow on the open slopes in heavy soil on the high hills at the head of Lumreau Creek on the west slope of the Greenhorn Range. The flowering spikes are atypically congested and dense (T7715, Farnsworth 183, 315).

Hemizonia arida Keck. Red Rock tar weed. A very restricted endemic that grows along the short subalkaline seep in Red Rock Canyon about 300 yards above the freeway bridge. In years with wet winters, the colony extends along the ephemeral streamlet and moist subsurface sand for perhaps a quarter mile, in dry years for as little as 20 yards.

The plant is closely related to *Hemizonia pallida,* common in late spring in light dry soil around the head of the San Joaquin Valley, and it is possible both species evolved from a common ancestor after separation and isolation by the uplift of the Tehachapi Mountains. *Hemizonia arida* may have been widespread in the Mohavean woodland of Pliocene time. Drying of what is now the desert would have eliminated favorable habitats, resulting in the near extinction of the species.

A factor in the survival of the small colony is its very long blooming season, from as early as mid-April until mid-November in favorable years. It can endure even the heaviest grazing and trampling. Before the advent of livestock transportation by truck, many thousands of sheep were driven through Red Rock Canyon each year in the late spring on the "Long Trail" from the San Joaquin Valley to the Bishop region. Mary Austin (The Flock, p. 76) gives a vivid description of the congested passage of both cattle and sheep through the canyon and their desperate crowding around its scant water. The recent increase in recreational use of the Red Rock region presents new hazards. The sheep did not spin wheels while making sharp turns in the sandy wash, which has become a favorite play area for motorcyclists, jeepsters, and other mechanized vandals who completely destroyed the plants that grew in 1965. If such activities are continued, Red Rock tarweed's future may be bleak.

The type collection is David Daniels Keck & Palmer Stockwell 3279, 11 May 1935. The author's collections, representing all stages of growth, are 7095, 10,150, 10,400, and 11,763.

Hemizonia Fitchii Gray. Fitch spikeweed. A rare weed in Kern County, known only from a recurrent roadside colony in the Upper Sonoran grassland 0.7 mile west of Granite Station (T4042). (*Centromadia Fitchii.*)

Hemizonia Heermanii. Greene. Tehachapi tarweed. A common summer annual at the middle elevations of the mountains from the Tehachapi region north; west to the borders of the valley. The plant is local and probably adventive along the highway on the east slope of Cottonwood Pass in the southernmost extension of the Diablo Ranges in extreme northwestern Kern County. The type was collected near Keene (Charles Christopher Parry and Edward L. Greene, 9 September 1881). (*Hemizonia virgata* var. *Heermannii, Holocarpha Heermannii.*)

Hemizonia Kelloggii Greene. Occasional on the east slope of the Greenhorn Range and around Kernville, scarce in Choice Valley in the northern Temblor Range, rare and possibly only a roadside adventive in Cummings Valley in the Tehachapi Mountains. (*H. Wrightii.*)

Hemizonia obconica Clausen & Keck. Common along the low east-side foothills south to Kern Canyon and on the mesas east of Bakersfield. (*Holocarpha obconica.*)

Hemizonia pallida Keck. San Joaquin tarweed. A late spring tarweed endemic to the arid upper San Joaquin Valley and the surrounding foothills. It is often abundant on the higher plains and lower hill slopes, coloring them a light yellow after the spring annuals have dried. The type was collected at the head of the San Joaquin Valley, 5.3 miles north of Grapevine (David Daniels Keck & Wm. McKinley Heusi (Hiesey) 2255, 6 May 1933).

Hemizonia pungens (Hooker & Arnott) Torrey & Gray. Spikeweed. Common in often dense colonies around vernal pools, winter-flooded places, along the borders of sloughs, canal banks, etc., in the valley and the Temblor Range. Colonies are occasional in the mountain valleys, especially in the marshy areas in South Fork Valley. (*Centromadia pungens.*)

Lagophylla glandulosa Gray. Collected at Cameron, in the lower Tehachapi Pass on the borders of the creosote bush association, by Marcus Jones in 1900 (POM). This is an extremely anomolous station for this plant of the Sierra Nevada foothills south to Fresno County. The very arid climate at Cameron, the disjunct locality, and the failure to rediscover it suggest that the Jones collection was a waif. Confusion in labelling is another possibility. However, the record of Jones's botanical explorations shows he collected at Cameron on July 24, 1900, and that none of his collecting that year was done in regions where *L. glandulosa* usually grows (Leafl. West. Bot., 10:220, 1965).

Lagophylla ramosissima Nuttall. Common hareleaf. Common, mostly in the Upper Sonoran grassland from the Temblor Range southeast to the Joshua tree groves in the northwestern Antelope Valley, thence through the foothills of the mountains to the Granite Station region and Davis Guard Station. This summer-blooming plant is remarkably well adapted to growth in seemingly dry soil, making most of its growth after the spring annuals have dried from lack of moisture.

Layia glandulosa (Hooker) Hooker & Arnott. White layia. Common in all associations to the borders of the yellow pine forest, including the valley floor and the desert, where it even grows in the dunes around the alkali dry lakes. Thus, it has one of the most extensive ranges of any of our native plants.

Layia heterotricha (DeCandolle) Hooker & Arnott. Scarce in the Upper Sonoran associations from Yeguas Peak in the Temblor Range southeast to the Mt. Pinos region; thence through the mountains northeast to Sand Canyon near Monolith. It is restricted to clay, usually the highly local beds of ultra-fine friable (dry bog) clay, and is often colorfully common in this unusual soil. (*L. graveolens.*)

Layia leucopappa Keck. Endemic to vernally wet usually whitish clay soil on small flats on the slopes from the Comanche Hills southeast of Arvin south to the neighboring Tejon Hills, apparently growing in only the most favorable years. It was once collected in what are now farmlands southeast of Mettlers Station. The great increase in cultivated farming around the head of the valley, especially on the plains below Comanche Point, may have eliminated some of the very few colonies of this Kern County rarity. The type was collected near Comanche Point, nearly 10 kilometers south of Arvin (E. Roy Weston 583, 3 April 1927).

Layia Munzii Keck. Munz layia. Occasional in the Upper and Lower Sonoran grasslands in the Temblor Range and the bordering high plains along the west side of the valley, growing mostly in swales and on low flats.

Layia pentachaeta Gray. Rare in the east-side foothills, mostly in the Upper Sonoran grassland on north slopes, south to the west slope of Breckenridge Mountain (mouth of Kern Canyon, T5945; near Parker Spring on Breckenridge Mountain, T10,519, where it is scarce in a scattered colony of ssp. *albida*).

Layia pentachaeta ssp. **albida** Keck. Common in the Temblor Range, the valley, and the east-side foothills; occasional to the yellow pine forest.

Layia platyglossa (Fischer & Meyer) Gray ssp. **campestris** Keck. Tidy-tips. Scarce in the central and northern Temblor Range, south to the pinyon woodland in the Mt. Abel region and the Upper Sonoran grassland at the southwest border of Antelope Valley along the Tehachapi Mountains (especially in the Mexican Mine region). It is usually restricted to heavy clay, especially in places where water collects briefly following winter storms, where it often grows in dense colonies.

Madia densifolia Greene. Occasional in waste places and along roadsides in Lynns Valley; a rare roadside weed in Tehachapi Valley (Highline Road at Water Canyon Road, T11,159). (*M. elegans* ssp. *densifolia.*)

Madia elegans D. Don. Common in the Douglas oak woodland in the Woody-Glennville region; rare in the pinyon woodland northwest of Mt. Abel.

Madia elegans ssp. **vernalis** Keck. Occasional in colonies in the mountains, mostly in the chaparral and the higher elevations of the Douglas oak woodland. The plant is a vernal form of the typical subspecies.

Madia exigua (Smith) Gray. A roadside colony grew in 1965 on the west slope of Basket Peak at 3,775 feet elevation in the Douglas oak wood land in the Greenhorn Range (T10,995).

Madia gracilis (Smith) Keck. Uncommon but widespread in the yellow pine forest and the Douglas oak woodland in the mountains south to Tunis Ridge in the Tehachapi Mountains.

Madia minima (Gray) Keck. Colonies are scarce in disturbed or clay soils in the yellow pine forest in the Greenhorn Range, on Breckenridge Mountain, and at Tecuya Ridge and Mt. Pinos. (*Hemizonella minima.*)

Madia radiata Kellogg. Rare in usually extensive colonies in ultra-fine friable (dry bog) clay in the Temblor Range (head of Bitterwater Creek in the Pinole Hills, T10,494; head of McClean Canyon, T1233).

Madia Wheeleri Gray. Occasional in colonies in the yellow pine forest, growing most often in dry loamy soil at the borders of meadows. (*M. elegans* ssp. *Wheeleri, Hemizonia Wheeleri.*)

Helenieae. Sneezeweed Tribe

Chaenactis carphoclinia Gray. Occasional in canyons or other rocky or gravelly places in the creosote bush association. The typical variety and var. *attenuata* are among the most drouth resistant of the desert annuals; occasional plants succeed even in the driest years.

Chaenactis carphoclinia var. **attenuata** (Gray) Jones. Scarce in canyons in the El Paso Mountains and on the sandy slopes east of Ridgecrest, becoming common on the broad sandy plains at the head of Salt Wells (Poison) Canyon in neighboring San Bernardino County. It is perhaps even more drouth tolerant than the typical variety.

Chaenactis Fremontii Gray. Pin-cushion flower. Common in sandy or light soils in the Lower Sonoran grassland and the creosote bush association. It is occasional in the foothills around the valley north to the head of Adobe Canyon in the Greenhorn Range. The plant usually grows in light or poor often gypseous soils.

Chaenactis glabriuscula DeCandolle. Golden girls. Common and colorful in the Temblor Range and from the high plains around the valley to the edge of the yellow pine forest.

Chaenactis glabriuscula var. **curta** (Gray) Jepson. Rare in the Douglas oak and pinyon woodlands and the chaparral from Kernville southwest through the mountains to the upper Cuyama Valley, usually growing on sandy flats, dry washes, etc.

Chaenactis glabriuscula var. **lanosa** (DeCandolle) Hall. Rare in chalky or sandy soil in the northern Temblor Range (0.5 mile northwest of the Choice Valley School, T6035).

Chaenactis glabriuscula var. **megacephala** Gray. Scarce on the high plains on the east side of the Valley and in the neighboring low foothills south to Comanche Point. Although it usually grows on sandy or gravelly arid slopes, it sometimes occurs in heavy clay (Long Tom Gulch, C. N. Smith 1073, JEPS).

Chaenactis macrantha D. C. Eaton. Rare in sand in the shadscale scrub along County Line Road on the plains east of Ridgecrest (T11,837).

Chaenactis rubricaulis Rydberg. Scarce on sandy slopes in the Jeffrey pine forest on the Kern Plateau and on the southwest flank of Lookout Point in the Piute Mountains. (*C. Douglasii* in part.)

Chaenactis santolinoides Gray. Perennial pin-cushion flower. Scarce but widespread in rocky or coarse soils in the yellow pine forest and the higher elevations of the arid Douglas oak woodland in the mountains. It occurs from the San Bernardino Mountains to Mt. Pinos, thence northeast to the southern Sierra Nevada and is a typical plant of what Willis Linn Jepson (Manual of the Flowering Plants of California, p. 12) called the Tehachapan Endemic Area.

Chaenactis stevioides Hooker & Arnott. Scarce (although almost common in unusually wet years) in sandy soils in the creosote bush association.

Chaenactis stevioides var. **brachypappa** (Gray) Hall. Occasional on the desert plains and slopes, scarce in sandy places in the mountains around the head of the San Joaquin Valley, north to the Temblor Range.

Chaenactis Xantiana Gray. Tejon chaenactis. Occasional on dry slopes in the mountains southwest to Mt. Pinos and in the Temblor Range, growing mostly in the arid Upper Sonoran associations. It is occasional on the desert plains, where some plants grow in even the driest years. The type was collected at Ft. Tejon (Xantus de Vesey in 1857–58).

Eatonella Congdonii Gray. Scarce in the valley, usually growing in wind-modified light soil or on sand dunes. It grows only in years of more than normal rainfall.

Eriophyllum ambiguum (Gray) Gray. Common from the Greenhorn Range south to Ft. Tejon, growing mostly in the Douglas oak woodland. It is occasional at the edge of the valley (Sand Ridge) and in the desert-facing canyons through the El Paso Range to the Rademacher Hills and the Rand Mountains. The type was collected at Ft. Tejon by George Henry Van Horn in 1863.

Eriophyllum ambiguum var. **paleaceum** (Brandegee) Ferris. A distinctive form with large flowering heads that is occasional on the desert slopes of the mountains and in the El Paso Range, west to the east slope of the Greenhorn Range and, according to Roxanna Ferris (Illustrated Flora of the Pacific States, 4:230), in Walker Basin. The type was collected at Kernville in 1892 by Townshend Stith Brandegee.

Eriophyllum confertiflorum (DeCandolle) Gray. Yellow yarrow. Common in the cismontane Upper Sonoran shrubby associations. It occurs east of the mountains at the head of canyons north of Walker Pass and along the high ridges north of Cameron. A form with loose inflorescences that has been called var. *laxiflorum* Gray grows at the Piute cypress grove on Black Mountain 2.9 miles west of Wofford (T8468).

Eriophyllum lanatum var. **Hallii** Constance. Tejon yarrow. A very rare and local form known from only two collections: in the mountains south of Ft. Tejon (Coville & Funston 1184, 7 July 1891, US) and the type from 3,500 feet elevation in the foothill woodland near Ft. Tejon [in the rocky canyon 0.5 mile west of the fort] (Harvey Monroe Hall 6262, June 16–19, 1905, UC).

Eriophyllum lanatum var. **obovatum** (Greene) Hall. A plant of the San Bernardino Mountains that is occasional on rocky slopes in the ponderosa pine forest in the Greenhorn Range, becoming rare in the southern Sierra Nevada of Tulare County.

Eriophyllum Pringlei Gray. Occasional in sandy places on the desert, southwest through the arid shrub association and the pinyon woodland to the Mt. Abel region, west to Kelso and South Fork valleys.

Eriophyllum Wallacei (Gray) Gray. Occasional in sandy places in the creosote bush associaton, usually growing in small scattered colonies.

Gaillardia aristata Pursh. Blanket flower. Known as a garden escape near Tehachapi (road to Old Town, about 2 miles west of Tehachapi, Ben Easley & W. V. Johnson, 12 August 1964). Native from Colorado to Oregon and British Columbia.

Helenium Bigelovii Gray. Occasional in wet places on the north and east slopes of Sunday Peak in the red fir and ponderosa pine forests, in the sunny cienegas at the head of Tejon Creek in the Tehachapi Mountains, and at the wet meadow at Iris Point in the Jeffrey pine forest on Mt. Pinos.

Helenium puberulum DeCandolle. Sneezeweed, rosilla. Scarce along Poso Creek near Poso Mine, on the Kern River flood plain above Lake Isabella, along Kern River at Hart Park, and on a shaded moist bank in the dense growth in the marsh below Ft. Tejon.

Lasthenia chrysantha (Greene) Greene. A San Joaquin Valley endemic that reaches its southern limits around vernal pools on the alkali plains northwest of Pond (Garces Road 2 miles west of Corcoran Highway, T4148). (*Crockeria chrysantha.*)

Lasthenia chrysostoma (Fischer & Meyer) Greene. Gold fields. Abundant in the valley and the surrounding hills through the Douglas oak woodland, occasional on sandy plains on the desert, often coloring broad expanses bright yellow in the spring. The plant tolerates a variety of edaphic situations and is abundant in dry light soils in the hills and in moist or even wet often subalkaline soils in the valley alkali sinks, around vernal pool-beds, and at other sunny wet places. Gold fields blooms in early spring; in dry years plants hardly an inch tall flower and set seed. (*Baeria chrysostoma, B. chrysostoma* ssp. *gracilis.*)

Lasthenia debilis (Greene) Ornduff. Scarce but widespread in the woody cismontane Upper Sonoran associations. The plant is most frequent south through the Greenhorn Range to Kern Canyon, usually growing in small colonies in protected shady places, often near patches of *L. chrysostoma* that occupy sunny areas in openings in the Douglas oak woodland. In unusually wet years it is much more common and has been collected in the shade of saltbush at such arid places as the McKittrick Hills and the foot of Grocer Grade near Maricopa. (*Baeria debilis.*)

Lasthenia Fremontii (Torrey) Gray. Scarce in colonies around vernal pools and at low places on the alkali plains from Semitropic to Pond and Delano. Semitropic is the type locality for the form called *Baeria Fremontii* var. *heterochaeta* Hoover (Robert F. Hoover 944, 11 April 1936), where it probably no longer occurs because of land leveling for farming. (*Baeria Fremontii.*)

Lasthenia Ferrisiae Ornduff. Alkali daisy. Occasional in dense colorful colonies in wet usually subalkaline soils, especially around vernal pools, in the valley and the Temblor Range. (Cf. Robert Ornduff. The Goldfields Genus *Lasthenia*. Univ. Calif. Pub. Bot., 40:74, 1966). (*L. glabrata* var. *Coulteri* as to almost all of our plants.)

Lasthenia glabrata Lindley ssp. **Coulteri** (Gray) Ornduff. Robert Ornduff (Univ. Calif. Pub. Bot. 40:74) cites Heller 7824 from Tehachapi, the only Kern County record for this plant of southern California.

Lasthenia microglossa (DeCandolle) Greene. Scarce in the cismontane Upper Sonoran associations; rare on the desert (Soledad Mountain, T4326). It usually grows in small dense colonies in the shade of shrubs or around rock outcrops. (*Baeria microglossa.*)

Lasthenia minor (DeCandolle) Ornduff. Common, often abundant, in winter-wet soil in the valley and surrounding hills. It is especially dense and colorful around subalkaline vernal pools. (*Baeria minor.*)

Monolopia lanceolata Nuttall. Hilltop daisy. Abundant in light arid soils on open hillsides in the Temblor and San Emigdio ranges and on the plains on the west side of the valley. It is occasional through the Upper Sonoran associations northeast to Adobe Canyon in the Greenhorn foothills and east to the western edge of the desert, where it is rare. The great fields of bright clear yellow in the early spring on the broad east slopes in the Temblor Range form one of California's notable wildflower displays. The small colony in fine clay at 4,050 feet elevation on the divide between Keene and Tehachapi is probably an altitude record for the plant. Although rare on the desert, in spring an extensive colony colors the northeast slope of Soledad Mountain, a landmark on the Mojave Desert's western border. (*M. major* var. *lanceolata.*)

Monolopia stricta Crum. Scarce in light arid often gypseous soils in the valley and the Temblor and San Emigdio ranges. It is rare in light arid soils below the Douglas oak woodland in the east-side foothills. The type was collected 2 miles west of Lost Hills (Herbert Louis Mason 9314, 2 May 1935).

Nicolletia occidentalis Gray. Hole-in-the-sand, Nicollet. Scarce in deep loose sand in the arid shrub association in Short Canyon north of Walker Pass (T4540, 7120) and in the creosote bush association in the sand dunes at Rogers (T10,816) and Buckhorn dry lakes (T10,684). The caudex and stems are usually well buried in the loose often wind-drifted sand.

Orochaenactis thysanocarpha (Gray) Coville. Occasional in granite sand on slopes in the sparse Jeffrey pine forest on the southwestern Kern Plateau. It is rare on the east slope of the Piute Mountains, growing on the open arid slopes where the pinyon woodland and the Jeffrey pine forest merge (1.7 mile east of Landers Meadow, T7442).

Perityle Emoryi Torrey. Rock daisy. Scarce in metamorphic talus at the base of a rocky canyon wall in Last Chance Canyon at 2,400 feet elevation in the El Paso Range (T4362, 11,863).

Pseudobahia Heermannii (Durand) Rydberg. Known from a disjunct occurrence on the east slope of the Greenhorn Range on the [old] Kernville-Glennville road at 5,000 feet elevation (Peirson 8855, RSA). The plant is infrequent in colonies in the western foothills of the Sierra Nevada south to the north slope of King George Mountain in southern Tulare County (T11,801). (*Eriophyllum Heermannii.*)

Pseudobahia Peirsonii Munz. Colonies are scarce in clay in the Upper Sonoran grassland and the Douglas oak woodland in the hills west of Woody south to Dry Bog Knoll at the head of Adobe Canyon.

Rigiopappus leptocladus Gray. Occasional in the Douglas oak woodland and chaparral in the mountains and the Temblor Range, usually growing in clay or other heavy soils, often in swales or disturbed places.

Syntrichopappus Fremontii Gray. Occasional on gravelly slopes and sandy flats in the creosote bush and shadscale scrub associations.

ANTHEMIDEAE. Mayweed Tribe

Achillea borealis Bongard ssp. **californica** (Pollard) Keck. California yarrow. Occasional from the Upper Sonoran grassland through the Douglas oak woodland and at wet places in the valley; scarce in the Jeffrey pine forest in the Tehachapi Mountains. (*A. Millefolium* var. *borealis, A. californica.*)

Achillea gigantea Pollard. Giant yarrow. A distinctive tall yarrow that is scarce in wet places in the valley (rather common along canals in the Buttonwillow region). (Philip A. Munz does not list the plant in synonymy in "A California Flora"; presumably it would be included in his concept of *A. borealis* ssp. *californica.* Malcom Nobs in Ferris & Abrams "Illustrated Flora of the Pacific States" gives *A. Millefolium* var. *gigantea.*)

Achillea lanulosa Nuttall. Mountain yarrow. Common in meadows in the yellow pine forest, growing from the very wet soil in the open areas to the dry often disturbed loam on the meadow borders. (*A. Millefolium* var. *lanulosa.*)

Anthemis Cotula Linnaeus. Mayweed. Common along roads, in waste places, and along creeks in the Greenhorn foothills; an infrequent but widespread attractive weed elsewhere, mostly in the Douglas oak woodland. Native of Europe.

Artemisia californica Lessing. California sagebrush, old man. Common in the woody Upper Sonoran associations in the Temblor and San Emigdio ranges. On very hot days, its aroma and that of California juniper combine to make a delicate and appealing scent in the late afternoon in canyons in the Temblor Range.

Artemisia Douglasiana Besser. California mugwort. Common in wet places through the Upper Sonoran associations in the mountains and the Temblor Range and along Kern River to Bakersfield. (*A. vulgaris* var. *heterophylla.*)

Artemisia Dracunculus Linnaeus. Dragon's head, tarragon. Widespread in the cismontane Upper Sonoran asociations, occasional in moist soil on the desert-facing mountain slopes and in the yellow pine forest. In Kern County, the plant is often invasive in disturbed places, such as meadows and open slopes subjected to years of heavy grazing, old logging depots, abandoned ranchgrounds, etc. Apparently this is not true elsewhere in California.

Artemisia ludoviciana Nuttall. Occasional in wet places in the valley and at lower elevations in the San Emigdio Range.

Artemisia spinescens D. C. Eaton. Bud sage. Occasional on the desert, mostly in the shadscale scrub or on the borders of alkali sinks.

Artemisia tridentata Nuttall. Big sage, sagebrush. Common in dry places in the Jeffrey pine forest and the pinyon woodland, often dominant on open slopes and in dry meadows. It occurs most abundantly on the east slope of the mountains southwest to Cuyama Valley and is common north to the Tulare County line on the east slope of the Greenhorn Range. Sagebrush is locally common on slopes in the Douglas oak woodland and chaparral at Poso Flat and along Rattlesnake Grade on the west slope of the Greenhorn Range, in canyons in the San Emigdio Range, and on flats at higher elevations on the west slope of the Tehachapi Mountains (where it is especially common at the southwest end of Cummings Valley and on Cummings Mountain). Unusually large almost arborescent plants grow on the sandy flats in the arid canyons of the upper Cuyama Valley. The extensive occurrence of this primarily transmontane plant in the Douglas oak woodland and chaparral shows that vegetation classification is a highly generalized rather than an exact science.

Most range management specialists consider sagebrush undesirable and feel it should be removed or restricted in favor of grasses and herbs; local livestock people consider it an excellent browse plant better adapted to the places where it grows than anything that might replace it. This conflict between academic and pragmatic approaches to the economics of vegetation is not uncommon; time proves one correct as often as the other.

Although the supporting evidence in Kern County is inconclusive, sagebrush is probably invasive in heavily grazed places here as it is elsewhere, especially the dry meadows in the Jeffrey pine forest. Seedlings are poor competitors; this may be a factor in its general restriction to light soils in arid regions. Cultivated plants grow and thrive even in heavy soil and benefit from occasional summer irrigation. A plant 12 years old grows in very stiff soil at the Dorothy Twisselmann Ranch garden in the Temblor Range; despite occasional pruning it is 10 feet tall.

Chrysanthemum Parthenium (Linnaeus) Bernhardi. Feverfew. A sparingly established garden escape at the head of Big Canyon just below the Ernest Still ranchgrounds near Annette in the northern Temblor Range (T5910); occasional in moist places in waste places in Bakersfield. Native of Europe.

Cotula australis (Sieber) Hooker. A scarce weed, usually in colonies in beaten places, at Kernville, the parking area at the mouth of Cow Flat Creek in Kern Canyon, and at Bakersfield. Native of Australia.

Cotula coronopifolia Linnaeus. Brass buttons. Colonies are occasional in sunny muddy places in the Temblor Range. In the swampy flats along Poso Creek near the Poso Station monument it grows in colonies several hundred yards long. Native of South Africa.

Dicoria canescens Torrey & Gray ssp. **hispidula** (Rydberg) Keck. Bug seed. Common on sandy plains, roadsides, etc., in the creosote bush association, especially in years with good summer rains.

Eclipta alba (Linnaeus) Hasskarl. Scarce in the valley in rice fields, along canal banks, and at Lake Woolloomes near Delano; rapidly becoming more common. Its prompt appearance when the dry subalkaline plains were flooded at the Kern and Pixley National Wildlife refuges suggests that the seed is carried by waterfowl. Although apparently native in California, it has become established in Kern County only recently.

Matricaria matricarioides (Lessing) Porter. Pineapple weed. A universal weedy plant of waste places, paths, beaten areas, etc., in the Lower and Upper Sonoran associations, asecending to the yellow pine forest at Shirley Meadows and in disturbed places on Breckenridge Mountain and in the Tehachapi Mountains. Pioneers used an infusion of the dried plants for stomach disorders. (*M. suaveolens.*)

Matricaria occidentalis Greene. Alkali pineapple weed. Uncommon in wet usually subalkaline soils in the valley, around Rosamond on the desert, and in Cummings Valley. It is most common in years with early heavy rains. This species can be distinguished from the similar *M. matricarioides* by its erect habit, large size, and the evident pappus crown of the achenes. (*M. matricarioides* in part.)

Calenduleae. Calendula Tribe

Dimorphotheca sinuata DeCandolle. African daisy. A garden escape commonly spontaneous and possibly sparingly established in and about Shafter (T4257). Native of South Africa.

Senecioneae. Groundsel Tribe

Arnica Chamissonis Lessing ssp. **foliosa** (Nuttall) Maguire. Scarce at Browns Meadow in the Jeffrey pine forest in the Piute Mountain (T3195, Breedlove 3992). Although this plant is not known from the Greenhorn Range, it is common in meadows on the southern Kern Plateau in Tulare County. (*A. foliosa.*)

Crocidium multicaule Hooker. Colonies are occasional in the Douglas oak woodland in the Greenhorn Range, usually growing in shaded old disturbed places, such as bulldozed slopes, or, according to Charlotte Smith (personal communication), on chilly wet north slopes. Extensive samplings indicate that all of our plants are pappose.

Lepidospartum squamatum (Gray) Gray. Broom scale. Common in washes, canyon bottoms, sandy disturbed places (such as old quarries and sand pits), etc., in the desert and the lower elevations of the mountains southwest to Cuyama Valley. It is scarce in the valley and very rare in the northern Temblor Range. Broom scale migrates freely and often appears promptly at freshly graded areas, road construction cuts and fills, etc., frequently in places some distance from possible parental colonies.

Psathyrotes annua (Nuttall) Gray. Desert velvet. Occasional in sandy or subalkaline clay flats on the plain on the north side of Koehn Dry Lake; rare in the east fork of Red Rock Canyon. While it frequently blooms in late spring, it is more common in the fall, especially after summers with heavy rains. Thus, the plant is probably a relict from a distinctly different Mohavean climate.

Senecio astephanus Greene. According to Philip A. Munz and John C. Roos (Aliso 3:129) the species occurs in the Tehachapi Mountains.

Senecio Breweri Davy. Occasional in the Douglas oak woodland in the mountains and the Temblor Range, usually growing in deep soil. Plants in flower are sometimes hard to find as the flowering stems are eagerly eaten by cattle and deer.

Senecio californicus DeCandolle. Known from the Breckenridge Mountain Road at 1,400 feet elevation (Benson 3056, DS). It has been collected on the west side of Cottonwood Pass in San Luis Obispo County not far from the Kern County line (Eastwood & Howell 2049).

Senecio Clarkianus Gray. Scattered colonies grow at wet places in the ponderosa pine and red fir forests on Sunday Peak (canyon 0.9 mile north of Tiger Flat, Howell 38,800, T6405; north slope of Sunday Peak, T7760).

Senecio Douglasii DeCandolle. Bush groundsel. A relatively short-lived subshrub that is common in cismontane Upper Sonoran associations, especially in sandy or light soils. In years with late wet springs the plants are often infested with mildew.

Senecio Douglasii var. **monoensis** (Greene) Jepson. A rather indistinct glabrous form that replaces the typical variety in the pinyon woodland on the east slope of the Piute Mountains and in the desert-facing slopes and canyons of the Sierra Nevada north of Walker Pass.

Senecio Douglasii var. **tularensis** Munz. Common in the Greenhorn Range from the Douglas oak woodland to the yellow pine forest, north to southern Tulare County. The regularly pinnate leaves readily distinguish it from the typical variety.

Senecio ionophyllus Greene. Scarce around rock outcrops in the vicinity of Piute Peak (T3035) and in deep loam in the yellow pine forest on the steep north slope of Tehachapi Peak (T11,117). The type was collected in the pine woods south of Tehachapi (Edward L. Greene, 25 June 1889), very likely from the Tehachapi Peak colony.

Senecio pseudaureus Rydberg. Known from Sunday Peak, where the plant reaches its southern Sierran limit (C. N. Smith 39, JEPS). (*S. pauciflorus* var. *jucundulus.*)

Senecio serra Hooker. A small dense colony grows in wet soil in the shade of willows at the north end of French Meadow in the Piute Mountains (T7478).

Senecio vulgaris Linnaeus. Common groundsel. A common weed of early spring in farm lands in the valley and occasional in gardens and moist places elsewhere (except the desert). It is infrequent around rock outcrops in range land in the Upper Sonoran grassland and Douglas oak woodland in the east-side foothills east to Woody and Caliente. Native of the Old World.

Tetradymia canescens DeCandolle. Horse brush. Occasional in the pinyon woodland on the east slope of the mountains southwest to the Frazier Park region. It is usually a large shrubby plant but grows as a matted subshrub among the granite outcrops and blocks on the southern Kern Plateau and around the high exposed metamorphic outcrops at Lookout Point in the Piute Mountains.

Tetradymia glabrata Gray. Desert horse brush. Scarce in the arid shrub and the creosote bush associations on the desert-facing slopes of the mountains and in canyons in the western El Paso Range; rare on the sand dunes in the Muroc dry lakes region.

Tetradymia spinosa Hooker & Arnott var. **longispina** Jones. Cotton thorn. Occasional on the hot arid desert-facing slopes of the mountains west to Kelso Canyon and on the rocky slopes on the north side of South Fork Valley west to Lake Isabella. (*T. axillaris.*)

Tetradymia stenolepis Greene. Rare in alkali sinks in the Boron and Muroc regions, west to Willow Springs and Mojave. It is also known from the slope of metamorphic rocks northeast of Lake Isabella (Howell 38,666). The type was collected between Mojave and Cameron Station by Katharine Layne Curran in 1884.

CYNAREAE. Thistle Tribe

Carthamus tinctorius Linnaeus. Safflower. Although only recently planted as a commercial crop in the valley for its volatile oils, safflower is already a widespread escape and appears to be established in moist places in the alkali sinks along Adobe Road southwest of Arvin. It is occasionally planted for quail food in the Temblor Range and may be sparingly established in Big Canyon below the Ernest Still Ranch at Annette and about the Temblor Ranch grounds. Native of Eurasia.

Centaurea Cyanus Linnaeus. Corn flower, bachelor's button. Naturalized for many years at Glennville, where a large and colorful colony grows around the rock outcrops just above the fire station. It is often an impurity in commercial range land grass seed and is frequently spontaneous in seedings of burns. However, the seeds have no drouth-adaptive germination mechanisms and these adventive introductions that are sometimes self-sustaining for several years do not survive the first extended winter or early spring drouth. Native of Europe.

Centaurea melitensis Linnaeus. Napa thistle, tocalote. Common in waste places, roadsides, and heavily grazed grasslands in the Temblor Range and the foothills of the mountains; especially common in the Poso Hills. The seedlings are poor competitors and light grazing in the early spring can usually control the plant in regions where it threatens to be a pest. Native of Europe.

Centaurea repens Linnaeus. Russian knapweed. Common in the valley, where it is often troublesome. It is occasional in low places and in farmlands on the western portion of the desert where it is an especially common waste place weed in Mojave. Russian knapweed is rare in the northern Temblor Range (Kenneth Davies Ranch, north of Annette). Native of the Caucasus.

Centaurea solstitialis Linnaeus. Yellow star-thistle, Barnaby's thistle. A noxious weed that is scarce and vigorously controlled in the valley; occasional in the Lebec region, in the Tehachapi Valley, and in alfalfa fields near Rosamond. Native of Europe.

Cirsium Andersonii (Gray) Petrak. Scarce in rocky places in the yellow pine forest in the Greenhorn Range and the Piute Mountains.

Cirsium californicum Gray. California thistle. Scarce in the Douglas oak woodland, chaparral, and shin oak brush south through the mountains to the Tehachapi Mountains.

Cirsium crassicaule (Greene) Jepson. Slough thistle. One of California's rarest thistles, known from sloughs and canals near Stockton and in Kern County from the Carmel Ranch east of Lost Hills north to the Kern National Wildlife Refuge. It grows at widely separated stations in often dense root-spreading colonies. The plant's casual appearance is that of a common canal bank weed; why it is so rare is difficult to comprehend. Plants at the Kern National Wildlife Refuge set viable seed in ample quantity. Plants from the Buena Vista Slough 7 miles southeast of Lost Hills have unusually large flowering heads (T1278).

Cirsium cymosum (Greene) J. T. Howell. Peregrine thistle. A plant of the Coast Ranges south to the Diablo Range, this thistle is rare in the low hills at the mouth of Water Canyon in the Tehachapi Mountains (T7251). (*C. Vaseyi.*)

Cirsium foliosum (Hooker) DeCandolle. Occasional at Mil Potreros, in Cuddy Valley, the cienegas at the head of Tejon Canyon, and around vernal pools in Bear Valley in the Tehachapi Mountains. (*C. Drummondii* in part.)

Cirsium occidentale (Nuttall) Jepson. Cobweb thistle. Occasional, usually in the Upper Sonoran grassland, from the Temblor Range south to the Mt. Abel region. (*C. Coulteri* in part.)

Cirsium proteanum J. T. Howell. Red thistle, Venus thistle. Occasional, mostly below the yellow pine forest, in the mountains and the high hills in the central Temblor Range. The brilliant crimson flowers on stems often six feet tall are striking in the late spring when the grass has dried. Especially fine plants grow on the ridge southeast of Reyes Station and on the road to Mt. Abel. (*C. Coulteri* in part.)

Cirsium vulgare (Savi) Tenore. Bull thistle. Occasional in moist places in the montane valleys (abundant on the west side of Walker Basin and at the south end of Castac Lake); rare in the Lake Isabella region and around Glennville. Native of Europe. (*C. lanceolatum.*)

Cnicus benedictus Linnaeus. Blessed thistle. A potentially noxious weed known from a roadside colony about four miles west of Wofford on the Greenhorn Pass highway (Myrick 1122 in 1965, SBBG).

Silybum Marianum (Linnaeus) Gaertner. Milk thistle. This weed with beautifully marbled leaves grows in small colonies in the Douglas oak woodland in Chanac and Tejon canyons in the Tehachapi Mountains and is scarce on canal banks in the Carmel Ranch region east of Lost Hills. Native of the Mediterranean region.

Cichorieae. Chicory Tribe

Agoseris glauca (Pursh) Greene var. **dasycephala** (Torrey & Gray) Jepson. Rare in dry sandy soil in the sagebrush flats in the Jeffrey pine forest at Woolstaff Meadow in the Piute Mountains (T7302).

Agoseris grandiflora (Nuttall) Greene. Occasional in the Upper Sonoran associations in the mountains.

Agoseris heterophylla (Nuttall) Greene. Occasional in the cismontane Upper Sonoran associations, especially the grasslands. The varieties *californica* (Nuttall) Jepson and *crenulata* (Hall) Jepson occur at random with the typical variety.

Agoseris retrorsa (Bentham) Greene. Occasional in the Douglas oak and pinyon woodlands to the yellow pine forest in the mountains; rare in the northern Temblor Range.

Anisocoma acaulis Torrey & Gray. Scale bud. Common in sandy places on the desert; occasional west to the east slope of the Greenhorn Range (occurring at the border of the ponderosa pine forest at Woodard Spring) and southwest to the borders of the open Jeffrey pine forest on Tecuya Ridge in the Mt. Pinos region. On the Kern Plateau it occurs in extreme southern Tulare County along Taylor Creek at 7,200 feet elevation (Howell & True 41,852).

Calycoseris Parryi Gray. Yellow tack-stem. Scarce on gravelly or rocky slopes in the creosote bush association.

Cichorium Intybus Linnaeus. Chicory. An infrequent weed at widely scattered cismontane stations below the yellow pine forest, mostly along roadsides and in waste places. As livestock are very fond of the plant, flowering stems rarely survive to set seed in pasture lands. The roots are used in many parts of the world as a coffee substitute or adulterant; Kern County plants are sometimes gathered for this purpose by immigrants from the southern states. Native of the Old World.

Crepis acuminata Nuttall. Hawk's beard (a common name for most species of the genus). Scarce in metamorphic rock at the lower levels of the Jeffrey pine forest on the north end of Piute Mountain (7.7 miles above Bodfish Gap, T8901, Breedlove 3889).

Crepis occidentalis Nuttall. Known from the summit of Mt. Pinos (Breedlove 521A, DS). Peter H. Raven writes (personal communication) that the collection is distinct from ssp. *pumila,* which also grows on Mt. Pinos.

Crepis occidentalis ssp. **pumila** (Rydberg) Babcock & Stebbins. Rare at the summit of Mt. Pinos (T3809), at the ridge east of Catskin Creek in the Tehachapi Mountains (Roy V. Parker, 2 June 1964), and in sagebrush in the open Jeffrey pine forest bordering Woolstaff Meadow in the Piute Mountains (T7298).

Glyptopleura setulosa Gray. Colonies are rare on sandy subalkaline flats from Rogers Dry Lake to the Boron region (Mojave Desert 11.3 miles west of Boron, T3388).

Hieracium albiflorum Hooker. White-flowered hawkweed. Occasional in the ponderosa pine forest in the Greenhorn Range, mostly around rocks and on gravel banks.

Hieracium horridum Fries. Scarce around granite outcrops in the northern Greenhorn Range, the southern Kern Plateau, and at Piute Peak and Lookout Point in the Piute Mountains.

Hypochoeris glabra Linnaeus. Smooth cat's ear. It has been well established for many years in a small area near the old branding corral at Carneros Rocks in the Temblor Range (T2639) and is occasional about rock outcrops in the hills south of White River in Kern County near the Tulare County line south to Poso Creek near Poso Mine. Native of Eurasia and north Africa.

Lactuca Serriola Linnaeus. Wild lettuce. A common weed of waste places and dry-farm wheat fields in the Temblor Range and the valley. It is occasional through the Douglas oak woodland in the mountains and is a rare waste place weed in damp soil on the desert. Livestock and deer probably graze wild lettuce more avidly than any other plant in the county; this severely limits its distribution. Native of Europe.

Lactuca Serriola var. **integrata** Grenier & Godron. This distinctive glabrous form with entire leaves is common in moist soils in the valley and rare elsewhere. It requires moister soil than the typical variety. Native of Europe.

Lygodesmia exigua Gray. Egbertia. A small colony grows in gravel below a low metamorphic outcrop on the ridge just east of the Salisbury Mine on the northwest slope of Government Peak in the Rand Mountains (T12,018).

Malacothrix altissima Greene. Tall marigold. Occasional in the Upper Sonoran associations from Breckenridge Mountain and the north end of Piute Mountain southwest to Cuyama Valley. The type was collected near Tehachapi Station by Katharine Layne Curran in 1884. (*M. saxatilis* var. *altissima.*)

Malacothrix californica DeCandolle. Wild marigold. Common on sandy plains in the valley, often forming extensive displays of light yellow in the early spring. Occasional in light sandy soil in the surrounding foothills, extending east to Kernville and the borders of the desert along the Tehachapi Mountains in the west end of Antelope Valley.

Malacothrix Clevelandii Gray. A rare plant of the chaparral that is perhaps most common following burns (Breckenridge Mountain, Hardham 2184; burn on the north slope of Black Mountain in the Greenhorn Range, Howell 38,199). It is also known from a very small dense colony by a large granite outcrop in the thin Jeffrey pine forest near Pine Flat on the southern Kern Plateau (T9872).

Malacothrix Coulteri Gray. Snake heads. Common in the valley and the surrounding foothills, mostly in very light often gypseous soils, and on sandy plains and slopes on the desert.

Malacothrix floccifera (DeCandolle) Blake. Scarce in light sandy soil in the chaparral on Messa Ridge in the Temblor Range, southwest to the pinyon woodland and Jeffrey pine forest at Tecuya Ridge in the Mt. Pinos region. (*M. obtusa.*)

Malacothrix glabrata Gray. Desert dandelion. Abundant in light sandy soils in the creosote bush association; occasional, sometimes in extensive colonies, on sandy flats west to Lake Isabella. It is common in wind-modified sand on the west side of the valley in the Belridge region and scarce in similar places east to Adobe Canyon in the Greenhorn foothills (C. N. Smith 1346, JEPS). On the desert plains, the plant frequently grows in great masses that in flowering season are one of the region's impressive floral displays.

Microseris campestris Greene. Occasional in the Upper Sonoran grassland in the Temblor Range, usually growing in colonies in heavy soil. (*M. aphantocarpha* var. *palealis.*)

Microseris Douglasii (DeCandolle) Schultz-Bipontinus. Occasional in the Upper Sonoran grassland; rare in the Lower Sonoran grassland in the valley.

Microseris elegans Greene. Kenton Chambers (Contr. Dud. Herb. 4:306) cites a collection from 8 miles northwest of Lokern [Belridge Plain] (Johannsen 1407, UC).

Microseris heterocarpa (Nuttall) Chambers. Common in the Temblor Range and the low hills and higher plains around the valley. (*Uropappus leucocarpus.*)

Microseris linearifolia (Nuttall) Schultz-Bopontinus. Uropappus. Common, mostly in the Upper Sonoran grassland and the Douglas oak woodland in the Temblor Range and the hills and high plains bordering the valley. It is rare near the summit of Walker Pass (Howell 37,235) and in the creosote bush association on the edge of the desert (Willow Springs, T6841). (*Uropappus linearifolius, U. Lindleyi.*)

Microseris montana (Greene) Hall. Rare in red clay in the chaparral on the northwest slope of Piute Mountain (T3623).

Microseris sylvatica (Bentham) Gray. Occasional in the chaparral and Douglas oak woodland on the west slope of the Greenhorn Range and on Breckenridge Mountain. According to Philip A. Munz (A California Flora, p. 1290), it also occurs in the Tehachapi Mountains. (*Scorzonella sylvatica.*)

Rafinesquia californica Nuttall. California chicory. Scarce in the cismontane Upper Sonoran associations, growing most often in poor light soil on steep slopes or around boulders.

Rafinesquia neomexicana Gray. Desert chicory. Scarce in dry washes, often in the shade of shrubs, and around rock outcrops in the El Paso Range and the Rand Mountains.

Sonchus asper (Linnaeus) Hill. Sow thistle. A common weed of gardens and in moist soils elsewhere. Native of Europe.

Sonchus oleraceus Linnaeus. Although widespread in California, this weed is rare in Kern County, known with certainty only from a logged slope south of Cow Creek in the Greenhorn Range (T9004 in 1963), along Tejon Creek at 3,325 feet elevation (T10,582 in 1965), and in the dense growth around Kern Hot Springs (T12,227). Native of Europe.

Stephanomeria cichoriacea Gray. Tejon milk-aster. Rare from the Mt. Pinos region east to Ft. Tejon (Bitter Creek on the south side of Mt. Pinos at 7,000 feet elevation, H. M. Hall, July, 1905, UC); among rocks in the canyon 0.5 mile west of Ft. Tejon, Hall 6299, June 16–19, 1905, UC). The type was collected at Ft. Tejon by George Henry Van Horn in 1863.

Stephanomeria exigua Nuttall. Common in the arid shrub and creosote bush associations on the desert, in dry mostly sandy places in the Douglas oak woodland in the mountains, and in all associations in the Temblor Range.

Stephanomeria Parryi Gray. Rare in loose sandy soil in the creosote bush association (Indian Wells Valley, 2 miles west of Inyokern, T4398).

Stephanomeria pauciflora (Torrey) Nelson. Common on the desert and in arid places through the Douglas oak and pinyon woodlands in the mountains; occasional in the valley and in light arid soils in the Temblor Range. (*S. runcinata.*)

Stephanomeria virgata Bentham. Common in the cismontane Upper Sonoran associations, especially in heavy soils, east to the Piute Mountains and South Fork Valley. Plants in the Glennville region are particularly tall and robust. It is relished by livestock and deer.

Taraxacum laevigatum (Willdenow) DeCandolle. Known from moist heavy black soil below an old spring near the Bernard Ranch about four miles southwest of the head of Eugene Grade in the Douglas oak woodland in the Greenhorn Range (C. N. Smith 1443, 11 April 1966, JEPS). Native of the Old World.

Taraxacum officinale Weber. Dandelion. Common and thoroughly naturalized in sunny moist mountain meadows and in moist places along streams in the foothills. It is a ubiquitous garden weed, especially in lawns. Naturalized from Europe.

Tragopogon pratensis Linnaeus. Meadow salsify. One of the rare California occurrences of this plant is at Pettit Meadow along Little Poso Creek on the west slope of the Greenhorn Range (T8533), where it is well established in the dryish soil on the borders of the wet areas of this secluded and isolated meadow. Native of Europe.

A colony of the more common salsify or oyster plant, *T. porrifolius* Linnaeus, has survived with no care or water at the old Bill Carter place, which was abandoned in 1943, in the northern Temblor Range in San Luis Obispo County, 200 feet from the Kern County line.

Excluded Records

Recent California floras specifically include Kern County in the range of the following plants, apparently erroneously:

Luzula parviflora (Ehrhart) Desvaux. The Kern County record is apparently based on a collection by Frank Peirson from Silver Lake, a tributary of the Little Kern River, at 10,000 feet elevation, Tulare County.

Brasenia Schreberi J. F. Gmelin. The Kern County record is based on a collection from Little Kern Lake, Kern Canyon at 6,200 feet, Jepson 4924 (Jeps. Fl., 1:513), a Tulare County station.

Lupinus Peirsonii Mason. While it is possible that this lupine will yet be collected in the Tehachapi Mountains, the Kern County record is apparently based on Los Angeles County collections attributed to the wrong county.

Lupinus Stiversii Kellogg. The southern Sierra Nevada limits seem to be at Cedar Grove in Tulare County. Perhaps confusion in places names led to it being attributed to Cedar Creek in Kern County (Bull. Torr. Club, 48:220, cited in Jeps. Fl., 2:278).

Ceanothus pinetorum Coville. The record is based on collections from Durrwood Meadow, Tulare County; the labels give the county as Kern. (The plant occurs in extreme southern Tulare County on the Kern Plateau and may eventually be collected in Kern County in that region.)

Camissonia bistorta (Nuttall) Raven. No specimen can be found to support the Kern County record of this evening primrose of southern California. (*Oenothera bistorta* var. *Veitchiana.*)

Linanthus breviculus (Gray) Greene. The Kern County references are based on misdetermined collections of *L. nudatus.*

Phacelia frigida Greene. A subalpine phacelia that reaches its southern limit at Olancha Peak (Munz 15,273, RSA).

Cryptantha Clevelandii Greene. The Kern County record is based on misdetermined collections of *C. nemaclada* from the Annette region in the northern Temblor Range.

Oreocarya confertiflora Greene. Attributed to Mt. Laura in the Piute Mountains, possibly because of a confusion of labels.

Diplacus longiflorus Nuttall. The Kern County record is based on misdetermined collections of *D. calycinus.* While these species often intergrade, apparently true *D. longiflorus* does not occur inland as far as Kern County.

Mimulus bicolor Hartweg. The Kern County record is apparently based on misdetermined collections of *M. montioides* and *M. Whitneyi.*

Encelia farinosa Gray. Although the plant has been attributed to Kern Canyon in Kern County, the canyon has been carefully searched for this desert species by the writer and others with no success. Possibly the record is based on a misdetermined collection of *E. virginensis* A. Nelson ssp. *actoni* (Elmer) Keck, which is common in lower Kern Canyon.

Rudbeckia californica Gray. No specimen can be found to support the Kern County occurence of this plant of the Sierra Nevada.

Hemizonia laevis (Keck) Keck. Attributed to southwest Kern County. No herbarium record has been found to support the Kern County reference for this tarweed of coastal southern California.

Hulsea vestita Gray. The Kern County record is apparently based on collections from Frazier Mountain in Ventura County.

INDEX TO THE CATALOGUE OF PLANTS

Abies, 168
Abronia, 231
Acamptopappus, 351
Acer, 280
Aceraceae, 280
Achillea, 373
Achyrachaena, 365
Aconitum, 236
Adder's-tongue family, 165
Adenostoma, 255
Aesculus, 280
Agastache, 327
Agoseris, 379
Agropyron, 182
Agrostis, 186
Ailanthus, 277
Ailanthus family, 277
Aira, 185
Aizoaceae, 232
Alchemilla, 255
Alder, white, 210
Alfalfa, 270
Alhagi, 250
Alisma, 174
Alismataceae, 174
Alkali weed, 306
Allenrolfea, 224
Allium, 200
Allocarya, 321
Allophyllum, 307
Almond, desert, 258
Alnus, 210
Alopecurus, 187
Althaea, 282
Alum-root, San Emigdio, 253
Alyssum, desert, 248
Amaranth family, 230
Amaranthaceae, 230
Amaranthus, 230
Amelanchier, 255
Ammannia, 290
Amole, 204
Amsinckia, 321
Anacardiaceae, 279
Anagallis, 302
Androsace, 303
Anemopsis, 215
Angelica, 296
Anise, 298
Anisocoma, 380
Antelope brush, 258
Antennaria, 361
Anthemis, 373
Antirrhinum, 334, 337
Apiastrum, 296
Apium, 296
Apocynaceae, 304
Apocynum, 304
Aquilegia, 236
Arabidopsis, 244
Arabis, 244
Arceuthobium, 214
Arctostaphylos, 301
Arenaria, 233
Argemone, 240
Aristida, 187
Arnica, 375
Arrow-grass, 174
Arrowhead, 175
Arrow-scale, 226
Arrow-weed, 225, 362
Artemisia, 373
Arundo, 175
Asclepiadaceae, 304
Asclepias, 304
Ash, Oregon, 303
Asparagus, 200
Aster, 351, 359
Astragalus, 260
Athel, 285
Athyrium, 165
Athysanus, 245
Atriplex, 225
Avena, 185
Avens, 256
Azolla, 167

Baby-blue-eyes, 316
Baccharis, 352
Bachelor's button, 378
Bacopa, 334
Baeria, 371
Balsamorhiza, 364, 365
Balsam-root, 364
Bamboo, 175
Barbarea, 245
Barberry, 239
    family, 239
Barley, 184
Barleys, wild, 183
Bassia, 227
Bay, California, 239
Bean, castor, 279
    horse, 270, 273
Beaver-tail, 289
Bed-straws, 346 to 347
Beech family, 211
Bee plant, 344
Beet, sugar, 227
Beggar ticks, 364
Bellflower family, 349
Berberidaceae, 239
Berberis, 239
Bergia, 285
Berula, 296
Beta, 227
Betulaceae, 210
Bidens, 364
Bignonia family, 344
Bignoniaceae, 344
Bindweed, field, 305
Birch family, 210
Bird-of-paradise tree, 270
Bird's beak, 336
    foot trefoil, 264
    nest, 220
Bitter brush, 258
    root, 232
Black bush, 256
Blackberries, 259
Bladder pod, 252
Blanket flower, 371
Blazing stars, 287, 288
Bleeding heart, 244
Blepharizonia, 365
Bloomeria, 200
Blow-wives, 365
Blue curl, 330
    dicks, 202
    flower, 317
Blue-eyed-Mary, 336
Boisduvalia, 291
Borage family, 321
Boraginaceae, 321
Botrychium, 165
Bottle cleaner, 291
Bowlesia, 296
Box elder, California, 280
    thorn, 332
Bracken, western, 166
Bramble, 259
Brasenia, 384
Brass buttons, 375
Brassica, 245
Brickell bushes, 351
Brickellia, 351
Brodiaea, 202
Bromus, 175
Brooklime, American, 344
Broom scale, 376
Broomrape, 345
    family, 345
Buck brush, 280
Buckeye, California, 280
    family, 280
Buck-thorn, 281
    family, 280

Buckwheat family, 215
Buckwheats, 217–222
Buffalo berry, 290
Bug seed, 375
Bulrushes, 197–198
Burhead, upright, 175
Bur-reed, 173
family, 173
Burro brush, 363
Bur-weed, 363
Bush hops, 363
Butter and eggs, 341
Buttercup, Bermuda, 275
California, 239
water, 238, 344
Butterfly bush, 327
Buttonwillow, 346

Cabbage, skunk, 205
squaw, 246
Cacao family, 284
Cactaceae, 288
Cactus family, 288
Cactuses, 288–290
Caesalpinia, 270
Calabazilla, 348
Calandrinia, 232
Calico flower, 340
Callitrichaceae, 279
Callitriche, 279
Calocedrus, 171
Calochortus, 202
Caltrops family, 276
Calycadenia, 365
Calycoseris, 380
Calyptridium, 232
Camas, death, 207
Camel thorn, 260
Camissonia, 291, 384
Campanulaceae, 349
Canaigre, 224
Canbya, 240
Canchalagua, 303
Cannabis, 210
Cantaloupe, 348
Canyon tia, 173
Caper family, 252
Capparidaceae, 252
Caprifoliaceae, 347
Capsella, 245, 248
Cardaria, 245
Careless weeds, 230
Carex, 193
Carpet weed family, 232
Carrot, common, 297
Carthamus, 377
Caryophyllaceae, 233
Cassia, 263
Castanopsis, 211
Castilleja, 334
Catalpa, 344
Catchfly, sleepy, 235
Cats-ears, 380
Cat-tail family, 173
Cat-tails, 173
Caucalis, 296
Caulanthus, 245
Ceanothus, 280, 384
Cedar, incense, 171
salt, 285
Celery, 296
mock, 296
Celtis, 214
Cenchrus, 191
Centaurea, 378
Centaurium, 303
Centromadia, 366
Cephalanthus, 346
Cerastium, 234
Ceratophyllaceae, 236
Ceratophyllum, 236
Cercis, 263
Cercocarpus, 255
Chaenactis, 369
Chaetochloa, 192
Chaetopappa, 353
Chamaebatia, 256
Chamise, 255
Chaparral white-thorn, 280
Charlock, 245
Checker bloom, 283–284
Cheese brush, 363
weed, 283
Cheilanthes, 165
Chenopodiaceae, 224
Chenopodium, 227
Cherries, wild, 257–258
Cherry, ground, 333
Chess, Australian, 175
soft, 176
Chia, 329
Chickweed, Indian, 232
Chickweeds, common, 236
Chicory, 380
California, 382
Chinaberry tree, 279
Chinese houses, 336
Chinquapin, 211
Chloris, 189
Chlorogalum, 204
Cholla, silver, 290
Chorizanthe, 215
Chrysanthemum, 374
Chrysopsis, 353
Chrysothamnus, 353, 358
Chufa, 196
Cichorium, 380
Cicuta, 296
Cinquefoil, 257
Circaea, 293
Cirsium, 378
Citrullus, 348
Clarkia, 293
Cleavers, 347
Clematis, 237
Cleomella, 252
Cliff brake, 165
Clotbur, spiny, 363
Clover, bur, 270
jackass, 252
mustang, 265, 313
owl's, 341, 342
Spanish, 265
Clover-fern family, 167
Clovers, 271–273
sweet, 270
Cnicus, 379
Cockle-bur, 363
Coffee berry, 281
Coldenia, 322
Coleogyne, 256
Collinsia, 335
Collomia, 307
Columbine, crimson, 237
Comandra, 215
Compositae, 351
Conium, 297
Convolvulaceae, 305
Convolvulus, 305
Conyza, 355
Coral root, 207
Corallorhiza, 207
Cordylanthus, 336
Corethrogyne, 355
Coreopsis, 364
Corn, domestic, 193
flower, 378
gyp, 193
Cornaceae, 300
Cornus, 300
Cotton, 283
Cotton-batting plant, 361
Cotton-thorn, 377
Cottonweed, 362
Cottonwoods, 208
Cotula, 375
Cotyledon, 253
Cranesbills, 275
Crassulaceae, 253
Cream cups, 243
Creambush, 256
Creosote bush, 276
Crepis, 380
Cress, hoary, 245
rock, 244
shield, 249

Cress—Contd.
water, 249
western yellow, 249
winter, 245
Cressa, 306
Crocidium, 376
Crockeria, 371
Croton, 277
Crowfoot family, 236
Cruciferae, 244
Cryptantha, 322, 384
Cucumis, 348
Cucurbita, 348
Cucurbitaceae, 348
Cudweed, California, 361
glandular, 355
Cupressaceae, 171
Cupressus, 171
Currants, 254
Cuscuta, 306
Cymopteris, 297
Cynodon, 190
Cynoglossum, 325
Cyperaceae, 193
Cyperus, 195
Cypress family, 171
Piute, 171
Cystopteris, 165

Dactylis, 177
Daisy, African, 375
alkali, 372
hill-top, 372
Philadelphia, 357
rock, 372
Dalea, 263
Damasonium, 174
Dandelion, 383
desert, 381
Datisca, 288
family, 288
Datiscaceae, 288
Datura, 331
Daucus, 297
Deer brush, 280
Delia, 303
Delphinium, 237
Dendromecon, 240
Dentaria, 246
Deschampsia, 185
Descurainia, 246
Desert calico, 312
candle, 246
hornbill, 275
star, 360
trumpet, 219
velvet, 376
Devil's lantern, 295
root, 231
Dicentra, 244
Dichelostemma, 202
Dicoria, 375
Digitaria, 191
Dimorphotheca, 375
Diplacus, 337, 384
Distichlis, 177
Ditch-grass family, 174
Dithyrea, 247
Docks, 224
Dodders, 306
Dodecatheon, 303
Dogbane family, 304
western, 304
Dogwood family, 300
Dogwoods, 300–301
Downingia, 349
Draba, 247
Dragon's head, 374
Dryopteris, 166
Duckweed family, 198
least, 198
Dudleya, 253
Durango root, 288

Eastwoodia, 356
Eatonella, 370
Echinocactus, 288
Echinochloa, 191
Echinocystis, 349
Echinodorus, 175
Eclipta, 375
Egbertia, 381
Eichhornia, 198
Elaeagnaceae, 290
Elatinaceae, 285
Elatine, 285
Elderberries, 348
Eleocharis, 196
Eleusine, 190
Ellisia, 316, 320
Elm family, 214
slippery, 284
Elms, 214
Elodea, 175
Elymus, 182
Emmenanthe, 315
Encelia, 364, 384
Ephedra, 172
Ephedraceae, 172
Epiliobium, 294
Epipactis, 208
Equisetaceae, 167
Equisetum, 167
Eragrostis, 178
Eremalche, 282
Eremocarpus, 278
Eriastrum, 307
Ericaceae, 301
Ericameria, 358
Erigeron, 355, 356
Eriochloa, 191
Eriodictyon, 315
Eriogonum, 216
Eriophyllum, 370, 373
Erodium, 273
Eruca, 247
Eryngium, 297
Erysimum, 247
Eschscholzia, 240
Eucalyptus, 290
Eucnide, 287
Eucrypta, 316
Eulophus, 299
Euphorbia, 278
Euphorbiaceae, 277
Eurotia, 228
Evax, 361
Evening-primrose family, 291
Evening snow, 313
Everlasting, fragrant, 361

Fagaceae, 211
Farewell-to-spring, 282, 293
Fennel, hog, 298
Fern, clover, 167
duckweed, 167
family, 165
grape, 165
mosquito, 167
Ferns, 165–166
moss, 167
Fescue, 178–179
Festuca, 178
Fever-few, 374
Ficus, 210
Fiddleneck, 321
Fiesta flower, 320
Fig, domestic, 210
Figwort family, 334
Filago, 361
Filarees, 273–275
Fimbristylis, 197
Fireweed, 321
Firs, 168
Five hook, 227
spot, desert, 282
spot, mountain, 316
Flannel bush, 284
Flaxes, 275–276
Fleabane, 356
marsh, 362
Flowering quillwort family, 174
Foeniculum, 298
Fog weed, 225

Forestiera, 303
Four-o'clock family, 231
Four wing, 231
Fox-tails, 183
Frankenia, 285
family, 285
Frankeniaceae, 285
Franseria, 263
Frasera, 303
Fraxinus, 303
Fremontia, 284
Fremontodendron, 284
Fringe pod, 251
Fritillaria, 204
Fritillaries, 204
Frogbit family, 175
Frying pans, 242
Fuchsia, humming-bird, 295
Fumariaceae, 244
Fumitory family, 244

Gaillardia, 371
Galium, 346
Gamble weed, 300
Garrya, 300
Garryaceae, 300
Gaura, 295
Gayophytum, 244
Gentian family, 303
hiker's, 303
Gentianaceae, 303
Gentians, green, 303
Geraniaceae, 273
Geranium, 275
family, 273
Geum, 256
Gilia, 307, 308
Gilias, 308–312
Giraffe's head, 328
Githopsis, 349
Glasswort, 229
Glinus, 232
Glycyrrhiza, 263
Glytopleura, 380
Gnaphalium, 361
Gold fields, 371
Golden bushes, 357–359
eardrops, 244
fleece, 358
girls, 369
head, 351
prince's plume, 250
rods, 360–361
stars, 200
Golden-top, 179
Gooseberries, 254
Goosefoot family, 224
Fremont's, 228
Gopher plant, 278
Gossypium, 283
Gourd family, 348
wild, 348
Gramineae, 175
Grape family, 281
Grapes, 281–282
Grass, arrow, 174
bear, 205
blue-eyed, 207
ditch, 174
family, 175
goose, 347
Whitlow, 247
yellow nut, 196
Grasses, 175–193
blue, 180–181
bent, 186–187
brome, 175–177
wheat, 182
Gratiola, 337
Grayia, 229
Greeneocharis, 322
Grindelia, 357
Ground-cherry, 333
Groundsels, 376–377
Gum weed, 357
Gutierrezia, 357
Guttiferae, 284
Gymnogramme, 166

Habenaria, 208
Hackberry, Arizona, 214
Hackelia, 325
Haloragidaceae, 296
Haplopappus, 357
Hareleaf, common, 367
Hawksbeard, 380
Hawkweed, 380
Hazardia, 359
Heath, alkali, 285
Heather family, 301
Hedgenettle, 330
Helenium, 371
Heleocharis, 196
Heleochloa, 187
Helianthella, 364
Helianthus, 364
Heliotropium, 325
Hellebore, 205
Hemicarpha, 197
Hemizonella, 368
Hemizonia, 365, 368, 384
Hemlock, poison, 297
water, 296
Hemp, Indian, 304
Henbit, 328
Herniaria, 231
Hesperochiron, 316
Hesperocnide, 213
Hesperomecon, 243
Heterocodon, 349
Heterodraba, 248
Heterogaura, 295
Heteromeles, 256
Heterotheca, 359
Heuchera, 253
Hieracium, 380
Himalaya berry, 259
Hippocastanaceae, 280
Hoffmannseggia, 263
Holcus, 183
Holodiscus, 256
Hole-in-the-sand, 372
Hollisteria, 222
Holly, desert, 225
Hollyhock, 282
Honeysuckle, chaparral, 347
family, 347
Hordeum, 183
Horehound, 328
water, 328
Horkelia, 256
Horned pondweed family, 174
Hornwort, 236
family, 236
Horse brush, 377
Horsetail, common, 167
family, 167
Hound's tongue, 325
Howelliella, 337
Hugelia, 307
Hulsea, 384
Hutchinsia, 248
Hyacinth, water, 198
Hydrocharitaceae, 175
Hydrocotyle, 298
Hydrophyllaceae, 315
Hydrophyllum, 316
Hymenoclea, 363
Hypericum, 284
Hypochoeris, 380

Idahoa, 248
Illecebraceae, 231
Imperata, 191
Indian warrior, 342
Indian-pipe family, 301
Indigo bush, 263
Innocence, 336
Iodine bush, 224
Ipomoea, 306
Ipomopsis, 312
Iridaceae, 207
Iris, 207
family, 207

Isocoma, 357
Isoetaceae, 168
Isoetes, 168
Isomeris, 252
Isopyrum, 238
Ithuriel's spear, 202
Iva, 363
Ivesia, 257

Jewel flowers, 250–251
Jimson weed, 331, 332
Johnny-tuck, 341
Joshua, Kern, 206
  tree, 205
Juglandaceae, 210
Juglans, 210
Juncaceae, 199
Juncaginaceae, 174
Juncus, 199
Junipers, 172
Juniperus, 172
Jussiaea, 295

Klamath weed, 285
Knapweed, Russian, 378
Knot-weeds, 223
Knotwort family, 231
Kochia, 229
Koeleria, 185

Labiatae, 327
Lactuca, 381
Lady's mantle, western, 255
  thumb, 223
Lagophylla, 367
Lamarckia, 179
Lamb's quarters, 227-228
Lamium, 328
Langloisia, 312
Larkspurs, 237–238
Larrea, 276
Lastarriaea, 215
Lasthenia, 371
Lathyrus, 264
Lauraceae, 239
Laurel family, 239
Layia, 367
Leguminosae, 260
Lemmonia, 316
Lemna, 198
Lemnaceae, 198
Lennoa family, 301
Lennoaceae, 301
Lepidium, 248
Lepidospartum, 376
Leptochloa, 190
Leptodactylon, 312, 314
Leptotaenia, 298
Lessingia, 359
Lettuce, Indian, 233
  miner's, 233
  wild, 381
Lewisia, 232
Libocedrus, 171
Licorice, common, 263
  wild, 263
Lilaea, 174
Lilaeaceae, 174
Liliaceae, 200
Lilium, 204
Lily, adobe, 204
  chocolate, 204
  corn, 205
  family, 200
  sand, 295
  Sierra, 204
  snake, 202
Limnanthes, 275
Limosella, 337
Linanthus, 312, 384
Linaria, 337
Lindernia, 337
Linum, 275
Lippia, 326
Lithophragma, 253
Live-forever, 253
Lizard-tail family, 215
Loasa family, 287
Loasaceae, 287
Lobelia family, 349
Lobeliaceae, 349
Loco weeds, 260–262
Locust, black, 271
Loeflingia, 234
Lolium, 184
Lomatium, 298
London rocket, 250
Lonicera, 347
Loose-strife, California, 290
  family, 290
Loranthaceae, 214
Lord's candle, 206
Lotus, 264
Ludwigia, 295
Lupines, 266–269
Lupinus, 266, 384
Luzula, 200, 384
Lycium, 332
Lycopersicon, 332
Lycopus, 328
Lygodesmia, 381
Lythraceae, 290
Lythrum, 290

Machaeranthera, 360
Machaerocarpus, 174
Maclura, 211
Madder family, 346
Madia, 368
Mahogany family, 279
  mountain, 255
Malacothamnus, 283
Malacothrix, 381
Mallow family, 282
Mallows, 282–284
Malva, 283
Malvaceae, 282
Malvastrum, 282
Man root, 349
Manzanitas, 301–302
Maple, big-leaf, 280
  family, 280
Marah, 349
Marigolds, wild, 381
Marijuana, 210
Marrubium, 328
Marsilea, 167
Marsileaceae, 167
Martynia, 345
  family, 345
Martyniaceae, 345
Matchweed, 357
Matricaria, 375
Matthiola, 249
Mayweed, 373
Meadow foam, 275
  rue, 239
Meconella, 243
Medic, black, 270
Medicago, 270
Melia, 279
Meliaceae, 279
Melica, 180
Melics, 180
Melilotus, 270
Melon, coyote, 348
  mush, 348
  water, 348
Mentha, 328
Mentzelia, 287
Mesquite, 270
Micropus, 362
Microseris, 382
Microsteris, 315
Mignonette family, 285
Milk-aster, Tejon, 383
Milkweed family, 304
Milkweeds, 304–305
Millet, broom corn, 192
  wild, 191
Milo, 193
Mimulus, 337, 384
Mint family, 327
  marsh, 328
Mirabilis, 231

Mission bells, 204
Mistletoe family, 214
Mistletoes, 214–215
Mohavea, 341
Mojave fish-hook, 288
Mollugo, 232
Monardella, 328
Monkey face, 350
Monkey flower, bush, 337
Monkey flowers, 338–341
Monkshood, 236
Monolepis, 229
Monolopia, 372
Monoptilon, 360
Montia, 232
Moraceae, 210
Morning-glory family, 305
Morning-glories, 305–306
Morus, 211
Moss-fern family, 167
Mountain misery, 256
Mouse ear, 234
  tail, 238
Mud-wort, northern, 337
Mugwort, California, 374
Muhlenbergia, 187
Muilla, 205
Mulberry family, 210
  white, 211
Mule ears, 365
  fat, 353
  weed, 337
Mullein, 344
  turkey, 278
Musk flower, 340
Mustard, California, 251
  family, 244
  oriental, 250
  tower, 244
  tumble, 250
Mustards, wild, 245–246
Myosotis, 325
Myosurus, 238
Myriophyllum, 296
Myrtaceae, 290
Myrtle, 304
  family, 290

Najadaceae, 174
Najas, 174
  family, 174
Nama, 316, 320
Nasturtium, 249
Navarretia, 314
Nemacladus, 350
Nemophila, 316, 320
Nettle, bull, 333
  family, 213
Nettle—Contd.
  rock, 287
  western, 213
Nettles, 214
Nicollet, 372
Nicolletia, 372
Nicotiana, 332
Nightshade, enchanter's, 293
  family, 331
Nightshades, 333–334
Nitrophila, 229
Nolina, 205
Noriega weed, 333
Notholaena, 166
Nyctaginaceae, 231

Oak, Jerusalem, 228
  poison, 279
Oaks, 211–213
Oat, cultivated, 185
Oats, side, 179
  wild, 185
Oenanthe, 299
Oenothera, 291, 295, 384
Old man, 373
Oleaceae, 303
Oleaster family, 290
Oligomeris, 285
Olive, desert, 303
  family, 303
Onagraceae, 291
Onions, wild, 200–201
Ophioglossaceae, 165
Opuntia, 289
Orange, osage, 211
Orchid family, 207
Orchidaceae, 207
Orchis, rein, 208
  stream, 208
Oreonana, 299
Orobanchaceae, 345
Orobanche, 345
Orochaenactis, 372
Orthocarpus, 341
Oryza, 191
Oryzopsis, 187
Osmorhiza, 299
Oxalis, 275
Oxypolis, 299
Oxytheca, 222

Paint brushes, 334–335
Palm, California fan, 198
  family, 198
Palmae, 198
Palo verde, Mexican, 270
Panicum, 192
Papaver, 243
Papaveraceae, 240
Parkinsonia, 270
Parrot's feather, 296
Parsley, button, 300
  family, 296
Parsleys, wild, 298–299
Parsnip, alkali, 298
  water, 296
  wild, 296
Parthenocissus, 281
Parvisedum, 253
Paspalum, 192
Pea, everlasting, 264
  family, 260
  garden, 270
Peach, desert, 257
Pea-weed, Austrian, 271
Pectocarya, 325
Pedicularis, 342
Pellaea, 166
Pellitory, Florida, 214
Penny-royals, 328–329
Penstemon, 342
Pentachaeta, 353
Pepper tree, 279
Pepper-grasses, 248
Perideridia, 299
Perityle, 372
Periwinkle, 304
Petalonyx, 288
Phacelia, 317, 384
Phlaris, 190
Phleum, 188
Phlox, 315
Phlox family, 307
  white, 315
Pholisma, 301
Pholistoma, 320
Phorodendron, 214
Photinia, 256
Phragmites, 180
Physalis, 333
Pickerel weed family, 198
Pig nut, 202
  weed, 222
  weed, rough, 230
Pigmy weeds, 253
Pimpernel, 302
Pinaceae, 168
Pin-cushion flowers, 369
Pine, desert, 285
  drops, 301
  family, 168
  mat, 280
Pineapple weeds, 375
Pines, 168–170
Pink family, 233

Pink family—Contd.
Indian, 235
windmill, 235
Pinus, 168
Pinyon, 169
Pisum, 270
Pityrogramma, 166
Plagiobothrys, 321, 326
Plantaginaceae, 345
Plantago, 345
family, 345
Plantain, 346
Plantains, water, 174
Platanaceae, 255
Platanus, 255
Platyspermum, 248
Platystemon, 243
Plectritis, 349
Pluchea, 362
Plums, wild, 258
Poa, 180
Pogogyne, 329
Poinciana, 270
Polemoniaceae, 307
Polemonium, 315
Polygonaceae, 215
Polygonum, 222
Polypodiaceae, 165
Polypodium, 166
Polypody, California, 166
valley, 287
Polypogon, 188
Pomegranate, 290
family, 290
Pond-weed family, 173
Pond weeds, 173–174
Pontederiaceae, 198
Pop-corn flowers, 326
Poppy, bush, 240
chaparral, 240
family, 240
Point-of-Rocks, 243
Prickly, 240
wind, 243
Poppies, California, 240–242
Populus, 208
Portulaca, 233
Portulacaceae, 232
Potamogeton, 173
Potamogetonaceae, 173
Potato, 334
tule, 175
Potentilla, 256
Poverty weed, 363
Pretty face, 202
Pride-of-the-mountains, 343
Primrose, desert, 295
Primrose—Contd.
evening, 295
family, 302
Primulaceae, 302
Proboscidea, 345
Prosopis, 270
Prunus, 257
Psathyrotes, 376
Pseudobahia, 373
Pseudotsuga, 170
Psilocarphus, 362
Psoralea, 271
Pteridium, 166
Pteris, 166
Pterospora, 301
Puccinellia, 181
Punica, 290
Punicaceae, 290
Puncture vine, 277
Purshia, 258
Purslane, 233
family, 232
horse, 232
marsh, 295
sea, 232
water, 233
Pyrola, 301
Pyrolaceae, 301

Quail brush, 225
Quercus, 211
Quillwort, 168
family, 168
flowering, 174

Rabbit brush, 353
Radicula, 249
Radish, wild, 249
Rafinesquia, 382
Ranunculaceae, 236
Ranunculus, 238
Raphanus, 249
Raspberry, western, 259
Rattlesnake weed, 278, 297
Rattleweed, Salinas Valley, 262
Redberries, 281
Redbud, western, 263
Red-scale, 226
Red-top, 186
Reed, giant, 175
Resedaceae, 285
Rhamnaceae, 280
Rhododendron, 302
Rhubarb, wild, 224
Rhus, 279
Ribes, 254
Ribwort, 346
Rice, 191
Ricinus, 279
Rigiopappus, 373
Rip-gut, 176
Robinia, 271
Rorippa, 249
Rosa, 259
Rosaceae, 255
Rose family, 255
wild, 259
Rosilla, 371
Rubiaceae, 346
Rubus, 259
Rudbeckia, 385
Rumex, 224
Ruppia, 174
Ruppiaceae, 174
Rush family, 199
wood, 200
Rushes, 199–200
scouring, 167
spike, 196
Rye, cultivated, 184
Rye-grasses, wild, 182–183

Sacaton, alkali, 188
Safflower, 377
Sage, big, 374
bladder, 329
bud, 374
bur, 363
hop, 229
thistle, 329
Sagebrush, 374
California, 373
Sages, 329–330
Sagina, 234
Sagittaria, 175
St. Johns-Wort, 285
family, 284
Salazaria, 329
Salicaceae, 208
Salicornia, 229
Salix, 209
Salsify, 383
Salsola, 229
Saltbush, common, 226
spiny, 227
Saltweed, Australian, 226
Salvia, 329
Salviniaceae, 167
Sambucus, 348
Sand cress, 232
Sandalwood family, 215
Sandbur, 191
Sandpaper plant, 288
Sand-worts, 233–234

Sanicle, 300
Sanicula, 300
Santalaceae, 215
Sarcodes, 301
Satin-tail, 191
Saucer plant, 222
Saururaceae, 215
Saxifraga, 255
Saxifragaceae, 253
Saxifrage family, 253
Scale bud, 380
Scarlet bugler, 342
Schinus, 279
Schismus, 185
Scirpus, 197
Scribneria, 184
Scrophularia, 344
Scrophulariaceae, 334
Scorzonella, 382
Scutellaria, 330
Secale, 184
Sedella, 253
Sedge family, 193
walking, 197
Sedum, 253
Selaginella, 167
Selaginellaceae, 167
Senecio, 376
Senna, desert, 263
Service berry, 255
Sesuvium, 232
Setaria, 192
Shadscale, 225
Shepherdia, 290
Shepherd's purse, 245
Shin-leaf, 301
Shooting star, 303
Sida, 283
Sidalcea, 283
Silene, 235
Silk-tassel bush, 300
family, 300
Silybum, 379
Simarubaceae, 277
Sisymbrium, 250
Sisyrinchium, 207
Sitanion, 184
Skeleton weed, 217
Skull cap, Sierra, 330
Sky rocket, 312
Sleepy catchfly, 235
Slim solomon, 205
Smart-weed, water, 224
Smilacina, 205
Smilo, 188
Snake heads, 387
Sneeze-weed, 371
Snow bush, 280
plant, 301
Snowberries, 348
Soap plant, 204, 228
Solanaceae, 331
Solanum, 333
Solidago, 360
Sonchus, 382
Sorghum, 193
Sorrel, sheep, 224
yellow, 275
Sparganiaceae, 173
Sparganium, 173
Spearmint, 328
Spectacle pod, 247
Speedwell, Persian, 324
purslane, 324
Spergularia, 235
Sphaeralcea, 283, 284
Sphaerophysa, 271
Sphenosciadium, 300
Spikeweeds, 366–367
Spirodela, 198
Sporobolus, 188
Spraguea, 232
Sprangle-top, 190
Spruce, big cone, 170
Spurge family, 277
Spurges, 278
Spurreys, 235–236
Squaw bush, 279
water brush, 353
Squirrel tail, 184
Stachys, 330
Stanleya, 250
Star flowers, 308
Stellaria, 236
Stenotopsis, 358
Stephanomeria, 383
Sterculiaceae, 284
Stick-seed, 325
shining, 287
valley, 287
Stillingia, 279
Stink weed, 252
Stipa, 188
Stock, night-scented, 249
Stone-crop family, 253
Pacific, 253
Streptanthella, 250
Streptanthus, 246, 250
Stylomecon, 243
Suaeda, 230
Sumuc family, 279
Suncups, 291–292
yellow, 295
Sunflower family, 351
mountain bush, 364
Sunflowers, 364
Sweet briar, 259
fennel, 298
Swertia, 303
Sycamore, 255
family, 255
Symphoricarpos, 348
Syntrichopappus, 373

Tack-stem, yellow, 380
Tamaricaceae, 285
Tamarisk family, 285
Tamarisks, 285
Taraxacum, 383
Tarragon, 374
Tarweeds, 365–366
Tauschia, 300
Tea, Mexican, 227
sheepherder, 172
Telegraph weed, 359
Tetradymia, 377
Thalictrum, 239
Thelypodium, 251
Thimble berry, 259
Thistle, blessed, 379
coyote, 297
Napa, 378
Russian, 229
sow, 382
Spanish, 363
yellow-star, 378
Thistles, 378–379
Thistle-sage, 329
Thorn-apple, 332
Thread stem, 350
Three awn, 187
square, 197
Thysanocarpus, 251
Tibinagua, 219
Tidy-tips, 368
Tillaea, 253
Timothy, 188
swamp, 187
Tinkers penny, 284
Toad-flax, Dalmatian, 337
Tobaccos, Indian, 332
Tocalote, 378
Tolguacha, 331
Tomato, 332
Toothwort, California, 248
Toyon, 256
Tree-of-heaven, 277
Trianthema, 232
Tribulus, 277
Tricardia, 320
Trichostema, 330
Trifolium, 271
Triglochin, 174
Trisetum, 186
Triticum, 184
Tropidocarpum, 252

Tule, common, 197
Tulips, mariposa, 202–203
Tumble weeds, 229, 230
Turpentine weeds, 330–331
Turricula, 320
Typha, 173
Typhaceae, 173

Ulmaceae, 214
Ulmus, 214
Umbelliferae, 296
Umbellularia, 239
Umbrella tree, 279
Unicorn plant, 345
Urtica, 214
Urticaceae, 213
Uropappus, 382

Valerian, 349
Valerianaceae, 349
Velaea, 300
Veratrum, 205
Verbascum, 344
Verbena, desert sand, 231
  sand, 231
  family, 326
Verbenaceae, 326
Verbenas, 327
Veronica, 344
Vervain, 327
Vetches, 273
Vicia, 273
Vinca, 304
Vinegar weed, 330
Viola, 286
Violaceae, 286
Violet, 286
  family, 286
Virginia creeper, 281
Virgin's bower, 237
Vitaceae, 327
Vitex, 327
Vitis, 281

Wagon wheels, 231
Wall flowers, wild, 247
Walnut family, 210
Wappato, 175
Washingtonia, 198
Water nymph, 174
Water-fern family, 167
Water-milfoil family, 296
Waterleaf, 136
  family, 315
Water-plantain family, 174
Water-starwort family, 279
Waterweed, Canadian, 175
  yellow, 295
Waterwort family, 285
Wheat, 184
  Indian
Wheat-grasses, 182
Whispering bells, 315
White heads, 300
  top, 245
Willow family, 208
Willow-herb, 294
Willow-weed, 223
Willows, 209–210
Winter-fat, 238, 352
Wire weed, 222
Wish bone bush, 231
Wislizenia, 252
Wolffiella, 198
Woodland star, 253
Woodwardia, 166
Woolly heads, 362
  pod, 262
Wyethia, 365

Xanthium, 363

Yarrow, yellow, 370
  California, 373
Yellow-flower, 287
Yerba golondrina, 278
  mansa, 215
  santa, 315
Yucca, 205

Zannichellia, 174
Zannichelliaceae, 174
Zauschneria, 295
Zea, 193
Zigadenes, 207
Zifiadenus, 207
Zygophyllaceae, 276
Zygophyllum, 277

METRIC 1 2 3 4 5 6 7 8 9 10

METRIC 1 2 3 4 5 6 7 8 9 10